T0140502

Studies in Computational Intelligence

Volume 613

Series editor

Janusz Kacprzyk, Polish Academy of Sciences, Warsaw, Poland
e-mail: kacprzyk@ibspan.waw.pl

About this Series

The series "Studies in Computational Intelligence" (SCI) publishes new developments and advances in the various areas of computational intelligence—quickly and with a high quality. The intent is to cover the theory, applications, and design methods of computational intelligence, as embedded in the fields of engineering, computer science, physics and life sciences, as well as the methodologies behind them. The series contains monographs, lecture notes and edited volumes in computational intelligence spanning the areas of neural networks, connectionist systems, genetic algorithms, evolutionary computation, artificial intelligence, cellular automata, self-organizing systems, soft computing, fuzzy systems, and hybrid intelligent systems. Of particular value to both the contributors and the readership are the short publication timeframe and the worldwide distribution, which enable both wide and rapid dissemination of research output.

More information about this series at http://www.springer.com/series/7092

Kurosh Madani · António Dourado
Agostinho Rosa · Joaquim Filipe
Janusz Kacprzyk
Editors

Computational Intelligence

Revised and Selected Papers
of the International Joint Conference,
IJCCI 2013, Vilamoura, Portugal,
September 20–22, 2013

 Springer

Editors
Kurosh Madani
Images, Signals and Intelligence Systems
  Laboratory
University PARIS-EST Créteil (UPEC)
Créteil
France

António Dourado
Departamento de Engenharia Informatica
  Polo II—Pinhal de Marrocos
University of Coimbra
Coimbra
Portugal

Agostinho Rosa
Lasseb-ISR-IST
Technical University of Lisbon (IST)
Lisbon
Portugal

Joaquim Filipe
Rua do Vale de Chaves, Estefanilha
Polytechnic Institute of Setúbal/INSTICC
Setúbal
Portugal

Janusz Kacprzyk
Systems Research Institute
Polish Academy of Sciences
Warsaw
Poland

ISSN 1860-949X   ISSN 1860-9503   (electronic)
Studies in Computational Intelligence
ISBN 978-3-319-79477-8   ISBN 978-3-319-23392-5   (eBook)
DOI 10.1007/978-3-319-23392-5

Printed on acid-free paper

Springer International Publishing AG Switzerland is part of Springer Science+Business Media
(www.springer.com)

Preface

The present book includes extended and revised versions of a set of selected papers from the Fifth International Joint Conference on Computational Intelligence (IJCCI 2013). Sponsored by the Institute for Systems and Technologies of Information, Control and Communication (INSTICC), IJCCI 2013 was held in Vilamoura, Algarve, Portugal, from 20 to 22 September, 2013, and was organized in cooperation with the AAAI—Association for the Advancement of Artificial Intelligence, APNNA—Asia Pacific Neural Network Assembly, and ACM SIGART—ACM Special Interest Group on Artificial Intelligence.

Since its first edition in 2009, the purpose of the International Joint Conference on Computational Intelligence (IJCCI) has been to bring together researchers, engineers and practitioners in computational technologies, especially those related to the areas of fuzzy computation, evolutionary computation and neural computation. IJCCI is composed of three co-located conferences, each one specialized in one of the aforementioned areas. Namely:

- International Conference on Evolutionary Computation Theory and Applications (ECTA)
- International Conference on Fuzzy Computation Theory and Applications (FCTA)
- International Conference on Neural Computation Theory and Applications (NCTA)

Their aim is to provide major forums for scientists, engineers and practitioners interested in the study, analysis, design, and application of these techniques to all fields of human activity.

In ECTA, modeling and implementation of bio-inspired systems namely on the evolutionary premises, both theoretically and in a broad range of application fields, is the central scope. Considered as a subfield of computational intelligence focused on combinatorial optimization problems, evolutionary computation is associated with systems that use computational models of evolutionary processes as the key elements in design and implementation, i.e., computational techniques which are inspired by the evolution of biological life in the natural world. A number of evolutionary computational models have been proposed, including evolutionary algorithms, genetic algorithms, evolution strategies, evolutionary programming, swarm optimization, and artificial life.

In FCTA, results and perspectives of modeling and implementation of fuzzy systems, in a broad range of fields, are presented and discussed. Fuzzy computation, based on the theory of fuzzy sets and fuzzy logic, is dedicated to the solution of information processing, system analysis, knowledge extraction from data, and decision problems. Fuzzy computation takes advantages of the powerful available technologies to find useful solutions for problems in many fields, such as medical diagnosis, automated learning, image processing, and understanding, and systems control.

NCTA is focused on modeling and implementation of artificial neural networks computing architectures. Neural computation and artificial neural networks have seen a continuous explosion of interest in recent decades, and are being successfully applied across an impressive range of problem domains, including areas as diverse as finance, medicine, engineering, geology, and physics, providing appealing solutions to problems as varied as prediction, classification, decision making, or control. Numerous architectures, learning strategies, and algorithms have been introduced in this highly dynamic field in the last couple of decades.

During the joint conference, IJCCI received 111 paper submissions from 30 countries, which demonstrates the global dimension of this conference. Of 111 papers, 24 papers were published as full papers (21.6 % of submissions) and 29 were accepted for short presentation (26 % of submissions). Moreover, 13 were accepted for poster presentation. These ratios denote a high level of quality which we aim to continue reinforcing in the next edition of this conference. This book includes revised and extended versions of a strict selection of the best papers presented at the conference.

On behalf of the Conference Organizing Committee, we would like to thank all participants. First of all, to the authors, whose quality work is the essence of the conference, and to the members of the Program Committee, who helped us with their expertise and diligence in reviewing the papers. As we all know, producing a post-conference book, within the high technical-level exigency, requires efforts of many individuals. We wish also to thank all the members of our Organizing Committee, whose work and commitment were invaluable.

December 2013   Kurosh Madani
                                                                    António Dourado
                                                                    Agostinho Rosa
                                                                    Joaquim Filipe

Contents

Organization

Conference Co-chairs

Joaquim Filipe, Polytechnic Institute of Setúbal/INSTICC, Portugal
Janusz Kacprzyk, Systems Research Institute, Polish Academy of Sciences, Poland

Program Co-chairs

António Dourado, University of Coimbra, Portugal
Kurosh Madani, University of PARIS-EST Créteil (UPEC), France
Agostinho Rosa, IST, Portugal

Organizing Committee

Marina Carvalho, INSTICC, Portugal
Helder Coelhas, INSTICC, Portugal
Vera Coelho, INSTICC, Portugal
Bruno Encarnação, INSTICC, Portugal
Ana Guerreiro, INSTICC, Portugal
André Lista, INSTICC, Portugal
Andreia Moita, INSTICC, Portugal
Raquel Pedrosa, INSTICC, Portugal
Vitor Pedrosa, INSTICC, Portugal
Susana Ribeiro, INSTICC, Portugal
Sara Santiago, INSTICC, Portugal
Mara Silva, INSTICC, Portugal

José Varela, INSTICC, Portugal
Pedro Varela, INSTICC, Portugal

ECTA Program Committee

Francisco Martínez Álvarez, Pablo de Olavide University of Seville, Spain
B.V. Babu, JK Lakshmipat University (JKLU), Jaipur, India
Thomas Baeck, Leiden University, The Netherlands
Pedro Ballester, European Bioinformatics Institute, UK
Michal Bidlo, Brno University of Technology, Faculty of Information Technology,
Czech Republic
Tim Blackwell, University of London, UK
Maria J. Blesa, Universitat Politècnica de Catalunya, Spain
Christian Blum, IKERBASQUE and University of the Basque Country, Spain
Indranil Bose, Indian Institute of Management Calcutta, India
William R. Buckley, California Evolution Institute, USA
David Cairns, University of Stirling, UK
Pei-Chann Chang, Yuan Ze University, Taiwan
Sung-Bae Cho, Yonsei University, Korea, Republic of
Antonio Della Cioppa, University of Salerno, Italy
David Cornforth, University of Newcastle, Australia
Lino Costa, Universidade do Minho, Portugal
Peter Duerr, Sony Corporation, Japan
Marc Ebner, Ernst-Moritz-Arndt-Universität Greifswald, Germany
El-SayedEl-Alfy, King Fahd University of Petroleum and Minerals, Saudi Arabia
Andries Engelbrecht, University of Pretoria, South Africa
Fabio Fassetti, DIMES, University of Calabria, Italy
Carlos M. Fernandes, University of Granada, Spain
Stefka Fidanova, Bulgarian Academy of Sciences, Bulgaria
Bogdan Filipic, Jozef Stefan Institute, Slovenia
Dalila B.M.M. Fontes, Faculdade de Economia and LIAAD-INESC TEC,
Universidade do Porto, Portugal
Girolamo Fornarelli, Politecnico di Bari, Italy
Marcus Gallagher, The University of Queensland, Australia
Aaron Garret, Jacksonville State University, USA
Steven Guan, Xian Jiaotong-Liverpool University, China
Jennifer Hallinan, Newcastle University, UK
Lutz Hamel, University of Rhode Island, USA
Wei-Chiang Hong, Oriental Institute of Technology, Taiwan
Seiya Imoto, University of Tokyo, Japan
Liu Jing, Xidian University, China
Colin Johnson, University of Kent, UK
Mark Johnston, California Institute of Technology, USA

Iwona Karcz-Duleba, Wroclaw University of Technology, Poland
Ahmed Kattan, AI Real-World Applications Research Laboratory, UK
Andy Keane, University of Southampton, UK
Ed Keedwell, University of Exeter, UK
Ziad Kobti, University of Windsor, Canada
Mario Köppen, Kyushu Institute of Technology, Japan
Ondrej Krejcar, University of Hradec Kralove, Czech Republic
Jiri Kubalik, Czech Technical University, Czech Republic
Dario Landa-Silva, University of Nottingham, UK
Antonio J. Fernández Leiva, Universidad de Málaga, Spain
Piotr Lipinski, University of Wroclaw, Poland
Wenjian Luo, University of Science and Technology of China, China
Penousal Machado, University of Coimbra, Portugal
Rainer Malaka, Bremen University, Germany
Euan William McGookin, University of Glasgow, UK
Jörn Mehnen, Cranfield University, UK
Juan J. Merelo, Universidad de Granada, Spain
Marjan Mernik, University of Maribor, Slovenia
Konstantinos Michail, Cyprus University of Technology, Cyprus
Sanaz Mostaghim, Karlsruhe Institute of Technology (KIT), Germany
Luiza de Macedo Mourelle, State University of Rio de Janeiro, Brazil
Pawel B. Myszkowski, Wroclaw University of Technology, Poland
Tomoharu Nakashima, Osaka Prefecture University, Japan
Kei Ohnishi, Kyushu Institute of Technology, Japan
Schütze Oliver, CINVESTAV-IPN, Mexico
Gary B. Parker, Connecticut College, USA
Aurora Pozo, Federal University of Parana, Brazil
Joaquim Reis, ISCTE, Portugal
José Risco-Martín, Universidad Complutense de Madrid, Spain
Mateen Rizki, Wright State University, USA
Katya Rodriguez, Instituto de Investigaciones en Matemáticas Aplicadas y en Sistemas (IIMAS), Mexico
Agostinho Rosa, IST, Portugal
Suman Roychoudhury, Tata Consultancy Services, India
Filipe Azinhais Santos, ISCTE-IUL, Portugal
Miguel A. Sanz-Bobi, Pontificia Comillas University, Spain
Emmanuel Sapin, University of Exeter, UK
Robert Schaefer, AGH University of Science and Technology, Poland
Adam Slowik, Koszalin University of Technology, Poland
Emilia Tantar, University of Luxembourg, Luxembourg
Jonathan Thompson, Cardiff University, UK
Krzysztof Trojanowski, Cardinal Stefan Wyszynski University, Poland
Elio Tuci, Aberystwyth University, UK
Massimiliano Vasile, University of Strathclyde, UK
Neal Wagner, Fayetteville State University, USA

Peter Whigham, University of Otago, New Zealand
Gary Yen, Oklahoma State University, USA

FCTA Program Committee

Sansanee Auephanwiriyakul, Chiang Mai University, Thailand
Daniel Antonio Callegari, PUC-RS Pontificia Universidade Catolica do Rio Grande
do Sul, Brazil
Heloisa Camargo, UFSCar, Brazil
Rahul Caprihan, Dayalbagh Educational Institute, India
João Paulo Carvalho, INESC-ID/Instituto Superior Técnico, Portugal
France Cheong, RMIT University, Australia
Francisco Chiclana, De Montfort University, UK
Mikael Collan, Lappeenranta University of Technology, Finland
Shuang Cong, University of Science and Technology of China, China
Martina Dankova, University of Ostrava, Czech Republic
Kudret Demirli, Concordia University, Canada
Scott Dick, University of Alberta, Canada
József Dombi, University of Szeged, Institute of Informatics, Hungary
António Dourado, University of Coimbra, Portugal
Didier Dubois, Institut de Recherche en Informatique de Toulouse, France
Yoshikazu Fukuyama, Meiji University, Japan
Robert Fullér, Obuda University, Hungary
Alexander Gegov, University of Portsmouth, UK
Brunella Gerla, University of Insubria, Italy
Maria Angeles Gil, University of Oviedo, Spain
Sarah Greenfield, De Montfort University, UK
Masafumi Hagiwara, Keio University, Japan
Susana Muñoz Hernández, Universidad Politécnica de Madrid (UPM), Spain
Francisco Herrera, University of Granada, Spain
Kaoru Hirota, Tokyo Institute of Technology, Japan
Katsuhiro Honda, Osaka Prefecture University, Japan
Chih-Cheng Hung, Southern Polytechnic State University, USA
Lazaros S. Iliadis, Democritus University of Thrace, Greece
Angel A. Juan, Open University of Catalonia, Spain
Cengiz Kahraman, Istanbul Technical University, Turkey
Frank Klawonn, Ostfalia University of Applied Sciences, Germany
Donald H. Kraft, Colorado Technical University, USA
Rudolf Kruse, Otto-von-Guericke University Magdeburg, Germany
Anne Laurent, Lirmm, University Montpellier 2, France
Kang Li, Queen's University Belfast, UK
Chin-Teng Lin, National Chiao Tung University, Taiwan
Ahmad Lotfi, Nottingham Trent University, UK

Edwin Lughofer, Johannes Kepler University, Austria
Francesco Marcelloni, University of Pisa, Italy
Corrado Mencar, University of Bari, Italy
Ludmil Mikhailov, University of Manchester, UK
Javier Montero, Complutense University of Madrid, Spain
Alejandro CarrascoMuñoz, University of Seville, Spain
Vesa Niskanen, University of Helsinki, Finland
Yusuke Nojima, Osaka Prefecture University, Japan
Vilém Novák, University of Ostrava, Czech Republic
Vasile Palade, Oxford University, UK
Valentina Plekhanova, University of Sunderland, UK
Daowen Qiu, Sun Yat-sen University, China
Jordi Recasens, Universitat Politècnica de Catalunya, Spain
Antonello Rizzi, University of Rome "La Sapienza", Italy
Roseli A. Francelin Romero, University of São Paulo, Brazil
Alessandra Russo, Imperial College London, UK
Hooman Tahayori, Ryerson University, Canada
Vicenc Torra, IIIA-CSIC, Spain
Dat Tran, University of Canberra, Australia
Wen-June Wang, National Central University, Taiwan
Thomas Whalen, Fromtline Health Care Workers Safety Foundation, USA
Jianqiang Yi, Institute of Automation, Chinese Academy of Sciences, China
Slawomir Zadrozny, Polish Academy of Sciences, Poland
Hans-Jürgen Zimmermann, European Laboratory for Intelligent Techniques
Engineering (ELITE), Germany

FCTA Auxiliary Reviewers

Michela Antonelli, University of Pisa, Italy
Christian Moewes, University of Magdeburg, Germany

NCTA Program Committee

Shigeo Abe, Kobe University, Japan
Francisco Martínez Álvarez, Pablo de Olavide University of Seville, Spain
Veronique Amarger, University PARIS-EST Créteil (UPEC), France
Gilles Bernard, Paris 8 University, France
Daniel Berrar, Tokyo Institute of Technology, Japan
Yevgeniy Bodyanskiy, Kharkiv National University of Radio Electronics, Ukraine
Antonio Padua Braga, Universidade Federal de Minas Gerais, Brazil

Ivo Bukovsky, Czech Technical University in Prague, Faculty of Mechanical Engineering, Czech Republic
Ning Chen, Instituto Superior de Engenharia do Porto, Portugal
Amine Chohra, University PARIS-EST Créteil (UPEC), France
Catalina Cocianu, The Bucharest University of Economic Studies, Faculty of Cybernetics, Statistics and Informatics in Economy, Romania
Shuang Cong, University of Science and Technology of China, China
Leonardo Franco, Universidad de Málaga, Spain
Josep Freixas, Escola Politècnica Superior d'Enginyeria de Manresa, Spain
Marcos Gestal, University of A Coruña, Spain
Vladimir Golovko, Brest State Technical University, Belarus
Michèle Gouiffès, Institut D'Electronique Fondamentale (IEF) CNRS 8622 University Paris-Sud 11, France
Barbara Hammer, Bielefeld University, Germany
Chris Hinde, Loughborough University, UK
Alexander Hošovský, Technical University of Kosice, Slovak Republic
Gareth Howells, University of Kent, UK
Yuji Iwahori, Chubu University, Japan
Magnus Johnsson, Lund University, Sweden
Juha Karhunen, Aalto University School of Science, Finland
Christel Kemke, University of Manitoba, Canada
Mohamed A. Khabou, University of West Florida, Pensacola, USA
Adnan Khashman, Near East University, Turkey
Dalia Kriksciuniene, Vilnius University, Lithuania
Adam Krzyzak, Concordia University, Canada
Honghai Liu, University of Portsmouth, UK
Jinhu Lu, Chinese Academy of Sciences, China
Jinwen Ma, Peking University, China
Hichem Maaref, IBISC Laboratory, France
Kurosh Madani, University of Paris-Est Créteil (UPEC), France
Yong Mao, UT MD Anderson Cancer Center, USA
Jean-Jacques Mariage, Paris 8 University, France
Mitsuharu Matsumoto, The University of Electro-Communications, Japan
David Gil Mendez, University of Alicante, Spain
Christo Panchev, University of Sunderland, UK
George Perry, University of Texas at San Antonio, USA
Manuel Roveri, Politecnico di Milano, Italy
Neil Rowe, Naval Postgraduate School, USA
Christophe Sabourin, IUT Sénart, University PARIS-EST Créteil (UPEC), France
Carlo Sansone, University of Naples, Italy
Gerald Schaefer, Loughborough University, UK
Alon Schclar, Academic College of Tel-Aviv Yaffo, Israel
Christoph Schommer, University Luxembourg, Campus Kirchberg, Luxembourg
Catherine Stringfellow, Midwestern State University, USA
Mu-Chun Su, National Central University, Taiwan

Johan Suykens, K.U. Leuven, Belgium
Norikazu Takahashi, Okayama University, Japan
Ah Hwee Tan, Nanyang Technological University, Singapore
Yi Tang, Yunnan University of Nationalities, China
Carlos M. Travieso, University of Las Palmas de Gran Canaria, Spain
Andrei Utkin, INOV INESC Inovação, Portugal
Brijesh Verma, Central Queensland University, Australia
Ricardo Vigário, Aalto University, Finland
Shuai Wan, Northwestern Polytechnical University, China
Fei Wang, IBM Almaden Research Center, USA
Yingjie Yang, De Montfort University, UK
Weiwei Yu, Northwestern Polytechnical University, China
Wenwu Yu, Southeast University, China
Cleber Zanchettin, Federal University of Pernambuco, Brazil

NCTA Auxiliary Reviewers

Peter Benes, Czech Technical University in Prague, Czech Republic
Giacomo Boracchi, Politecnico di Milano, Italy
Matous Cejnek, CVUT, Czech Republic
Gualberto Asencio Cortés, Universidad Pablo de Olavide, Spain
Antonio Marullo, Politecnico di Milano, Italy
Cyril Oswald, Faculty of Mechanical Engineering, Czech Technical University in
Prague, Czech Republic
Jorge Reyes, NT2 Labs, Chile
Thiago Rodrigues, Cefet-MG, Brazil

Invited Speakers

Kevin Warwick, University of Reading, UK
Leslie Smith, University of Stirling, UK
Juan J. Merelo, Universidad de Granada, Spain
Eyke Hüllermeier, Philipps University of Marburg, Germany
Belur V. Dasarathy, Information Fusion, USA
Alexandre Castro-Caldas, Portuguese Catholic University, Portugal

Part I
Evolutionary Computation Theory
and Applications

Incremental Hough Transform: A New Method for Circle Detection

A. Oualid Djekoune, Khadidja Messaoudi and Mahmoud Belhocine

Abstract The circle Hough transform (CHT) is a fundamental issue in image processing applications of industrial parts or tools. Because of its drawbacks, various modifications have been suggested to increase its performance. Most of them have met the problem of implicit evaluation of trigonometric functions that makes the implementation difficult. The CORDIC algorithm is used to simplify the trigonometric calculations when the basic CHT algorithm is implemented into a digital device such as FPGA. Although, this solution require computation time and device resources consumption for the CORDIC IP implementation. This paper presents a modified CHT method, called Incremental circle Hough transform (ICHT), suitable for hardware implementation. This method is mainly used to get around the implementation of CORDIC IP. This paper provides also the errors analysis of the proposed method against the basic CHT method to illustrate that it can replace the basic CHT method for small values of the resolution ε of the angle θ.

Keywords Hough transform · Incremental hough transform · Circle hough transform · Circle detection

1 Introduction

Shape recognition is one of the most important tasks in the image processing and pattern recognition. Many methods for detecting geometric primitives have been proposed. The Hough transform (HT) and its extensions constitute a popular and robust method for extracting analytic curves. It was first applied to the recognition of

A.O. Djekoune (✉) · M. Belhocine
Robotics and Industrial Automation, Advanced Technologies Development Centre,
Lotissement 20 Août 1956 Baba Hassen, BP17 Algiers, Baba Hassen, Algeria
e-mail: odjekoune@cdta.dz

K. Messaoudi
Systems Architectures and Multimedia, Advanced Technologies Development Centre,
Lotissement 20 Août 1956 Baba Hassen, BP17 Algiers, Baba Hassen, Algeria

© Springer International Publishing Switzerland 2016
K. Madani et al. (eds.), *Computational Intelligence*,
Studies in Computational Intelligence 613,
DOI 10.1007/978-3-319-23392-5_1

straight lines [3] and later extended to circles [4], ellipses [21] and arbitrarily shaped objects [15].

The principal concept of the HT is to define a mapping between an image space and a parameter space. Each feature point (or a set of feature points) in an image is mapped to the parameter space to vote for the parameters whose associated curves pass through the data point(s). The votes for each curve are accumulated, and after all the points in an image have been considered, local maxima in the parameter space correspond to the parameters of the detected curves. The curves detection in the image space therefore become a peak detection problem in the parameter space. The advantages of the HT include robustness to noise, shape distortions and to occlusions/missing parts of an object. Its main disadvantage is the fact that computational and storage requirements of the algorithm increase as a power of the dimensionality of the curve. This means that the computational complexity and storage requirements are $O(n^2)$ for straight lines, $O(n^3)$ for circles and $O(n^5)$ for ellipses [5].

To overcome of this disadvantage, we introduce in this paper a new modified CHT method, called Incremental circle Hough transform (ICHT), that is aimed at improving the voting process. For each input point in the image, the new method computes incrementally the circle point coordinates passing through the input point using new formulation of the parametric representation of the circle. By using approximations on cosine and sine in the parametric representation of the circle, the new formulation is: easy to use because the point coordinates of the circle at the iteration n is computed from the coordinates point of the circle of the iteration n − 1 by using simple equations; provides a solution to the use of trigonometric functions that causes problems in digital device implementations such as FPGA; can be seen as a solution to the use of the CORDIC (Cordinate Rotation Digital Computer) algorithm, and finally very suitable for parallelization in the calculation of circles point coordinates that passing through the input point in the image.

In this paper, we show in detail the feasibility and the simplicity of the introduced method which can replace the basic CHT method both in software and hardware applications.

The paper is organized as follow: In Sect. 2, we position our work relative to some published CHT methods existing in the literature. In Sect. 3, we present an overview of the basic CHT algorithm followed by its hardware implementation with or without the CORDIC algorithm. The details, algorithm development, the errors analysis and the software implementation of the proposed method are shown in the Sect. 4. Finally, conclusion and future work are given in the Sect. 5.

2 Related Works

Extracting circles from images has received more attention for several decades because an extracted circle can be used to yield the location of circular object in many industrial applications. Many variations on original CHT method have been proposed to increase its performance. One type of method address issues of efficiency to reduce significantly the amount of computation and storage required to implement the Hough transform [14], Another type of method replaces the formal parameterization of the target object with a look-up table, the Generalized Hough Transform (GHT), allowing the Hough approach to be used to detect arbitrary shapes [1]. A third type of method uses the probabilistic interpretation of the Hough approach [13, 16]. Other type of method uses the randomized selection of edge points and geometrical properties of the circle [2, 10, 20], and the edge orientation information of each edge pixel to reduce the computing time or the requirement of the accumulator [12]. And finally the type of method proposes a variety of voting scheme used in the Hough transform [17]. An excellent reviews of a number of circle detection methods based on variations of the Hough transform can be found in [22].

Other than the software solution of the CHT drawbacks, we also find in the literature, the hardware solution which provides an attractive solution to computationally intensive applications in real-time whilst maintaining the flexibility of a software solution [7, 18]. The Hough Transform has traditionally been implemented using complex processor architecture. These are either slow or complicated due to the transform's intensive calculations of trigonometric, multiplication and addition operations [6]. To overcome this major setback, the CORDIC algorithm is used [6, 8]. The CORDIC algorithm can be used to calculate elementary trigonometric functions such as sine, cosine, tangent, and arctangent as well as *ln* and *exp*.

In this context, we present in this work a new ICHT method aimed at improving the voting process. This method fully both exploits the software and the hardware solution advantages because it doesn't use any trigonometric calculations, simple to use, easily fitted into digital device, such as FPGA, without consuming too device resources, and very suitable for a parallel implementation.

3 The Basic CHT Method

3.1 An Overview

A circle with centre *(a, b)* and radius r, in a binary image, is specified by the parameters *(a, b, r)* in the equation:

$$(x - a)^2 + (y - b)^2 = r^2 \tag{1}$$

with *(x, y)* the set edge pixels that make up the circumference of this circle.

The parametric representation of the circle is:

$$\begin{cases} x = a + r \cos\theta \\ y = b + r \sin\theta \end{cases} \tag{2}$$

For each edge pixel, the basic Hough transform method constructs a circular cone, in the (a, b, r) parameter space (or Hough space), resulting from the voting process of the (a, b, r) parameters whose associated circles pass through the considered pixel by using a fourfold loop over x, y, a and b (Fig. 1). This operation runs slowly because it is mainly due to the both use a large number of mathematical operations (1) and trigonometric calculations (2). This raises the computational cost of the transform, often to unacceptable levels.

For simplicity, some works in the literature set the radius to a constant value (hard coded) or provide the user with the option of setting a range (maximum and minimum) prior to running the application, or use the edge direction information to limit voting to a section of the cone, or use the CORDIC algorithm to overcome to intensive calculations of trigonometric, multiplication and addition operations of the CHT [8].

3.2 Hardware Implementation

The CORDIC algorithm is one of the existing hardware solutions of the CHT (or HT) implementation to overcome its intensive calculation of trigonometric, multiplication and addition operations. In follows we will show how the CORDIC algorithm works, its hardware implementation on a FPGA circuit alone and with the basic CHT algorithm.

3.2.1 The CORDIC Algorithm

The CORDIC algorithm, proposed by Volder in 1959 [19], is used to calculate elementary trigonometric functions such as sine, cosine, tangent, and arctangent as well as *ln* and *exp*. It provides an iterative method of performing vector rotations by arbitrary angles using only shifts and adds. Let two points (x, y) and (x', y') resulting of a rotation of a vector by an angle φ (Fig. 2). We have then:

$$\begin{cases} x = R \sin\beta \\ y = R \cos\beta \end{cases} \tag{3}$$

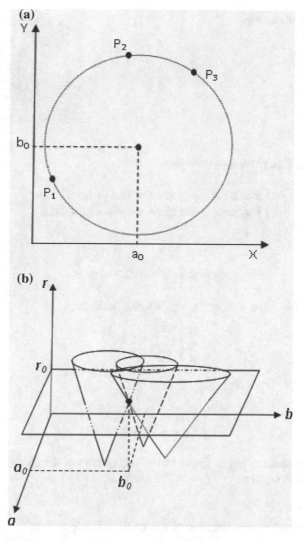

Fig. 1 Relationship between a binary image plan and the Hough space. **a** P_1, P_2 and P_3 are edge pixels belonging to a same imaginary circle with r_0 the radius and (a_0, b_0) the coordinates of its centre. **b** Each edge pixel from the binary image generates a circular cone in the Hough space. The cones in the Hough space intersect at (a_0, b_0, r_0) corresponding to the parameters of the circle formed by the edge pixels P_1, P_2 and P_3

and

$$\begin{cases} \acute{x} = R\,sin(\beta + \varphi) \\ \acute{y} = R\,cos(\beta + \varphi) \end{cases} \qquad (4)$$

Fig. 2 Rotation of vector by
the angle φ

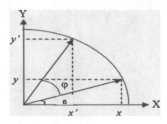

combining the Eqs. (3) and (4), we get:

$$\begin{cases} \acute{x} = x\,cos\varphi - y\,sin\varphi = (x - y\,tan\varphi)\,cos\varphi \\ \acute{y} = y\,cos\varphi + x\,sin\varphi = (y + x\,tan\varphi)\,cos\varphi \end{cases} \tag{5}$$

we have:

$$\varphi_0 = atan2^{-j} \approx 2^{-j},\ j > 3 \tag{6}$$

if we choose $j = 4$, so $\varphi \ll$ and $cos\varphi \approx 1$, we get then:

$$\begin{cases} \acute{x} = x - y\,2^{-4} \\ \acute{y} = y + x\,2^{-4} \end{cases} \tag{7}$$

to generalize for all points:

$$\begin{cases} X_{i+1} = X_i - Y_i\,2^{-4} \\ Y_{i+1} = Y_i - X_i\,2^{-4} \end{cases} \tag{8}$$

where i is the iteration index. So, to sweep one quarter of the circle, the number of micro rotation is equal to ${}^{(\pi/2)}/_{2^{-4}} \approx 25$ iterations.

3.2.2 Hardware Implementation of the CORDIC Algorithm

A hardware implementation of the CORDIC algorithm was presented in [9]. In this work the CORDIC IP was implemented on Agility RC10 board witch contains the Xilinx Spartan 3L XC3S1500L-4-FG320 FPGA circuit, in order to compute the following functions: sinus, cosine, and arctangent and vector magnitude (Fig. 3). The implementation results are given in the Table 1. Implemented alone, it consumes a significant amount of device resources.

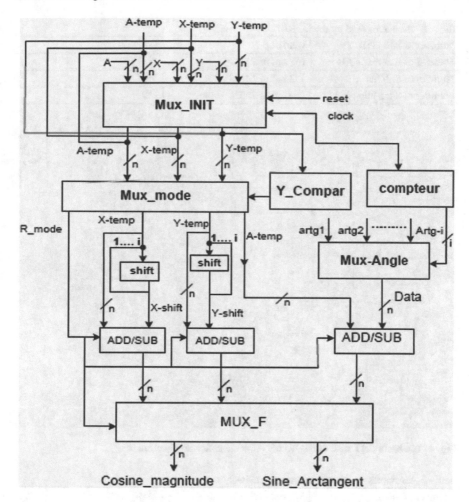

Fig. 3 CORDIC IP architecture [9]

3.2.3 Hardware Implementation of the CORDIC with the Basic CHT Algorithms

The hardware implementation on a FPGA circuit of the basic CHT with the CORDIC algorithms was presented in [8] (Fig. 4). In this work, the CORDIC IP is used in rotation mode in order to compute the sine and cosine functions. The resulting architecture was implemented on the V2MB1000 board witch contains the Xilinx Virtex-II, XC2V1000-4fg456C FPGA device. This implementation results in a complication of the final architecture and a significant consumption of the device resources. The device resources consumption are given in the Table 2.

Table 1 Resources used by the CORDIC IP

Number of BUFGMUXs	1 out of 8	12%
Number of External IOBs	162 out of 221	73%
Number of SLICEs	352 out of 13312	2%
Number of Slice Flip Flop	166 out of 26624	1%

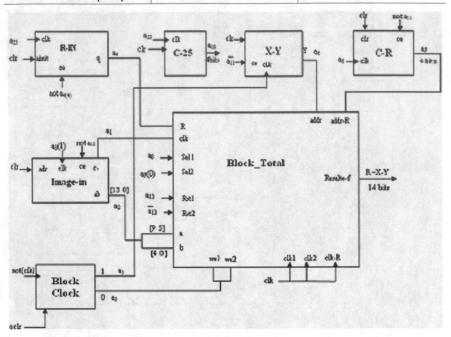

Fig. 4 The basic CHT with the CORDIC implementation architecture [8]

Table 2 Resources used of the proposed architecture

Number of Slices	578 out of 5120	11%
Number of ExternalIOBs	19 out of 324	5%
Number of GCLKs	3 out of 16	18%
Number of BRAMs	8 out of 40	20%

Despite the implementations above are not performed on the same Xilinx device board, we can estimate that the CORDIC device resources occupation regarding the number of slices is almost half (578/352).

Furthermore, the basic CHT is generally not implemented alone. It is a module often used with other modules performing a given task. Save device resources in a

FPGA circuit is giving the possibility to integrate other modules in the same circuit thus avoiding the use of a additional FPGA circuits. Hence our motivation, find a trick to avoid the use of the CORDIC algorithm when implementing the basic CHT.

4 The New ICHT Method

4.1 Algorithm Development

The main goal of this work is to try to improve the basic CHT to make it simple to use, and easily adapted and fitted into the digital device without consuming too device resources. Thus combining both advantages of hard and soft solutions described above.

Our improvement mainly concerns the voting process, it uses new equations, or formulation, of the parametric representation of a circle. These equations compute incrementally the coordinates point of a circle, such that each coordinates point of a circle at the iteration n is computed from the coordinates point of the same circle of the iteration n − 1. We proceeded as follows:

For discrete values of the angle, (2) is written as follows we have used the same notation as in [7, 18]:

$$
\begin{cases}
x_n = a + r \cos \theta_n \\
y_n = b + r \sin \theta_n \\
x_0 = a + r \\
y_0 = b \\
\theta_n = n\,\varepsilon \\
n_\theta = \frac{2\pi}{\varepsilon} \\
0 \leqslant \theta_n < 2\pi \\
0 \leqslant n
\end{cases}
\tag{9}
$$

with n, ε and n_θ are, respectively, the angle index, the angle resolution and the number of angle values in the θ interval.

To make (3) as incremental, it must be written in the following form:

$$
\begin{cases}
x_{n+1} = f(x_n, y_n) \\
y_{n+1} = f(x_n, y_n)
\end{cases}
\tag{10}
$$

i.e., the point coordinates of the circle (x_{n+1}, y_{n+1}) at the iteration $n + 1$ is only computed from the point coordinates of the circle (x_n, y_n) of the iteration n.

By replacing n by $n + 1$ in (9), we will have:

$$\begin{cases} x_{n+1} = a + r\cos\theta_{n+1} \\ \quad\quad = a + r\cos[(n+1)\varepsilon] \\ \quad\quad = a + r[\cos n\varepsilon \cos\varepsilon - \sin n\varepsilon \sin\varepsilon] \\ y_{n+1} = b + r\sin\theta_n \\ \quad\quad = b + r\sin[(n+1)\varepsilon] \\ \quad\quad = b + r[\sin n\varepsilon \cos\varepsilon - \cos n\varepsilon \sin\varepsilon] \end{cases} \quad (11)$$

Making the approximation, in the expression above, on cosine and sine for the small values of the angle by assuming $\cos\varepsilon = 1$ and $\sin\varepsilon = \varepsilon$. The Eq. (11) becomes:

$$\begin{cases} x_{n+1} = a + r[\cos n\varepsilon - \varepsilon \sin n\varepsilon] \\ \quad\quad = x_n - r\varepsilon \sin n\varepsilon \\ y_n = b + r[\sin n\varepsilon + \varepsilon \cos n\varepsilon] \\ \quad\quad = y_n - r\varepsilon \cos n\varepsilon \end{cases} \quad (12)$$

Note that from (9), we can get:

$$\begin{cases} r\cos n\varepsilon = x_n - a \\ r\sin n\varepsilon = y_n - b \end{cases} \quad (13)$$

By replacing (13) in (12), then rearranging the obtained expression to get the following general expression of our new ICHT method:

$$\begin{cases} x_{n+1} = x_n - \varepsilon y_n + \varepsilon b \\ y_{n+1} = y_n + \varepsilon x_n - \varepsilon a \\ x_0 = a + r \\ y_0 = b \\ 0 \le n < n_\theta \\ n_\theta = \frac{2\pi}{\varepsilon} \end{cases} \quad (14)$$

We can note that (2) and (14) are almost similar except that (2) is highly dependent to the trigonometric functions, which is not the case with (14). We can therefore conclude that (14) is:

- Purely incremental,
- Doesn't use any trigonometric calculations,
- Simple to use,
- Can be seen as an alternative to the CORDIC algorithm,
- Can be easily fitted into digital device such as FPGA,
- Can be very suitable for parallelization.

Fig. 5 The circles
Circle$_{ICHT}$ and Circle$_{CHT}$
drawn at a fixed position with
radius $= 100$ and $\varepsilon = 0.5°$

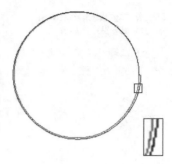

4.2 Error Analysis

In the following, we show the errors, if exist, caused by the above approximations when using (2) and (14) to draw circles.

When drawing two circles (Circle$_{ICHT}$ and Circle$_{CHT}$) with the same parameters using (2) and (14), we note that the points of the two circles overlap for small values of θ and diverge for larger values of θ (Fig. 5).

There are many criteria which can be considered to evaluate this divergence, but in our study the most important point relates to errors analysis. The errors analysis is measured using: the average error ($E_{average}$) and the quadratic error ($E_{quadratic}$) between the radii resulting from the generated points using the two above equations; the difference area ($Diff_{area}$) and the Jaccard coefficient ($Coef_{Jaccard}$) to compare the similarity of the two generated circles. The Jaccard coefficient measures the ratio of the intersection area of two sets divided by the area of their union [11].

These errors are computed from different values of the radius R and the resolution ε of the θ angle. They are expressed as follows:

$$E_{average} = \frac{1}{n_\theta} \sum_{n=0}^{n_\theta-1} (R_{ICHT}[n] - R_{CHT}[n]) \tag{15}$$

$$E_{quadratic} = \frac{1}{n_\theta} \sum_{n=0}^{n_\theta-1} (R_{ICHT}[n] - R_{CHT}[n])^2 \tag{16}$$

with $R_{CHT}[n]$ and $R_{ICHT}[n]$ the radii computed from the coordinates point at the nth value of θ using (2) and (14).

It is interesting to note that our new ICHT method achieves very small errors for small values of the resolution ε of the angle θ which remain within a narrow tolerance despite the high that can have the radius R. But these errors increase considerably when the resolution ε increases with high value of the radius R (Figs. 6, 7, 8 and 9).

The Figs. 6, 7, 8 and 9 show that for values of the resolution ε less than 1°, the errors $E_{average}$, $E_{quadratic}$ and $Diff_{area}$ are very small, and consequently the $Coef_{Jaccard}$ value reach the one value. The one value means that the two circles are substantially

Fig. 6 The average error in
a 2D and **b** 3D version

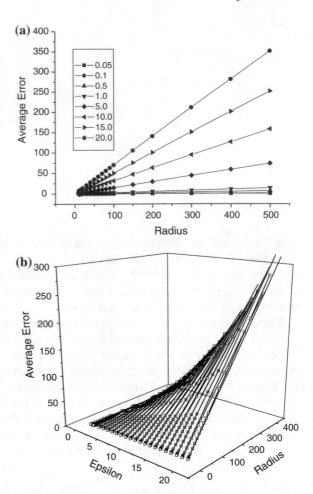

similar. Beyond the value $1°$ and for small values of the radius R, the $Coef_{Jaccard}$ value decreases giving rise to significant divergences.

The Figs. 6 and 7 show that the average error and the difference area between $Circle_{ICHT}$ and $Circle_{CHT}$ circles increases linearly with a slope depending on the values of the resolution ε of the angle θ. This gives us an idea of how the computed points, from expression (14), diverge from those computed from (2).

In conclusion, our new ICHT method can replace the basic CHT method for small values of the resolution ε of the angle θ with the advantage of does'nt using any trigonometric calculations, It then compete the CORDIC algorithm when implemented into digital device.

Fig. 7 The quadratic error
in **a** 2D and **b** 3D version

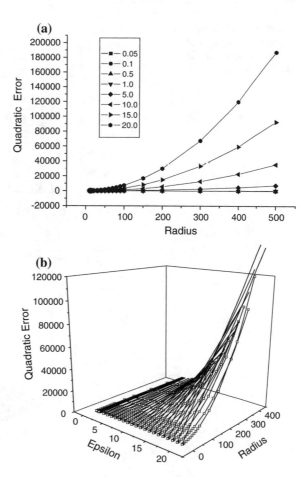

4.3 Implementation

The new ICHT method is tested against the basic CHT method. This will be done
to illustrate the consequences of the used approximations in the parametric representation of the circle in the processing time of the voting process and to see the
computational efficiency of the Hough space of the two methods. The new ICHT
method and the basic CHT method were implemented in the programming language
Matlab v.7. The implementation was performed using a laptop PC equipped with 2.6
GHz i5 processor and 6 GB RAM. A real grey scale image, of size 225×220 pixel

Fig. 8 The difference area between $Circle_{ICHT}$ and $Circle_{CHT}$ circles in **a** 2D and **b** 3D version

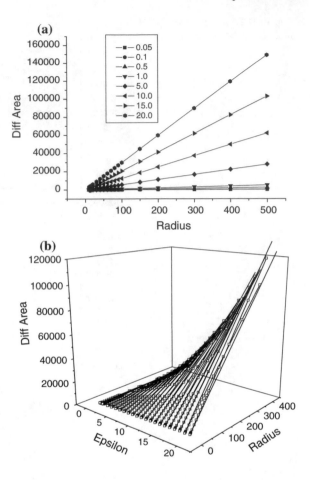

is used (Fig. 10). The binary edge points shown in (Fig. 8) are obtained by using the Matlab Canny operator.

The voting process algorithm of the new ICHT and the basic CHT methods, applied in a binary edge image, are performed using $r_{min} = 10$ and $r_{Max} = \sqrt{225^2 + 220^2} \cong 315$. The ε resolution value of the θ angle is initially set by the user.

The Table 3 assess the time processing of the two methods, where the time of the voting process, the time required to process one binary edge pixel, and the time ratio are presented. The binary edge image of the Fig. 11 is used using different values of the resolution ε of the θ angle. The processing time per pixel, in milliseconds, is obtained by dividing the time of the voting process, expressed in second, by the

Fig. 9 The *Jaccard* coefficient in **a** 2D and **b** 3D version

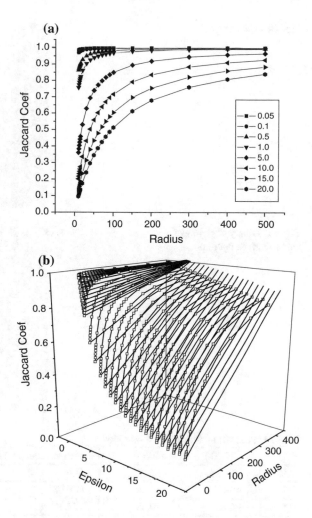

number of the binary edge point contained in the Fig. 11, in our case this number is equal to 2508. The time ratio is obtained, in this case, by dividing the time of the voting process of the basic CHT method by the time of the voting process of the new ICHT method. The Table 3 not only show that the new ICHT method is fast more than the basic CHT method but it is more than two time faster.

After evaluating the processing time of the two methods, now we try to show the Hough space obtained from these two methods. The Figs. 12 and 13 show the plans of the Hough space with the radius $R = 10$, and confirm the conclusions done above. The Hough spaces obtained from these two methods are the same for values of the resolution less than $1°$, and diverge significantly beyond the value $1°$.

Fig. 10 The test image

Fig. 11 Binary edge image
using *Deriche* operator

Table 3 Processing time of the voting process

$\varepsilon(°)$	CHT(s)	ICHT(s)	Time per pixel(ms) CHT/ICHT	Time ratio
0.05	862.39	412.51	343.85/164.48	2.09
0.1	438	211.60	174.64/84.37	2.07
0.5	96.69	49.55	38.54/19.76	1.95
1.0	50.87	26.24	20.28/10.46	1.94
5.0	10.5	5.15	4.19/2.05	2.04
10.0	5.32	2.51	2.12/1.0	2.12
15.0	3.6	1.62	1.43/0.64	2.22
20.0	2.73	1.20	1.09/0.48	2.27

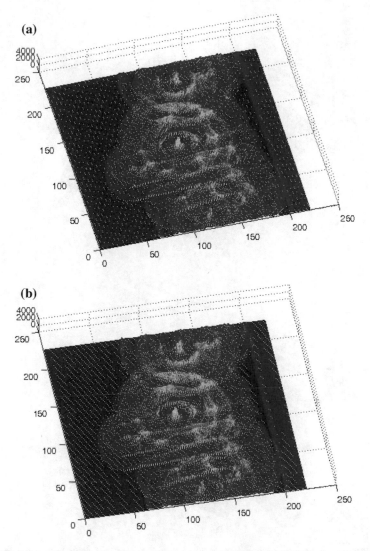

Fig. 12 Hough space plan with radius $= 20$ and $\varepsilon = 0.05°$ obtained from both **a** the basic CHT method and **b** our new ICHT method

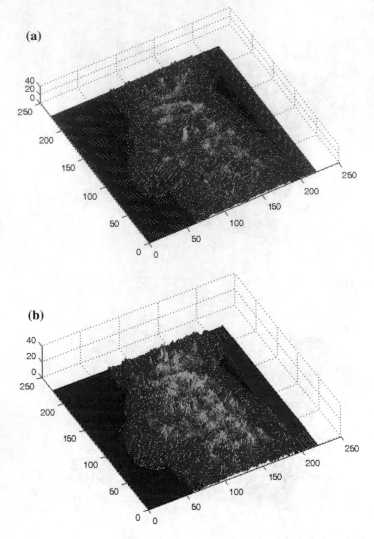

Fig. 13 Hough space plan with radius = 20 and $\varepsilon = 5°$ obtained from both **a** the basic CHT method and **b** our new ICHT method

5 Conclusions

We presented a new modified CHT method with enhanced formulation for improving the computational performance and efficiency of the voting process of the basic CHT. Called Incremental circle Hough transform (ICHT), the method fully both exploits the software and the hardware solution advantages with no trigonometric calculations, it can be seen as an alternative to the CORDIC algorithm and consequently easily fitted into digital device, such as FPGA, without consuming too device resources, and very suitable for a parallel implementation.

We have presented theoretical and errors analysis of our method, and have shown experimentally that, for small values of angle, the new method has the same accuracy as the basic CHT method.

We are currently trying to further improve the time of the voting process of the proposed ICHT by changing if possible the expression (14).

Acknowledgments The authors would like to thank all those who helped to achieve this modest work as well as their useful discussions and comments.

References

1. Ballard, D.: Generalised Hough transform to detect arbitrary shapes. IEEE Trans. Pattern Anal. Mach. Intell. **13**(2), 111–122 (1981)
2. Bandera, A., Perez-Lorénzo, J.M., Bandera, J.P., Sandoval, F.: Mean shift based clustering of Hough domain for fast line segment detection. Pattern Recognit. Lett. **27**(6), 578–586 (2006)
3. Duda, R., Hart, P.: Use of the Hough transform to detect lines and curves in pictures. Commun. ACM **15**(1), 11–15 (1972)
4. Davies, E.: A modified Hough scheme for general circle location. Pattern Recognit. Lett. **7**, 37–43 (1987)
5. Ioannou, D., Huda, W., Laine, A.F.: Circle recognition through a 2D Hough Transform and radius histogramming. Image Vis. Comput. **17**, 15–26 (1999)
6. Deng, D.D.S., ElGindy, H.: High-speed parameterisable hough transform using reconfigurable hardware, In: Proceedings of the Pan-Sydney Area Workshop on Visual Information Processing. VIP01, vol.11, pp. 51–57. Sydney, May 2001
7. Djekoune, A.O., Achour, K.: Incremental hough transform: an improved algorithm for digital device implementation. Real-Time Imaging **10**(6), 351–363 (2004)
8. Ferhat-taleb Alim, F., Messaoudi, K., Seddiki, S., Kerdjidj, O.: Modified circular hough transform using FPGA. In: ICM, Algiers (2012)
9. Ferhat-taleb Alim, F., Messaoudi, K., Ait Mohamed L., Kerdjidj, O., Seddiki, S.: CORDIC IP description with hande 1-C and VHDL languages. In: Signal PROCESSING Algorithms, Architectures, Arrangements, and Applications SPA13, Poznań, 26–28 September 2013
10. Ho, Chun, Chen, L.: A fast ellipse/circle detector using geometric symmetry. Pattern Recognit. **28**(1), 117–124 (1995)
11. Jaccard, P.: The distribution of flora in the alpine zone. New Phytol. **11**(2), 37–50 (1912)
12. Kimme, C., Ballard, D., Sklansky, J.: Finding circles by an array of accumulators. Proc ACM **18**, 120–122 (1975)
13. Kälviäinen, H., Hirvonen, P., Xu, L., Oja, E.: Probabilistic and non-probabilisticHough transforms: overview and comparisons. Image Vis. Comput. **13**(4), 239–252 (1995)

14. Li, H., Lavin, M.A., LeMaster, R.J.: Fast Hough transform: a hierarchical approach. Comput. Vis. Graph. Image Process **36**, 139–161 (1986)
15. Pao, D.C.W., Li, H.F., Jayakumar, R.: Shapes recognition using the straight line Hough transform: theory and generalizaion. IEEE Trans. Pattern Anal. Mach. Intell. **14**, 1076–1089 (1992)
16. Stephens, R.S.: A probabilistic approach to the Hough transform. Proc. British Mach. Vis. Con. 55–60 (1990)
17. Guo, S., Pridmore, T., Kong, Y., Zhang, X.: An improved Hough transform voting scheme utilizing surround suppression. Pattern Recognit. Lett. **30**, 1241–1252 (2009)
18. Tagzout, S., Achour, K., Djekoune, O.: Hough transform algorithm for FPGA implementation. Signal Process. **81**(6), 1295–1301 (2001). Elsevier
19. Volder, J.E.: The CORDIC trigonometric computing technique. IRE Trans. Electron. Comput **EC–8**(N° 3), 330–334 (1959)
20. Xu, L., Oja, E., Kultanan, P.: A new curve detection method: randomized Hough transform (RHT). Pattern Recognit. Lett. **11**(5), 331–338 (1990)
21. Yip, R., Tam, P., Leung, D.: Modification of Hough transform for circles and ellipses detection using a 2-dimensional array. Pattern Recognit. **25**, 1007–1022 (1992)
22. Yuen, H., Princen, J., Illingworth, J., Kittler, J.: A comparative study of Hough transform methods for circle finding. Image Vis. Comput. **8**(1), 71–77 (1990)

Self-adaptive Evolutionary Many-Objective Optimization Based on Relation ε-Preferred

Nicole Drechsler

Abstract Many real-world optimization problems consist of several mutually dependent subproblems. If more than three optimization objectives are involved in the optimization process, the so-called *Many-Objective Optimization* is a challenge in the area of multi-objective optimization. Often, the objectives have different levels of importance that have to be considered. For this, relation ε-*Preferred* has been presented, that enables to compare and rank multi-dimensional solutions. ε-*Preferred* is controlled by a parameter ε that has influence on the quality of the results. In this paper for the setting of the epsilon values three heuristics have been investigated. To demonstrate the behavior and efficiency of these methods an *Evolutionary Algorithm* for the multi-dimensional *Nurse Rostering Problem* is proposed. It is shown by experiments that former approaches are outperformed by heuristics that are based on self-adaptive mechanisms.

Keywords Many-objective optimization · Nurse rostering problem · Relation ε-preferred · User preferences

1 Introduction

During the last 20 years solving *Multi-Objective Optimization* (MOO) problems is getting more and more important. Many real-world problems consist of multiple competing subproblems that have to be optimized in parallel. For *Evolutionary Algorithms* (EAs) many approaches have been presented that cope with MOO problems [1–4]. If more than three objectives are involved in the optimization process the corresponding problems are called many-objective optimization problems. Especially, if real-world optimization problems are of interest, more than three objectives are considered during the optimization process [5–7]. Furthermore, in industrial problems optimization criteria have often different levels of importance. These user

N. Drechsler (✉)
Institute of Computer Science, University of Bremen, 28359 Bremen, Germany
e-mail: nd@informatik.uni-bremen.de

© Springer International Publishing Switzerland 2016
K. Madani et al. (eds.), *Computational Intelligence*,
Studies in Computational Intelligence 613,
DOI 10.1007/978-3-319-23392-5_2

preferences have to be taken into account during optimization. Considering both, many-objective optimization and user preferences, there is a need for optimization models that combine these properties [8–12].

To overcome these problems one classical method to combine multiple optimization criteria with user preferences is the *Weighted Sum* approach. Here, a single value is computed by a linear combination of the considered criteria. By the choice of the weights for each criterion the influence of the user preference can be controlled. It is often used in industrial applications, because it is easy to implement and at a first view scales well. Further examinations have shown that it is a challenge to adjust the weights such that the search is guided in the desired direction [5, 13]. A disadvantage of the weighted sum approach is that it is incapable to find compromise solutions of concave *Pareto* fronts. A further classical approach is the use of *Non-dominated Sets* that is based on the *Pareto-Dominance* relation [14]. Using the *Dominance* relation a ranking between multi-dimensional solutions can be required. If EAs are used for MOO, the method NSGA-II [3] is a basic approach that is based on non-dominated sorting. It is suitable in low dimensions, but for more than three objectives more sophisticated approaches are required [15, 16]. As an alternative, the hypervolume indicator is proposed, i.e. to each candidate solution an indicator value is assigned, but due to the computational complexity it can only be applied in low dimensions. An approximation of the hypervolume for higher dimensions is presented in [4].

In further developments relation ε-*Preferred* has been proposed for many-objective optimization [17]. Using relation ε-*Preferred* a ranking between solutions can be determined and solutions that are incomparable using *Dominates* can be distinguished. Thus ε-*Preferred* is a refinement of relation *Dominates*. In [12] the model based on ε-*Preferred* has been enlarged such that it can also handle user preferences (priorities). For this model, the influence of parameter ε has been investigated and method AEP, that determines ε automatically, has been presented. In this context the *Nurse Rostering Problem* (NRP) has been considered, i.e. a scheduling problem where a working plan for employees in a hospital has to be computed. The proposed method is compared to the well-known NSGA-II approach, because the modeling of user preferences as proposed in this article can easily be used within this method. A comparison to more sophisticated approaches, like e.g. the MOEA/D [18] or the hypervolume approach [4] can not directly be performed: Taking the user defined priorities into account, the comparison to these approaches without user preference modeling is not meaningful, because the usage of user preferences is not provided.

In this paper an *Evolutionary Algorithm* that makes use of the ε-*Preferred* relation including user preferences is applied to the NRP.[1] In contrast to [12] the full potential of the presented model has been exploited. To model the user preferences in [12] only two types of priorities are used, the soft constraints and hard constraints. The hard constraints have a higher priority during optimization than the soft constraints. The hard constraints map the rules of the nurse station. Each soft constraint itself

---

[1]For the investigation of this approach the NRP has been used as application, because it consists of many objectives with different levels of priorities. There, in contrast to standard benchmarks for MOO (DTLZ [19]), the priorities are provided in the benchmark files.

consists of up to 90 rules that have different user preferences. These user preferences are directly given as weights in the benchmark examples [20]. Then, a weighted sum is constructed to compute the constraints. Using the soft and hard constraints, a multi-dimensional fitness function is computed, such that the hard constraint and for each employee the corresponding soft constraints are provided. Following this, benchmarks with up to 17 optimization criteria are considered. In contrast, in this approach each rule of the soft constraints is treated as a separated optimization criterion. This leads to fitness functions with up to 90 objectives. For each rule a priority is calculated dependent on the weight that is specified in the benchmark. In the experiments it is shown that the results from [12] can be further improved if the advanced model as described above is used.

Furthermore, the justification of the epsilon values is examined. Two self-adapting methods for epsilon adaptation are presented. It is shown in our experiments that AEP [12] can be further improved. Additionally, an approach based on *Weighted Sums* is outperformed, where previously published methods fail.

2 Preliminaries

First, we give a short introduction into the basic techniques of multi-objective optimization and relations used for comparison.

2.1 Multi-objective Optimization

A multi-objective optimization problem is defined as follows: Given a search space Ω, an evaluation function $f : \Omega \to \mathbb{R}^m$ is defined to calculate the fitness vector $F(A) : \forall A \in \Omega$ of size m. Then we have to minimize (or maximize) the elements of $F(A)$. In the following we assume, without loss of generality, that F has to be minimized for all objectives. According to [14] it holds:

Definition 1 Let $A, B \in \Omega$.

$$A \prec_{dominates} B :\Leftrightarrow \exists j : F_j(A) < F_j(B) : F_i(A) \leqslant F_i(B), 1 \leqslant i \leqslant m. \quad (1)$$

Based on this, we can describe the Pareto set as

$$\chi : \forall p \in \chi : \nexists q \in \Omega : q \prec_{dominates} p. \quad (2)$$

It can be directly seen from the definition that for $A, B \in \Omega$ element A dominates B only if A is better than B in at least one component and equal or better in all components. Relation *Dominates* is a partial order. In evolutionary multi-objective optimization relation *Dominates* is used to perform *Non-dominated Sorting* [3]: All elements of a population are compared using *dominates* and the non-dominated elements are computed. This set is called *Non-dominated Set*. Disregarding the *Non-dominated Set* the next level of non-dominated elements is considered. This is repeated, until all elements are classified. The elements $A \in \Omega$ in the *Non-dominated Set* are equal or not comparable and hence, the designer is interested in solutions from the *Non-dominated Set*.

2.2 Relation Preferred

In [5] a refinement of relation *Dominates* has been presented. The approach is well-suited for problems in many-objective optimization, i.e. if more than three optimization criteria are considered. In [3] it has been shown that in higher dimensions more than 90 % of the population are *Non-dominated* elements and thus, a ranking used for selection mechanisms cannot be performed. To overcome these problems relation *Preferred* is defined as follows:

Definition 2 Let $A, B \in \Omega$ and $1 \leq i, j, \leq m$.

$$A \prec_{preferred} B :\Leftrightarrow |\{i : F_i(A) < F_i(B)\}| > |\{j : F_j(B) < F_j(A)\}|. \qquad (3)$$

Relation *Preferred* considers the number of different objectives of A and B. A is *preferred* to B if $i\,(i < m)$ objectives of A are smaller or equal than the corresponding objectives in B and only $j\,(j < i)$ objectives of B are smaller or equal than the corresponding objectives in A.

Using relation *Preferred* the solutions in a population are classified in so-called *Satisfiability Classes* (SCs) [5]. All solutions $A \in \Omega$ are compared using relation *Preferred*. Then the relation graph is constructed, where each element is a node and preferences are represented by edges. *Preferred* is not a partial order, because the relation graph can have cycles, and thus it is not transitive.

To overcome this property the relation graph is modified such that cycles are eliminated. The main idea is that elements that are included in a cycle should be ranked equally. For this the *Strongly Connected Components* (SCC) of the graph are computed by a linear time DFS-based algorithm [21]. Then the relation graph is modified such that each SCC is replaced by a new node representing all elements in the corresponding cycle. Doing so, all cycles in the relation are eliminated. The relation that is represented by the acyclic relation graph is transitive and antisymmetric, which is sufficient for our purposes. Level sorting of the nodes in the acyclic relation graph determines a ranking of SCCs, where each level defines a SC. This is illustrated in the following example:

Example 1 Consider some solution vectors from $\mathbb{R}^3$, i.e. each vector is a solution consisting of three objectives (m_1, m_2, m_3):

$$(0, 1, 2) \quad (1, 1, 2) \quad (2, 1, 1) \quad (7, 0, 9) \quad (8, 7, 1) \quad (1, 9, 6)$$

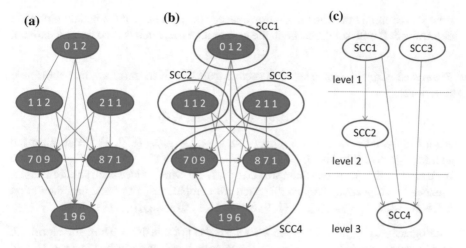

Fig. 1 Relation graph and *Satisfiability Classes*

The relation graph of these elements and relation *Preferred* is given in Fig. 1a. Elements (0, 1, 2), (1, 1, 2) and (2, 1, 1) are preferred to the remaining elements, but (0, 1, 2) and (2, 1, 1) ((1, 1, 2) and (2, 1, 1)) are not comparable. Additionally, element (0, 1, 2) is preferred to (1, 1, 2). The remaining three vectors (8, 7, 1), (1, 9, 6), and (7, 0, 9) are pairwise comparable. But as can be seen in the relation graph they describe a "cycle". Thus relation *Preferred* is not transitive. For more details see [5].

2.3 Relation ε-Preferred

In [17] an enlargement for many-objective optimization of relation *Preferred* has been introduced. For the proposed relation *ε-Preferred* fitness limits ε_i, $1 \leq i \leq m$, for each dimension are defined.

Definition 3 Let $A, B \in \Omega$ and ε_i, $1 \leqslant i \leqslant m$.

$$A \prec_{\varepsilon-exceed} B \Leftrightarrow |\{i : F_i(A) < F_i(B) \wedge |F_i(A) - F_i(B)| > \varepsilon_i\}| \quad (4)$$
$$> |\{j : F_j(A) > F_j(B) \wedge |F_j(A) - F_j(B)| > \varepsilon_j\}| \quad (5)$$

ε-exceed counts how often a solution exceeds the given limits ε_i. Then solution A is better than solution B with respect to the limits ε_i, if A has more exceeding than B. Using *ε-exceed* the extension *ε-Preferred* is defined as follows:

Definition 4 Given two solutions $A, B \in \Omega$.

$$A \prec_{\varepsilon-preferred} B \Leftrightarrow A \prec_{\varepsilon-exceed} B \vee (B \not\prec_{\varepsilon-exceed} A \wedge A \prec_{preferred} B) \quad (6)$$

First it is counted how often a solution exceeds the ε-limits and the better solution is determined. If both solutions are in the given range *Preferred* is used for comparison.

Example 2 Consider some solution vectors from $\mathbb{R}^3$, i.e. the results of three objective functions:

$$(7, 0, 9) \quad (8, 7, 1) \quad (1, 9, 6)$$

Additionally, let $\varepsilon_i = 5, 1 \leq i \leq 3$. $(7, 0, 9) \prec_{\varepsilon-preferred} (8, 7, 1)$, because for the second objective it holds $|0 - 7| > \varepsilon_2$, where solution $(7, 0, 9)$ "wins", and for the third it holds $|9 - 1| > \varepsilon_3$, where solution $(8, 7, 1)$ "wins". Since each solution has an ε-exceeding objective, *Preferred* is used for comparison. The same argumentation holds for $(8, 7, 1) \prec_{\varepsilon-preferred} (1, 9, 6)$ and $(1, 9, 6) \prec_{\varepsilon-preferred} (7, 0, 9)$.

Analogously to *Preferred* relation ε-*Preferred* is not transitive. Thus, the algorithm that computes the SCCs is applied to the relation graph as described in Sect. 2.2 and in [5].

2.4 Relation Prio-ε-Preferred

In many real world applications the optimization criteria have user specific preferences that have to be modeled during the optimization process. To model priorities of optimization objectives in [12] relation *Prio-ε-Preferred* is defined. It is a combination of relation ε-*Preferred* and a lexicographic ordering of the objectives.

Let us assume that priorities $1, 2, \ldots, k$ are assigned to the objectives in an ascending ordering, i.e. the lower the index i, $1 \leq i \leq k$, the higher the priority.

Definition 5 Let $p = (p_1, \ldots, p_k)$ be a priority vector. p_i determines the number of objectives that have priority i. The priority of an objective is calculated by the function:

$$pr : \{1, \ldots, m\} \rightarrow \{1, \ldots, k\} \tag{7}$$

The subvector of objectives $c|_i$ of priority i is defined as

$$c|_i \in \mathbb{R}^{p_i}, c|_i = (c_r, \ldots, c_s) \tag{8}$$

where

$$r = \sum_{j=1}^{i-1} p_j + 1 \wedge s = \sum_{j=1}^{i} p_j. \tag{9}$$

Fig. 2 Sketch of basic
algorithm

```
Prio-ε-Preferred (population) {
    for_all (individuals)
    {
        calculate_pairwise_Prio-ε-Preferred () ;
        construct_relation_graph () ;
        calculate_strongly_connected_components () ;
        perform_level_sorting () ;
    }
    return level_of_individuals ;                    //ranking
}
```

For $A, B \in \Omega$ the relation $\prec_{\varepsilon-priopref}$ (*Prio-ε-Preferred*) is defined by

$$A \prec_{\varepsilon-priopref} B : \Leftrightarrow \exists j \in \{1, \ldots, k\} : A|_j \prec_{\varepsilon-preferred} B|_j \tag{10}$$

$$\wedge \ (\forall h < j : A|_h \not\prec_{\varepsilon-preferred} B|_h \wedge B|_h \not\prec_{\varepsilon-preferred} A|_h) \tag{11}$$

To perform a ranking of a set of elements, analogously to Sect. 2.2 the *Satisfiability Classes* are computed. For this a set of elements is pairwise compared using *Prio-ε-Preferred* and the relation graph is constructed. Then the algorithm for finding the *Strongly-Connected Components* (SCC) is applied to eliminate cycles in the relation graph. A sketch of the algorithm is given in Fig. 2.

To give an impression on the properties of relation *Prio-ε-Preferred* an example is considered.

Example 3 Let us consider a problem with 5 objectives with 3 different priorities. Let $c = (c_1, c_2, c_3, c_4, c_5)$ a solution vector and $p = (1, 3, 1)$ a priority vector, i.e. one objective has priority 1 (i.e. $p_1 = 1$), three objectives have priority 2 ($p_2 = 3$) and one objective has priority 3 ($p_3 = 1$). This leads to the function pr with $pr(1) = 1$, $pr(2) = 2, pr(3) = 2, pr(4) = 2$, and $pr(5) = 3$ what means that the first objective has priority 1, the second objective priority 2, and so on. For priority 2 the projection is $c|_2 \in \mathbb{R}^3, c|_2 = (c_2, c_3, c_4)$, since $r = 1 + 1 = 2$ and $s = 1 + 3 = 4$.

Now, let us consider two solution vectors, $A = (2, 7, 0, 9, 15)$ and $B = (2, 1, 9, 6, 5)$. Then it holds, that $B \prec_{\varepsilon-priopref} A$. For this, first the objectives with priority 1 are compared. Since they are equal, next the objectives with priority 2 are compared with relation $\prec_{\varepsilon-preferred}$, i.e. $(1, 9, 6) \prec_{\varepsilon-preferred} (7, 0, 9)$ (see Example 2) which leads to the statement $B \prec_{\varepsilon-priopref} A$. The last objective has not to be considered anymore, because it has lowest priority and the decision, which solution is better with respect to relation $\prec_{\varepsilon-priopref}$, has already been made.

3 Nurse Rostering Problem

In this section a description of the *Nurse Rostering Problem* (NRP) and the algorithm for evolutionary many-objective optimization is given.

3.1 Problem Description

Since several years the NRP is of high interest and many approaches for optimization have been presented [20, 22]. The NRP is a utilization planning problem, where a working plan for employees in a hospital has to be determined. Working shifts, like e.g. day shift, night shift, stand-by shift, long shift or vacations have to be assigned to each employee and working day, such that sufficient employees are on duty and the working contracts of the employees are fulfilled.

For optimization solution schedules have to be evaluated by a fitness function. The fitness function consists of multiple criteria, that can be categorized in hard constraints that have to be fulfilled and soft constraints that improve the fitness function. A solution can still be valid, even though a soft constraint is not fulfilled. But indeed, a solution that does not fulfill soft constraints can be rejected by the planner. The constraints are given as rules that are specified in the benchmarks [20]. The benchmarks are available from [23]. The rules can be categorized into the following main areas:

1. Rules of the nurse station, e.g. sufficient nurses per shift
2. Restrictions by law, e.g. maximal hours of work per day or maximal working days per month
3. Rules resulting from ergonomics, e.g. having ergonomic shift pattern

Following the benchmarks from [20] the rule of the first category is modeled as hard constraint, whereas the rules of item 2. and 3. are given as soft constraints. The influence of the rules in the fitness function is controlled by weights that are given in the benchmarks. Concerning the weights in the benchmarks it is assumed to use a *Weighted Sum* for the calculation of the fitness function. In this application schedules for up to 16 employees for a planning period of 30 days are considered. Instead of using a weighted sum, the weights of the rules determine a priority that is used by relation *Prio-ε-Preferred*. The details are described in Sect. 3.2.

Example 4 In Fig. 3 an example of a schedule for the NRP is given. To each employee A–H and day (02nd–29th) a shift is assigned, where the shifts are labeled as follows: Day shift (D), night shift (N). In the vertical columns the rules of the nurse station are evaluated. In this example for shift D three employees and for shift N one employee have to be on duty. This hard constraint is fulfilled for each day. For the maximal working days per month the vertical rows have to be evaluated. Notice, that employees E–H have half time positions, thus they have less working days. It can be seen that the ergonomic rules of the shift pattern (soft constraints) are fulfilled, i.e. desired patterns like e.g. DDDNN are given in the solution. Undesired patterns like single shifts are not scheduled. For more details about the NRP see [20, 22, 23].

	1							2							3							4						
	02	03	04	05	06	07	08	09	10	11	12	13	14	15	16	17	18	19	20	21	22	23	24	25	26	27	28	29
	M	T	W	T	F	S	S	M	T	W	T	F	S	S	M	T	W	T	F	S	S	M	T	W	T	F	S	S
A	D	D	D	D				D	D	D	D	D					D	D	D	D	D	D				N	N	N
B				D	D	D	D	N	N		D	D	D	D	N	N							D	D	D	D	D	D
C	D	D	N	N				D	D	D	D	D					D	D	D	D	N	N				D	D	D
D	N	N			D	D	D	D	D	D	D	D			D	D	D	D	D	D		D	D	N	N			
E	D	D	D								D	D						N	N	N				D	D			
F				N	N	N					D	D			D	D	D					D	D					
G		D	D								N	N	N				D	D	D					D	D			
H				D	D	D			N	N					N	N										D	D	D

Fig. 3 Example nurse rostering schedule for benchmark GPost

3.2 Proposed Model for Many-Objective Optimization

In this application an *Evolutionary Algorithm* (EA) is used to optimize the schedules. Details of representation of the individuals and evolutionary operators are left out due to page limitation.

The fitness function $F(A)$, $A \in \Omega$, consists of m functions F_i, $1 \leq i \leq m$, that are directly derived from the benchmark under consideration. For each hard and soft constraint in the benchmark an objective F_i is defined that has to be minimized. Following Definition 5 the priority function pr for the objectives is calculated such that the higher the weight of an objective i the higher is the priority $pr(i)$. First, the objectives that correspond to hard constraints get the highest priority value 1. The remaining objectives that correspond to soft constraints get priority values depending on its weights. Objectives that have the highest weight value given in the benchmark lead to priority 2, i.e. $pr(i) = 2$, $\forall i$ with maximum weight. Then, the objectives with maximum weight are disregarded and again the objectives with maximum weight are considered, they get priority 3. Following this, the next weights are considered one after another. This is repeated, until a priority is assigned to each objective. The number of priorities of the considered benchmarks ranges from 2 to 10.

For the ranking of the solutions relation *Prio-ε-Preferred* is used. The solutions are compared using relation *Prio-ε-Preferred*. Then, the relation graph is constructed and the SCCs of the directed graph are computed as described in Sect. 2. For the determination of the epsilon values several methods are examined that adapt the epsilon values automatically. In Sect. 4 several methods for the justification of the epsilon values are proposed.

4 Approaches for ε-Adaptation

The justification of the epsilon values for relation *Prio-ε-Preferred* is an interesting task. For this in [12] the influence of the epsilon values in relation *Prio-ε-Preferred* has been investigated. It has been shown that the choice of the epsilon values influences the quality of the optimization. A method called AEP (*Adapted Epsilon Preferred*) has been proposed that adapts the epsilon values automatically. AEP is a straight forward method that computes the same epsilon value for all objectives. In this section more sophisticated methods for the adaptation of the epsilon values are presented. The methods examined in the experiments are described in the following:

Adapted Epsilon Preferred (AEP) [12]. In method AEP for all objectives one epsilon value is determined. Therefore, one individual out of the best *Satisfiability class* (SC) derived by relation *Prio-ε-Preferred* is randomly chosen. The new epsilon value is determined by the average value of all objectives of that individual:

$$\varepsilon = \sum_{j=1}^{m} \frac{Ind_{best,j}}{m} \tag{12}$$

where $Ind_{best,j}$ is the jth objective of an randomly chosen individual out of the best *SC*. The epsilon value is updated in each generation. The idea behind this method is that individuals can be distinguished by relation *Prio-ε-Preferred*, if the difference of the individuals exceeds the calculated average range.

Median Epsilon Preferred (MEP). In method MEP one separated epsilon value for each objective is calculated. For this all individuals in a population are considered and for each objective the epsilon value is set to the median of each objective:

$$\varepsilon_j = median(\{Ind_{i,j} | 1 \le i \le |P|\}), 1 \le j \le m \tag{13}$$

where m is the number of objectives, $|P|$ is the size of the population and $Ind_{i,j}$ is the jth objective of the ith individual in population P.

Self-adaptation 1 (SA1). For each objective a separated epsilon value is calculated. First, the epsilon values are initialized using method AEP. Then in each generation a randomly chosen epsilon value is decremented. If this reduces the number of SCs, this step is revised. The idea is that a higher number of SCs leads to a meaningful ranking of the solutions.

Self-adaptation 2 (SA2). Again, for each objective a separated epsilon value is calculated. The epsilon values are initialized using method AEP. Then in each generation for a randomly chosen objective the epsilon is bisected, if the set of best elements has not changed for 100 generations. The idea is to give more restriction in the ranking mechanism, if the optimization is in progress.

Table 1 Properties of benchmarks for the NRP

Benchmark	Rules/objectives	Priorities	Employees	Days
Millar-2Shift-DATA1	11	2	8	14
WHPP	12	2	30	14
Valouxis-1	15	4	16	28
GPost	42	6	8	28
ORTEC01	92	10	16	31

Table 2 Comparison of standard methods

Benchmark	Objectives	Weighted Sum	MOO model [12] AEP	Proposed MOO model NSGA-II	AEP
Millar-2Shift-DATA1	11	1310	1590	1390	1190
WHPP	12	29	-	133	27
Valouxis-1	15	13986	16692	148500	13542
GPost	42	7159	7557	26528	11830
ORTEC01	92	9132	11672	36740	10040

5 Experimental Results

In this section the experimental results of the presented approaches are described. The benchmarks for the NRP are taken from [23]. and its properties are summarized in Table 1. The optimization rules given in the benchmark directly correspond to the objectives (column *Rules/Objectives*). The number of different priorities of the objectives are given in column *Priorities*. Columns *Employees* and *Days* show the benchmarks' number of employees and the planning period, respectively. For each benchmark and for each method presented in Sect. 4 the EA is run 10 times with different random seeds. Then, the average value over these 10 runs is calculated. The population size is set to 50 and the EA runs for 5000 generations. The average values of the presented approaches are given in Tables 2 and 3. The methods are compared using the *Weighted Sum*. For this, the objectives are transferred into a single objective fitness function, such that the weights in the benchmarks are taken to weight each objective.[2]

In a first series of experiments the proposed model for MOO, where for each rule an objective is defined, is compared to the restricted model from [12]. There only hard and soft constraints are considered as optimization objectives. For both models method AEP from [12] is applied to the NRP. The average values can be seen in columns *AEP* of Table 2. The results can be improved, if a refinement of the model for MOO as proposed in this paper is performed.

---

[2]Originally the benchmarks are designed for optimization using a *Weighted Sum*. Thus, the weights are justified by a planner and directly given in the benchmark.

Table 3 Comparison of ε-adaptation methods for the proposed MOO model

Benchmark	Objectives	Weighted Sum	NSGA-II	AEP	MEP	SA1	SA2
Millar-2Shift-DATA-1	11	1310	1390	1190	1540	1280	1220
WHPP	12	29	133	27	32	27	24
Valouxis-1	15	13986	148500	13542	12890	11852	13132
GPost	42	7159	26528	11830	12300	13006	11401
ORTEC01	92	9132	36740	10040	10927	11549	10618

Additionally, AEP is compared to NSGA-II [3], which is a basic method in evolutionary multi-objective optimization (see column *NSGA-II*). For comparison, analogously to *Prio-ε-Preferred* NSGA-II is extended such that it can also handle priorities.[3] Thus, it is comparable to the methods that are based on relation *Prio-ε-Preferred*. Furthermore, it can be seen that AEP outperforms NSGA-II for each considered benchmark. Especially for *Valouxis-1* an improvement of more than 90 % can be observed. Furthermore, a comparison to an approach that is based on *Weighted Sums* is given. It is a single objective evolutionary algorithm, where the weights in the benchmarks are used to calculate the fitness function. The comparison shows that for most benchmarks the overall quality has been improved.

In a next series of experiments the approaches for adaptation of the epsilon values presented in this paper are compared. The results are summarized in columns *MEP*, *SA1* and *SA2* of Table 3. A comparison to NSGA-II in Table 2 shows that MEP fails only for one example (*Millar-2Shift-DATA1*), whereas the self-adaptive approaches SA1 and SA2 improve NSGA-II for all benchmarks. For three out of the considered benchmarks both methods compute better results than the weighted sum approach. Notice, the benchmarks are designed such that optimization with *Weighted Sums* can easily be performed, i.e. the weights are specified in the benchmark. Thus, even these results can be improved, if the full potential of the proposed model for many-objective optimization is used. Only for benchmarks *GPost* and *ORTEC01* SA1 and SA2 fail to calculate the best results. Both benchmarks consist of 42 and more objectives. A comparison shows that AEP can be improved using the self-adapting techniques SA1 and SA2. It is focus of current work to investigate the self-adapting techniques such that also problems with a higher number of objectives are solved sufficiently.

To give an impression on the quality of the priority based optimization presented above a solution element out of the best SC derived by SA1 is compared to an element from the non-dominated set derived by NSGA-II. In Figs. 4 and 5 a comparison for benchmarks *Valouxis-1* and *WHPP* is shown. The objectives and its priorities are

---

[3]The main reason for using NSGA-II for comparison is that it can easily be enlarged such that it can handle priorities as described in this paper. Other methods like e.g. Hype [4] are more suitable for *Many-Objective Optimization*, but it is not obvious how to incorporate the priorities. This is an interesting task for further developments.

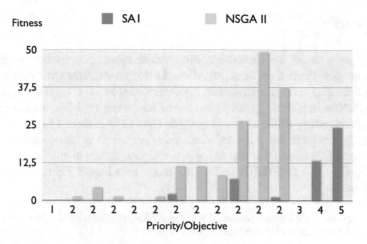

Fig. 4 Comparison of solutions: benchmark Valouxis-1

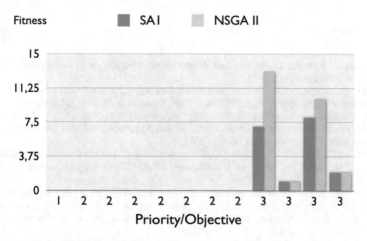

Fig. 5 Comparison of solutions: benchmark WHPP

specified at the x-axis. For benchmark *Valouxis-1* it can be seen, that the solution obtained by SA1 has only 5 objectives over zero, whereas NSGA-II has 10 objectives over zero. This means that the solution obtained by SA1 is *Preferred* to the solution obtained by method NSGA-II. Additionally, if the absolute values of the objectives are compared, SA1 performs better than NSGA-II. The same observation holds for benchmark *WHPP*. For this example the objectives with priorities 1 and 2 are solved optimally. For objectives with priority 3 it can be observed that the solution obtained by SA1 even *Dominates* the solution from NSGA-II.

6 Conclusions

In this paper a model for *Many-Objective Optimization*, i.e. optimization problems
with more than three objectives, based on the *Prio-ε-Preferred* relation has been
investigated. For this, heuristics for the determination of the epsilon values are
presented. The model is applied to the *Nurse Rostering Problem*, a resource planning problem where different working shifts have to be assigned to the nurses in a
hospital. To compare the proposed methods experiments on benchmark examples
are performed. It turned out that using self-adapting mechanisms for the adaption
of the epsilon value NSGA-II and an approach based on *Weighted Sums* can be
outperformed.

Acknowledgments I'd like to thank André Sülflow and Rolf Drechsler for helpful discussions and
comments and for their contributions to previous work.

References

1. Fonseca, C., Fleming, P.: An overview of evolutionary algorithms in multiobjective optimization. Evol. Comput. **3**(1), 1–16 (1995)
2. Zitzler, E., Thiele, L.: Multiobjective evolutionary algorithms: a comparative case study and
   the strength pareto approach. IEEE Trans. Evol. Comput. **3**(4), 257–271 (1999)
3. Deb, K.: Multi-objective Optimization using Evolutionary Algorithms. Wiley, New York (2001)
4. Bader, J., Zitzler, E.: HypE: an algorithm for fast hypervolume-based many-objective optimization. Evol. Comput. **19**(1), 45–76 (2011)
5. Drechsler, N., Drechsler, R., Becker, B.: Multi-objective optimisation based on relation favour.
   In: International Conference on Evolutionary Multi-Criterion Optimization, pp. 154–166
   (2001)
6. Hughes, E.: Radar waveform optimization as a many-objective application benchmark. In:
   International Conference on Evolutionary Multi-Criterion Optimization, pp. 700–714 (2007)
7. Pizzuti, C.: A multiobjective genetic algorithm to find communities in complex networks. IEEE
   Trans. Evol. Comput. **16**(3), 418–430 (2012)
8. Schmiedle, F., Drechsler, N., Große, D., Drechsler, R.: Priorities in multi-objective optimization
   for genetic programming. In: Genetic and Evolutionary Computation Conference, pp. 129–136
   (2001)
9. Wickramasinghe, U., Li, X.: A distance metric for evolutionary many-objective optimization
   algorithms using user-preferences. In: 22nd Australasian Joint Conference on Advances in
   Artificial Intelligence (AI'09), pp. 443–453 (2009)
10. Auger, A., Bader, J., Brockhoff, D., Zitzler, E.: Articulating user preferences in many-objective
    problems by sampling the weighted hypervolume. In: Genetic and Evolutionary Computation
    Conference, pp. 555–562 (2009)
11. Wagner, T., Trautmann, H.: Integration of preferences in hypervolume-based multiobjective
    evolutionary algorithms by means of desirability functions. IEEE Trans. Evol. Comput. **14**(5),
    688–701 (2012)
12. Drechsler, N., Sülflow, S., Drechsler, R.: Incorporating user preferences in many-objective opti-
    mization using relation ε-preferred. In: International Conference on Evolutionary Computation
    Theory and Applications (2013)
13. Geiger, M.: Multi-criteria curriculum-based course timetabling—a comparison of a weighted
    sum and a reference point based approach. In: International Conference on Evolutionary Multi-
    Criterion Optimization, pp. 290–304 (2009)

14. Goldberg, D.: Genetic Algorithms in Search Optimization and Machine Learning. Addison-Wesley Publisher Company, Inc, Reading (1989)
15. Corne, D., Knowles, J.: Techniques for highly multiobjective optimization: theorie and applications. In: Genetic and Evolutionary Computation Conference, pp. 773–780 (2007)
16. Ishibuchi, H., Tsukamoto, N., Nojima, Y.: Evolutionary many-objective optimization: a short review. In: IEEE Congress on Evolutionary Computation, pp. 2424–2431 (2008)
17. Sülflow, A., Drechsler, N., Drechsler, R.: Robust multi-objective optimization in high-dimensional spaces. In: International Conference on Evolutionary Multi-Criterion Optimization, pp. 715–726 (2007)
18. Zhang, Q., Li, H.: Moea/d: a multiobjective evolutionary algorithm based on decomposition. IEEE Trans. Evol. Comput. **11**(6), 712–731 (2007)
19. Deb, K., Thiele, L., Laumanns, M., Zitzler, E.: Scalable test problems for evolutionary multi-objective optimization. Technical Report 112, Computer Engineering and Networks Laboratory (TIK), Swiss Federal Institute of Technology (ETH), Zurich, Switzerland (2001)
20. Burke, E., Curtois, T., Qu, R., Vanden-Berghe, G.: Problem model for nurse rostering benchmark instances. Technical report, ASAP, School of Computer Science, University of Nottingham, UK (2012)
21. Cormen, T., Leierson, C., Rivest, R.: Introduction to Algorithms. MIT Press, Cambridge (1990)
22. Burke, E., Causmaecker, P.D., Berghe, G., Landeghem, H.: The state of the art of nurse rostering. J. Sched. **7**, 441–499 (2004)
23. Benchmarks: Employee scheduling benchmark data set: Technical report, ASAP, School of Computer Science, The University of Nottingham, UK (2012). http://www.cs.nott.ac.uk/~tec/nrp/

Automated Graphical User Interface Testing Framework—Evoguitest—Based on Evolutionary Algorithms

Gentiana Ioana Latiu, Octavian Augustin Cret and Lucia Vacariu

Abstract Software testing has become an important phase in software applications' lifecycle. Graphical User Interface (GUI) components can be found in a large number of desktops and web applications and also in a wide variety of mobile devices. In the last years GUIs have become more and more complex and interactive, their testing process requiring an interaction with the GUI components, mainly by generating mouse, keyboard and touch events. Given their increased importance, GUIs verification for correctness contributes to the establishment of the correct functionality of the corresponding software application. The current research on GUI testing methodologies primarily focuses on automated testing. This paper presents EvoGUITest, a novel automated GUI testing framework based on evolutionary algorithms which tests the GUI independently from the application code itself. The framework is designed for testing GUIs of web applications. Results have been compared, based on specific metrics, with others existing frameworks.

Keywords Graphical user interface testing · Evolutionary algorithms · Automated testing framework

1 Introduction

GUI is a specification for the look and feel of the software application [1]. A GUI consists of graphical elements such as windows, icons, menus, buttons, textboxes. A well designed GUI must be intuitive and user friendly, being the image of the

G.I. Latiu (✉) · O.A. Cret · L. Vacariu
Computer Science Department, Technical University of Cluj-Napoca,
26-28 Baritiu Street, Cluj-napoca, Romania
e-mail: Gentiana.Latiu@cs.utcluj.ro

O.A. Cret
e-mail: Octavian.Cret@cs.utcluj.ro

L. Vacariu
e-mail: Lucia.Vacariu@cs.utcluj.ro

© Springer International Publishing Switzerland 2016   39
K. Madani et al. (eds.), *Computational Intelligence*,
Studies in Computational Intelligence 613,
DOI 10.1007/978-3-319-23392-5_3

application. A good quality of the GUI is necessary and the diminishing of the testing cost becomes an important requirement. The GUI's set of components can be a crucial point in the users' decisions to either use or not use that specific software application [2].

While GUIs have become ubiquitous and increasingly complex, their testing remains largely ad-hoc. Due to its complexity, the testing process is problematic and time-consuming [3].

During the manual GUI testing process, each test case needs a long time to execute (tens of seconds, for a medium complexity GUI). The manual checking process of the results needs another time spent by the human tester, which is also of a few tens of seconds. If for instance there is a suite of 10,000 test cases to be applied, then the total testing time becomes enormous (hundreds of hours) [4].

If the test cases are executed automatically, it takes around 3 seconds for each test case to be executed, and another 1 second for checking the output results. 10,000 test cases need around 10 hours to be executed, which shows an acceleration of one order of magnitude compared to the manual testing process [4]—that is why the research mainly focuses on automatic GUI testing.

Different frameworks were built to automate the testing process for Web applications GUIs, to eliminate the human tester involvement, etc., but many of these were made either for some particulary GUI software systems, or for the systems at a very general level.

A survey by Al-Zain et al. on automation testing tools for web applications shows, using different criteria (the effectiveness of recorder/playback tools, handling of page waits, cross browser compatibility, technical support, and the number of different techniques available to programmatically locate elements on web pages), that free and simple tools can be more powerful and time saving, compared to commercially sophisticated and expensive tools [5]. The authors have also summarized the best practices and guidelines to be considered when adopting automated GUI functional tests for web applications [5].

A comparative study of automated testing tools was conducted in [6]. Based on criteria such as the efforts involved with generating test scripts, capability to playback the scripts, result reports, speed and cost, Mercury QuickTest Professional and the AutomatedQA TestComplete have been compared. Analyzing the features supported by these two functional testing tools, which help minimizing the resources in script maintenance and increasing the efficiency of script reuse, the authors conclude that both tools are good, but for data security needs the QuickTest Professional is better.

Some years ago, test cases were generated randomly during the automatic GUI testing process. Because the coverage of random input testing was very weak, the scientific community started studying the usage of the Evolutionary Algorithms (EA) for automating the GUI testing process.

To only mention some of the most spectacular applications of EA in real life, we could say that in the last years the Evolutionary Art was used in a lot of applications, with interactive EAs in which the user assigns scores to images based on their suitability [7]; also, the EvoSpace framework is used for developing interactive algorithms for artistic design [8].

The rest of this paper is organized as follows: Sect. 2 describes the automatic process for GUI testing, Sect. 3 provides a detailed description of the EA process, Sect. 4 describes our novel proposed Web GUI testing framework (EvoGUITest). In this Section the framework architecture and the experimental results are also presented. Section 5 concludes the paper, summarizing the future work planned.

2 Automatic GUI Testing

The GUI testing is a process which aims at testing the software application's user interface and detecting if the GUI is functionally correct. GUI testing includes checking the way the software application handles mouse and keyboard events [9].

The automatic GUI testing process includes automatic manual testing tasks performed by human testers. By the automatic testing process, a software program executes the testing tasks and analyzes if the GUI under test is functionally correct.

Automatic GUI testing can be executed using different techniques.

2.1 Capture/Replay Tools

These tools have two modes of functioning: capture and replay. In capture (record) mode, the tool is able to record testers' actions while they are interacting with the GUI. The set of actions is recorded inside test scripts. These tools provide a scripting language which can be used by engineers for maintaining the test scripts.

In replay mode, the recorded test scripts are executed. During the execution of each test script, some mouse or keyboard events are executed on the GUI. The test scripts' execution process is automatic and can be repeated several times.

The most important disadvantage of these GUI testing tools is the lack of structure of the test scripts, which makes the maintenance process difficult. These tools don't provide any support to design and evaluate test cases based on coverage criteria.

Three examples for these tools are: Selenium [10], WinRunner [11] and Rational Robot [12].

2.2 Random Input Testing

This testing technique is also referred in the literature as stochastic testing or monkeys testing [13]. Random input testing refers to the idea that somebody seats in front of a software application and interacts randomly with it, by sending keyboard and mouse events.

The goal of monkeys testing is to crash the GUI of the software application under test. They generate tests cases randomly without knowing anything about the software application. The biggest problem of this testing technique is that monkeys cannot recognize software errors. There is a smarter category of monkeys called "smart monkeys" which have some knowledge about the software application under test. These monkeys can find more bugs, but they are more expensive to be developed [2].

Even if random input testing tools have a weak coverage, one of the biggest software companies has reported that 10–20 % of the bugs in their software applications were found by using random input testing method [13].

2.3 Unit Testing Frameworks

Unit testing technique for GUI testing requires programming the test cases. Unit testing frameworks like NUnit [14] can be used for executing GUI test cases.

These tools are helpful in case many bugs can only be discovered through a particular sequence of actions. With these tools the tester has to write code to simulate user interaction with the GUI under test. After executing the test cases the tester should check if the result obtained is the one expected.

In order to be effective, the GUI testing process using unit testing frameworks needs a lot of programming effort. There are some GUI libraries such as Abbot [15] which provide methods to simulate user interaction.

2.4 Model-Based Testing

Model-based testing requires that GUI states and events are described with a certain type of model. Having these models in place, the test cases can be generated automatically, either randomly or according to some particular coverage criteria.

The model-based testing process is presented in Fig. 1.

The model based testing process starts with the construction of the GUI's model. The model is used to generate test cases which are then executed over the GUI. In the last step, the obtained results are compared to the expected results described in the model.

The most important existing testing models used for model based testing are the following ones [4]:

- *Event Sequence Graph* (ESG)—a directed graph which contains a finite set of nodes and a finite set of edges. Each node represents a GUI event and the sequence of nodes represents the sequence of GUI events [16].
- *Event Flow Graph* (EFG) and *Event Interaction Graph* (EIG)—inside the EFG, each node represents a GUI event and all events which can be executed immediately after one event are directly linked with directed edges from this event. A path inside

Fig. 1 Model based testing

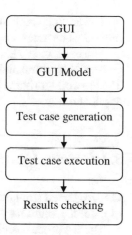

the EFG represents a sequence of GUI events and can be considered a test case. EIG is the later version of the EFG. The EIG's structure is composed by all the GUI events which represent the GUI nodes and all relationships between events which represent the graph edges.

The model-based testing technique is usually used to test the structural representation of a GUI [17].

Some of the frameworks used in the testing process of the GUIs of Web applications are WebGuitar [18], Artemis [19], Atusa [20] and Kudzu [21]. Web Guitar uses the ESG model; Atusa and Kudzu use the EIG model, all being based on functional testing. Artemis has a grey-box testing style, both structural and functional.

The EvoGUITest framework that was developed by our team uses in the beginning of testing process a random input testing method for generating the first set of test cases. Then the test cases evolve using an evolutionary process. The aim of the EvoGUITest framework is to determine the longest sequence of events which tests as many GUI controls as possible. The EvoGUITest framework will be further detailed in Sect. 4.

3 Evolutionary Algorithms

EAs are software programs that attempt to solve complex problems by mimicking the processes of Darwinian evolution [22]. They operate on a population of possible solutions by applying the principle called *survival of the fittest* to produce better approximations to a solution [23].

During the EA process a big number of artificial individuals search the solution over the problem space.

The artificial individuals are usually represented by vectors of binary values. Each individual encodes a possible solution for the problem which needs to be solved.

The most widely known EA is the Genetic Algorithm (GA). In the following, both Genetic Algorithm and the Simulated Annealing (SA) algorithm will be presented. These two algorithms were used for generating test cases inside the EvoGUITest application.

3.1 Genetic Algorithms

GA originated from the work of John Holland. They are the most obvious mapping of natural evolutionary process into a software application [24].

The GA process begins with a set of candidate solutions which is called population. A population is composed of individuals who are constituted from one or more genes. A population's individuals are used to form a new population by using crossover and mutation operators. During the GA process there is an expectation that the newly generated individuals are better than their parents.

GAs are well known and widely used in scientific and technical research because of their parallel nature, of their design space exploration and also due to their ability to solve non-linear problems [25].

A GA has four important phases:

- *Evaluation*—during this phase each individual is evaluated by the evaluation method. The *fitness function* is used for evaluation. It calculates how good the individual is to satisfy the test criteria;
- *Selection*—during this phase individuals are chosen randomly from the current population for creating new individuals in the next generation. The main idea of the selection methods is that fittest individual has the biggest probability of survival; therefore he has a greater probability to be picked for reproduction;
- *Crossover*—during this phase, recombination reproduces the chosen individuals and pair wise information will be exchanged and will result in a new population [25]. The crossover process joins two selected individuals at a crossover point, thus producing two new offsprings. During crossover, for instance the first parent's right half genes can be exchanged with the subsequent right half of the second parent. After crossover is performed, each parent pair will result in two offsprings. Crossover is the operator which is responsible for improving the individuals;
- *Mutation*—during this phase a randomly chosen bit is changed from '0' to '1' or from '1' to '0'. Each bit inside an individual has the same probability to mutate. Mutation is the operator which is responsible for introducing variety inside the population.

3.2 Simulated Annealing

SA is a probabilistic method for finding the global minimum of a cost function that may possess several local minima [26]. This algorithm emulates the physical process whereby a solid is slowly cooled until its structure becomes frozen. This happens at a minimum energy configuration.

The SA algorithm has four basic elements [27]:

- *Configurations*—these represent the possible problem solutions over which the process will search for the problem solution;
- *Move Set*—this set represents the computations performed to move from one configuration to another, as annealing proceeds;
- *Cost Function*—measures how "good" a particular configuration is;
- *Cooling Schedule*—anneal the problem from a randomly generated possible solution to a good solution. Usually the schedule needs a starting hot temperature and different rules for establishing when the current temperature should be decreased, by which amount temperature should be lowered and when the process should take end.

The most important feature of the SA algorithm is that it is a probabilistic method where during the search process the moves that increase the cost function are accepted in addition to moves which decrease the cost function [28]. This feature is the central point of the algorithm which enables the search process to locate the global minimum among all the other local minima.

The most important challenge in improving the performance of the SA algorithm is to decrease the temperature and in the same time to ensure that the process does not stop in a local minimum.

The goal of the SA algorithm is to find the quickest annealing schedule that achieves a value for finding the global minimum equal to unity [28].

The SA algorithm is suitable for solving large scale optimization problems inside which the global minimum is located among many local minima values.

4 EvoGUITest

EvoGUITest is a novel GUI automatic testing framework based on evolutionary algorithms. It automatically generates test cases which are used afterwards for testing the GUI. The test cases suite is generated automatically by an EA-based process. EvoGUITest's objective is to find the sequence of events which produces the biggest number of changes inside the GUI in a minimum amount of time. A bigger number of changes inside the GUI guarantee a better coverage of the search space, i.e. capturing a greater number of situations for testing the GUI's functionality.

4.1 The EvoGUITest Framework Architecture

The EvoGUITest application is a GUI testing framework which uses EAs for generating GUI test cases. It is developed in JavaScript and it runs on client side. Being developed in JavaScript it is very easy to be extended without any need of extra tools to write JavaScript. EvoGUITest is able to generate test cases for Web applications which have a GUI component already developed.

The testing process with this GUI testing framework consists of the following main steps:

- *Analysis*—the GUI state together with each GUI controls' states are analyzed. The result of this step is the list of HTML properties and events which correspond with each control located inside GUI;
- *Test Cases Generation*—generate test cases by using the specific EAs methods;
- *Test Cases Execution*—executes test cases;
- *Results Verification*—verifies the results after the execution of the test cases.

Figure 2 presents the main components of the EvoGUITest framework.

The most important part of the framework is the module which generates test cases using EAs. Each test case is represented by an individual. The first population of individuals is randomly generated Fig. 3 shows such an initial population for the classical Calculator application running under Windows.

Each individual consists of an array of genes, each corresponding to a GUI control. In Fig. 3 the array of genes for each individual corresponds to an array of ids which correspond to each GUI control. Each GUI control which appears inside an individual is linked with a user action on the GUI. After the first population of individuals is generated, the individuals are evolved by means of the EA process. After each generation, the new individuals are displayed together with their objective, age and fitness function. Figure 4 shows the individuals from the first generation. The population of individuals is generated for testing the GUI of a complex application. The individuals are classified so that the first one is the best individual from the current generation. As it can be easily observed, the first individual is the one which

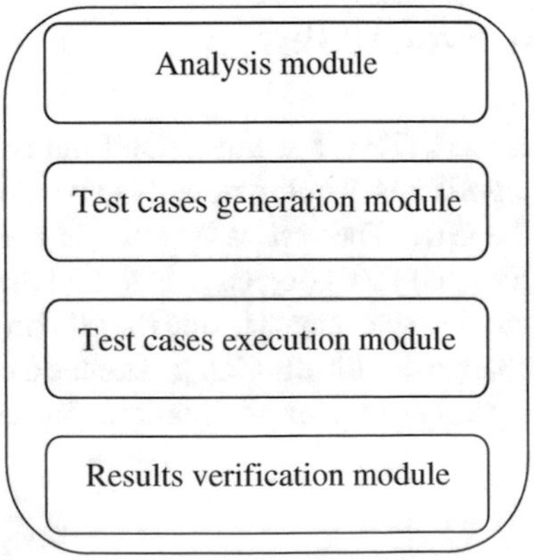

Fig. 2 The EvoGUITest
architecture

Analysis module

Test cases generation module

Test cases execution module

Results verification module

id	genes
1	btn1,div11,btn2,div11,btn7,tf13
2	div1,btn10,btn2,i1,div3,btn10,btn7,tf31,p1
3	i2,header,p1,div8,tf13,btn5
4	div1,div11,btn5,tf21,div11,div10,div11,p3
5	tf21,tf31,btn6,span2,div7,tf13
6	tf36,tf31,span4,tf34,div2,span1,btn4,tf1,span3
7	span4,div9,div11,div9,tf31
8	div9,div3,tf38,div8,i1,tf18,i1,tf18
9	div8,btn8,div10,tf21,btn4,btn4,tf34,btn1
10	div6,div5,div7,tf13,btn9,tf19,i1,btn3,p2

Fig. 3 Randomly generated individuals for testing GUI of a calculator application

id	genes	obj	age	fitness
1	btn4,div2,div4,tf38,btn2,btn10,btn2,tf31,tf19,div3	0.011364	1	0.0506
2	div6,tf18,btn10,btn9,span2,tf19,span2,btn7	0.012346	1	0.0494
3	btn6,tf34,tf31,btn1,btn8,btn7,span3	0.014706	1	0.0481
4	tf36,tf21,btn1,btn7,div10,btn6	0.014706	1	0.0469
5	div4,span3,btn6,div3,tf13	0.015385	1	0.0456
6	btn4,tf38,btn7,tf38,span3,i2,btn9,div11,tf18	0.015873	1	0.0444
7	div1,divhdn,div6,btn10,tf31,div11,i2,span1,i2	0.016667	1	0.0431
8	div7,div1,btn8,tf36,btn6,tf31	0.016667	1	0.0419
9	btn3,div9,btn4,span2,tf34,btn7,i1,span1,divhdn	0.016667	1	0.0406
10	div9,tf38,btn2,btn10,tf36	0.018868	1	0.0394
11	tf1,span3,btn7,tf18,tf21,btn10,header,span1	0.019608	1	0.0381
12	tf1,btn3,tf21,span5,btn9,div4	0.02	1	0.0369
13	div10,btn9,btn3,tf21,p1,tf34,btn7,btn3,btn9	0.020833	1	0.0356
14	btn10,p3,div3,div10,div6,div3	0.020833	1	0.0344
15	div8,btn3,span1,span3,p2,btn4,p3,tf19	0.022222	1	0.0331
16	span5,btn9,p1,div5,btn3	0.022222	1	0.0319
17	div4,btn9,div2,div7,btn2,span1,div3,div9,div1	0.025	1	0.0306
18	btn7,span4,tf36,div1,span2,btn2,btn4,div3,span3,tf34	0.026316	1	0.0294
19	btn1,tf13,div7,div5,div11,btn9,span4	0.028571	1	0.0281
20	tf31,div8,p1,tf31,tf18,btn4,p3,div4	0.028571	1	0.0269

Fig. 4 First generation of individuals for testing a complex GUI component

contains more button controls; therefore it is the one which produces the biggest number of changes inside the GUI. The age represents the current generation number. The objective column contains the objective value for each individual, and the fitness column contains the fitness value assigned to each individual. The objective attribute represents the performance of the individuals, while the fitness value represents rang of individuals inside the hierarchy.

For example, if we have the following objective values:

Individual 1: 2
Individual 2: 1000
Individual 3: 65536

if the *roulette wheel selection* will be applied on the above population of individuals the last individual won't have any chance to be selected for reproduction. If we assign a fitness function for each individual, who have the following values:

Individual1 : 2   Fitness : 0.5
Individual2 : 1000   Fitness : 0.3
Individual3 : 65536   Fitness : 0.2

then the last individual has a small chance to be selected for crossover.
The objective function which evaluates each individual is presented in formula (1):

$$Objective = (1/no\_of\_changes)+$$
$$1/(100 \times no\_of\_similar\_states)+ \qquad (1)$$
$$1/(100 \times no\_of\_useless\_states)$$

Each individual should produce the greatest number of changes and the smallest number of similar states and useless actions. A *useless action* is an action which doesn't produce any change inside the GUI. A *similar state* is a state which has already appeared earlier inside the set of states produced by the same individual.

The EvoGUITest framework contains a separate section where the user can set values for the most important parameters used by the GA and SA algorithms. For each one of these two algorithms, the user can select the values for the parameters presented in Table 1. The variables that affect the outcome of the SA algorithm are:

Table 1 Parameters list for GA and SA algorithms

GA	Values	SA	Values
Number of individuals	40	Initial temperature	100
Number of genes (min, max)	Min: 10 Max: 25	Epsilon	0.001
Number of selected pairs for crossover	20	Alpha	0.999
Mutation probability	0.2	–	
Mutation addition probability	0.5	–	
Mutation removal probability	0.5	–	
Number of generations	50	–	

the initial temperature, the rate at which the temperature decreases (alpha) and the stopping condition of the algorithm (epsilon).

The number of individuals indicates how many individuals exist in each population while the number of generations represents the generations for which the GA algorithm will be performed. The number of genes represents the minimum and the maximum length of each individual from the first population. The number of selected pairs for crossover represents how many individuals will be selected for reproduction. The mutation probability refers to the application of the mutation operator. Mutation can be applied in two ways: either by removing a gene from an individual or by adding a new gene.

Figure 5 displays the section which consists of the GA parameters list for the EvoGUITest application.

Fig. 5 GA parameters settings area

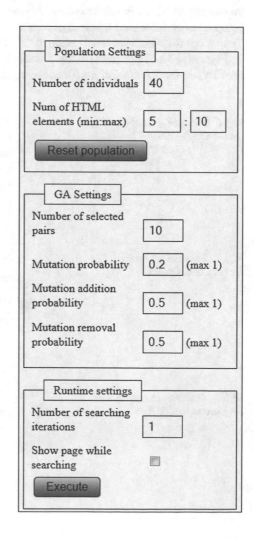

4.2 *The EvoGUITest Experimental Results*

All the experiments were performed on a computing system having the following configuration: Intel I3 processor, 2.2 GHz, Windows 7 Operating System. Three GUIs were tested: the first one is a simple GUI which consists of two buttons and two textboxes, the second one is the GUI of the classic Calculator application from Windows and the last one is a complex GUI which consists of more than twenty user controls.

For test cases generation we used both the GA and the SA algorithms. The selection method used for GA algorithm was the roulette wheel method. For each specific parameter, for each algorithm, the values presented in Table 1 were used in order to generate the test cases. These values were chosen to be used for running EAs based on our empirical studies done before. All the EAs' specific parameters' values were setup after we have tried hundred of runs with different values for these parameters. The values for which we have obtained the best results were chosen.

Figures 6, 7 and 8 present the test results obtained for each of the three GUIs using the GA and the SA algorithms for evolving the test cases suite.

Fig. 6 Test case generation for the simple GUI

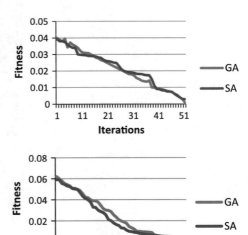

Fig. 7 Test case generation for the Calculator GUI

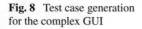

Fig. 8 Test case generation for the complex GUI

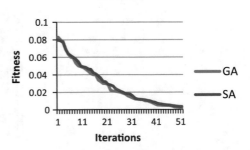

Table 2 Best individuals' performance for the GA and SA algorithms

GUI	GA Performance	No. of GUI Changes	SA Performance	No. of GUI Changes
Best individual for simple GUI testing	0.001	14	0.0029	11
Best individual for Calculator GUI testing	0.0012	19	0.0019	15
Best individual for complex GUI testing	0.0015	27	0.0034	23

Table 3 Convergence time(s) for the GA and SA algorithms

GUI Type	GA Convergence (s)	SA Convergence (s)
Simple GUI testing	30	46
Calculator GUI testing	40	57
Complex GUI testing	60	78

The performance of the best individuals for both GA and SA algorithms is presented in Table 2.

From Figs. 6, 7, 8 and Table 2 one can notice that the GA is able to find better test data compared to the SA algorithm. GA manages to find out the sequence of events which produces more changes inside GUI in comparison with SA. The individual which produces the biggest number of changes inside the GUI is the one which has the smallest value of its fitness function, because the testing problem is transformed into a minimization problem. It shows that individuals have evolved from the first generation to the last one. The best individual from the last generation produces the biggest number of changes inside the GUI; therefore, it has the smallest value of the fitness function.

The mean value of convergence time (in seconds) obtained from ten runs of each algorithm is presented in Table 3.

The convergence time for GA algorithm is smaller than the convergence time obtained for SA algorithm.

4.3 Performance Metrics

To evaluate the testing results obtained with the EvoGUITest framework we made a series of tests using other available open source frameworks.

To facilitate frameworks comparisons and provide information about the frameworks' performances, some metrics had to be defined. The defined metrics are as general as possible, in order to be applicable on any testing software that might be used for testing the GUIs of Web applications.

The metrics we defined are:

- Metrics #1: Number of HTML content errors *per* number of source code lines (NECL)

This metric represents the number of HTML content errors found over the number of source code lines of the tested application. The metric offers an idea about the density of HTML errors in the application source code. Measuring this parameter gives an image of the application's quality.

- Metrics #2: Average number of HTML content errors *per* number of source code lines (ANECL)

This metric represents the average number of HTML content errors over the number of source code lines of the tested applications. This parameter gives an idea about the density of HTML errors compared to the average size of the tested application. Measuring this parameter also gives an image of the tested application's quality.

- Metrics #3: Number of HTML content errors *per* test suite (NETS)

This metric indicates the number of HTML content errors discovered after testing the software application over the number of tests in the test suite. It can be extended for any other applications and test suites. Its extension refers to the total number of HTML errors over the total number of tests run for all the tested applications.

- Metrics #4: Average number of HTML content errors *per* test suite (ANETS)

The metric represents the average number of HTML content errors found over the number of tests in the test suite. This parameter offers an idea about the density of HTML errors discovered by each test from the test suites. It gives an image of the quality of the tests used in the testing process.

- Metrics #5: Number of HTML content errors *per* test (NET)

The metric represents the number of HTML content errors found after testing the software application with one single test from the suite of tests.

- Metrics #6: Average number of HTML content errors *per* test (ANET)

This metric represents the average number of HTML content errors found by a certain testing scenario. It shows the average abilities of a testing scenario to find errors in the tested graphical interface.

4.4 Experimental Results

Figure 9 presents a comparison between four test suites composed of ten test cases each. The test suites were generated with EvoGUITest, Selenium [8], WinRunner [9] and Rational Robot [10]. They were used in the regression testing phase for detecting errors inside a set of benchmark Web applications.

From Fig. 9 one can notice that the test suite generated using EvoGUITest is able to find more defects in comparison with other test suites based on the two functioning modes (i.e. capture and replay), even if they contain the same number of tests. This illustrates the fact that the test suite generated with EvoGUITest is better than those generated with Selenium, WinRunner and Rational Robot frameworks.

Artemis [17], Atusa [18] and Kudzu [19] frameworks for testing the GUIs of Web applications were used to observe the similarities and differences with our EvoGUITest framework.

The number of benchmark applications that were tested in order to make comparisons between EvoGUITest and Artemis, respective Atusa frameworks, was 30, and the number of tests from the tests suite was 100. These Web applications were selected among popular Web applications available on the Internet.

Figure 10 shows the differences between EvoGUITest and Artemis when they were used for finding out errors in different Web graphical interfaces.

The number of errors discovered in the HTML code by EvoGUITest and Artemis are presented in Table 4. In case of just two applications out of 30, the Artemis framework had better results than our EvoGUITest framework.

Using the NECL, NETS and NET metrics defined before, in Table 5 we present the better results obtained by EvoGUITest, compared to those obtained using the Artemis framework. The results clearly show better results for EvoGUITest.

Figure 11 shows the differences between EvoGUITest and Atusa when looking for errors in the 30 benchmark applications tested.

The results of the number of errors discovered in the HTML code by EvoGUITest and Atusa are presented in Table 6. In all the 30 of tested cases, EvoGUITest framework had better results than Atusa framework.

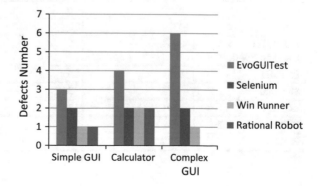

Fig. 9 Number of defects discovered by different testing frameworks

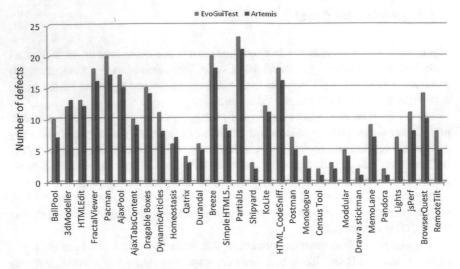

Fig. 10 Errors find out by EvoGUITest and Artemis

Table 4 Errors in HTML code discovered by EvoGUITest and Artemis

Tested Applications	Lines code number	EvoGuiTest (HTML errors number)	Artemis (HTML errors number)
3dModeller	393	12	13
BallPool	256	10	7
FractalViewer	750	18	16
HTMLEdit	568	13	12
Pacman	1857	20	17
AjaxPool	250	17	15
AjaxTabsContent	156	10	9
DragableBoxes	697	15	14
DynamicArticles	156	11	8
Homeostasis	2037	6	7
Qatrix	1712	4	3
Durandal	2159	6	5
Breeze	14730	20	18
Simple HTML5 Drawing Application	439	9	8
PartialJs	5857	23	21
Shipyard	73	3	2
KoLite	381	12	11
HTML CodeSniffer	1433	18	16
Postman	199	7	5

(continued)

Table 4 (continued)

Tested Applications	Lines code number	EvoGuiTest (HTML errors number)	Artemis (HTML errors number)
Monologue	215	4	2
Census Tool	1567	2	1
Computer Language Benchmarks Game	1678	3	2
Moddular	945	5	4
Draw a stickman	785	2	1
MemoLane	567	9	7
Pandora	1128	2	1
Lights	957	7	5
jsPerf	589	11	8
BrowserQuest	1368	14	10
RemoteTilt	688	8	5
Average	1486	10	8

Table 5 Metrics results for EvoGUITest and Artemis

Applications	EvoGuiTest			Artemis		
	NECL	NETS	NET	NECL	NETS	NET
3dModeller	0.03	12	0.12	0.033	13	0.13
BallPool	0.039	10	0.1	0.027	7	0.07
FractalViewer	0.024	18	0.18	0.021	16	0.16
HTMLEdit	0.022	13	0.13	0.021	12	0.12
Pacman	0.01	20	0.2	0.009	17	0.17
AjaxPool	0.068	17	0.17	0.06	15	0.15
AjaxTabsContent	0.064	10	0.1	0.057	9	0.09
DragableBoxes	0.021	15	0.15	0.02	14	0.14
DynamicArticles	0.07	11	0.11	0.05	8	0.08
Homeostasis	0.002	6	0.06	0.003	7	0.07
Qatrix	0.002	4	0.04	0.002	3	0.03
Durandal	0.003	6	0.06	0.002	5	0.05
Breeze	0.001	20	0.2	0.001	18	0.18
Simple HTML5 Drawing Application	0.02	9	0.09	0.018	8	0.08
PartialJs	0.004	23	0.23	0.003	21	0.21
Shipyard	0.04	3	0.03	0.027	2	0.02
KoLite	0.031	12	0.12	0.028	11	0.11
HTML CodeSniffer	0.012	18	0.18	0.011	16	0.16

(continued)

Table 5 (continued)

Applications	EvoGuiTest			Artemis		
	NECL	NETS	NET	NECL	NETS	NET
Postman	0.0351	7	0.07	0.025	5	0.05
Monologue	0.0186	4	0.04	0.009	2	0.02
Census Tool	0.001	2	0.02	0.001	1	0.01
Computer Language Benchmarks Game	0.001	3	0.03	0.001	2	0.02
Moddular	0.005	5	0.05	0.004	4	0.04
Draw a stickman	0.002	2	0.02	0.001	1	0.01
MemoLane	0.015	9	0.09	0.012	7	0.07
Pandora	0.001	2	0.02	0.001	1	0.01
Lights	0.007	7	0.07	0.005	5	0.05
jsPerf	0.018	11	0.11	0.0135	8	0.08
BrowserQuest	0.010	14	0.14	0.007	10	0.1
RemoteTilt	0.011	8	0.08	0.007	5	0.05
	ANECL	**ANETS**	**ANET**	**ANECL**	**ANETS**	**ANET**
Average for NECL, NETS, NET	0.0195	10	0.1	0.0159	8	0.084

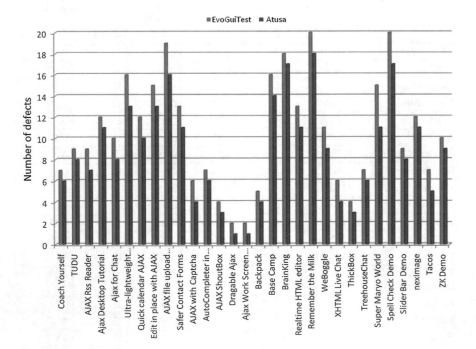

Fig. 11 Errors discovered by EvoGUITest and Atusa

Table 6 Errors in HTML code find out by EvoGUITest and Atusa

Tested applications	Lines code number	EvoGuiTest (HTML errors number)	Atusa (HTML errors number)
TUDU	580	9	8
Coach Yourself	150	7	6
AJAX Rss Reader	600	9	7
Ajax Desktop Tutorial	856	12	11
Ajax for Chat	678	10	8
Ultra-lightweight charts for AJAX	1234	16	13
Quick calendar using AJAX	453	12	10
Edit in place with AJAX	745	15	13
AJAX file upload tutorial	897	19	16
Safer Contact Forms without Captcha	1145	13	11
Using AJAX with Captcha	459	6	4
AutoCompleter in AJAX	239	7	6
AJAX ShoutBox	712	4	3
Dragable Ajax	324	2	1
Making Ajax Work with Screen Readers	123	2	1
Backpack	890	5	4
Base Camp	1256	16	14
BrainKing	2367	18	17
Realtime HTML editor	543	13	11
Remember the Milk	967	20	18
WeBoggle	1145	11	9
XHTML Live Chat	678	6	4
ThickBox	564	4	3
TreehouseChat	343	7	6
Super Maryo World	756	15	11
Spell Check Demo	897	20	17
Slider Bar Demo	235	9	8
nexImage	123	12	11
Tacos	378	7	5
ZK Demo	456	10	9
Average	693	10	8

58
G.I. Latiu et al.

Table 7 Metrics results for EvoGUITest and Atusa

Aplications	EvoGuiTest			Atusa		
	NECL	NETS	NET	NECL	NETS	NET
TUDU	0.015	9	0.09	0.013	8	0.08
Coach Yourself	0.046	7	0.07	0.04	6	0.06
AJAX Rss Reader	0.015	9	0.09	0.011	7	0.07
Ajax Desktop Tutorial	0.014	12	0.12	0.013	11	0.11
Ajax for Chat	0.0147	10	0.1	0.0118	8	0.08
Ultra-lightweight charts for AJAX	0.013	16	0.16	0.010	13	0.13
Quick calendar using AJAX	0.0264	12	0.12	0.022	10	0.1
Edit in place with AJAX	0.020	15	0.15	0.017	13	0.13
AJAX file upload tutorial	0.021	19	0.19	0.0178	16	0.16
Safer Contact Forms without Captcha	0.011	13	0.13	0.009	11	0.11
Using AJAX with Captcha	0.0130	6	0.06	0.008	4	0.04
AutoCompleter in AJAX	0.0292	7	0.07	0.0251	6	0.06
AJAX ShoutBox	0.005	4	0.04	0.004	3	0.03
Dragable Ajax	0.006	2	0.02	0.003	1	0.01
Making Ajax Work with Screen Readers	0.0162	2	0.02	0.008	1	0.01
Backpack	0.005	5	0.05	0.004	4	0.04
Base Camp	0.0127	16	0.16	0.011	14	0.14
BrainKing	0.007	18	0.18	0.007	17	0.17
Realtime HTML editor	0.024	13	0.13	0.020	11	0.11
Remember the Milk	0.008	20	0.2	0.007	18	0.18
WeBoggle	0.009	11	0.11	0.007	9	0.09
XHTML Live Chat	0.008	6	0.06	0.005	4	0.04
ThickBox	0.007	4	0.04	0.005	3	0.03
TreehouseChat	0.020	7	0.07	0.017	6	0.06
Super Maryo World	0.019	15	0.15	0.014	11	0.11
Spell Check Demo	0.022	20	0.2	0.018	17	0.17
Slider Bar Demo	0.038	9	0.09	0.034	8	0.08
nexImage	0.097	12	0.12	0.089	11	0.11
Tacos	0.018	7	0.07	0.013	5	0.05
ZK Demo	0.021	10	0.1	0.019	9	0.09
	ANECL	ANETS	ANET	ANECL	ANETS	ANET
Average for NECL, NETS, NET	0.0193	10	0.1	0.0154	8	0.088

Using the NECL, NETS and NET metrics in Table 7 we present the results obtained by EvoGUITest and compare them to those obtained by Atusa framework. The results show that EvoGUITest had better results than Atusa for all metrics.

The third comparison was made with the Kudzu framework. For 28 benchmark applications that were tested, Fig. 12 shows the errors obtained by EvoGUITest and by Kudzu frameworks.

The number of errors discovered in the HTML code by EvoGUITest and Kudzu for 28 benchmark applications are presented in Table 8. In the tested cases when errors were discovered, EvoGUITest framework had better or same results than Kudzu framework. In just one single case (the TVGuide application), Kudzu discovered one error while EvoGUITest was unable to find any error.

Using the NECL, NETS and NET metrics in Table 9 we present the results obtained by EvoGUITest framework, compared to Kudzu framework. The results show that EvoGUITest had better results than Kudzu.

All the comparisons made until now show that the results obtained with EvoGUI-Test framework in automated testing of graphical user interfaces for Web applications have a better quality. The EvoGUITest framework managed to find out more errors in the HTML code and to generate more performing test suites than others similar frameworks.

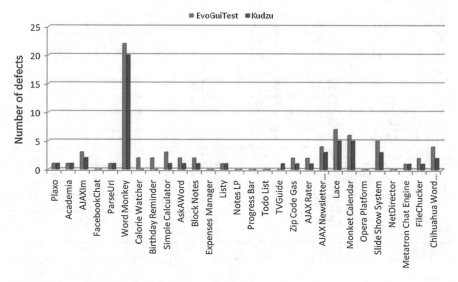

Fig. 12 Errors discovered by EvoGUITest and Kudzu

Table 8 Errors in HTML code discovered by EvoGUITest and Kudzu

Tested Applications	Lines Code Number	Tests Suite Dimension	EvoGuiTest(HTML errors number)	Kudzu (HTML errors number)
Plaxo	17854	178	1	1
Academia	1604	16	1	1
AJAXIm	9328	93	3	2
FacebookChat	15789	144	–	–
ParseUri	601	6	1	1
Word Monkey	5437	93	22	20
Calorie Watcher	2808	28	2	–
Birthday Reminder	1678	16	2	–
Simple Calculator	2340	93	3	1
AskAWord	2690	93	2	1
Block Notes	1798	28	2	1
Expenses Manager	3608	32	–	–
Listy	2564	26	1	1
Notes LP	3516	30	–	–
Progress Bar	1439	15	–	–
Todo List	2076	20	–	–
TVGuide	3789	32	–	1
Zip Code Gas	5587	54	2	1
AJAX Rater	4312	47	2	1
AJAX Newsletter Signup	4789	87	4	3
Lace	3912	76	7	5
Monket Calendar	2945	34	6	5
Opera Platform	12347	120	–	–
Slide Show System	5489	67	5	3
NetDirector	3675	104	–	–
Metatron Chat Engine	4012	81	1	1
FileChucker	2934	72	2	1
Chihuahua Word Puzzle	5476	55	4	2
Average	4799	62	3	2

Table 9 Metrics results for EvoGUITest and Kudzu

Aplications	EvoGuiTest			Kudzu		
	NECL	NETS	NET	NECL	NETS	NET
Plaxo	0	1	0.005	0.00005	1	0.005
Academia	0.0006	1	0.062	0.0006	1	0.062
AJAXIm	0.0003	3	0.032	0.0002	2	0.021
FacebookChat	0	0	0	0	0	0
ParseUri	0.0016	1	0.166	0.0016	1	0.166
Word Monkey	0.004	22	0.23	0.0036	20	0.21
Calorie Watcher	0.0007	2	0.071	0	0	0
Birthday Reminder	0.0011	2	0.12	0	0	0
Simple Calculator	0.0012	3	0.032	0.0004	1	0.01
AskAWord	0.0007	2	0.021	0.0003	1	0.01
Block Notes	0.0011	2	0.071	0.0005	1	0.035
Expenses Manager	0	0	0	0	0	0
Listy	0.0003	1	0.038	0.0003	1	0.038
Notes LP	0	0	0	0	0	0
Progress Bar	0	0	0	0	0	0
Todo List	0	0	0	0	0	0
TVGuide	0	0	0	0.0002	1	0.031
Zip Code Gas	0.0003	2	0.037	0.0001	1	0.018
AJAX Rater	0.0004	2	0.042	0.0002	1	0.021
AJAX Newsletter Signup	0.0008	4	0.045	0.0006	3	0.034
Lace	0.0017	7	0.092	0.0012	5	0.065
Monket Calendar	0.002	6	0.176	0.0016	5	0.147
Opera Platform	0	0	0	0	0	0
Slide Show System	0.0009	5	0.074	0.0005	3	0.044
NetDirector	0	0	0	0	0	0
Metatron Chat Engine	0.0002	1	0.012	0.0002	1	0.012
FileChucker	0.0006	2	0.0277	0.0003	1	0.013
Chihuahua Word Puzzle	0.0007	4	0.0727	0.0003	2	0.0363
	ANECL	**ANETS**	**ANET**	**ANECL**	**ANETS**	**ANET**
Average for NECL, NETS, NET	0.0006	3	0.050	0.0004	2	0.034

5 Conclusions and Future Work

This paper presents EvoGUITest, an original framework for automatically testing graphical user interfaces of Web applications based on EAs techniques. The main features of the EvoGUITest framework are the following:

- It tests the GUI separately from the application source code itself;
- It automatically generates and executes the test suite;
- It is able to find the sequence of events which produces the biggest number of changes inside the GUI, so it checks the biggest possible number of controls inside the GUI.

The EvoGUITest framework is original because it runs on client side, being developed in Javascript and it tests the GUI of the application separately from the software application itself. To the best of our knowledge, it is the first GUI testing application developed only using JavaScript. The advantage of using JavaScript is that it is platform-independent and it can test GUI components developed in any programming language. The extension of the framework is very easy to make because there is no need of any extra tools to write JavaScript code. This can be done using any plain text or HTML editor.

EvoGUITest has the objective to find out the most important sequence of events which produces the biggest number of changes inside the GUI. By producing the biggest number of changes, the sequence is able to verify as many components as possible inside the GUI.

EvoGUITest is able to discover the most important sequence of GUI events.

Future work will involve using EvoGUITest framework for testing larger projects. We will also focus on using EvoGUITest for regression testing. The test cases suite will be used to check if the GUI still functions correctly after each development change is performed. The framework will be extended with other evolutionary algorithms: Particle Swarm Optimization (PSO) and Ant Colony Optimization (ACO) algorithms.

A complete automated testing framework based on EAs could be designed and implemented, for completely automating the GUI testing process.

Acknowledgments This work was supported by a grant of the Romanian National Authority for Scientific Research, CNDI-UEFISCDI, project number 47/2012.

References

1. Jansen, B.J.: The Graphical User Interface: an introduction. In: Seminal Works in Computer Human Interaction, vol. 30(2), pp. 24-26. ACM, New York (1998)
2. Pimenta, A.: Automated Specification-based Testing of Graphical User Interfaces. Ph.D. Thesis, Department of Electrical and Computer Engineering, Porto University, Portugal (2006)
3. Ganov, S., Killmar, C., Khurshid, S., Perry, D.: Test generation for graphical user interfaces based on symbolic execution. In: Proceedings of the 3rd International Workshop on Automation of Software Test , pp. 35-40. ACM, New York (2008)
4. Yang, X.: Graphic User Interface Modelling and Testing Automation. Ph.D. Thesis, School of Engineering and Science, Victoria University Melbourne, Australia (2011)
5. Al-Zain, S., Eleyan, D., Hassouneh, Y.: Comparing GUI automation testing tools for dynamic web applications. Asian J. Comput. Inf. Syst. **01**(02), 38–48 (2013)
6. Kaur, M., Kumari, R.: Comparative study of automated testing tools: testcomplete and quicktest pro. Int. J. Comput. Appl. **24**(1), 1–7 (2011)

7. Bergen, S., Ross, J.: Evolutionary art using summed multi-objective ranks. In: Genetic Programming Theory and Practice VIII, vol. 8, pp. 227–244. Springer Science, Berlin (2011)
8. Valdez-Garcia, M., Trujillo, I., Fernandez de Vega, F., Guervos, J.M., Olague, G.: EvoSpace-Interactive: a framework to develop distributed collaborative-interactive evolutionary algorithms for artistic design. In: Evolutionary and Biologically Inspired Music, Design, Sound Art and Design, LNCS, vol. 7834, pp. 121–132. Springer (2013)
9. Prabhu, J., Malmurugan, N.: A survey on automated GUI testing procedures. Eur. J. Sci. Res. **64**(3), 456–462 (2011)
10. Selenium Framework. http://seleniumhq.org
11. WinRunner Framework. http://mercury.com
12. Rational Robot Framework. http://www-01.ibm.com/software/awdtools/tester/robot/
13. Nyman, N.: In defense of monkey testing. In: Software Testing and Quality Engineering Magazine, pp. 18–21 (2000)
14. NUnit Framework. http://nunit.org
15. Abbot GUI libraries. http://abbot.sourceforge.net
16. Belli, F.: Finite-State testing and analysis of Graphical User Interfaces. In: Proceedings of the 12th International Symposium on Software Reliability Engineering, pp. 34–43. IEEE Xplore (2001)
17. Qureshi, I.A., Nadeem, A.: GUI testing techniques: a survey. Int. J. Future Comput. Commun. vol. 2(2), pp. 142–146 (2013)
18. Web Guitar. http://www.cs.umd.edu/~atif/GUITAR-Web
19. Artzi, S., Dolby, J., Jensen, S. H., Moller, A., Tip, F.: A framework for automated testing of JavaScript web applications. In: Proceedings of the 33rd International Conference on Software Engineering, pp. 571–580. ACM, New York (2011)
20. Mesbah, A., Van Deursen, A.: Invariant-Based automatic testing of AJAX user interfaces. In: Proceedings of the 31st International Conference on Software Engineering, pp. 210–220. IEEE Computer Society Washington, DC (2009)
21. Saxena, P., Akhawe, D., Hanna, S., McCamant, S., Song, D., Mao, F.: A symbolic execution framework for JavaScript. In: Proceedings of 31st IEEE Symposium on Security and Privacy, pp. 513–528. IEEE Computer Society Washington, DC (2010)
22. Jones, G.: Genetic and evolutionary algorithms. In: Encyclopedia of Computational Chemistry, pp.1–10. Wiley (1990)
23. Pohlheim, H.: Evolutionary algorithms: overview, methods and operators. In: Geneticand Evolutionary Algorithm Toolbox for Matlab (2006)
24. Streichert, F.: Evolutionary Algorithms in Multi-Modal and Multi-Objective Environments, Ph.D. Thesis, Department of Computer Architecture, University of Tubingen, Germany (2007)
25. Rauf, A.: Coverage Analysis for GUI Testing, Ph.D. Thesis, Department of Computer Science, National University of Computer and Emerging Sciences Islamabad, Pakistan (2010)
26. Bertsimas, D., Tsitsiklis, J.: Simulated annealing. Stat. Sci. **8**(1), 10–15 (1993)
27. Ruthenbar, R.: Simulated Annealing algorithms: an overview. IEEE Circuits Devices Mag. **5**, 19–26 (1989)
28. Nascimento, V., Carvalho, V., Castilho, C., Soares, E., Bittencourt, C., Woodruff, D.: The simulated annealing global search algorithm applied to the crystallography of surfaces by Leed. In: Surface Review and Letters, vol. 6(5), pp. 651–661 (1999)

Evolving Protection Measures for Lava Risk Management Decision Making

Giuseppe Filippone, Donato D'Ambrosio, Davide Marocco
and William Spataro

Abstract Many volcanic areas around the World are densely populated and urbanized. For instance, Mount Etna (Italy) is home to approximately one million people, despite being the most active volcano in Europe. Mapping both the physical threat and the exposure and vulnerability of people and material properties to volcanic hazards can help local authorities to guide decisions about where to locate a priori critical infrastructures (e.g. hospitals, power plants, railroads, etc.) and human settlements and to devise for existing locations and facilities appropriate mitigation measures. We here present the application of Parallel Genetic Algorithms for optimizing earth barriers construction by morphological evolution, to divert a case study lava flow that is simulated by the numerical Cellular Automata model Sciara-fv2 at Mt Etna volcano (Sicily, Italy). The devised area regards Rifugio Sapienza, a touristic facility located near the summit of the volcano, where the methodology was applied for the optimization of the position, orientation and extension of an earth barrier built to protect the zone. The study has produced extremely positive results, providing insights and scenarios for the area representing, to our knowledge, the first application of morphological evolution for lava flow mitigation.

Keywords Evolutionary computation · Genetic algorithms · Parallel computing · Decision support system · Cellular automata · Morphological evolution

1 Introduction

When dealing with lava flow risk assessment, the use of thematic maps of volcanic hazard is of fundamental relevance to support policy managers and administrators in effective land use planning and taking proper actions that are required during an

G. Filippone (✉) · D. D'Ambrosio · W. Spataro
Department of Mathematics and Computer Science, University of Calabria,
Rende, Italy
e-mail: G.Filippone@ed.ac.uk

D. Marocco
Centre of Robotics and Neural Systems, School of Computing and Mathematics,
Plymouth University, Plymouth, UK

© Springer International Publishing Switzerland 2016   65
K. Madani et al. (eds.), *Computational Intelligence*,
Studies in Computational Intelligence 613,
DOI 10.1007/978-3-319-23392-5_4

emergency phase. In particular, hazard maps are a key tool for emergency management by describing the threat that can be expected at a certain location for future eruptions. At Mt. Etna (Italy), the most active volcano in Europe, the majority of events that have occurred in the last four centuries report damage to human properties in numerous towns on the volcano flanks [1]. Current efforts for hazard evaluation and contingency planning in volcanic areas depend heavily on hazard maps and numerical simulations for the purpose of individuating affected areas in advance.

Although many computational modeling methods [2–4] for lava flow simulation and related techniques for the compilation of susceptibility maps are already known to the international scientific community, the problem of defining a standard methodology for the construction of protection works, in order to mitigate volcanic risk, remains open. Techniques to slow down and divert lava flows, caused by collisions with protective measures such as artificial barriers [5, 6] or dams [7], are now to be considered empirical, exclusively based on past experiences. The proper positioning of protective measures in the considered area may depend on many factors (viscosity of the magma, output rates, volume erupted, steepness of the slope, topography, economic costs). As a consequence, in this context one of the major scientific challenges for volcanologists is to provide efficient and effective solutions.

Morphological Evolution (ME) is a recent development within the field of engineering design, by which evolutionary computation techniques are used to tackle complex design projects. This branch of evolutionary computation is also known as evolutionary design and it is a multidisciplinary endeavour that integrates concepts from evolutionary algorithms, engineering, and complex systems to solve engineering design problems [8]. Morphological evolution has been largely explored in evolutionary robotics, both for the design of imaginary 3D robotics bodies [9] and for the efficient and autonomous design of adaptive moving robots [10]. Principles of evolutionary design have been also applied in structural engineering at different level of the design process, from the structural design itself to the logistic involved in the construction [11].

This paper describes the application of ME by Parallel Genetic Algorithms (PGAs), for the first time to our knowledge, for optimizing earth barriers construction to divert a case study lava flow that is simulated by the latest release fv2 of the SCIARA Cellular Automata lava flow model [12]. Cellular Automata (CA) were introduced in 1947 by John von Neumann [13], quickly gaining the attention of the Scientific Community both as powerful parallel computational models and as a convenient apparatus for modeling and simulating several types of complex physical phenomena. CA have been applied to a variety of fields and their major interest regard their pratical use in Complex Systems modelling in Physics, Biology, Earth Sciences and Engineering (e.g., see [14–17]).The GA fitness evaluation, which was adopted for evaluating the "goodness" of the protective works of the CA model generated lava flow scenarios, has implied a massive use of the numerical simulator that runs thousands of concurrent simulations for every GA generation computation. Therefore, a GPGPU (General Purpose computation with Graphic Processor Units) library was developed to accelerate the GA execution. A visualization system [18] was also implemented, thereby allowing interactive analysis of the results.

Eventually, a study of GA dynamics, with reference to emergent behaviors, is also discussed later. In the following, after the description of the case study adopted for the experiments (Sect. 2), the main characteristics of the implemented algorithm, framework and results are presented (Sect. 3). Section 4 concludes the paper with final comments and future works.

2 The Case Study: The 2001 Mt Etna Eruption

The 2001 eruption of Mt. Etna began on July 17, characterized by lava emission from several vents on the southern flank of the volcano, at elevations of 2100, 2550, 2600, 2700, 2950, 3050 m, the latter four being directly connected to the conduit of the SE crater [19] (see Fig. 1). Lava flows emitted from the lowermost vents (2100, 2550, 2600–2700 m) caused damage and threatened some important facilities and infrastructure, which were protected by earthen barriers. Effusion rates at the main eruptive vents were estimated daily by [1] from the volume/time ratio and were obtained by careful mapping of the flow area and estimating its mean thickness. The facilities of the Sapienza zone were undoubtedly at risk because of their short distance from the 2700 and 2550 m effusive vents (respectively 3 and 2.5 km). The most probable path for the lava flow emitted from the 2100 m fissure was simulated (Crisci et al. 2001 and M.T. Pareschi, unpublished reports to Civil Protection) and was considered for the carried out experiments presented in the next sections. Thirteen artificial barriers were built during the July August 2001 Mt. Etna eruption. Their locations, together with investigated area here considered, are shown in the map of Fig. 1. The flow emitted from the lower vent, the 2100 m fissure, immediately interrupted the road SP92 and invaded a part of the adjacent wide parking area located between Mts. Silvestri and the Sapienza zone (1900 m a.s.l.). Starting on 21 July, a large barrier was progressively built on the eastern flank of the flow to protect two tourist facilities. This barrier worked properly and the two buildings were saved. The lava flow emitted from the 2100 m fracture descended about 6 km southwards (Fig. 1) and after the SP92 road near Mts. Silvestri it cut some other minor rural roads and destroyed a few isolated country houses. Had the lava advanced further, it would have re-crossed the SP92 road at a lower elevation, causing the complete isolation of the upper part of Mt. Etna. Workers and machines were moved to a possible critical point on the western front ready to build a diversion barrier to protect the road. An intervention plan was also set up for the protection of the Nicolosi and Belpasso villages, located on the most probable path of the lava, at only 4 km distance from its lowermost front. Eventually, the rate effusion decrease beginning in the last days of July prevented any further advance of the flow and thus the planned interventions were not necessary.

Fig. 1 Set of interventions carried out during the 2001 eruption event to divert the lava flow away from the facilities. The *green* perimeters represent the Rifugio Sapienza and other facilities (security area), which delimitates the area that has to be protected by the flow for the study. The *red* perimeter (work area), specifies the area in which the earth barrier can be located (Base figure taken from [5])

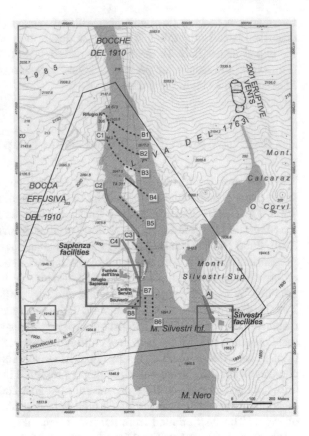

3 Morphological Evolution of Protective Works Through Parallel Genetic Algorithms

Genetic Algorithms (GAs) [20] are general-purpose iterative search algorithms inspired by natural selection and genetics. Among other applications, GA have been applied to combinatorial problems [21] in the study of the interaction between evolution and learning [22], evolutionary robotics [23, 24], for improving the performance of CA in resolving difficult computational tasks (e.g. [25]). GAs based methods have also been applied to CA for modelling bioremediation of contaminated soils [26] and for the optimisation of lava and debris flow simulation models (e.g., [27–30]).

GAs simulate the evolution of a population of candidate solutions, called phenotypes, to a specific problem by favouring the reproduction of the best individuals. Phenotypes are codified by genotypes, typically using strings, whose elements are called genes. In order to determine the best possible solution of a given problem, the GA must explore the so-called search (or solution) space, defined as the set of all possible values that the genotype can assume. The members of the initial population evaluated by means of a "fitness function", determining the individuals "adaptivity"

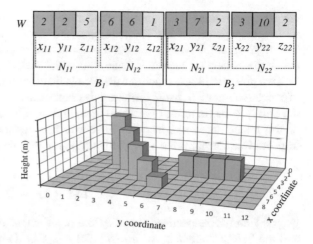

Fig. 2 Example of barriers encoding into a GA genotype. The height of the intermediate points of each barrier is obtained by connecting the work protections extremes through a linear function

value (also called fitness value), i.e. a measure of its goodness in resolving the problem. Best individuals are chosen by means of a "selection" operator and reproduced by applying random "genetic" operators to form a new population of offspring. Typical genetic operators are "crossover" and "mutation": they represent a metaphor of sexual reproduction and of genetic mutation, respectively. The overall sequence of fitness assignment, selection, crossover, and mutation is repeated over many generations (i.e. the GA iterations) producing new populations of individuals. According to the individual's probability of selection, any change that actually increases the individual's fitness will be more likely to be preserved over the selection process, thus obtaining better generationsas stated by the fundamental theorem of genetic algorithms [20]. For a complete overview of GAs, see [31, 32].

While GAs have been applied several times in the past for optimizing CA models, as the ones previously reported, by considering the 2001 Nicolosi case study, in this work GAs were adopted in conjunction with the SCIARA-fv2 CA model for the morphological evolution of protective works to control lava flows. The numerical model finite set of states was extended by introducing two substates defined as:

$$Z \subseteq R \tag{1}$$

where Z is the set of cells of the cellular automaton that specifies the Safety Zone, which delimitates the area that has to be protected by the lava flow and

$$P \subseteq R, P \cap Z = \oslash \tag{2}$$

where P is the set of CA cells that identifies the Protection Measures Zone identifying the area in which the protection works are to be located.

The Protection work $W = B_1, B_2, \ldots, B_n$ was represented as a set of barriers, where every barrier $B_i = N_{i1}, N_{i2}$ is composed by a pair of nodes $N_{ij} = x_{ij}, y_{ij}, z_{ij}$, where x_{ij}, y_{ij} represent CA coordinates for the generic node j of the barrier i, and z_{ij} the height (expressed in m). The solutions were encoded into a GA genotype, directly as integer values (Fig. 2) and a population of 100 individuals, randomly generated inside the Protection Measures Zone, was considered.

Two different fitness functions were considered to suitably evaluate the goodness of a given solution: f_1, based on the areal comparison between the simulated event and the Safety Zone (in terms of affected area) and f_2, which considers the total volume of the protection works in order to reduce intervention costs and environmental impact. More formally, the f_1 objective function is defined as:

$$f_1 = \frac{\mu(S \cap Z)}{\mu(S \cup Z)} \tag{3}$$

where S and Z respectively identify the areal extent of the simulated lava event and the Safety Zone area, with $\mu(S \cap Z)$ e $\mu(S \cup Z)$ being the measures of their intersection and union. The function f_1, assumes values within the range [0, 1] where 0 occurs when the simulated event and Safety Zone Area are completely disjointed (best possible simulation) and 1 occurs when simulated event and Safety Zone Area perfectly overlap (worst possible simulation).

The f_2 objective function is defined as:

$$f_2 = \frac{\sum_{i=1}^{|W|} p_c \cdot d(B_i) \cdot h(B_i)}{V_{max}} \tag{4}$$

where $d(B_i)$ and $h(B_i)$ represent the length (in meters) and the average height of the ith barrier, respectively. The parameter p_c is the cell side and $V_{max} \in R$ is a threshold parameter (i.e., the maximum building volume) given by experts, for the function normalization. Since the barriers are composed of two nodes, the function can be written as:

$$f_2 = \frac{\sum_{i=1}^{|W|} p_c \cdot d(N_{i1}, N_{i2}) \cdot \bar{h}(N_{i1}, N_{i2})}{V_{max}} \tag{5}$$

where $\bar{h}(N_{i1}, N_{i2}) = \frac{|z_{i1} + z_{i2}|}{2}$ is considered as the average height value between two different nodes and $d(N_{i1}, N_{i2}) = \sqrt{(x_{i1} - x_{i2})^2 + (y_{i1} - y_{i2})^2}$ identifies the Euclidean distance between them. The final fitness function f_2 is thus:

$$f_2 = \frac{\sum_{i=1}^{|W|} p_c \cdot \sqrt{(x_{i1} - x_{i2})^2 + (y_{i1} - y_{i2})^2} \cdot \frac{|z_{i1} + z_{i2}|}{2}}{V_{max}} \tag{6}$$

The function f_2, assumes values within the range [0, 1]: it is nearly 0 when the work protection is the cheapest possible, 1 otherwise.

For the genotype fitness evaluation, a composite (aggregate) function f_3 was also introduced as follows:

$$f_3 = f_1 \cdot \omega_1 + f_2 \cdot \omega_2 \tag{7}$$

where $\omega_1, \omega_2 \in R$ and $(\omega_1 + \omega_2) = 1$, represent weight parameters associated to f_1 and f_2. Several different values where tested and the considered ones in this work chosen on the basis of trial and error techniques. The goal for the GA is to find a solution that minimizes the considered objective function $f_3 \in [0, 1]$.

In order to classify each genotype in the population, at every generation run, the algorithm executes the following steps:

1. CA cells elevation a.s.l. are increased/decreased in height on the basis of the genotype decoding (i.e., the barrier cells). In addition, an extending Bresenham's original algorithm [33] is applied to determine the cells inside the segment between the work protection extremes and f_2 subsequently computed.
2. A SCIARA-fv2 simulation is performed (about 40000 calculation steps) and the impact of the lava thickness on Z area (f_1 computation) is evaluated.
3. f_3 is computed and individuals are sorted according to their fitness.

The adopted GA is a rank based and elitist model, as at each step only the best genotypes generate off-spring. The 20 individuals which have the highest fitness generate five off-spring each and the $20 \times 5 = 100$ offspring constitute the next generation. After the rank based selection, the mutation operator is applied with the exception of the first 5 individuals.

The complete list of GA characteristics and parameters is reported in Table 1. Each gene mutation probability depends on its representation: p_{mc} for genes corresponding to coordinates value and p_{mh} viceversa. Therefore, if during the mutation process, a coordinate gene is chosen to be modified, the new value will depend on the parameters x_{max} and y_{max} which represent the cell radius within the node, the position of which can vary. The interval $[h_{min}, h_{max}]$ is the range within which the values of height nodes are allowed to vary (Fig. 3). This strategy ensures the possibility for the GA to provide, as output, either protective barriers or ditches.

To ensure a better exploration of the search space and to avoid a fast convergence of solutions to local optima a n point crossover operator has been introduced. Two parent individuals are randomly chosen from the mating pool and two different cutting points for each parents are selected. Cut points always coincide with the first gene of a sub-solution and after the selection portions of the sub-solution chosen in the genotype, they are exchanged. The crossover operator is applied according to a prefixed probability, p_c, for each sub-solution encoded in the genotype.

3.1 Parallel Implementation and Performance

The fitness evaluation of a GA individual consists in an entire CA simulation, followed by a comparison of the obtained result with the actual case study. This phase

Table 1 List of parameters of the adopted GA

GA parameters	Specification	Value
g_l	Genotypes length	6
p_s	Population size	100
n_g	Number of generations	100
p_{mc}	Coord. gene mutation probability	0.5
x_{max}	Gene x position variation radius	10
y_{max}	Gene y position variation radius	10
p_{mh}	Height gene mutation probability	0.5
h_{min}	height min variation range	−5
h_{max}	height max variation range	10
p_c	crossover probability	0.05
c_{h+}	Cost to build	1
c_{h-}	Cost to dig	1
ω_{f1}	f_1 weight parameter	0.90
ω_{f2}	f_2 weight parameter	0.10

Fig. 3 Graphical representation of the genotype mutation phase. Each gene, representing a CA coordinate, can vary within a variation radius $[x_{max}, y_{max}]$

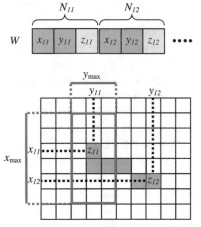

may require several seconds, or even several hours: for example, on a 2-Quadcore Intel Xeon E5472, 3.00 GHz CPU such evaluation requires approximately 10 min, as at least 40,000 CA steps are required for a simulation. For instance, if the GA population is composed of 100 individuals, the time required to run one seed test (100 generation steps) exceeds 69 days. Moreover, the GA execution can grow, depending on both the extent of the considered area and the number of different tests to run.

As a consequence, a *CPU/GPU* library was developed to accelerate the GA running. Specifically, a "Master-Slave" model was adopted in which the Host-CPU (Master) executes the GA steps (selection, population replacement, mutation and crossover), while GPU cores (slaves) evaluate the individuals fitness (i.e., a complete SCIARA-fv2 simulation).

Since the most intensively computational work is needed for this latter phase, a *multi-simulator* was devised to efficiently exploit the considered GPGPU hardware, in order to permit the execution of more simulations in parallel. For this purpose, two different implementation strategies were implemented and a landscape benchmark case study was considered, modeled through a Digital Elevation Model composed of 200×318 square cells with a side of 10 m. In addition, a set of 50 hypothetical barriers placed with 2 different inclinations ($135, 225°$) to the lava flow direction was considered leading to a total of 100 simulations to be performed. Four CUDA devices were used in the experiments: a nVidia Tesla C2075 and three nVidia Geforce graphic cards, namely the GTX480, GTX 580 and the GTX 680. Also, in order to quantify the achieved parallel speedup, sequential versions of the same GPU strategies were run on a workstation equipped with a 2-Quadcore Intel Xeon E5472 (3.00 GHz) CPU.

Starting from previous research in CA modelling by means of GPGPU (e.g. [34–37]), a first straightforward parallel implementation, labeled as WCSI (Whole Cellular Space Implementation) was considered where the CUDA kernels operate on the whole cellular space. However, since the transition function of the currently active cells (i.e., cells containing lava) is invoked, simulating only one simulation at a time would imply a high percentage of uselessly scheduled threads. In addition, given the limited extension of most simulations (on average, 20 % of cells of the entire automaton are active during a single simulation), the number of active threads would be too low to allow the GPU to effectively activate the latency-hiding mechanism [38] of CUDA. To increase thread occupancy, in the WCSI approach more than a single lava episode are simultaneously executed. This means that the main CUDA kernel is executed over a number of simulations which are simultaneously executed at the same CA step. In particular, each simulation performed is mapped on a different value of z and on a grid of threads composed of 16×16 blocks. That is, the grid of threads used for the CA transition function is three-dimensional, with the base representing the considered CA space and the vertical dimension corresponding to the different launched simulations.

Using the adopted GPU devices, the algorithm was implemented with the WCSI approach and execution times evaluated for a variable number of simultaneous lava simulations. For a fair comparison, the sequential version of the same algorithm was used and the elapsed time achieved by the CPU was 26039 s. According to the results shown in Fig. 4a, the GTX 680 achieved the lowest elapsed time of 650,96 s, concurrently simulating 50 lava events. The gain provided by the parallelisation in terms of computing time was significant and corresponded to a parallel speedup of over 40 for the used CPU (Fig. 4).

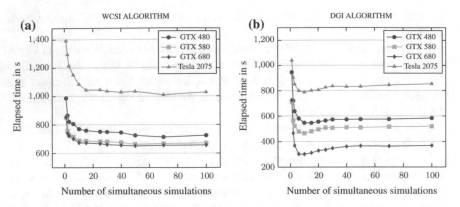

Fig. 4 Elapsed time as a function of simultaneous lava events usign the WCSI (**a**) and DGI (**b**) approaches on different considered GPGPU hardware

For CA lava flow models, the application of the transition function can be restricted to the only active cells where computation is actually taking place. Thus, the CA space can be confined within a rectangular bounding box (RBB). This optimization drastically reduces execution times, since the sub-rectangle is usually quite smaller than the original CA space. This may result in having a high percentage of computationally inactive threads in the CUDA grid, as in the case of the WCSI CA implementation. For these reasons, a second approach was developed in which the grid of threads is dynamically computed during the simulation in order to keep low the number of computationally irrelevant threads. In such an approach, labelled as DGI (Dynamic Grid Implementation), a number of lava flow simulations are simultaneously executed as in the WCSI procedure.

In addition, at each CA step the procedure involves the computation of the smallest *common* rectangular bounding box (CRBB) that includes any active cells in every concurrent simulation. Figure 5 shows all kernels required by the CA step that are

Fig. 5 Mapping of the CA transition function into a CUDA grid of threads (*right*) in case of the simultaneous lava flows (*left*)

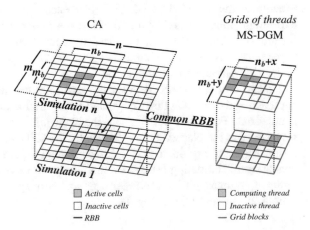

mapped on such CRBB, thus reducing the number of useless threads and significantly improving the computational performance.

An analogous strategy based on the bounding box has been developed for the sequential version of the program. for a fair comparison. Using the reference CPU, such sequential procedure required 20180 s on the same case study. Figure 4b shows times taken by the parallel DGI approach as a function of the number of concurrent simulations. As seen, the GTX 680 achieved the lowest elapsed time of 301,18 s, giving rise to a parallel speedup of 67.

3.2 Experiment and Results

By considering the Nicolosi lava flow event (barriers uphill from Sapienza Zone) and by adopting the parallel multi-simulator described in the previous section, ten GA runs (based on different random seeds) of 100 generation steps each were carried out, each one with a different initial population. The elapsed time achieved for the ten GA runs was less than nine hours of computation on a 10 multi-GPU GTX 680 GPU Kepler Devices Cluster (note that the same experiment, on a sequential machine, would had lasted more than seven months). Furthermore, during the running, a Visualization System Software [18], based on OpenGL and C++ and integrated into Qt interface, allowed the interactive visualization and analysis phases of the results.

For this preliminary experiment, only solutions with two nodes were considered ($|W| = 1$), while Z and P were chosen as in Fig. 1. The cardinality of W (Protection work) and the gene values in which they are allowed to vary (depending of Z area), define the search space S_r for the GA:

$$S_r = \{[P_{x_{min}}, P_{x_{max}}] \times [P_{y_{min}}, P_{y_{max}}] \times [(h_{min} \cdot n_g), (h_{max} \cdot n_g)]\}^{2|W|} \qquad (8)$$

The temporal evolution of the f_3 fitness is graphically reported in Fig. 7a, in terms of average results over the ten considered experiments. GA experiment parameters values are also listed in Table 1. The related CA simulation, obtained by adopting the best individual is shown in Fig. 6.

The study, though preliminary, has produced quite satisfying results. Among different best individuals generated by the GA for each seed test, the best one (Table 2) consists of a barrier with an average height of 7,5 and 410 m in length with an inclination angle of 141° with respect to the direction of the lava flow. The barrier (its properties are shown in Table 2) completely deviates the flow avoiding that the lava reaches the inhabited and building facilities areas. It is worth to note that, the best solution provided by GA (Fig. 6) in this work is approximately five times more efficient (in term of total m^3 volume used to keep safe tha safety areas) respect to the one applied in the real case (Fig. 1), consisting of thirteen earthen barriers.

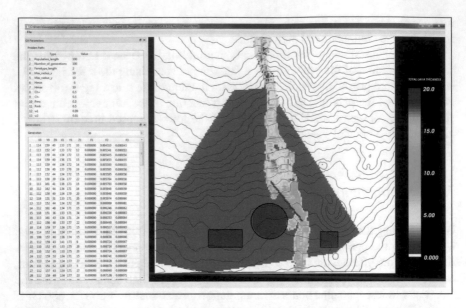

Fig. 6 SCIARA—fv2 simulation visualization adopting the GA best solution. As seen, the devised barrier (*blue*) completely diverts the lava flow from the Safety Areas (*red*)

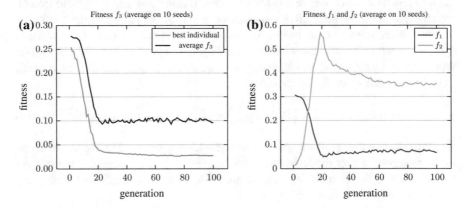

Fig. 7 Temporal evolution of composite f_3 fitness of best individual (*in black*) and of average fitness of whole population (*in gray*) (**a**). Temporal evolution of average fitness f_1 (*in red*) and f_2 (*in green*) of whole population (**b**). Fitness values were obtained as an average of 10 GA runs, carried out by adopting different seeds for generation of random numbers

3.3 Considerations on the GA Dynamics and Emergent Behaviors

In the GA experiments that have been performed, individuals with high fitness evolved rapidly, even if the initial population was randomly generated and the search

Table 2 Properties of the best barrier evolved by GA run

Barrier Properties	Length (m)	Height (m)	Base Width (m)	Volume (m^3)	Inclination (°)
[206,96,2] [238,122,13]	410	7,5	10	24750	141

Fig. 8 Nodes distribution of the best 100 solutions generated by the GA. Scale values indicates occurrence of nodes

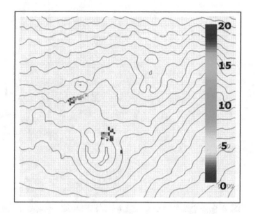

space was quite large (Eq. 8). By analyzing several individuals evolved in ten different GA executions, similar solutions were observed. This behavior is due to the presence of problem constraints (e.g. morphology, lava vent, emission rate, Z and P areas) that lead the GA to search in a "region" of the solution space characterized by a so called "local optimum". In particular, f_1 reaches the minimum value (0) around the twentieth GA generation and the remaining 80 runs are used by GA for the f_2 optimization (cf. Fig. 7b).

In any case, the evolutionary process has shown, in accordance with the opinion of the scientific community [5, 39], the ineffectiveness of barriers placed perpendicular to the lava flow direction despite diagonally oriented solutions (130–160°).

Furthermore, a systematic exploitation of morphological characteristics by GA, during the evolutionary process, has emerged. To better investigate such GA emergence behaviour, a study of nodes distribution was conducted (Fig. 8). By considering the best 100 solutions provided by GA, each node was classified on the basis of the *slope proximity* calculation, as an average of altitude differences between node neighborhood cells (with radius 10) and the central cell. More formally, the function that assigns to each generic node j a *slope proximity* value is defined as:

$$sp_j = \frac{\sum_{i=1}^{|X|} \bar{z}_i - \bar{z}_0}{|X|} \qquad (9)$$

where X is the set of cells that identifies the neighborhood of j and $\bar{z}_i \in Q_z$ is the topographics altitude (index 0 represents the central cell). As shown in Fig. 9, starting from the tenth GA generation, the evolutionary process has shown an increase

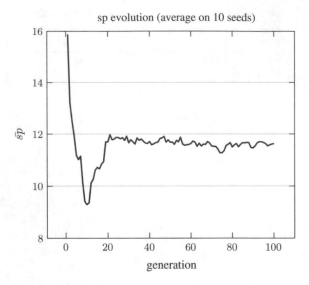

Fig. 9 Temporal evolution of average slope proximity values for the best individuals

in slope proximity values. Therefore, after the f_1 optimization (cf. Fig. 7b), in order to minimize f_2, there is a specific evolutionary temporal phase (i.e., up to the 25th generation) where the algorithm generates solutions that are located in the proximity of elevated slopes.

4 Conclusions and Future Works

This paper has presented a novel approach for devising protective measures to divert lava flows. Starting from the problem of the high computational complexity of the GA algorithm, a library was developed for executing a large number of concurrent lava simulations using GPGPU. The parallel speedups attained through the proposed approaches and by considering GPGPU hardware, were indeed significant. In fact, the adoption of PGAs permitted to perform, in reasonable times, a greater number of tests shortening the execution by a factor of 67. In addition to the GA algorithm acceleration implementation, an interaction visualization system was also developed for the analysis phases of the results.

In this preliminary release of the algorithm only two nodes based solutions were considered and evaluated on the basis of two fitness functions. The first fitness function guarantees the goodness of the solution in terms of security; the second one minimizes the environmental impact.

First observations of the GA results permitted to conjecture the presence of a local optima in the search space, probably due to problem constraints. To better investigate GA dynamic characteristics, a study of nodes distribution was also conducted and a systematic exploitation of morphological characteristics by GA during the evolutionary process emerged.

PGAs experiments, carried out by considering the Nicolosi case-study, demonstrated that artificial barriers can successfully change the direction of lava flow in order to protect predefined point of interests.

In particular, by performing extensive experiments, simulations demonstrated that protective works are more effective when placed nearly parallel to the flow direction, while a barrier placed perpendicular to the flow direction can only stop the flux temporarily, ultimately allowing the solidified crust to accumulate and cause the following mass to go over the barrier.

Though preliminary, the study has produced extremely positive results and simulations have demonstrated that GAs can represent a valid tool to determine protection works construction in order to mitigate the lava flows risk. However, considering two-nodes barriers is a strong limitation and a critical aspect of GA implementations that can significantly improve the efficiency of the final solution is the the possibility to provide multi-barrier protection measures. For this reason, future work will firstly consider the investigation of solutions consisting of multiple protective interventions and the introduction of lava cooling by water jets as parameter of the methodology. By considering the hazard evaluation context, an important application of the methodology could be to take into account as an event to be mitigated a grid of hypothetical vents defined as the source for the simulations to be carried out. In this case, protection measures provided by the GA can represent a preventive solution to assess the effect of possible human interventions. Furthermore, it could be very important to evaluate the extension of this method to other different complex natural phenomena such as a debris flow models.

Acknowledgments Authors gratefully acknowledge the support of NVIDIA Corporation for this research. The work was partially funded by the European Commission European Social Fund (ESF) and by the Regione Calabria (Italy).

References

1. Behncke, B., Neri, M.: The July-August 2001 eruption of Mt. Etna (Sicily). Bull. Volcanol. **65**(7), 461–476 (2003)
2. Miyamoto, H., Sasaki, S.: Simulating lava flows by an improved cellular automata method. Comput. Geosci. **23**, 283–292 (1997)
3. Avolio, M.V., Crisci, G.M., Di Gregorio, S., Rongo, R., Spataro, W., D'Ambrosio, D.: Pyroclastic flows modelling using Cellular Automata. Comput. Geosci. **32**, 897–911 (2006)
4. Del Negro, C., Fortuna, L., Herault, A., Vicari, A.: Simulations of the 2004 lava flow at Etna volcano using the magflow cellular automata model. Bull. Volcanol. **70**(7), 805–812 (2008)
5. Barberi, F., Brondi, F., Carapezza, M., Cavarra, L., Murgia, C.: Earthen barriers to control lava flows in the 2001 eruption of Mt. Etna. J. Volcanol. Geoth. Res. **123**, 231–243 (2003)
6. Colombrita, R.: Methodology for the construction of earth barriers to divert lava flows: the Mt. Etna 1983 eruption. Bull. Volcanol. **47**(4), 1009–1038 (1984)
7. Barberi, F., Carapezza, M., Valenza, M., Villari, L.: The control of lava flow during the 1991–1992 eruption of Mt. Etna. J. Volcanol. Geoth. Res. **56**, 1–34 (1993)

8. Bentley, P.: An introduction to evolutionary design by computers. In: Bentley, P.J. (ed.) Evolutionary Design by Computers, ch. 1, pp. 1–73. Morgan Kaufman, San Francisco (1999)
9. Sims, K.: Evolving 3d morphology and behavior by competition. In: Proceedings of Artificial Life IV, pp. 28–39. MIT Press (1994)
10. Bongard, J.: Morphological change in machines accelerates the evolution of robust behavior. In: Proceedings of the National Academy of Sciences, vol. 108, pp. 1234–1239 (2011)
11. Kicinger, R., Arciszewski, T., Jong, K.D.: Evolutionary computation and structural design: a survey of the state-of-the-art. Comput. Struct. **83**, 1943–1978 (2005)
12. Spataro, W., Avolio, M.V., Lupiano, V., Trunfio, G.A., Rongo, R., D'Ambrosio, D.: The latest release of the lava flows simulation model SCIARA: First application to Mt Etna (Italy) and solution of the anisotropic flow direction problem on an ideal surface. In: Proceedings of International Conference on Computational Science, vol. 1, pp. 17–26. Procedia Computer Science (2010)
13. Neumann, J.V.: Theory of Self-Reproducing Automata. University of Illinois Press, Champaign (1966)
14. Chopard, B., Droz, M.: Cellular Automata Modeling of Physical Systems. Cambridge University Press, Cambridge (2000)
15. Trunfio, G.A., D'Ambrosio, D., Rongo, R., Spataro, W., Di Gregorio, S.: A new algorithm for simulating wildfire spread through cellular automata. ACM Trans. Model. Comput. Simul. **22**, 6:1–6:26 (2011)
16. Succi, S.: The Lattice Boltzmann Equation for Fluid Dynamics and Beyond. Clarendon Press, Oxford (2001)
17. Crisci, G.M., Gregorio, S.D., Rongo, R., Spataro, W.: Pyr: a cellular automata model for pyroclastic flows and application to the 1991 mt. pinatubo eruption. Future Gen. Comput. Syst. **21**(7), 1019–1032 (2005)
18. Filippone, G., D'Ambrosio Spataro, D., Marocco, D.: An interactive visualization system for lava flows cellular automata simulations using CUDA. In: Poster Presented at GPU Technology Conference. San Jose, California (2013)
19. Barberi, F., Carapezza, M.L.: Mt. Etna: Volcano Laboratory, ch. The Control of Lava Flows at Mt. Etna, pp. 357–369. American Geophysical Union, Washington (2004)
20. Holland, J.H.: Adaptation in Natural and Artificial Systems: An Introductory Analysis with Applications to Biology, Control, and Artificial Intelligence. The MIT Press, Cambridge (1992)
21. Goncalves, J.F., Resende, M.G.: Biased random-key genetic algorithms forcombinatorial optimization. J. Heuristics **17**(5), 487–525 (2011)
22. Hinton, G.E., Nowlan, S.J.: How learning can guide evolution. Complex Syst. pp. 495–502 (1987)
23. Nolfi, S., Marocco, D.: Evolving robots able to integrate sensory-motor information over time. Theory Biosci. **120**, 287–310 (2001)
24. ElSayed, A., Kongar, E., Gupta, S., Sobh, T.: A robotic-driven disassembly sequence generator for end-of-life electronic products. J. Intell. Rob. Syst. **68**(1), 43–52 (2012)
25. Piwonska, A., Seredynski, F., Szaban, M.: Learning cellular automata rules for binary classification problem. J. Supercomput. **63**(3), 800–815 (2013)
26. Di Gregorio, S., Serra, R., Villani, M.: Applying cellular automata to complex environmental problems: the simulation of the bioremediation of contaminated soils. Theoret. Comput. Sci. **217**(1), 131–156 (1999)
27. Iovine, G., D'Ambrosio, D., Di Gregorio, S.: Applying genetic algorithms for calibrating a hexagonal cellular automata model for the simulation of debris flows characterised by strong inertial effects, Geomorphology, vol. 66, no.14, pp. 287–303 (2005)
28. Rongo, R., Spataro, W., D'Ambrosio, D., Avolio, M.V., Trunfio, G.A., Di Gregorio, S.: Lava flow hazard evaluation through cellular automata and genetic algorithms: an application to Mt Etna volcano. Fundam. Inf. **87**, 247–267 (2008)
29. D'Ambrosio, D., Rongo, R., Spataro, W., Trunfio, G.A.: Meta-model assisted evolutionary optimization of cellular automata: an application to the sciara model. In: Proceedings of the 9th International Conference on Parallel Processing and Applied Mathematics - Volume Part II, PPAM'11, pp. 533–542. Springer, Berlin (2012)

30. D'Ambrosio, D., Rongo, R., Spataro, W., Trunfio, G.: Optimizing Cellular Automata through a Meta-model Assisted Memetic Algorithm. In: Proceedings of Parallel Problem Solving from Nature - PPSN XII, Lecture Notes in Computer Science, vol. 7492, pp. 317–326. Springer, Berlin (2012)
31. Genetic Algorithms in Search, Optimization and Machine Learning, 1st edn. Addison-Wesley Longman Publishing Co. Inc., Boston (1989)
32. Mitchell, M.: An introduction to Genetic Algorithms. MIT Press, Cambridge (1996)
33. Bresenham, J.: Algorithm for computer control of a digital plotter. IBM Syst. J. 4(1), 25–30 (1965)
34. Filippone, G., Spataro, W., Spingola, G., D'Ambrosio, D., Rongo, R., Perna, G., Di Gregorio, S.: GPGPU programming and cellular automata: Implementation of the SCIARA lava flow simulation code. In: 23rd European Modeling and simulation Symposium (WMSS), pp. 12–14. Rome, September 2011
35. Di Gregorio, S., Filippone, G., Spataro, W., Trunfio, G.A.: Accelerating wildfire susceptibility mapping through GPGPU. J. Parallel Distrib. Comput. 73(8), 1183–1194 (2013)
36. D'Ambrosio, D., Filippone, G., Marocco, D., Rongo, R., Spataro, W.: Efficient application of gpgpu for lava flow hazard mapping. J. Supercomput. 65(2), 630–644 (2013)
37. D'Ambrosio, D., Filippone, G., Rongo, R., Spataro, W., Trunfio, G.A.: Cellular automata and GPGPU: an application to lava flow modeling. Int. J. Grid High Perform. Comput. 4, 30–47 (2012)
38. NVIDIA Corporation, CUDA C Best Practices Guide. NVIDIA Corporation, 2701 San Tomas Expressway, Santa Clara 95050, USA, 5.0 ed. (2012)
39. Fujita, E., Hidaka, M., Goto, A., Umino, S.: Simulations of measures to control lava flows. Bulletin of Volcanology 71, 401–408 (2009)

A Targeted Estimation of Distribution Algorithm Compared to Traditional Methods in Feature Selection

Geoffrey Neumann and David Cairns

Abstract The Targeted Estimation of Distribution Algorithm (TEDA) introduces into an EDA/GA hybrid framework a 'Targeting' process, whereby the number of active genes, or 'control points', in a solution is driven in an optimal direction. For larger feature selection problems with over a thousand features, traditional methods such as forward and backward selection are inefficient. Traditional EAs may perform better but are slow to optimize if a problem is sufficiently noisy that most large solutions are equally ineffective and it is only when much smaller solutions are discovered that effective optimization may begin. By using targeting, TEDA is able to drive down the feature set size quickly and so speeds up this process. This approach was tested on feature selection problems with between 500 and 20,000 features using all of these approaches and it was confirmed that TEDA finds effective solutions significantly faster than the other approaches.

Keywords Estimation of distribution algorithms · Feature selection · Genetic algorithms · Hybrid algorithms

1 Introduction

Classification problems concern the task of sorting samples, defined by a set of features, into two or more classes. Feature Subset Selection (FSS) is the process by which redundant or unnecessary features are removed from consideration [5]. Reducing the number of redundant features used is vital as it may improve classification accuracy, allow for faster classification and enable a human expert to focus on the most important features [10]. We therefore approach the problem of FSS with

G. Neumann (✉) · D. Cairns
Computing Science and Mathematics, University of Stirling, Stirling, UK
e-mail: gkn@cs.stir.ac.uk
url: http://www.cs.stir.ac.uk/

D. Cairns
e-mail: dec@cs.stir.ac.uk

© Springer International Publishing Switzerland 2016
K. Madani et al. (eds.), *Computational Intelligence*,
Studies in Computational Intelligence 613,
DOI 10.1007/978-3-319-23392-5_5

83

two objectives: to develop a FSS algorithm that is able to find feature subsets that are as small as possible while also enabling samples to be classified with as great an accuracy as possible.

Evolutionary Algorithms (EAs) have often been applied to FSS problems. An EA is a heuristic technique where a random population of potential solutions is generated and then combined based on a fitness score to produce new solutions. Due to their population based nature they are able to investigate multiple possible sets of features simultaneously. They are implicitly parallel and investigate features as sets rather than individually.

Due to the fact that features are often correlated [2] this can make them more effective than techniques that only consider features individually. Estimation of Distribution Algorithms (EDAs) are a class of EA that build an estimated model of the population using several fit individuals. This is in contrast to Genetic Algorithms (GAs) that produce new individuals by crossing over genes (in this case features) between two parent individuals.

EDAs can offer advantages over GAs as they are able to discover patterns within the population and they are able to explicitly model relationships between features. Cantu-Paz [3] demonstrated that both GAs and EDAs were equally capable of solving FSS problems but that a simple GA was able to find good solutions faster than EDAs.

Inza [10] introduced the concept of using EDAs for feature selection. He compared an EDA to both traditional hill climbing approaches (Forward Selection and Recursive Feature Elimination, a form of Backward Selection) and GAs and found that an EDA was able to find significantly more effective feature sets than any of the techniques that it was compared against [11]. This work is motivated by the goal of combining the beneficial properties of both GAs and EDAs, enabling good solutions to be found rapidly while using a probability distribution model to make the most informed choice of features to select.

Many investigations of FSS problems, including those of Cantu-Paz and of Inza et al., looked at problems with fewer than 100 features. Many real world problems involve much larger feature sets. We therefore explore applying EAs to problems with between 500 and 20,000 features. For these problems the initial number of features is so large that complex EDA approaches are impractical [11]. Many of these problems are very noisy and only a small proportion of the features are useful [9]. For problems which are so noisy that only a tiny proportion of features are useful, driving down the size of the feature set is an important part of the optimization process.

To achieve this objective, techniques such as constraining the number of features and then iteratively removing features to fit within this constraint have been explored [21]. In this approach a decision needs to be made as to what value to set the constraint to. This is problematic as it requires previous knowledge of the problem. In this work, we demonstrate a GA/EDA hybrid that is specifically designed to drive down the number of features to consider before attempting to build an EDA model.

Targeted EDA (TEDA) was initially developed to solve time series problems by finding the optimal number of actions or 'interventions', required to solve these problems (a process referred to as 'Intervention Targeting' or 'Feature Targeting').

Previous work on time series problems has shown that TEDA is effective at solving 'bang bang control' problems [16] and other problems [17] where there is a concept of parameters or features being either 'on' or 'off' and where a key consideration is the total number of variables that are 'on' in a solution. TEDA works by making a prediction as to how many features or interventions should set to 'on' in an ideal solution. This is achieved by examining the number of features set to 'on' in two or more parent solutions, before deciding which features should be used.

TEDA transitions over time from initially operating like a GA to operating like an EDA. The transition occurs as the population starts to converge and the probability distribution becomes more reliable. This paper addresses whether TEDA can use this capability to determine the number of features needed to solve a FSS problem and so effectively find both small and accurate feature subsets. In applying TEDA to FSS problems we are also testing TEDA's ability to solve problems much larger than those that it has previously been tested with and we introduce a new modification to TEDA which makes its transitioning process more efficient and less dependent on prior knowledge of the problem.

We also investigate how Fitness Directed Crossover (FDC) [7] performs on these problems. FDC is a precursor to TEDA that introduced intervention targeting as a part of the crossover operation. FDC provides a useful performance baseline for the intervention targeting component of TEDA but does not include the EDA elements that are introduced by TEDA.

We begin this paper with a discussion of the background to this research area, introducing existing FSS and classification techniques. We then introduce TEDA in Sect. 3. The final three sections are used for explaining our methodology (Sect. 4), presenting our results (Sect. 5), and exploring any conclusions drawn (Sect. 6).

2 Background

2.1 Classification

A typical classification problem will involve constructing a classifier based on samples in a training set where the class that a given sample belongs to is already known. New samples are then classified based on the information extracted from the training set. Popular approaches include K Nearest Neighbour (KNN) [12] and Support Vector Machines (SVM). In KNN the k individuals in the training set that are most similar to the new sample are used to determine the new sample's class. SVM is a classification technique where two classes are distinguished by drawing between them the hyperplane that separates the instances of each class by the greatest possible margin. New samples are then classified based on which side of the hyperplane they fall on. For the experiments detailed here we have used the LIBSVM library for SVMs [4].

2.2 Feature Subset Selection

Feature Subset Selection (FSS), an NP hard problem in machine learning, involves the identification of the minimum number of features that will most accurately classify a given set of samples. As there are 2^n possible subsets of a feature set of length n an exhaustive search is not possible and so various search heuristics have been developed [5]. Techniques can be divided into filter and wrapper methods [13]. Filters build feature sets by calculating the capacity of features to separate classes whereas wrappers use the final classifier to assess complete feature sets. Wrapper methods can be more powerful than filter methods because they consider multiple features at once and yet they tend to be more computationally expensive [9].

Forward Selection and Backward Selection [13] are also popular search methods as they have been described as being less prone to over fitting than wrapper methods such as TEDA [8]. In Forward Selection the most informative feature is selected to begin with. After this a greedy search is carried out and the second most informative feature is added. This process is repeated until a feature set of size L, a pre-specified limit, is reached or no further improvement can be achieved by adding any of the remaining features. In Backward Selection an SVM initially attempts to carry out classification using the entire feature set. The SVM assigns a weight to each feature and the least useful features are eliminated. An alternative to the greedy search that does not require as many iterations is to order the features by usefulness and then use the top n features as the feature set. A decision then needs to be made as to where to draw the line between features used in the feature set and features that are to be discarded. One method that has been discussed before [22] is to introduce a random dummy feature or *probe* to the feature set and to discard any features that perform less well at the classification task than this dummy feature.

2.3 Genetic Algorithms

In GAs, new solutions are generated by exchanging genetic information between two fit solutions via a crossover process. Following crossover each new solution may, with a small probability, be mutated to a different value. The aim of this step is to introduce variability into the gene pool.

The main varieties of crossover are distinguished by the method in which genes from the two parents are selected. In *One Point Crossover* a single index is selected within the genome to be the position where the parents are to be crossed over. A new child will be produced that combines the genes taken from before the index in one parent with the genes taken from after the index in the other parent. *Two Point Crossover* is similar except that the genome is split at two separate points, dividing each parent into three parts. When using *Uniform Crossover* to generate new solutions a separate decision is made for each individual gene as to which parent it should be copied from. In this work *One Point Crossover* is used as a benchmark as it has previously been shown to be reasonably effective at feature selection problems [3].

2.4 Fitness Directed Crossover

FDC can be used in optimal control problems to provide a GA based crossover method that will drive intervention selection towards levels used by the fittest individuals that have been selected for breeding [7]. FDC was used with fixed length encodings where every gene corresponded to an intervention and either had a value greater than 0 (on) or had a value of 0 (off). FDC starts with a population of solutions that have been randomly initialised to either an on or off setting. To produce the next generation, two parents are selected via tournament selection and used to derive a target number of interventions I_T. New children are produced with exactly I_T genes set to 'on' as using the FDC rule given in Eq. (1).

$$I_T = I_F + (2T - 1)(I_1 - I_2)(F_1 - F_2) \tag{1}$$

The number of interventions set in parent one and parent two are denoted as I_1 and I_2 respectively. F_1 and F_2 are the normalised fitnesses of the above two parent solutions compared against the current population and I_F is the number of interventions in the fitter solution. T is 0 for a minimisation problem and 1 for a maximisation problem. The effect of this equation is to set the target number of interventions I_T such that if the fitter parent has more interventions than the less fit parent then I_T will be greater than the number in the fitter parent and vice versa. The level of overshoot is determined by the difference in fitness between the two parents.

Once I_T has been determined, we need to choose which particular interventions to set. We start by placing all interventions set in both parent solutions in the set S_{dup} and all interventions set in only one parent in the set S_{single}. Interventions to set are then selected randomly from S_{dup} until either I_T interventions have been set or S_{dup} is empty. If more interventions are needed then interventions will be selected randomly from S_{single} until it is empty or I_T has been reached.

2.5 Estimation of Distribution Algorithms

EDAs use a set of relatively fit solutions to build a probability model indicating how likely it is that a given gene has a particular value. They sample this model to produce new solutions that are centred around the derived probability distribution. Univariate EDAs treat every gene as independent whereas multivariate approaches also model interdependencies between genes. Multivariate EDAs are essential in many problems where genes are highly interrelated but they have the disadvantage that, as the number of interactions increases, there is a substantial increase in computational effort required to model these interdependencies [14].

One of the simplest and best known univariate EDAs is the Univariate Marginal Distribution Algorithm (UMDA) [15]. For a binary problem, Eq. (2) shows how UMDA calculates the marginal probability, ρ_i, that the gene at index i is set.

$$\rho_i = \frac{1}{|B|} \sum_{x \in B, x_i = 1} 1 \tag{2}$$

$$\rho_i = \frac{1}{\sum_{x \in B} f_x} \sum_{x \in B, x_i = 1} f_x \tag{3}$$

Here B is a subset of fit solutions selected from the current population. ρ_i is the proportion of members of B in which x_i is true. Alternatively, we can weight the probability based on the normalised fitness f of each solution where x_i is true, as shown in Eq. (3).

Once the probabilities for each gene being set have been calculated, new solutions are generated by sampling this distribution according to probability ρ_i. Using ρ directly to provide the probability of setting genes in new solutions is the simplest method of probability distribution sampling and the method that is used by UMDA, but a number of alternative sampling techniques have been explored [14]. In Population Based Incremental Learning (PBIL) [1] a sampling vector is produced where every gene has an initial probability 0.5 and each generation a learning rate is used to move each of these values in the direction of ρ.

2.6 Hybrid Algorithms

TEDA falls into the category of hybrid algorithms that use both GAs and EDAs. These approaches are useful as neither EDAs nor GAs perform better than the other approach on all problems. On some problems EDAs become trapped in local optima while on other problems they produce faster convergence than GAs. It can be difficult to predict whether an EDA or a GA will perform better for a particular problem [18]. One hybrid approach developed by Pena is called GA-EDA [18]. This approach generates two populations, one through an EDA and one through a GA. Individuals are then selected from these populations to form the next generation based on their fitness. Zhang proposed an approach called Guided Mutation that produces new solutions that contain a mixture of genes copied from a promising individual and sampled from an EDA style probability distribution [23]. Aside from these approaches, most of the work done on hybridising EDAs or GAs with other techniques appears to have focussed on integrating local search or simulated annealing techniques into either GAs or EDAs instead of combining the two [18]. We have not found any techniques that dynamically transition over time from a GA to an EDA based on population convergence, allowing the GA to move towards the global optima before applying an EDA.

3 TEDA

The main principle behind TEDA is that it should use feature targeting in a similar manner to FDC and that it should transition from behaving like a GA before the population has converged to behaving like an EDA after it has converged. Specifically, the pre-convergence behaviour of TEDA should match that of FDC as this proved effective when using the targeting principle. This transitioning process is important as dictating exactly how many features solutions should have risks causing a loss of diversity in the population. It is therefore necessary to use an explorative method early on. In addition to this, a number of authors have noted the tendency of simple Gaussian EDAs to prematurely converge [14, 19].

TEDA is described in detail in Algorithms 1–3. The main steps of the process are as described below:

1. The population is randomly initialized.
2. The fitness of the population is evaluated.
3. A set of parents from the population is selected to form a 'breeding pool', B, of size b. The transitioning process controls how this set of parents is selected (described in more detail below).
4. The fittest and least fit individuals are selected from B and used as parents I_1 and I_2 in the FDC rule (Eq. 1) to decide on the number of features to set in offspring (I_T).
5. A probability model is built from B according to Eq. (3). This is used instead of Eq. (2) because the transitioning process means that TEDA initially uses a very small breeding pool and so the frequency of features is not always a useful measure.
6. b new solutions are created from B. For each new solution, features are selected and set until I_T features have been set. Features found in all members of the breeding pool are preferentially selected and then the remaining features are selected at random and set based on their marginal probability, as with standard UMDA.
7. Steps 3–6 are repeated until a new population has been generated.
8. Steps 2–7 are repeated until a specified number of generations has passed.

3.1 Controlling TEDA Transitioning

The TEDA transitioning process controls whether TEDA behaves like an EDA or a GA by managing the size of two sets:

- S is the 'selection pool', this consists of the fittest s solutions in the population.
- The breeding pool B contains the b parents that are used to build the probability model and so generate new solutions. B is selected from S using tournament selection.

The sizes of B and S are limited to between b_{min} and b_{max} and between s_{min} and s_{max} respectively. To begin with s is equal to s_{max} where s_{max} is set to the size of the whole population. B will initially contain b_{min} parents where b_{min} is 2. In this initial configuration, TEDA operates as a standard GA, selecting 2 parents for breeding from the whole population with tournament selection. The crossover mechanism is equivelent to that used by FDC.

The decision on whether to add another parent to B is made with a probability p, as calculated according to Eq. (4). This measures the similarity between two parents.

$$p = \frac{|f_1 \cap f_2|}{|f_1 \cup f_2|} \tag{4}$$

In this equation, f_1 and f_2 are the feature sets of the two parents that are being compared. These two parents are the last two that were added to B. Initially they will be the first two parents in the pool. If a parent is added according to this rule, the process is repeated until the first occurrence of a parent not being added or b reaches b_{max}.

When a new parent is added, s is decreased. The result is that as the level of variance within the population decreases, the selection pressure increases. s may be decreased until it reaches s_{min}. We recommend that b_{min} and s_{max} should be equal in value. If this is the case then TEDA is likely to ultimately use the fittest b individuals in the population to build a probability model, and therefore behave like an EDA.

This method of transitioning is a recent innovation for TEDA. Previously, the level of difference between chromosomes was calculated by randomly selecting a set of pairs from the population and obtaining $\bar{D}$, the average Hamming distance within this set of pairs.

$$b = b_{max} - (\bar{D}/d_{max})(b_{max} - b_{min})$$
$$s = s_{min} + (\bar{D}/d_{max})(s_{max} - s_{min}) \tag{5}$$

Equation (5) shows how $\bar{D}$ was used to move b and s. d_{max} was a tuneable value set by the user with the role of controlling the speed of transitioning. The new approach has an advantage over this method in that by not using d_{max} it eliminates a parameter that previously needed tuning. In order to ensure smooth transitioning, the value that d should be set to had to be carefully chosen based on the size of the genome and how the size of the solutions that TEDA generates varied over time. Given that in this problem we are working with feature sets that range in size from 500 to 20,000 genes, we cannot make a prediction as to what size of valid solutions TEDA will uncover. It would therefore be difficult to accurately tune this parameter. The probabilistic aspect of the new method ensures that the transitioning process responds smoothly and appropriately to the current population diversity. The new method is also more efficient than the previous method as there is no need to select a large number of individuals in order to estimate the diversity across the whole population.

All methods use genome similarity between solutions to measure population diversity. This should be a more reliable indicator than using the variance in fitness across the population. Previous work [16] has shown that for some problems the fitness function is volatile, leading to situations where a sharp drop in fitness variance may not necessarily mean that the population has converged and the probability distributions can be relied upon.

Algorithm 1. TEDA Pseudocode- Main Evolution Loop.

---

function EVOLVE
    $P_0 \leftarrow$ InitialisePopulation()
    $s \leftarrow s_{max}$ ▷ normally $s_{max} = popSize$
    for $g = 0 \rightarrow generations$ **do**
        $\forall P_{g_i} \in P_g$ AssessFitness(P_{g_i})
        $P_{g+1} \leftarrow$ Elite(P_g)
        while $|P_{g+1}| < popSize$ **do**
            $B \leftarrow$ GetBreedingPool(l, b, P_g)
            $i_t \leftarrow$ GetTargetNumOfFeats(B)
            $\rho \leftarrow$ BuildUMDAProbabilityModel(B)
            $S_{all} \leftarrow \forall i \in \rho$ where $\rho_i = 1$
            $S_{some} \leftarrow \forall i \in \rho$ where $0 < \rho_i < 1$
            for b times **do**
                $I \leftarrow$ Mutate(Breed($S_{all}, S_{some}, \rho, i_t$))
                $P_{g+1} \leftarrow P_{g+1} \cup I$

3.2 Genome Encoding

Two versions of TEDA have been explored. Initially TEDA was designed for fixed length integer strings that were treated as binary strings [16]. Later, TEDA was modified to best perform with a routing problem that featured variable length chromosomes [17]. In this paper a variable length encoding is used due to FSS problems tending towards large and sparse chromosomes. In this case, chromosomes consist of an ordered list of all the selected features. Despite this internal representation, the chromosomes are treated as fixed length binary strings with each feature being either 'on' (present in the chromosome) or 'off' (not present), creating behaviour that is functionally the same as standard TEDA.

4 Experimental Method

This section demonstrates the performance of TEDA compared with EAs and sequential feature selection methods on a set of test feature selection problems. In addition to TEDA, the algorithms that have been included are forward selection, FDC, a standard EDA using UMDA and a standard GA using one point crossover, previously shown

Algorithm 2. TEDA Pseudocode- Selection and Transitioning.

function GETBREEDINGPOOL
    $S \leftarrow$ bestSelection(s)
    $b \leftarrow b_{min}$ ▷ normally $b_{min} = 2$
    $B_1, B_2 \leftarrow$ tournamentSelectionFromSet(S)
    $p \leftarrow$ getOverlap(B_b, B_{b-1})
    while random(1) $< p$ **do**
        $b \leftarrow b + 1$
        $s \leftarrow s - 1$
        $S \leftarrow$ bestSelection(s)
        $B_b \leftarrow$ tournamentSelectionFromSet(S)
        if $b = b_{max}$ **then**
            $p \leftarrow 0$
        else
            $p \leftarrow$ getOverlap(B_b, B_{b-1})

function GETOVERLAP(B_1, B_2)
    $\bar{f}_1 \leftarrow$ all features in B_1
    $\bar{f}_2 \leftarrow$ all features in B_2
    return size($f_1 \cap f_2$) / size($f_1 \cup f_2$)

Algorithm 3. TEDA Pseudocode- Breeding.

function GETTARGETNUMOFFEATS(B, t)
    $Q \leftarrow$ Fittest(B) $\cup$ LeastFit(B)
    $I_f \leftarrow$ NumberOfFeaturesIn(Fittest(B))
    $I_1 =$ NumberOfFeaturesIn(Q_1)
    $F_1 =$ Fitness(Q_1)
    $I_2 =$ NumberOfFeaturesIn(Q_2)
    $F_2 =$ Fitness(Q_2)
    $I_t \leftarrow I_f + (2t - 1)(I_1 - I_2)(F_1 - F_2)$
    return I_t

function BREED($S_{all}, S_{some}, \rho, i_t$)
    $A \leftarrow \{\}$ ▷ Make new individual
    while $I_t > 0$ and $S_{all} \neq \{\}$ **do**
        $r \leftarrow$ random feature $\in S_{all}$
        $A \leftarrow A \cup r$
        $I_t \leftarrow I_t - 1$
        remove S_{all_r} from S_{all}
    while $I_t > 0$ and $S_{some} \neq \{\}$ **do**
        $r \leftarrow$ random feature $\in S_{some}$
        if $\rho_r >$ random(1.0) **then**
            $A \leftarrow A \cup r$
            $I_t \leftarrow I_t - 1$
            remove S_{some_r} from S_{some}
    return A

Table 1 Datasets

Name	Domain	Type	Feat.
Arcene	Mass spectrometry	Dense	10,000
Dexter	Text classification	Sparse	20,000
Madelon	Artificial	Dense	500
Name	Train (pos.)	Test (pos.)	
Arcene	100 (44)	100 (44)	
Dexter	300 (150)	300 (150)	
Madelon	2000 (1000)	600 (300)	

to be effective at FSS problems [3]. *UMDA1* is a configuration of UMDA that uses parameters common in literature. As such, it does not use mutation and builds a probability model using Eq. (2) from a breeding pool consisting of the top 50 % of the population. *UMDA2* is a configuration of UMDA with parameters that match those used in TEDA. As such, it uses the same mutation rate as used in TEDA and builds a probability model using Eq. (3) from a breeding pool consisting of the top 10 % of the population.

4.1 The Datasets

Tests were carried out on three datasets taken from the NIPS 2003 feature selection challenge [9]. These were all binary classification problems. The only preprocessing and data formatting steps applied to the datasets are those described in [9].

Table 1 provides the domain, type, training set size and the test set size for each dataset. All information is from [9] except for the the numbers of positive samples which can be found in [6]. Madelon is an artificial dataset designed to feature a high level of interdependency between features, and so by using it we are able to demonstrate how well TEDA performs in a highly multivariate environment [9]. All three datasets are relatively balanced, and so a simple accuracy score is used to assess how successful the classifiers that we use are.

4.2 The Fitness Function

The basis for the fitness function is the accuracy, calculated as the percentage of samples in the test set that are correctly classified. A penalty is subtracted from this to reflect the fact that smaller numbers of features are preferable. Given an accuracy value of a, a feature set of size l and a penalty of p, the fitness function f is calculated as $f = a - lp$. LIBSVM, A Support Vector Machine produced by [4] is used as the classifier with all parameters kept at their default values.

Table 2 Evolutionary
parameters

Parameter	Value
Population size	100
Crossover probability (for GAs)	1
Mutation probability	0.05
Generations	100
Tournament size	5
Elitism	1
Penalty (p)	$10/n$
TEDA: s_{min} and b_{max}	10
TEDA: s_{max}	100
TEDA: b_{min}	2

4.3 Evolutionary Parameters

All algorithms were implemented using the ECJ toolkit and were all tested with the parameters given in Table 2, in which n is the maximum number of features for each problem.

Generational replacement was used with one elite individual being kept from one generation to the next. Tournament selection of size 5 was used for all genetic algorithms.

The same mutation technique was applied to every algorithm. For each solution mutation is attempted a number of times equal to the current size of the feature set. Each time mutation is carried out with a probability of 0.05. If mutation is to be carried out, then with a probability of 0.5, a feature currently not used will be picked at random and added to the feature set, otherwise a feature will be picked at random and removed from the feature set.

For each algorithm, every individual in the starting population was initialised by first choosing a size k between 1 and n. Features are then chosen at random until k features have been selected. The only constraint placed on which solutions may be produced, either through initialisation, breeding or mutation, is that all solutions must have between 1 and n features and no feature may appear more than once in the same solution.

The following section shows the results for 50 runs of each algorithm on each of the three problems. For each problem three graphs are provided, showing the following metrics:

- The accuracy achieved by the fittest individual in the population on the y axis against the number of generations on the x axis. This accuracy is given as a percentage of test set samples correctly classified.
- The number of features used by the fittest individual in the population on the y axis against the number of fitness evaluations on the x axis.

- The accuracy achieved by the fittest individual in the population on the y axis against time on the x axis. This is the mean of the times that each solution in the population took to complete the classification task. This is important as classification can be time consuming for these large problems that use a lot of features. The mean for each population at each generation is then added to a running total and it is this total that is plotted. From this data we also present, in tabular form, the length of time that each algorithm took to reach a given accuracy level.

Each test was run 50 times and, for each graph, the value plotted is the median of the 50 runs with first and third quartiles given by the variance bars. The median was judged to be more reliable than the mean due to the fact that the variance in accuracy and feature set sizes do not follow a normal distribution.

5 Results

5.1 TEDA Compared to Sequential Selection

The results shown in Fig. 1 compare classification accuracy over time for TEDA against forward selection using both greedy search and probe based methods.

Both variants of forward selection are slower at reaching an optimum than TEDA as they require more fitness function evaluations. Using a probe, every feature is independently evaluated in order for them to be ranked. In this problem this corresponds to 20,000 evaluations. By contrast, a 100 generation run of TEDA with a population of 100 corresponds to only 10,000 evaluations, only a fraction of which TEDA actually needs to achieve a high fitness.

Even though evaluating a single feature is less time consuming than evaluating a large feature set these results show that the total amount of time taken is still greater. Using the greedy search method a fitness of 92.73 was ultimately achieved after the search eventually terminated after 179,963 evaluations. This number is much higher than the other approaches as the greedy search involves making multiple iterations across the population. This method has a worst case complexity of $O(n^2/2)$. Where the number of features is 20,000, as in this problem, this means that the maximum number of evaluations that it may take is $20,000^2/2$ or 2×10^8.

Both methods ultimately reach an optimum that is lower than that reached by TEDA. This may be because of a previously documented limitation of both forward and backward selection [20]. In Forward Selection, a selected feature cannot later be eliminated and in Backward Selection an eliminated feature cannot later be selected. This prevents the techniques from carrying out further exploration once a potential solution has been discovered. In such a large search space the need to carry out further exploration to ultimately find the optimal feature set is especially important.

Backward selection is expected to take even longer than forward selection using a greedy search as it too involves iteratively making changes to the feature set, with each change consisting of the removal of one feature. The tests here show that the

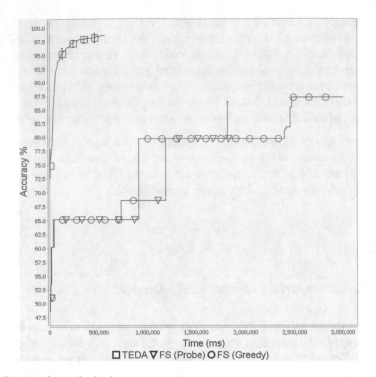

Fig. 1 Dexter—forward selection

most effective solutions that TEDA is able to discover usually contain less than 1,000 features. To discover solutions of this size starting with the entire feature set of 20,000, backward selection would take $19,000^2/2$ evaluations. Each of these evaluations would be much more time consuming than those carried out through forward selection as much larger candidate feature sets would be tested each time, starting with the entire population. Backward Selection is therefore likely to be prohibitively expensive.

Given these results, it would seem that sequential methods do not scale effectively for problems of this size and therefore the rest of this paper will concentrate on the comparison between TEDA and other evolutionary algorithms.

5.2 TEDA Compared to Evolutionary Algorithms

Classification Task: Dexter. The results for the *Dexter* classification problem are shown in Figs. 2, 3 and 4. The results in Fig. 2 show that TEDA is consistently able to find better solutions than any of the other techniques up until at least the 50th generation. UMDA1 performs worse than any other technique throughout the test.

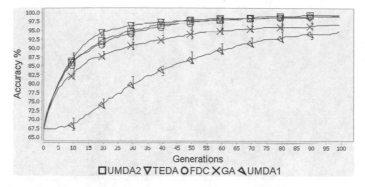

Fig. 2 Dexter—accuracy versus generations

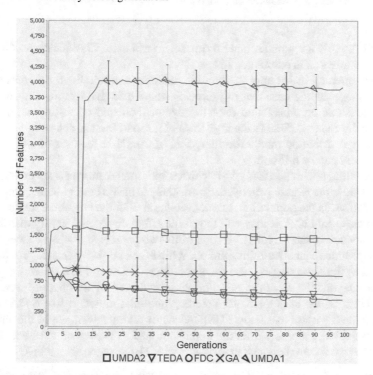

Fig. 3 Dexter—features versus generations

The graph in Fig. 3 indicates that algorithms that are most effective at finding accurate feature sets also tend to be more effective at finding smaller feature sets. The exception is FDC, which finds feature sets that are of an accuracy similar to those found by UMDA2 but tend be smaller. When we compare performance against time (Fig. 4) rather than against number of evaluations, the margin of difference between TEDA and UMDA2, the GA and UMDA1 is greater. This is because the feature sets

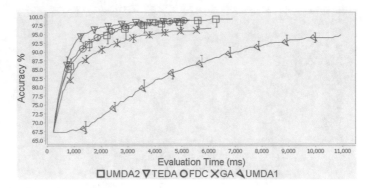

Fig. 4 Dexter—accuracy versus classification time

that TEDA finds are smaller and so quicker to evaluate. Classification with these smaller feature sets is completed in less time.

It is interesting that it appears that this problem is unsuitable for a conventional EDA. It might be the case that in problems where effective feature sets are small, fit solutions can only be found once the size of the explored feature set has been substantially reduced. Due to the high level of noise in Dexter, determining a useful probability distribution model for a large set of candidate features of which only a few are valid can be difficult.

In the initial population it is possible that some small feature sets are generated by chance. Due to the feature penalty, these are likely to have a better fitness compared to other solutions in the population. In a conventional EDA the large breeding pool may obscure these solutions as they will have little effect on the probability distribution. A GA may select such solutions as one of its two parents and when it does so it is likely to produce a smaller child solution. Whilst GAs might by chance produce new solutions of the same size as these small solutions, TEDA and FDC do this explicitly and drive beyond the size of these solutions to find even smaller feature sets.

UMDA2, which uses a smaller breeding pool and mutation like a GA, is able to overcome the noise that affects UMDA1 while taking advantage of the ability of EDAs to exploit patterns within the population and so proves more effective. This advantage that EDAs demonstrate explains why TEDA outperforms FDC.

Classification Task: Arcene. The accuracies obtained by selecting features for the Arcene classification task are shown in Fig. 5. From these results, it can be seen that FDC and TEDA both find better solutions early on than the other approaches. UMDA2 starts to perform slightly better than these approaches from around generation 25 onwards but for the first 10 generations it is completely unable to improve upon the fittest individual in the initial population. UMDA1 is only able to start improving after about generation 70. The standard GA is also slower at finding good solutions than TEDA and FDC, even though it is more effective early on than UMDA.

By looking at the number of features used (Fig. 6) we can see that for both UMDAs the fittest solution in the initial population has a median size of 75 and that for a period

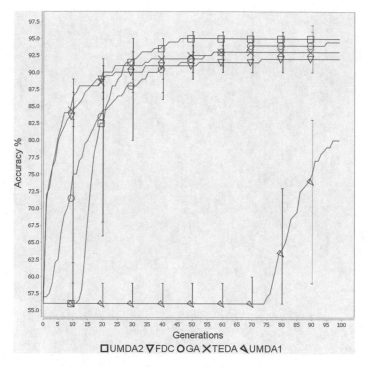

Fig. 5 Arcene—accuracy versus generations

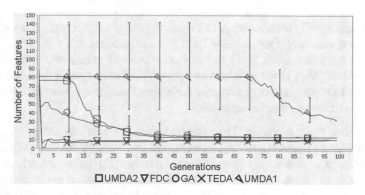

Fig. 6 Arcene—features versus generations

of time both techniques are unable to improve upon this. This is considerably smaller than the maximum feature set size of 10,000 features. We can assume that the sizes of solutions in the initial population is evenly distributed across the range 1 to 10,000. Small individuals would be effectively invisible to the probability model.

It would appear that the situation is the same for both *Arcene* and *Dexter*. Initial high levels of noise mean that until an algorithm starts to explore smaller solutions all

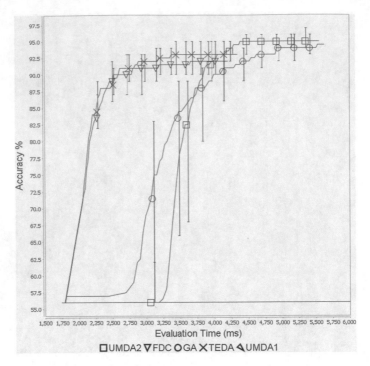

Fig. 7 Arcene—accuracy versus classification time

solutions are equally ineffective. A GA might by chance select a small solution and breed a new, similarly sized solution but TEDA accelerates this process by making it explicit. As with *Dexter*, Fig. 7 shows that these small solutions can be classified more efficiently than larger solutions and so, when plotted against time, we see that TEDA and FDC have almost completed a 100 generation run before UMDA and the GA start to discover effective solutions.

Classification Task: Madelon. The results for the Madelon classification task are shown in Figs. 8, 9 and 10. In the Madelon problem both TEDA and UMDA2 find good feature sets quicker than the other techniques but UMDA1 eventually overtakes both techniques. Both FDC and the GA are less effective. A traditional EDA is more effective at this problem than the other problems possibly because the need to dramatically reduce the size of feature set does not apply in this case. The initial feature set size is considerably smaller and there is less noise, so feature sets that use a large proportion of the available features can be very effective. Figure 9 confirms this, showing no steep declines or sudden drops in feature set size as seen in the other problems. TEDA and FDC show the greatest reduction in the size of feature set and UMDA1 shows the least reduction, as with the other problems. Despite not reducing the feature set size as fast or as far as for the other problems, plotted against time (Fig. 10), TEDA is still able to find good solutions earlier than the other techniques (Table 3).

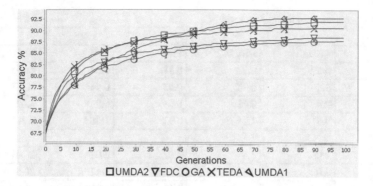

Fig. 8 Madelon—accuracy versus generations

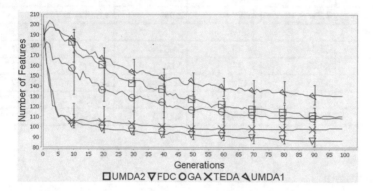

Fig. 9 Madelon—features versus generations

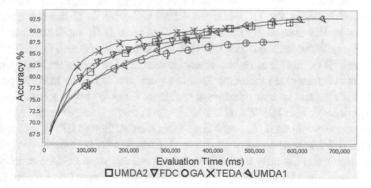

Fig. 10 Madelon—accuracy versus classification time

Table 3 Seconds to reach accuracy level

Dexter

Acc.	TEDA	UMDA2	FDC	GA	UMDA1
70.0	0.29	0.3	0.31	0.34	1.65*
76.0	0.4	0.43	0.43	0.54*	2.73*
82.0	0.6	0.63	0.65	0.82*	4.02*
88.0	0.81	1.1*	0.92*	1.46*	6.16*

Arcene

70.0	2.06	3.44*	2.07	3.03*	26.42*
74.0	2.08	3.49*	2.11	3.11*	26.53*
78.0	2.12	3.49*	2.16	3.27*	26.68*
82.0	2.18	3.58*	2.21	3.38*	–
86.0	2.28	3.69*	2.35	3.63*	–

Madelon

70.0	23.54	24.79	24.02	23.52	30.67
74.0	39.98	35.74	50.0	52.19*	50.01*
78.0	52.43	67.28*	69.27	92.89*	96.97*
82.0	74.17	123.1*	106.58*	168.88*	175.93*
86.0	136.32	210.82*	200.73*	343.32*	270.93*

6 Conclusions

In this work we have shown the benefits of applying TEDA to feature selection problems. We have tested TEDA on three FSS problems from literature and in all three cases it was able to find feature sets that were both small and accurate in comparably quicker time and less effort than standard EDAs and GAs. The speed with which TEDA finds these small solutions enables it to complete fitness function evaluations at a faster rate than comparable algorithms. It is able to solve problems that are of a size and sparsity that pose problems for other approaches. Tests were run with problems with 10,000 and 20,000 features and sequential selection methods do not appear to scale well for problems of this size. Although EDAs are capable of selecting a suitable feature set, they are unable to effectively optimise a feature set until the size is significantly reduced. From our results, we therefore conclude that TEDA is a suitable algorithm for problems that contain a sparse set of useful features within a large number of possible features and where fitness function evaluations are time consuming.

References

1. Baluja, S.: Population-based incremental learning: A method for integrating genetic search based function optimization and competitive learning. Technical Report CMU-CS-94-163, Computer Science Department, Carnegie Mellon University (1994)
2. Bo, T., Jonassen, I.: New feature subset selection procedures for classification of expression profiles. Genome Biol. **3**(4), 1–17 (2002)
3. Cantu-Paz, E.: Feature subset selection by estimation of distribution algorithms. In Proceedings of Genetic and Evolutionary Computation Conference MIT Press, pp. 303-310 (2002)
4. Chang, C.C., Lin, C.J.: Libsvm: a library for support vector machines. ACM Trans. Intell. Syst. Technol. (TIST) **2**(3), 27 (2011)
5. Dash, M., Liu, H.: Feature selection for classification. Intell. Data Anal. **1**, 131–156 (1997)
6. Frank, A., Asuncion, A.: UCI machine learning repository (2010)
7. Godley, P., Cairns, D., Cowie, J., McCall, J.: Fitness directed intervention crossover approaches applied to bio-scheduling problems. In: IEEE Symposium on Computational Intelligence in Bioinformatics and Computational Biology, pp 120-127 (2008)
8. Guyon, I., Elisseeff, A.: An introduction to variable and feature selection. J. Mach. Learn. Res. **3**, 1157–1182 (2003)
9. Guyon, I., Gunn, S., Ben-Hur, A., Dror, G.: Result analysis of the NIPS 2003 feature selection challenge. Adv. Neural Inf. Process. Syst. **17**, 545–552 (2004)
10. Inza, I., Larranaga, P., Etxeberria, R., Sierra, B.: Feature subset selection by bayesian networks based on optimization. Artif. Intell. **123**(1), 157–184 (2000)
11. Inza, I., Larranaga, P., Sierra, B.: Feature subset selection by bayesian networks: a comparison with genetic and sequential algorithms. Int. J. Approx. Reason. **27**(2), 143–164 (2001)
12. Keller, J., Gray, M., Givens, J.: A fuzzy k-nearest neighbor algorithm. In: IEEE Transactions on Systems, Man and Cybernetics, vol. 4, pp. 580–585 (1985)
13. Lai, C., Reinders, M., Wessels, L.: Random subspace method for multivariate feature selection. Pattern Recognit. Lett. **27**(10), 1067–1076 (2006)
14. Larranaga, P., Lozano, J.A.: Estimation of Distribution Algorithms: A New Tool For Evolutionary Computation, vol 2. Springer (2002)
15. Muhlenbein, H., Paass, G.: Recombination of genes to the estimation of distributions. PPSN, pp. 178–187. Springer, Berlin (1996)
16. Neumann, G., Cairns, D.: Targeted eda adapted for a routing problem with variable length chromosomes. In: IEEE Congress on Evolutionary Computation (CEC), pp. 220–225 (2012)
17. Neumann, G.K., Cairns, D.E.: Introducing intervention targeting into estimation of distribution algorithms. In: Proceedings of the 27th ACM Symposium on Applied Computing, pp. 334 - 341 (2012)
18. Pena, J., Robles, V., Larranaga, P., Herves, V., Rosales, F., Perez, M.: GA-EDA: Hybrid evolutionary algorithm using genetic and estimation of distribution algorithms. Innovations in Applied Artificial Intelligence, pp. 361–371. Springer, Berlin (2004)
19. Posik, P.: Preventing premature convergence in a simple eda via global step size setting. In: Proceedings of the 10th International Conference on PPSN X (2008)
20. Pudil, P., Novovicova, J., Kittler, J.: Floating search methods in feature selection. Pattern Recognit. Lett. **15**(11), 1119–1125 (1994)
21. Saeys, Y., Degroeve, S., Aeyels, D., de Peer, Y.V., Rouz, P.: Fast feature selection using a simple estimation of distribution algorithm: a case study on splice site prediction. Bioinformatics **19**(suppl 2), 179–188 (2003)
22. Stoppiglia, H., Dreyfus, G., Dubois, R., Oussar, Y.: Ranking a random feature for variable and feature selection. J. Mach. Learn. Res. **3**, 1399–1414 (2003)
23. Zhang, Q., Sun, J., Tsang, E.: Combinations of estimation of distribution algorithms and other techniques. Int. J. Autom. Comput. **4**(3), 273–280 (2007)

Genetic Programming Model Regularization

César L. Alonso, José Luis Montaña and Cruz Enrique Borges

Abstract We propose a tool for controlling the complexity of Genetic Programming models. The tool is supported by the theory of Vapnik-Chervonekis dimension (VCD) and is combined with a novel representation of models named straight line program. Experimental results, implemented on conventional algebraic structures (such as polynomials), show that the empirical risk, penalized by suitable upper bounds for the Vapnik-Chervonekis dimension, gives a generalization error smaller than the use of statistical conventional techniques such as Bayesian or Akaike information criteria.

Keywords Genetic Programming · Straight Line Program · Pfaffian Operator · Symbolic Regression

1 Introduction

Inductive inference from examples is one of the most studied problems in Artificial Intelligence and has been addressed for many years using different techniques. Among them are included statistical methods such as inference techniques, regression and decision trees and other machine learning methods like neuronal networks and support vector machines [3, 12, 18, 20].

C.L. Alonso
Centro de Inteligencia Artificial, Campus de Gijón, Universidad de Oviedo,
33271 Gijón, Spain
e-mail: calonso@aic.uniovi.es

J.L. Montaña (✉)
Departamento de Matemáticas, Estadística y Computación, Universidad de Cantabria,
39005 Santander, Spain
e-mail: montanjl@unican.es

C.E. Borges
DeustoTech (Energy Unit), Universidad de Deusto, 48007 Bilbao, Spain
e-mail: cruz.borges@alumnos.unican.es

© Springer International Publishing Switzerland 2016
K. Madani et al. (eds.), *Computational Intelligence*,
Studies in Computational Intelligence 613,
DOI 10.1007/978-3-319-23392-5_6

In the last two decades, genetic programming (GP) has been applied to solving problems of inductive learning with some remarkable success [15–17, 19]. The general procedure involves the evolution of populations of data structures that represent models for the target function. In the evolutive process, the fitness function for evaluating the population measures some empirical error between the empirical value of the target function and the value of the considered individual over the sample set. Usually this fitness function must be regularized with some term that depends on the complexity of the model. Identifying optimal ways to measure the complexity of the model is one of the main goals in the process of regularization.

Most of the work devoted to develop GP strategies for solving inductive learning problems makes use of the GP-trees as data structures for representing programs [14]. We have proposed a new data structure named straight line program (slp) to deal with the problem of learning by examples in the framework of genetic programming. The slp has a good performance in solving symbolic regression problem instances as shown in (see [2]). A slp consists of a finite sequence of computational assignments. Each assignment is obtained by applying some function (selected from a given set) to a set of arguments that can be variables, constants or pre-computed results. The slp structure can describe complex computable functions using a few amount of computational resources than GP-trees. The key point for explaining this feature is the ability of slp's for reusing previously computed results during the evaluation process. Another advantage with respect to trees is that the slp structure can describe multivariate functions by selecting a number of assignments as the output set. Hence one single slp has the same representation capacity as a forest of trees. We study the practical performance of ad-hoc recombination operators for slps and we apply the slp- based GP approach to regression. In addition we study the Vapnik-Chervonekis dimension of slps representing models. We consider families of slp's constructed from a set of Pfaffian functions. Pfaffian functions are solutions of triangular systems of first order partial differential equations with polynomial coefficients. As examples, polynomials, exponential functions, trigonometric functions on some particular intervals and, in general, analytic algebraic functions are Pfaffian. The main outcome of this work is a penalty term for the fitness function of a genetic programming strategy based on slp's to solve inductive learning problems. Experimental results point out that the slp structure, if suitably regularized, may result in a robust tool for supervised learning.

2 Supervised Learning and Regression

Genetic Programming can be seen as a direct evolution method of computer programs for inductive learning. Inductive GP can be considered as a specialization of GP, in that it uses the framework of the last one in order to solve inductive learning problems. These problems are, in general, searching problems where the target is to construct some prediction model from a finite set of observed data. Providing a framework for studying inductive learning problems is one of the goals of Statistical

Learning Theory. In this sense the inclusion of methods from Statistical Learning Theory into the GP paradigm is a relevant contribution to inductive GP. Inductive inference process consists of three steps: to observe a phenomenon, to construct a model of that phenomenon and to make predictions by using this model. Given some sample set obtained by means of the observation step, it could seem that the best model might fit exactly the data, but this situation could lead to a poor performance on unseen instances in the presence of noise. Hence, the general idea is to look for a model that fits well the data set, being at the same time as simple as possible. This immediately raises the question of how to measure the complexity of a model. There are many ways to do this. For example we would prefer models with a small number of free parameters, which corresponds to simple mathematical formulas. In other cases where the model is represented by a program, we would consider the length of the program as a complexity measure. Usually, for tree structures, a measure of the complexity is the height or the width of the tree. There is no universal way of measuring the complexity of the model and the choice of a specific measure inherently depends on the problem at hand.

We will consider symbolic regression formulation under the general setting for predictive learning (see, [7, 21, 22]). The goal is to estimate an unknown real-valued function that fits a given finite sample set of data points. More formally, we consider an input space $X = \mathbb{R}^n$ and an output space $Y = \mathbb{R}$. We are given a sample of m pairs $z = (x_i, y_i)_{1 \le i \le m}$. These examples are drawn according to an unknown probability measure ρ on the product space $Z = X \times Y$ and they are generated according to an independent identically distributed (i.i.d.) process. As usual, probability measure ρ factorizes through its marginal distribution in X, $\rho(x)$, and the conditional distribution in Y, $\rho(y|x)$, that is:

$$\rho(x, y) = \rho(x)\rho(y|x) \tag{1}$$

The goal is to construct a function $f : X \to Y$ which predicts the value $y \in Y$ from a given $x \in X$. To choose the function f we use a criterion of a low probability of error. The best estimation of the function is the mean of the output conditional probability:

$$g(x) = \int y\rho(y|x) \tag{2}$$

A learning method selects the best model $f \in \mathcal{H}$, where $\mathcal{H}$ is some class of functions. In general, the error of the estimator f, $\varepsilon(f)$, is written as

$$\varepsilon(f) = \int Q(x, f, y)d\rho, \tag{3}$$

where Q measures some notion of loss between $f(x)$ and the target value y, and ρ is the distribution from which examples (x, y) are drawn to the learner. For regression tasks one usually takes $Q(x, f, y) = (y - f(x))^2$.

For a class of functions $\mathcal{H}$ of finite bounded complexity (for instance trees with bounded size or height), the model can be chosen minimizing the empirical error, also known as empirical risk:

$$\varepsilon_m(f) = \frac{1}{m} \sum_{i=1}^{m} Q(x_i, f, y_i) \tag{4}$$

Obviously, this method will have a good performance when the optimal model belongs to the complexity bounded class of functions considered. Nevertheless, usually we are not able to make such an assumption and a large class $\mathcal{H}$ must be considered. In this case, the problem of regression estimation requires optimal selection of model complexity in addition to model estimation via minimization of the empirical risk.

Analytical models selection criteria estimate the real error (Eq. 3) as a function of the empirical error (Eq. 4) with a penalty term related with some measure of model complexity:

$$\varepsilon(f) = \varepsilon_m(f) * pen(h, m); \quad * \in \{+, \cdot\} \tag{5}$$

where f is the model, h is the model complexity and m is the size of the sample set. In the above equation, there exists a degree of freedom which is the selection of the measure h for the model complexity. This measure always depends on the considered class $\mathcal{H}$ and more exactly on the representation structure for the models in $\mathcal{H}$. For example if the functions are described by multivariate polynomials, h is usually the number of monomials or a linear function involving the degree of the polynomial. In other cases, when the elements of $\mathcal{H}$ are represented by programs or trees, some typical complexity measures are the length of the programs or the size of the trees.

In this work, as we will see in the next sections, the classes $\mathcal{H}$ are families of programs named straight line programs, that are constructed from a set of operators F and a set of terminals T. The elements of F are Pfaffian functions, that is, a more general class of functions than polynomials or rational functions. Pfaffian functions include, for example, the analytic algebraic functions as for instance square root extraction.

Motivated by the concept of degree of polynomials, in our programs we will only consider the non-scalar instructions for measuring the complexity of the model. The non-scalar instructions are those in which the selected operator in F is different from $\{+, -\}$. The main theoretical result in this work is the computation of a polynomial upper bound for a new complexity measure of our straight line programs with Pfaffian instructions, that does not involve the length of the corresponding program, but only the number of the non-scalar instructions. A simplification of this complexity measure will be considered in the Eq. (5) for the fitness regularization in GP with our structure.

3 Straight Line Program Genetic Programming

Straight line programs have a large history in the field of Computational Algebra. A particular class of straight line programs, known in the literature as arithmetic circuits, constitutes the underlying computation model in Algebraic Complexity Theory [6]. Arithmetic circuits with the standard arithmetic operations $\{+, -, *, /\}$ are the natural model of computation for studying the computational complexity of algorithms solving problems with an algebraic flavor. They have been used in linear algebra problems [4], in quantifier elimination [13] and in algebraic geometry [9, 10]. Also, slp's constitute a promising alternative to the trees in the field of Genetic Programming (see [2]). The formal definition of the straight line program structure is as follows: Let $F = \{f_1, \dots, f_n\}$ be a set of functions, where f_i has arity a_i, for $1 \le i \le n$, and let $T = \{t_1, \dots, t_m\}$ be a set of terminals. A *straight line program (slp)* over F and T is a finite sequence of computational instructions $\Gamma = \{I_1, \dots, I_l\}$ where

$$I_k \equiv u_k := f_{j_k}(\alpha_1, \dots, \alpha_{a_{j_k}}); \text{ with } f_{j_k} \in F,$$

$\alpha_i \in T$ for all i if $k = 1$ and $\alpha_i \in T \cup \{u_1, \dots, u_{k-1}\}$ for $1 < k \le l$.

Terminal set T is of the form $T = V \cup C$, where $V = \{x_1, \dots, x_n\}$ is a finite set of variables and $C = \{c_1, \dots, c_q\}$ is a finite set of constants. The number of instructions l is the *length* of Γ.

Note that if we consider the *slp* Γ as the code of a program, then a new variable u_i is introduced at each instruction I_i. We will denote by $\Gamma = \{u_1, \dots, u_l\}$ a slp. Each of the non-terminal variables u_i can be considered as an expression over the set of terminals T constructed by a sequence of recursive compositions from the set of functions F. The set of all slp's over F and T is denoted by $SLP(F, T)$.

An output set of a slp $\Gamma = \{u_1, \dots, u_l\}$ is any set of non-terminal variables of Γ, that is, $O(\Gamma) = \{u_{i_1}, \dots, u_{i_t}\}$, $i_1 < \cdots < i_t$. Provided that $V = \{x_1, \dots, x_p\} \subset T$ is the set of terminal variables, the function computed by Γ, denoted by $\Phi_\Gamma : I^p \to O^t$, is defined recursively in the natural way and satisfies $\Phi_\Gamma(a_1, \dots, a_p) = (b_1, \dots, b_t)$, where b_j stands for the value of the expression over V of the non-terminal variable u_{i_j} when we replace the variable x_k with a_k; $1 \le k \le p$.

Example 1 Let F be the set given by the three binary standard arithmetic operations, $F = \{+, -, *\}$ and let $T = \{1, x_1, x_2\}$ be the set of terminals. In this situation any slp over F and T is a finite sequence of instructions where each instruction represents a polynomial in two variables with integer coefficients. If we consider the following slp Γ of length 5 with output set $O(\Gamma) = \{u_5\}$:

$$\Gamma \equiv \begin{cases} u_1 := x_1 + 1 \\ u_2 := u_1 * u_1 \\ u_3 := x_2 + x_2 \\ u_4 := u_2 * u_3 \\ u_5 := u_4 - u_3 \end{cases} \tag{6}$$

Fig. 1 Directed graph
representing a slp

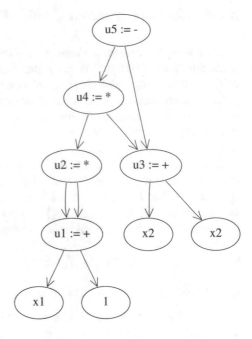

then the function computed by Γ is the polynomial

$$\Phi_\Gamma = 2x_2(x_1 + 1)^2 - 2x_2$$

Every slp, $\Gamma = \{u_1, \ldots, u_l\}$, over F and T can be represented by a directed acyclic graph $G_\Gamma = (V, E)$. The set of vertices is $V = T' \cup \{u_1, \ldots, u_l\}$, where T' contains all terminals involved in the computation. The set of edges E is constructed as follows: for every k, $1 \le k \le l$, we draw an edge (u_k, α_i) for each $i \in \{1, \ldots, a_{j_k}\}$. Note that T' is the set of leaves of G_Γ and it is a subset of the set T of terminals. Figure 1 is a directed graph representing the slp described in Eq. (6)

For computing the initial population, the well known methods for trees (see [14]) can be easily adapted to slp's. In order to compute the fitness function in a GP process to solve a particular problem, the computation of the function Φ_Γ, considering its own definition, would be often necessary.

3.1 SLP Crossover and Mutation

For slp-GP, 1-point crossover and in general k-point crossover are easily defined. However, a new specific crossover operation that produces another type of information exchange between the two selected parents, has been designed. The objective is to carry subexpressions from one parent to the other. A subexpression is captured by an instruction u_i and all the instructions that are used to compute the expression over the set of terminals represented by u_i. Now follows a description of this crossover with a clarifying example.

Given two slp's, Γ_1 and Γ_2, first a position k en Γ_1 is randomly selected. Let S_{u_k} be the set of all instructions related to the computation of the subexpression associated to node u_k. We obtain the first offspring by randomly selecting an allowed position t in Γ_2 and making the substitution of a part of its instructions by those instructions in S_{u_k} suitably renamed. For the second offspring we symmetrically repeat the strategy.

Example: Consider $F = \{*, +\}$, $L = 5$ and $T = \{x, y\}$. Let Γ_1 and Γ_2 be the following two slp's:

$$\Gamma_1 \equiv \begin{cases} u_1 := x + y \\ u_2 := u_1 * u_1 \\ u_3 := u_1 * x \\ u_4 := u_3 + u_2 \\ u_5 := u_3 * u_2 \end{cases} \qquad \Gamma_2 \equiv \begin{cases} u_1 := x * x \\ u_2 := u_1 + y \\ u_3 := u_1 + x \\ u_4 := u_2 * x \\ u_5 := u_1 + u_4 \end{cases}$$

If $k = 3$ then $S_{u_3} = \{u_1, u_3\}$, and t must be selected in $\{2, \ldots, 5\}$. If for instance $t = 3$, then the first offspring is as follows.

$$\Gamma_1' \equiv \begin{cases} u_1 := x * x \\ \mathbf{u_2} := \mathbf{x + y} \\ \mathbf{u_3} := \mathbf{u_2 * x} \\ u_4 := u_2 * x \\ u_5 := u_1 + u_4 \end{cases}$$

For the second offspring, if the selected position in Γ_2 is $k' = 4$, then $S_{u_4} = \{u_1, u_2, u_4\}$. Now if $t = 5$, then the offspring will be

$$\Gamma_2' \equiv \begin{cases} u_1 := x + y \\ u_2 := u_1 * u_1 \\ \mathbf{u_3} := \mathbf{x * x} \\ \mathbf{u_4} := \mathbf{u_3 + y} \\ \mathbf{u_5} := \mathbf{u_4 * x} \end{cases}$$

The mutation operation in the slp structure consists of a change in one of the instructions. This change can be either the substitution of the complete instruction by another one randomly generated, or a little modification of just one of the arguments of the function in F that defines the instruction.

4 Pfaffian Functions and VCD of Formulas

In this section we introduce some tools concerning the geometry of sets defined by boolean combinations of sign conditions over Pfaffian functions (semi-Pfaffian sets in the mathematical literature). A complete survey on the subject is due to Gabrielov and Vorobjov [8].

Definition 1 Let $U \subset \mathbf{R}^n$ be an open domain. A *Pfaffian chain* of length $q \geq 1$ and degree $D \geq 1$ in U is a sequence of real analytic functions $f_1, \ldots, f_q$ in U satisfying a system of differential equations

$$\frac{\partial f_i}{\partial x_j} = P_{i,j}(\mathbf{x}, f_1(\mathbf{x}), \ldots, f_i(\mathbf{x})) \tag{7}$$

for $i = 1, \ldots q$ where $P_{i,j} \in \mathbf{R}[\mathbf{x}, y_1, \ldots, y_i]$ are polynomials of degree at most D and $\mathbf{x} = x_1, \ldots, x_n$.

A function f on U is called a Pfaffian of order q and degree (D, d) if

$$f(\mathbf{x}) = P(\mathbf{x}, f_1(\mathbf{x}), \ldots, f_q(\mathbf{x})) \tag{8}$$

where $P \in \mathbf{R}[\mathbf{x}, y_1, \ldots, y_q]$ is a polynomial of degree at most $d \geq 1$ and $f_1, \ldots, f_q$ is a Pfaffian chain of length q and degree D.

The following functions are Pfaffian: $sin(x)$, defined on the interval $(-\pi + 2\pi r, \pi + 2\pi r)$; $tan(x)$, defined on the interval $(-\pi/2 + \pi r, \pi/2 + \pi r)$; e^x defined in $\mathbf{R}$; $log\ x$ defined on $x > 0$; $1/x$ defined on $x \neq 0$. $\sqrt{x}$ defined on $x \geq 0$. More generally, analytic algebraic functions are Pfaffian.

Definition 2 Let $\mathcal{F}$ be a class of subsets of a set X. We say that $\mathcal{F}$ shatters a set $A \subset X$ if for every subset $E \subset A$ there exists $S \in \mathcal{F}$ such that $E = S \cap A$. The VCD of $\mathcal{F}$ is the cardinality of the largest set that is shattered by $\mathcal{F}$.

Next we announce an upper bound for the VCD of a family of concept classes whose membership tests are computed by straight line programs involving Pfaffian operators over the real numbers. An important new issue is that we do not consider an upper bound for the length of the slp's. In previous results about VCD of programs or families of computation trees, a time bound approximated by the number of steps of the program execution or by the height of the computation tree is needed [11]. In our case we only need a bound for the number of the non-scalar slp's instructions. Those are instructions involving operations which are not in $\{+, -\}$.

A rough estimation of the VC dimension of slps using Pfaffian operators can be obtained computing the number of free parameters in families of slps with bounded non-scalar complexity. To do this let $T = \{t_1, \ldots, t_n\}$ be a set of terminals and let $F = \{+, -*, /, sign\} \cup \{f_1, \ldots, f_q\}$ be a set of functions, where the elements f_i constitute a Pfaffian chain of length q with arities bounded by A and the $sign$ function is defined as $sign(x) = 1$ if $x > 0$ and 0 otherwise.

Let $\Gamma_{n,L}$ be the collection of $slp's$ Γ over F and T using at most L non-scalar operations and a free number of scalar operations. Then, the number of free parameters of a universal slp Γ_U that parameterizes the elements of the family $\Gamma_{n,L}$ is exactly:

$$N := L[3 + q + A(n + \frac{L-1}{2}) + 1] + n \qquad (9)$$

The proof is as follows. Introduce a set of parameters α, β and γ taking values in Z^k for a suitable natural number k, such that each slp in the family can be obtained specializing the parameters. For this purpose we define $u_{-n+m} = t_m$, for $1 \le m \le n$. Note that any non-scalar assignment u_i, $1 \le i \le L$ in a slp Γ belonging to $\Gamma_{n,L}$ is a function of $t = (t_1, \ldots, t_n)$ that can be parameterized as follows:
$u_i = U_i(\alpha, \beta, \gamma)(t) =$

$$\gamma_{-n}^i[\alpha_{-n}^i(\sum_{j=-n+1}^{i-1} \alpha_j{}^{i_1} u_j) * (\sum_{j=-n+1}^{i-1} \alpha_j{}^{i_2} u_j)+$$

$$+(1 - \alpha_{-n}^i)[\beta_{-n}^i \frac{\sum_{j=-n+1}^{i-1} \alpha_j{}^{i_1} u_j}{\sum_{j=-n+1}^{i-1} \alpha_j{}^{i_2} u_j}+$$

$$+(1 - \beta_{-n}^i)sgn(\sum_{j=-n+1}^{i-1} \alpha_j{}^{i_1} u_j)]]+$$

$$+(1 - \gamma_{-n}^i)[\sum_{k=1}^{q} \gamma_k^i f_k(\sum_{j=-n+1}^{i-1} \alpha_j{}^{i_1} u_j, \ldots, \sum_{j=-n+1}^{i-1} \alpha_j{}^{i_A} u_j)]$$

Now considering the last assignment as the output set of the slp Γ, this last assignment is parameterized as:

$$U = \sum_{j=-n+1}^{L} \alpha_j u_j$$

where u_j, $1 \le j \le L$ are the non-scalar assignments.

Finally counting the number of introduced parameters we will obtain Eq. (9). The estimation given in that equation can be converted, after certain algebraic manipulations, into an upper bound using theory of Pfaffian operators (see [8]). We omit the proof due to lack of space.

Main Theorem. Let $\Gamma_{L,n}$ the set of slps with n variables, at most L non-scalar operations, using operators in F that contains the operations $\{+, -, *, /, sign\}$ and Pfaffian operations f, where each f belongs to a fixed Pfaffian chain $\{f_1, \ldots, f_q\}$ of length q and degree $D \geq 2$. Let N be as in Eq. (3). Then, the Vapnik-Chervonenkis dimension of $\mathcal{C}_{k,n}$ is in the class:

$$O((q(N+n))^2 + (N+n)(L+q)log_2((N+n)(L+1)(4+D))) \qquad (10)$$

Simplification. If we consider as constants parameters n, q, D and d, the VCD of the class is at most $O(L^4)$. This quantity gives an idea of the asymptotic maximum order of VCD of common classes of GP-models. We point out that this quantity is an upper bound and, possibly, far from being an optimal bound, but it can be used as starting point in further experimental developments.

5 Model Selection Criterion

In supervised learning problems like regression and classification a considerable amount of effort has been done for obtaining good generalization error bounds. The results by Vapnik (see [22]) state the following error bound:

$$\varepsilon(f) \leq \varepsilon_m(f) + \sqrt{\frac{h(log(2m/h)+1) - log(\eta/4)}{m}}, \qquad (11)$$

where h must be substituted by the upper bound of the VCD of the hypothesis class that contains the model f, η is the probability that the error bound is violated and m is the sample size. As usual in this context $\varepsilon(f)$ and $\varepsilon_m(f)$ stand, respectively, for the true mean square error and the empirical mean square error of the model f.

In our case, f will be represented by a straight line program $\Gamma \in SLP(F, T)$ where T contains n variables and F contains the operations on real numbers $\{+, -, *, /, sign\}$ and Pfaffian operations over the reals. Note that the sets F and T are invariants throughout the model selection process. Hence, the search space of models forms a nested structure:

$$C_1 \subset C_2 \subset \cdots \subset C_L \subset \cdots$$

where C_L represents the class of slp's in $SLP(F, T)$ that have at most L non-scalar instructions. In this situation we will finally choose the model that minimizes the right side of Eq. (5).

6 Experimental Results

In this section we present the obtained results after an experimental phase in which symbolic regression problem instances were solved using the selection criterion described in the previous section. Our proposal is to consider straight line programs with Pfaffian instructions as the structure that represents the model. Then a GP algorithm is executed considering the recombination operators for slp's described in Sect. 2 and with fitness regularization function expressed in Eq. (11). So we propose a model estimation via structural risk minimization (SRM). For the complexity measure h of the model, we will use the VCD bound in 4.

We will consider additive gaussian noise in the sample set $z = (x_i, y_i)_{1 \leq i \leq m}$. Hence, for a target function g, the sample set verifies: $y_i = g(x_i) + \epsilon$, where ϵ is independent and identically distributed (i.i.d.) zero mean random error.

We will compare the effectiveness of the VCD fitness regularization method (VCD-SRM) with two well known representative statistical methods with different penalization terms:

• Akaike Information Criterion (AIC) which is as follows (see [1]):

$$\varepsilon(f) = \varepsilon_m(f) + \frac{2h}{m}\sigma^2 \tag{12}$$

• Bayesian Information Criterion (BIC) (see [5]):

$$\varepsilon(f) = \varepsilon_m(f) + (ln\ m)\frac{h}{m}\sigma^2 \tag{13}$$

In the above expressions h stands for the number of free parameters of the model (Eq. 9).

For measuring the quality of the final selected model, we have considered a new set of unseen points, generated without noise from the target function. This new set of examples is known as the test set or validation set. So, let $(x_i, y_i)_{1 \leq i \leq n_{test}}$ a validation set for the target function $g(x)$ (i.e. $y_i = g(x_i)$) and let $f(x)$ be the model estimated from the training data. Then the prediction risk $\varepsilon_{n_{test}}$ is defined by the mean square error between the values of f and the true values of the target function g over the validation set:

$$\varepsilon_{n_{test}} = \frac{1}{n_{test}} \sum_{i=1}^{n_{test}} (f(x_i) - y_i)^2 \tag{14}$$

For the first experiment we have considered a set of 500 multivariate polynomials with real coefficients whose degrees are bounded by 5. The number of variables varies from 1 to 5 with 100 polynomials for each case.

A second experiment was performed considering some well known real benchmark problems. In all cases, when the GP process finishes, the best individual is selected as the proposed model for the corresponding target function.

We shall denote the set of polynomials as $P_n^R[X]$ with $X = (x_1, \ldots, x_n)$, $1 \leq n \leq 5$ and $x_i \in [-1, 1]$ $\forall i$. The individuals are slp's over $F = \{+, -, *, /, sqrt, sin, cos, exp\}$. In order to avoid errors generated by divisions by zero, instead of the traditional division we will use in our computation the operation usually named "protected division", that returns 1 if the denominator is zero. Besides the variables x_i, the terminal set also includes five constants c_i, $1 \leq i \leq 5$, randomly generated in $[-1, 1]$. Observe that although the target functions are polynomials, our set F not only contains the operators of sum, difference and product, but also contains other Pfaffian functions. This situation increments considerably the search space. Nevertheless, note that in a real problem situation usually we do not know if the target function is a polynomial or not.

The parameters for the GP process are the following: population size $M = 200$, probability of crossover $p_c = 0, 9$, probability of mutation $p_m = 0, 05$, and tournament selection of size 5. The real length of the slp's in the population is bounded by 40. Elitism and a particular generational replacement are used. In this sense, the offsprings do not necessarily replace their parents. After a crossover we have four individuals: two parents and two offsprings. We select the two best individuals with different fitness values. The motivation is to prevent premature convergence and to maintain diversity in the population.

As we are considering multivariate polynomials as target functions, the difficulty of the problem instance increases with the number of variables. Hence, to vary the size of the sample set as a function of the number of variables is a reasonable decission. Note that an upper bound for the number of monomials in a polynomial with n variables and degree d is $4 \cdot d^{n+1}$ and this is also a quite good estimation for a lower bound of the size of the sample set. Thus, in our case we have considered sample sets of size $4 \cdot 5^{n+1}$, $1 \le n \le 5$. In this experiment one execution for each strategy has been performed over the 500 generated target functions. In every execution the process finishes after 250 generations were completed. Finally, the validation set consists of a number of unseen points that is equal to two times the size of the sample set (Table 1).

Figures 2 and 3 represent the empirical distribution of the executions of the three compared strategies over the sets of polynomials. We have separated the polynomial sets by the number of variables, from one to five. These empirical distributions are displayed using standard box plot notation with marks at the best execution, 25%,

Table 1 Values of means and variances

$P_1{}^R[X]$	μ	σ	$P_2{}^R[X]$	μ	σ
AIC	0.39	0.42	AIC	1.17	0.86
BIC	0.38	0.41	BIC	1.25	0.85
VCD	0.24	0.27	VCD	0.82	0.54
$P_3{}^R[X]$	μ	σ	$P_4{}^R[X]$	μ	σ
AIC	2.62	1.28	AIC	4.79	1.49
BIC	2.89	1.57	BIC	5.08	1.74
VCD	2.24	0.85	VCD	4.76	1.43
$P_5{}^R[X]$	μ	σ			
AIC	8.51	2			
BIC	8.63	2.16			
VCD	8.63	2.17			

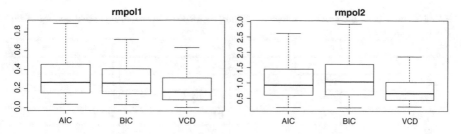

Fig. 2 Empirical distribution of the executions, for the univariate and bivariate polynomials

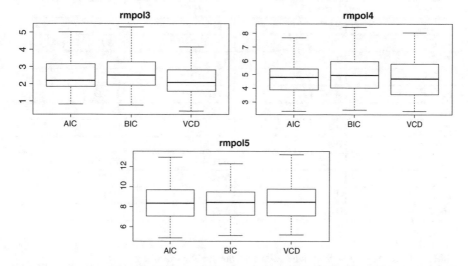

Fig. 3 Empirical distributions of the executions, for the multivariate polynomials with 3, 4, and 5 variables

50%, 75% and the worst execution, always considering the prediction risk of the selected model, represented on the y-axis and defined by the mean square error between the values of the model and the true values of the target function over the validation set. We also include tables that show means and variances as well as the prediction risk of the best obtained model for each method.

As we can see from the above figures and tables, it seems that VCD regularization performs better than the well known regularization methods AIC and BIC. This is more clear for the polynomials up to three variables and not so clear for the rest of the polynomial sets. This could be because for polynomials with four and five variables, as they constitute more complex problem instances, it would be necessary a large number of generations in the evolutive process. In order to confirm the comparative results of the studied strategies we have made crossed statistical hypothesis tests. The obtained results are showed in Table 2. Roughly speaking, the null-hypothesis in each test with associated pair (i, j) is that strategy i is not better than strategy j.

Table 2 Results of the crossed statistical hypothesis tests about the comparative quality of the studied strategies

$P_1{}^R[X]$	AIC	BIC	VCD
AIC		0.91	1
BIC	0.30		1
VCD	$3.93 \cdot 10^{-4}$	$1.1 \cdot 10^{-3}$	
$P_2{}^R[X]$	AIC	BIC	VCD
AIC		$3.91 \cdot 10^{-2}$	1
BIC	0.96		1
VCD	$1.86 \cdot 10^{-5}$	$1.12 \cdot 10^{-7}$	
$P_3{}^R[X]$	AIC	BIC	VCD
AIC		$7.7 \cdot 10^{-2}$	0.99
BIC	0.99		1
VCD	$1.2 \cdot 10^{-2}$	$3.93 \cdot 10^{-4}$	
$P_4{}^R[X]$	AIC	BIC	VCD
AIC		0.12	0.90
BIC	0.77		0.99
VCD	0.27	$2.7 \cdot 10^{-3}$	
$P_5{}^R[X]$	AIC	BIC	VCD
AIC		0.44	0.52
BIC	0.77		0.78
VCD	0.61	0.37	

Table 3 Prediction risk of the model obtained from the best execution

Instance	AIC	BIC	VCD
$P_1{}^R[X]$	$3.40 \cdot 10^{-2}$	$3.44 \cdot 10^{-2}$	$4.32 \cdot 10^{-3}$
$P_2{}^R[X]$	0.20	0.19	0.22
$P_3{}^R[X]$	0.82	0.77	0.42
$P_4{}^R[X]$	2.33	2.43	2.33
$P_5{}^R[X]$	4.88	5.13	5.20

Hence if value a_{ij} in Table 2 is less than a significance value α, we can reject the corresponding null-hypothesis (Table 3, Fig. 4).

Taking into account the results of the crossed statistical hypothesis tests with a significance value $\alpha = 0.05$, we can confirm that our proposed regularization method based on the VC dimension of families of SLP's is the best of the studied strategies for the considered sets of multivariate polynomials.

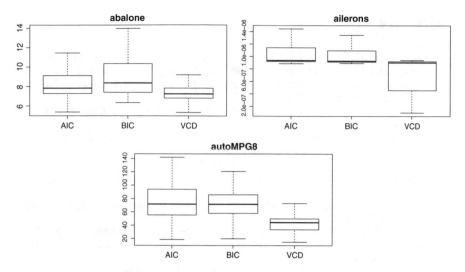

Fig. 4 Empirical distribution of the executions, for functions associated to real problems

7 Conclusions

Straight line programs constitute a promising structure for representing models in the Genetic Programming framework. Indeed, as it was published in a previous work, slp's outperform the traditional tree structure when GP strategies are applied for some kind of regression problems. In this paper we try to control the complexity of populations of slp's while they evolve in order to find good models for solving symbolic regression problem instances. The evolving structure is constructed from a set of functions that contains Pfaffian operators. We have considered the Vapnik Chervonenkis dimension as a complexity measure and we have found a theoretical upper bound of the VCD of families of slp's over Pfaffian operators as an important generalization of similar results for more simple sets of operators including rational functions. This theoretical upper bound is polynomial in the number of the non-scalar instructions of the family of the slp's. As a consequence of the main result, we propose a regularized fitness function included in a evolutionary strategy for solving symbolic regression problem instances. We have compared our fitness function based on the VCD upper bound with two well known statistical penalization criteria. The experimental results obtained after the execution of the compared strategies over two different groups of target functions, show that our proposed complexity measure and its corresponding penalization criterion is better than the others in over the group of the three real problem instances where the VCD regularization method is clearly the best.

Acknowledgments This work is partially supported by spanish grant TIN2011-27479-C04-04.

References

1. Akaike, H.: Statistical prediction information. Ann. Inst. Stat. Math. **22**, 203–217 (1970)
2. Alonso, C.L., Montaña, J.L.. Puente, J.: Straight line programs: a new linear genetic programming approach. In:Proceedings of the 20th IEEE International Conference on Tools with Artificial Intelligence (ICTAI), pp. 571–524 (2008)
3. Angluin, D., Smith, C.H.: Inductive inference: theory and methods. ACM Comput. Surv. **15**(3), 237–569 (1983)
4. Berkowitz, S.J.: On computing the determinant in small parallel time using a small number of processors. Inf. Process. Lett. **18**, 147–150 (1984)
5. Bernadro, J., Smith, A.: Bayesian Theory. Willey, New York (1994)
6. Burguisser, P., Clausen, M., Shokrollahi, M.A.: Algebraic Complexity Theory. Springer, New York (1997)
7. Cherkassky, V., Yunkian, M.: Comparison of model selection fo regression. Neural Comput. **15**(7), 1691–1714 (2003)
8. Gabrielov, A.N., Vorobjov, N.: Complexity of Computations with Pfaffian and Noetherian Functions, Normal Forms, Bifurcations and Finiteness Problems in Differential Equations. Kluwer, Dordrecht (2004)
9. Giusti, M., Heinz, J.: La Détermination des Points Isolés et la Dimension dúne Varieté Agebrique Peut se Faire en Temps Polynomial, Computational Algebraic Geometry and Commutative Algebra, Symposia Matematica XXXIV, Eisenbud, D, Robbiano, L. (eds.), pp. 216–256. Cambridge University Press, Cambridge (1993)
10. Giusti, M., Heintz, J., Morais, J., Morgentern, J.E., Pardo, L.M.: Straight line programs in geometric elimination theory. J. Pure Appl. algebra **124**, 121–146 (1997)
11. Goldberg, P., Jerrum, M.: Bounding the vapnik-chervonenkis dimension of concept classes parametrizes by real numbers. Mach. Learn. **18**, 131–148 (1995)
12. Gori, M., Maggini, M., Martinelli, E., Soda, G.: Inductive inference from noisy examples using the hybrid finite state filter. IEEE Trans. Neural Networks **9–3**, 571–575 (1998)
13. Heintz, J., Roy, M.F., Solerno, P.: Sur la Complexité du Principe de Tarski-Seidenberg. Bulletin de la Societé Mathematique de France **118**, 101–126 (1990)
14. Koza, J.: Genetic Programming: on the Programming of Computers by Means of Natural Selection. The MIT Press, Cambridge (1992)
15. Nikolaev, N.Y., Iba, H.: Regularization approach to inductive genetic programming. IEEE Trans. Evol. Comput. **5**(4), 359–375 (2001)
16. Okley, H.: In: Kinnear, K. (ed.) Advances in Genetic Programming. Two scientific applications of Genetic Programming: Stack filters and nonlinear fitting to chaotic data, pp. 369–389. MIT Press, Cambridge (1994)
17. Poli, R., Cagnoni, S.: In: Koza, J.R., Deb, K., Dorigo, M., Fogel, D.B., Garzon, M., Iba, H., Riolo, R.L. (eds.) Evolution of Pseudo-coloring Algoritms for Image Enhancement with Interactive Genetic Programming, pp. 269–277. MIT Press, Cambridge (1997)
18. Shaoning, P., Kasabov, N.: Inductive vs transductive inference. global vs local models: SVM, TSVM and SVMT for gene expression classification problems. In: Proceedings IEEE International Joint Conference on Neural Networks, vol. 2, pp. 1197–1202 (2004)
19. Tackett, W.A., Carmi, A.: In: Kinnear, K. (ed.) The Donut Problem: Scalability and Generalization in Genetic Programming, Advances in Genetic Programming. MIT Press, Cambridge (1994)
20. Tenebaum, J.B., Griffiths, T.L., Kemp, C.: Theory Based Bayesian Models of Inductive Learning and Reasoning. Trends in Cognitive Sciences, Kingston, vol. 10(7), pp. 309–318 (2006)
21. Vapnik, V., Chervonenkis, A.: Ordered risk minimization. Autom. Remote Control **34**, 1226–1235 (1974)
22. Vapnik, V.: Statistical Learning Theory. Willey, New York (1998)

A Radial Basis Function Neural Network-Based Coevolutionary Algorithm for Short-Term to Long-Term Time Series Forecasting

E. Parras-Gutierrez, V.M. Rivas and J.J. Merelo

Abstract This work analyzes the behavior and effectiveness of the L-Co-R method using a growing horizon to predict. This algorithm performs a double goal, on the one hand, it builds the architecture of the net with a set of RBFNs, and on the other hand, it sets a group of time lags in order to forecast future values of a time series given. For that, it has been used a set of 20 time series, 6 different methods found in the literature, 4 distinct forecast horizons, and 3 distinct quality measures have been utilized for checking the results. In addition, a statistical study has been done to confirms the good results of the method L-Co-R.

Keywords Time series forecasting · Co-evolutionary algorithms · Neural networks · Significant lags

1 Introduction

Formally defined, a time series is a set of observed values from a variable along time in regular periods (for instance, every day, every month or every year) [25]. Accordingly, the work of forecasting in a time series can be defined as the task of predicting successive values of the variable in time spaced based on past and present observations.

E. Parras-Gutierrez · V.M. Rivas
Department of Computer Sciences, University of Jaen, Campus Las Lagunillas s/n,
23071 Jaen, Spain
e-mail: eparrasg@vrivas.es

V.M. Rivas
e-mail: vrivas@vrivas.es

J.J. Merelo (✉)
Department of Computers, Architecture and Technology, University of Granada,
C/ Periodista Daniel Saucedo s/n, 18071 Granada, Spain
e-mail: jmerelo@geneura.ugr.es

© Springer International Publishing Switzerland 2016
K. Madani et al. (eds.), *Computational Intelligence*,
Studies in Computational Intelligence 613,
DOI 10.1007/978-3-319-23392-5_7

For many decades, different approaches have been used for to modelling and forecasting time series. These techniques can be classified into three different areas: descriptive traditional technologies, linear and nonlinear modern models, and soft computing techniques. From all developed method, ARIMA, proposed by Box and Jenkins [3], is possibly the most widely known and used. Nevertheless, it yields simplistic linear models, being unable to find subtle patterns in the time series data.

New methods based on artificial neural networks, such as the one used in this paper, on the other hand, can generate more complex models that are able to grasp those subtle variations.

The L-Co-R method [24], developed inside the field of ANNs, makes jointly use of Radial Basis Function Networks (RBFNs) and EAs to automatically forecast any given time series. Moreover, L-Co-R designs adequate neural networks and selects the time lags that will be used in the prediction, in a coevolutive [7] approach that allows to separate the main problem in two dependent subproblems. The algorithm evolves two subpopulations based on a cooperative scheme in which every individual of a subpopulation collaborates with individuals from the other subpopulation in order to obtain good solutions.

While previously work [24] was focused on 1-step ahead prediction, the main goal of this one is to analyze the effectiveness of the L-Co-R method in the medium and long-term horizon, using the own previously predicted values to perform next predictions. Thus, 6 different methods used in time series forecasting have been selected in order to test the behavior of the method.

The rest of the paper is organized as follows: Sect. 2 introduces some preliminary topics related to this research; Sect. 3 describes the method L-Co-R; and finally Sect. 4 presents the experimentation and the statistical study carried out.

2 Preliminaries

Approaches proposed in time series forecasting can be mainly grouped as linear and nonlinear models. Methods like exponential smoothing methods [34], simple exponential smoothing, Holt's linear methods, some variations of the Holt-Winter's methods, State space models [29], and ARIMA models [3], have stand out from linear methods, used chiefly for modelling time series. Nonlinear models arose because linear models were insufficient in many real applications; between nonlinear methods it can be found regime-switching models, which comprise the wide variety of existing threshold autoregressive models [31] as: self-exciting models [32], smooth transition models [8], and continuous-time models [4], among others. Nevertheless, soft computing approaches were developed in order to save disadvantages of nonlinear models like the lack of robustness in complex model and the difficulty to use [9].

ANNs have also been applied successfully [17] and recognized as an important tool for time-series forecasting. Within ANNs, the utilization of RBFs as activation functions were considered by works as [5] and [27], and applied to time series by Carse and Fogarty [6], and Whitehead and Choate [33]. Later works like the ones by Harpham and Dawson [13] or Du [10] focused on RBFNs for time series forecasting.

On the other hand, an issue that must be taken into account when working with time series is the correct choice of the time lags for representing the series. Takens' theorem [30] establishes that if d, a d-dimensional space where d is the minimum dimension capable of representing such a relationship, is sufficiently large is possible to build a state space using the correct time lags and if this space is correctly rebuilt also guarantees that the dynamics of this space is topologically identical to the dynamics of the real systems state space.

Many methods are based in Takens' theorem (like [19]) but, in general, the approaches found in the literature consider the lags selection as a pre or post-processing or as a part of the learning process [1, 23]. In the L-Co-R method the selection of the time lags is jointly faced along with the design process, thus it employs co-evolution to simultaneously solve these problems.

Cooperative co-evolution [26] has also been used in order to train ANNs to design neural network ensembles [12] and RBFNs [18]. But in addition, cooperative co-evolution is utilized in time series forecasting in works as the one by Xin [20].

3 Description of the Method

This section describes L-Co-R [24], a co-evolutionary algorithm developed to min-imize the error obtained for automatically time series forecasting. The algorithm works building at the same time RBFNs and sets of lags that will be used to predict future values. For this task, L-Co-R is able to simultaneously evolve two populations of different individual species, in which any member of each population can coop-erate with individuals from the other one in order to generate good solutions, that is, each individual represents itself a possible solution to the subproblem. Therefore, the algorithm is composed of the following two populations:

- Population of RBFNs: it consists of a set of RBFNs which evolves to design a suitable architecture of the network. This population employs real codification so every individual represent a set of neurons (RBFs) that composes the net. During the evolutionary process neurons can grow or decrease since the number of neurons is variable. Each neuron of the net is defined by a center (a vector with the same dimension as the inputs) and a radius. The exact dimension of the input space is given by an individual of the population of lags (the one chosen to evaluate the net).

- Population of lags: it is composed of sets of lags evolves to forecast future values of the time series. The population uses a binary codification scheme thus each gene indicates if that specific lag in the time series will be utilized in the forecasting

Fig. 1 General scheme of
method L-Co-R

```
Trend preprocessing
t = 0;
initialize P_lags(t);
initialize P_RBFNs(t);
evaluate individuals in P_lags(t);
evaluate individuals in P_RBFNs(t);
while termination condition not satisfied do
begin
  t = t+1;
  /* Evolve population of lags */
  for i=0 to max_gen_lags do
  begin
    set threshold;
    select P_lags'(t) from P_lags(t);
    apply genetic operators in P_lags'(t);
    /* Evaluate P_lags'(t) */
      choose collaborators from P_RBFNs(t);
      evaluate individuals in P_lags'(t);
      replace individuals P_lags(t) with P_lags'(t);
    if threshold < 0
    begin
      diverge P_lags(t);
    end
  end
  /* Evolve population of RBFNs */
  for i=0 to max_gen_RBFNs do
  begin
    select P_RBFNs'(t) from P_RBFNs(t);
    apply genetic operators in P_RBFNs'(t);
    /* Evaluate P_RBFNs'(t) */
      choose collaborators from P_lags(t);
      evaluate individuals in P_RBFNs'(t);
      replace individuals with P_RBFNs'(t);
  end
end
train models and select the best one
forecast test values with the final model
Trend post-processing
```

process. The length of the chromosome is set at the beginning corresponding with
the specific parameter, so that it cannot vary its size during the execution of the
algorithm.

As the fundamental objective, L-Co-R forecasts any time series for any horizon
and builds appropriate RBFNs designed with suitable sets of lags, reducing any hand
made preprocessing step. Figure 1 describes the general scheme of the algorithm
L-Co-R.

L-Co-R performs a process to automatically remove the trend of the times series to work with, if necessary. This procedure is divided into two main phases: preprocessing, which takes places at the beginning of the algorithm, and post-processing, at the end of co-evolutionary process. Basically, the algorithm checks if the time series includes trend and, in affirmative case, the trend is removed.

The performance of L-Co-R starts with the creation of the two initial populations, randomly generated for the first generation; then, each individual of the populations is evaluated. The L-Co-R algorithm uses a sequential scheme in which only one population is active, so the two population take turns in evolving. Firstly, the evolutionary process of the population of lags occurs: the individuals which will belong to the subpopulation are selected; following the CHC scheme [11], genetic operators are applied; the collaborator for every individual is chosen from the population of RBFNs; and the individuals are evaluated again and assigned the result as fitness. After that, the best individuals from the subpopulation will replace the worst individuals of the population. During the evolution, the population of lags checks that al least one gene of the chromosome must be set to one because necessarily the net needs one input to obtained the forecasted value.

In the second place, the population of RBFNs starts the evolutionary process. For the first generation, every net in the population has a number of neurons randomly chosen which may not exceed a maximum number previously fixed. As in population of lags, the individuals for the subpopulation are selected, the genetic operators are applied, every individual chooses the collaborator from the population of lags, and then, the individuals are evaluated and the result is assigned as fitness. Fitness function is defined by the inverse of the root mean squared error At the end of the co-evolutionary process, two models formed by a set of lags (from the first population) and a neural network (from the second population) are obtained. On the one hand, a model is composed of the best set of lags and its best collaborator, and on the other hand, the other model is composed of the best net found and its best collaborator. Then, the two models are trained again and the final model chosen is the one that obtains the best fitness. This final model obtains the future values of the time series used for the prediction, and then, forecasted data will be used to find next values.

The collaboration scheme used in L-Co-R is the best collaboration scheme [26]. Thus, every individual in any population chooses the best collaborator from the other population. Only at the beginning of the co-evolutionary process, the collaborator is selected randomly because the population has not been evaluated yet.

The method has a set of specific operators specially developed to work with individuals from every population. The operators used by L-Co-R are the followings:

- Population of RBFNs: tournament selection, x_fix crossover, four operators to mutate randomly chosen (C_random, R_random, Adder, and Deleter) and replacement of the worst individuals by the best ones of the subpopulation.
- Population of lags: elitist selection, HUX crossover operator, replacement of the worst individuals, and diverge (the population is restarted when it is blocked).

4 Experimentation and Statistical Study

The main goal of the experiments is to study the behavior of the algorithm
L-Co-R using 4 different and growing horizons, and to compare the results with
other 6 methods found in the literature and for 3 different quality measures.

4.1 Experimental Methodology

The experimentation has been carried out using 20 data bases taken from the INE.[1]
The data represent observations from different activities and have different nature,
size, and characteristics. The data bases have been labeled as: Airline, WmFranc-
fort, WmLondon, WmMadrid, WmMilan, WmNewYork, WmTokyo, Deceases,
SpaMovSpec, Exchange, Gasoline, MortCanc, MortMade, Books, FreeHouPrize,
Prisoners, TurIn, TurOut, TUrban, and HouseFin.

To compare the effectiveness of L-Co-R it has used, on the one hand, 6 methods
found within the field of time series forecasting: Exponential smoothing method
(ETS), Croston, Theta, Random Walk (RW), Mean, and ARIMA [16], and on the
other hand, 4 different horizons in order to test the effectiveness when the horizon
rises: 1, 6, 12, and 24.

An open question when dealing with time series is the measure to be used in order
to calculate the accuracy of the obtained predictions. Mean Absolute Percentage Error
(MAPE) [2] was the first measure employed in the M-competition [21] and most
textbooks recommended it. Later, many other measures as Geometric Mean Relative
Absolute Error, Median Relative Absolute Error, Symmetric Median and Median
Absolute Percentage Error (MdAPE), and Symmetric Mean Absolute Percentage
Error, among others, were proposed [22]. However, a disadvantage was found in
these measures, they were not generally applicable and can be infinite, undefined or
can produce misleading results, as Hyndman and Koehler explained in their work
[15]. Thus, they proposed Mean Absolute Scaled Error (MASE) that is less sensitive
to outliers, less variable on small samples, and more easily interpreted.

In this work, the measures used are MAPE (i.e., $mean(|\ p_t\ |)$), MASE (defined
as $mean(|\ q_t\ |)$), and MdAPE (as $median(|\ p_t\ |)$), taking into account that Y_t is the
observation at time $t = 1, ..., n$; F_t is the forecast of Y_t; e_t is the forecast error (i.e.
$e_t = Y_t - F_t$); $p_t = 100e_t/Y_t$ is the percentage error, and q_t is determined as:

$$q_t = \frac{e_t}{\frac{1}{n-1}\sum_{i=2}^{n}|\ Y_i - Y_{i-1}\ |}$$

[1]National Statistics Institute (http://www.ine.es/).

Table 1 Results of the methods L-Co-R, ETS, Croston, Theta, RW, Mean, and ARIMA, with respect to horizon 1 and MAPE

Time series	L-Co-R	ETS	ARIMA	CROSTON	THETA	MEAN	RW
Airline	30.380*	274.770	53.636	72.606	141.452	49.965	137.986
WmFrancfort	16.423	17.393	12.136*	40.544	22.745	64.632	25.169
WmLondres	2.860*	5.383	5.212	27.682	10.136	51.852	13.397
WmMadrid	20.101	27.035	12.930*	44.285	25.505	64.326	27.034
WmMilan	30.529*	34.858	34.823	49.750	34.078	59.840	34.823
WmNuevayork	8.259	7.182*	7.536	30.297	14.669	60.812	18.073
WmTokio	4.764*	12.807	12.591	20.556	10.575	42.627	12.591
Deceases	5.981*	8.002	8.040	7.472	7.264	9.663	8.040
SpaMovSpec	53.788*	217.978	88.197	78.648	70.500	63.288	78.935
Exchange	43.044	46.025	45.254	31.121	39.138	24.217*	33.631
Gasoline	1.654*	7.986	9.359	9.587	6.701	18.460	7.974
MortCanc	1.137*	12.979	5.440	32.489	5.889	46.655	6.256
MortMade	3.931*	13.526	31.000	46.362	40.272	42.120	12.800
Books	13.787*	23.588	23.476	23.122	22.360	24.895	22.640
FreeHouPrize	3.424*	8.540	10.227	29.271	5.215	48.746	9.220
Prisoners	8.392	3.103*	3.150	14.220	6.888	35.839	9.474
TurIn	1.357*	7.074	6.377	11.234	7.084	30.424	7.110
TurOut	8.133*	13.261	9.634	12.159	15.238	34.781	13.226
TUrban	2.734*	11.957	9.291	9.067	8.949	16.884	10.116
HouseFin	16.452*	22.296	19.555	21.548	19.947	42.314	22.887

Due to its stochastic nature, the results yielded by L-Co-R have been calculated as the average errors over 30 executions with every time series. For each execution, the following parameters are used in the L-Co-R algorithm: lags population size = 50, lags population generations=5, lags chromosome size = 10 %, RBFNs population size = 50, RBFNs population generations=10, validation rate=0.25, maximum number of neurons of first generation=0.05, tournament size = 3, replacement rate=0.5, crossover rate=0.8, mutation rate=0.2, and total number of generations=20.

Tables 1, 2, 3, 4, 5, and 6, show the results of the L-Co-R and the utilized methods to compare (ETS, Croston, Theta, RW, Mean, and ARIMA), for measures MAPE, MASE, and MdAPE, for horizons 1, and 6, respectively. Due to space limitations, this paper only shows results of the horizons 1 and 6, the results of the rest horizons, 12 and 24, can be accessed at https://goo.gl/frHK7z.

Table 2 Results of the methods L-Co-R, ETS, Croston, Theta, RW, Mean, and ARIMA, with respect to horizon 1 and MASE

Time series	L-Co-R	ETS	ARIMA	CROSTON	THETA	MEAN	RW
Airline	1.913	12.707	1.441*	2.738	5.853	2.045	5.664
WmFrancfort	3.578*	3.608	7.988	7.984	4.673	12.341	5.159
WmLondres	1.648	1.603*	3.484	8.410	3.099	15.566	4.119
WmMadrid	4.442*	5.686	8.625	9.126	5.362	13.050	5.685
WmMilan	5.967*	6.684	19.327	9.263	6.534	10.986	6.678
WmNuevayork	2.667	1.837*	6.228	7.982	3.942	15.620	4.879
WmTokio	2.791	2.443	1.628*	3.935	2.129	8.364	2.402
Deceases	1.059	1.059	1.144	0.952*	0.955	1.274	1.064
SpaMovSpec	1.027	2.027	1.933	1.009	1.023	0.997*	1.010
Exchange	41.181	44.039	70.734	30.448	37.807	23.911*	32.825
Gasoline	1.198*	1.543	1.698	1.864	1.274	3.533	1.541
MortCanc	0.646	1.618	0.277*	4.098	0.725	5.917	0.796
MortMade	1.314	1.303*	1.712	4.500	3.869	4.068	1.315
Books	0.762	0.965	1.147	0.936	0.894	1.040	0.759*
FreeHouPrize	3.339*	5.642	6.805	19.468	3.487	32.371	6.183
Prisoners	14.482	5.485	4.031*	23.979	11.934	58.935	16.305
TurIn	1.903	1.902	1.950	3.151	1.824*	8.328	1.916
TurOut	2.005	2.000	2.241	2.088	2.239	5.826	1.996*
TUrban	0.886	0.978	0.897	0.772	0.744*	1.576	0.887
HouseFin	1.319	1.283	1.502	1.234	1.095*	2.426	1.322

As mentioned before, every result indicated in the tables represent the average of 30 executions for each time series. Best result per database is marked with character '*'. Considering every horizon tested:

- Horizon 1: the L-Co-R algorithm obtains the best results in most of the time series. With respect to MAPE, the L-Co-R algorithm obtains the best results in 15 of 20 time series used, as can be seen in Table 1. Regarding MASE, L-Co-R stands out yielding the best results for 5 time series as can be observed in Table 2. And concerning MdAPE, L-Co-R acquires better results than the other methods in 12 of 20 time series, as Table 3 shows.
- Horizon 6: the L-Co-R has better results than all the other methods using MAPE and MdAPE, as can be seen in Tables 4 and 6, and the best results in 15 o the 20 time series for MASE, as can be observed in Table 5.
- Horizon 12: the L-Co-R yields the best results in 19, 17, and 18 of the 19 time series (MortCanc has not enough values to use with this horizon) respecting MAPE, MASE, and MdAPE, respectively.

Table 3 Results of the methods L-Co-R, ETS, Croston, Theta, RW, Mean, and ARIMA, with respect to horizon 1 and MdAPE

Time series	L-Co-R	ETS	ARIMA	CROSTON	THETA	MEAN	RW
Airline	15.057*	233.934	15.212	54.657	119.754	31.012	118.090
WmFrancfort	14.610	14.603	11.026*	39.259	19.960	63.868	22.750
WmLondres	3.498*	5.430	5.099	30.550	10.474	53.761	15.722
WmMadrid	22.718	28.116	11.446*	45.817	26.787	65.307	28.116
WmMilan	30.476*	34.685	34.643	50.040	33.872	60.072	34.643
WmNuevayork	9.114	4.598*	5.712	35.253	16.505	63.598	23.137
WmTokio	5.517*	9.864	9.556	18.782	9.075	40.967	9.556
Deceases	4.267*	5.464	5.458	6.121	4.440	7.144	5.458
SpaMovSpec	17.669*	107.283	54.033	51.653	53.104	54.045	51.568
Exchange	44.368	46.597	45.961	34.121	38.832	27.517	36.521
Gasoline	1.792*	7.587	8.923	9.045	6.429	18.825	7.563
MortCanc	11.25	9.694	5.116	30.568	4.047*	44.528	5.339
MortMade	3.459*	12.111	28.374	45.704	41.989	41.482	15.629
Books	4.868*	18.111	18.093	17.230	16.566	20.509	11.567
FreeHouPrize	1.803*	5.222	6.572	29.683	5.201	49.044	9.748
Prisoners	6.766	1.512*	1.621	12.651	5.287	34.665	7.817
TurIn	2.945*	6.627	4.605	11.696	4.779	31.502	6.669
TurOut	5.289*	11.331	7.689	11.518	10.873	36.500	11.392
TUrban	5.290	8.262	6.374	6.822	4.922*	17.828	8.900
HouseFin	18.286	22.623	17.297*	21.279	18.845	43.533	23.684

- Horizon 24: the L-Co-R algorithm obtains better results than the other methods in 17, 16, and 16 of the 17 time series (MortCanc, MortMade, and FreeHouPrize have not enough values to use with this horizon) with regard to MAPE, MASE, and MdAPE, respectively.

Thus, the L-Co-R algorithm is able to achieve a more accurate forecast in the most time series for any of the horizons and quality measures considered.

4.2 Analysis of the Results

To analyze in more detail the results and check whether the observed differences are significant, two main steps are performed: firstly, identifying whether exist differences in general between the methods used in the comparison; and secondly,

Table 4 Results of the methods L-Co-R, ETS, Croston, Theta, RW, Mean, and ARIMA, with respect to horizon 6 and MAPE

Time series	L-Co-R	ETS	ARIMA	CROSTON	THETA	MEAN	RW
Airline	28.740*	277.892	48.025	63.178	128.199	44.240	123.817
WmFrancfort	0.531*	19.056	13.004	43.264	25.102	66.250	27.844
WmLondres	0.113*	5.281	5.074	29.310	10.699	52.935	14.427
WmMadrid	0.312*	29.678	13.565	46.994	27.928	66.061	29.678
WmMilan	1.203*	38.440	38.403	52.914	37.562	62.369	38.403
WmNuevayork	0.140*	7.553	7.961	32.490	16.318	62.045	20.251
WmTokio	0.232*	13.255	13.052	20.777	10.908	42.825	13.052
Deceases	0.508*	8.266	8.309	7.385	7.440	10.085	8.309
SpaMovSpec	24.791*	235.399	93.095	82.501	72.432	64.599	82.821
Exchange	0.320*	46.431	33.296	30.949	39.226	24.028	33.465
Gasoline	0.205*	7.985	9.439	9.709	6.656	18.833	7.972
MortCanc	0.135*	12.562	5.963	36.563	5.829	51.164	6.334
MortMade	0.008*	15.078	34.276	55.378	49.472	50.875	12.375
Books	5.831*	23.590	21.059	23.026	22.118	25.159	21.274
FreeHouPrize	1.863*	12.678	15.393	30.282	5.416	49.478	10.517
Prisoners	0.204*	3.357	3.423	15.034	7.516	36.448	10.333
TurIn	0.042*	7.110	6.758	11.858	7.076	30.956	7.170
TurOut	0.603*	39.240	10.230	12.386	14.984	35.319	12.836
TUrban	2.052*	11.764	8.811	8.591	8.408	17.084	9.832
HouseFin	6.729*	21.571	18.953	20.797	19.092	42.674	22.177

determining if the best method is significant better than the rest of the methods. To do this, first of all it has to be decided if is possible to use parametric o non-parametric statistical techniques. An adequate use of parametric statistical techniques reaching three necessary conditions: independency, normality and homoscedasticity [28].

Owing to the former conditions are not fulfilled, the Friedman and Iman-Davenport non-parametric tests have been used. Tables with results of these tests are available at https://goo.gl/frHK7z. They show, from left to right, the Friedman and Iman-Davenport values (χ^2 and F_F, respectively), the corresponding critical values for each distribution by using a level of significance $\alpha = 0.05$, and the *p-value* obtained for the measures utilized. Finally, the critical values of Friedman and Iman-Davenport are smaller than the statistic, it means that there are significant differences among the methods in all cases.

Table 5 Results of the methods L-Co-R, ETS, Croston, Theta, RW, Mean, and ARIMA, with respect to horizon 6 and MASE

Time series	L-Co-R	ETS	ARIMA	CROSTON	THETA	MEAN	RW
Airline	1.595	12.290	1.585*	2.278	5.133	1.772	4.921
WmFrancfort	1.247*	3.946	2.360	8.548	5.133	12.831	5.675
WmLondres	1.317*	1.637	1.570	9.141	3.369	16.408	4.553
WmMadrid	1.302*	6.827	3.025	10.680	6.427	14.922	6.827
WmMilan	1.181*	7.915	7.908	10.710	7.735	12.537	7.908
WmNuevayork	1.235*	1.884	1.968	8.317	4.231	15.633	5.265
WmTokio	1.531*	2.459	2.423	3.862	2.150	8.182	2.423
Deceases	0.956*	1.113	1.119	0.963	0.997	1.348	1.119
SpaMovSpec	0.958	2.114	1.037	0.983	0.966	0.939*	0.984
Exchange	1.147*	44.047	32.240	30.039	37.574	23.546	32.399
Gasoline	0.051*	1.567	1.860	1.913	1.286	3.647	1.565
MortCanc	0.918	1.077	0.527	3.202	0.483*	4.497	0.533
MortMade	1.077*	1.689	3.876	6.335	5.641	5.810	1.370
Books	1.020	0.979	0.838	0.948	0.900	V1.062	0.730*
FreeHouPrize	1.214*	8.940	10.874	21.782	3.917	35.550	7.606
Prisoners	0.484*	5.684	5.795	24.350	12.457	57.773	17.012
TurIn	1.047*	1.863	1.728	3.225	1.769	8.237	1.882
TurOut	0.966*	5.986	1.556	2.131	2.200	5.912	1.943
TUrban	0.951	1.028	0.806	0.788	0.751*	1.705	0.928
HouseFin	1.026*	1.328	1.035	1.275	1.121	2.565	1.369

In addition, Friedman provides a ranking of the algorithms, so that the method with a lowest result is taken as the control algorithm. For this reason, and according to Tables 7, 8, 9, and 10, the L-Co-R algorithm results to be the control algorithm for all horizons considered and the three quality measures used.

In order to check if the control algorithm has statistical differences regarding the other methods used, the Holm procedure [14] is used. Tables 11, 12, 13, and 14 presents the results of the Holm's procedure since shows the adjusted p values from each comparison between the algorithm control and the rest of the methods for MAPE, MASE, and MdAPE, and for horizons 1, 6, 12, and 24 considering a level of significance of $alpha = 0.05$.

As can be seen in Tables 11, 12, 13, and 14, there are significant differences among L-Co-R and all the rest of the methods in the most of the cases. Analyzing more specifically for every horizon:

Table 6 Results of the methods L-Co-R, ETS, Croston, Theta, RW, Mean, and ARIMA, with respect to horizon 6 and MdAPE

Time series	L-Co-R	ETS	ARIMA	CROSTON	THETA	MEAN	RW
Airline	10.574*	223.150	13.222	31.476	89.154	26.266	85.401
WmFrancfort	2.332*	18.331	12.116	41.042	22.880	64.928	25.017
WmLondres	0.214*	5.394	5.078	30.658	11.247	53.833	15.853
WmMadrid	0.544*	28.484	12.129	46.094	27.270	65.484	28.484
WmMilan	0.519*	35.445	35.406	50.623	34.852	60.538	35.406
WmNuevayork	0.681*	4.909	5.755	36.166	18.312	64.112	24.221
WmTokio	1.207*	10.732	10.701	19.139	9.307	41.390	10.701
Deceases	0.513*	5.187	5.288	5.913	4.161	7.295	5.288
SpaMovSpec	7.824*	168.443	56.282	53.019	51.034	53.070	52.883
Exchange	0.011*	47.169	35.914	33.658	39.600	27.009	36.075
Gasoline	0.000*	7.353	8.995	9.394	6.365	19.412	7.329
MortCanc	0.152*	8.130	6.145	31.959	1.844	46.068	2.630
MortMade	2.582*	13.471	35.704	56.770	51.644	52.227	12.953
Books	1.849*	18.596	14.479	18.871	16.656	20.948	11.838
FreeHouPrize	1.547*	9.491	12.482	31.042	6.549	50.029	11.493
Prisoners	0.178*	1.786	1.906	14.123	6.422	35.766	9.371
TurIn	0.561*	6.781	5.482	12.795	4.614	32.355	6.671
TurOut	0.232*	35.128	8.219	11.965	10.860	37.078	10.784
TUrban	1.707*	8.341	6.054	6.431	4.729	17.642	8.694
HouseFin	3.028*	21.422	17.257	20.053	18.059	43.246	22.495

Table 7 Friedman's test ranking

MAPE		MASE		MdAPE	
L-Co-R	1.55	L-Co-R	2.63	L-Co-R	1.90
Theta	3.30	Theta	2.85	ARIMA	2.98
ARIMA	3.32	RW	3.60	Theta	3.10
RW	4.28	ETS	3.62	RW	4.15
ETS	4.40	Croston	4.60	ETS	4.23
Croston	5.00	ARIMA	4.65	Croston	5.30
Mean	6.15	Mean	6.05	Mean	6.35

Control algorithms are located in first row

- Horizon 1: significant differences exist between L-Co-R and the rest of the method for MAPE. With respect to MASE, there exist significant differences between the L-Co-R algorithm and Mean, ARIMA, and Croston, although it is not appropriate to assure that with methods ETS, RW, and Theta. Regarding MdAPE, L-Co-R has significant differences with all methods except ARIMA, as can be seen Table 11.

Table 8 Friedman's test ranking

MAPE		MASE		MdAPE	
L-Co-R	1.00	L-Co-R	1.70	L-Co-R	1.00
ARIMA	3.33	ARIMA	3.13	Theta	3.25
Theta	3.45	Theta	3.30	ARIMA	3.33
RW	4.35	RW	4.10	RW	4.10
ETS	4.68	ETS	4.63	ETS	4.48
Croston	5.05	Croston	5.00	Croston	5.45
Mean	6.15	Mean	6.15	Mean	6.40

Control algorithms are located in first row

Table 9 Friedman's test ranking

MAPE		MASE		MdAPE	
L-Co-R	1.00	L-Co-R	1.26	L-Co-R	1.05
ARIMA	3.40	ARIMA	3.24	Theta	2.53
Theta	3.42	Theta	3.42	ARIMA	2.55
RW	4.37	RW	4.26	RW	4.11
ETS	4.61	ETS	4.61	ETS	4.39
Croston	5.11	Croston	5.05	Croston	5.10
Mean	6.11	Mean	6.16	Mean	6.26

Control algorithms are located in first row

Table 10 Friedman's test ranking

MAPE		MASE		MdAPE	
L-Co-R	1.00	L-Co-R	1.18	L-Co-R	1.18
ARIMA	3.26	ARIMA	3.02	ARIMA	2.91
Theta	3.59	Theta	3.65	Theta	3.59
RW	4.44	RW	4.41	RW	4.29
ETS	4.76	ETS	4.74	ETS	4.74
Croston	4.88	Croston	4.94	Croston	5.24
Mean	6.05	Mean	6.05	Mean	6.06

Control algorithms are located in first row

- Horizon 6: L-Co-R has significant differences with all methods used, for every measure considered, as Table 12 shows.
- Horizon 12: there are significant differences among the control algorithm, L-Co-R, and the rest of the methods in all cases, as can be observed in Table 13.
- Horizon 24: as with horizons 6 and 12, there are also significant differences between L-Co-R and other methods, as Table 14 shows.

Table 11 Adjusted p values of Holm's procedure between the control algorithm (L-Co-R) and the other methods for MAPE, MASE, and MdAPE with respect to horizon 1

MAPE		MASE		MdAPE	
Mean	1.654E-11	Mean	5.340E-07	Mean	7.311E-11
Croston	4.412E-07	ARIMA	3.034E-03	Croston	6.454E-07
ETS	3.020E-05	Croston	3.839E-03	ETS	6.654E-04
RW	6.635E-05	ETS	1.432E-01	RW	9.890E-04
ARIMA	9.367E-03	RW	1.535E-01	Theta	7.898E-02
Theta	1.041E-02	Theta	7.419E-01	ARIMA	1.156E-01

Values lower than $alpha = 0.05$ indicate significant differences between L-Co-R and the corresponding algorithm

Table 12 Adjusted p values of Holm's procedure between the control algorithm (L-Co-R) and the other methods for MAPE, MASE, and MdAPE with respect to horizon 6

MAPE		MASE		MdAPE	
Mean	4.742E-14	Mean	7.311E-11	Mean	2.684E-15
Croston	3.055E-09	Croston	1.361E-06	Croston	7.311E-11
ETS	7.463E-08	ETS	1.854E-05	ETS	3.640E-07
RW	9.395E-07	RW	4.427E-04	RW	5.681E-06
Theta	3.352E-04	Theta	1.917E-02	ARIMA	6.654E-04
ARIMA	6.654E-04	ARIMA	3.698E-02	Theta	9.889E-04

Values lower than $alpha = 0.05$ indicate significant differences between L-Co-R and the corresponding algorithm

Table 13 Adjusted p values of Holm's procedure between the control algorithm (L-Co-R) and the other methods for MAPE, MASE, and MdAPE with respect to horizon 12

MAPE		MASE		MdAPE	
Mean	3.238E-13	Mean	2.874E-12	Mean	1.051E-13
Croston	4.704E-09	Croston	6.417E-08	Croston	7.372E-09
ETS	2.690E-07	ETS	1.856E-06	ETS	1.856E-06
RW	1.540E-06	RW	1.866E-05	RW	1.328E-05
Theta	5.517E-04	Theta	2.078E-03	ARIMA	3.611E-04
ARIMA	6.337E-04	ARIMA	4.862E-03	Theta	4.165E-04

Values lower than $alpha = 0.05$ indicate significant differences between L-Co-R and the corresponding algorithm

In conclusion, it is possible to confirm that the L-Co-R method is able to achieve a better forecast in majority of cases even when the horizon grows, comparing with the other 6 methods utilized and concerning to 3 different quality measures.

Table 14 Adjusted p values of Holm's procedure between the control algorithm (L-Co-R) and the other methods for MAPE, MASE, and MdAPE with respect to horizon 24

MAPE		MASE		MdAPE	
Mean	8.646E-12	Mean	4.421E-11	Mean	4.421E-11
Croston	1.609E-07	Croston	3.357E-07	Croston	4.306E-08
ETS	3.757E-07	ETS	1.563E-06	ETS	1.563E-06
RW	3.414E-06	RW	1.263E-05	RW	2.581E-05
Theta	4.775E-04	Theta	8.551E-04	Theta	1.134E-03
ARIMA	2.240E-03	ARIMA	1.240E-02	ARIMA	1.918E-02

Values lower than $alpha = 0.05$ indicate significant differences between L-Co-R and the corresponding algorithm

Acknowledgments This work has been supported by the regional projects TIC-3928 and -TIC-03903 (Feder Funds), the Spanish project TIN 2012-33856 (Feder Founds), TIN 2011-28627-C04-02 (Feder Funds).

References

1. Araújo, R.: A quantum-inspired evolutionary hybrid intelligent apporach fo stock market prediction. Int. J. Intell. Comput. Cybern. **3**(10), 24–54 (2010)
2. Bowerman, B., O'Connell, R., Koehler, A.: Forecasting: Methods and Applications. Thomson Brooks/Cole, Belmont, CA (2004)
3. Box, G., Jenkins, G.: Time series analysis: forecasting and control. Holden Day, San Francisco (1976)
4. Brockwell, P., Hyndman, R.: On continuous-time threshold autoregression. Int. J. Forecast. **8**(2), 157–173 (1992)
5. Broomhead, D., Lowe, D.: Multivariable functional interpolation and adaptive networks. Complex Syst. **2**, 321–355 (1988)
6. Carse, B., Fogarty, T.: Fast evolutionary learning of minimal radial basis function neural networks using a genetic algorithm. In: Proceedings of Evolutionary Computing. LNCS, vol. 1143, pp. 1–22 Springer, Heidelberg (1996)
7. Castillo, P., Arenas, M., Merelo, J., and Romero, G.: Cooperative co-evolution of multilayer perceptrons. In: Mira, J., lvarez, J.R. (eds.) Computational Methods in Neural Modeling, LNCS, vol. 2686, pp. 358–365. Springer, Heidelberg (2003)
8. Chan, K., Tong, H.: On estimating thresholds in autoregressive models. J. Time Ser. Anal. **7**(3), 179–190 (1986)
9. Clements, M., Franses, P., Swanson, N.: Forecasting economic and financial time-series with non-linear models. Int. J. Forecast. **20**(2), 169–183 (2004)
10. Du, H., Zhang, N.: Time series prediction using evolving radial basis function networks with new encoding scheme. Neurocomputing **71**(7–9), 1388–1400 (2008)
11. Eshelman, L.: The chc adptive search algorithm: how to have safe search when engaging in nontraditional genetic recombination. In: Proceedings of 1st Workshop on Foundations of Genetic Algorithms, pp. 265–283 (1991)
12. García-Pedrajas, N., Hervas-Martínez, C., Ortiz-Boyer, D.: Cooperative coevolution of artificial neural network ensembles for pattern classification. IEEE Trans. Evol. Comput. **9**(3), 271–302 (2005)

13. Harpham, C., Dawson, C.: The effect of different basis functions on a radial basis function network for time series prediction: a comparative study. Neurocomputing **69**(16–18), 2161–2170 (2006)
14. Holm, S.: A simple sequentially rejective multiple test procedure. Scand. J. Stat. **6**(2), 65–70 (1979)
15. Hyndman, R., Koehler, A.: Another look at measures of forecast accuracy. Int. J. Forecast. **22**(4), 679–688 (2006)
16. Hyndman, R.J., Khandakar, Y.: Automatic time series forecasting: the forecast package for r. J. Stat. Softw. **27**(3), 1–22 (2008)
17. Jain, A., Kumar, A.: Hybrid neural network models for hydrologic time series forecasting. Appl. Soft Comput. **7**(2), 585–592 (2007)
18. Li, M., Tian, J., Chen, F.: Improving multiclass pattern recognition with a co-evolutionary rbfnn. Pattern Recogn. Lett. **29**(4), 392–406 (2008)
19. Lukoseviciute, K., Ragulskis, M.: Evolutionary algorithms for the selection of time lags for time series forecasting by fuzzy inference systems. Neurocomputing **73**(10–12), 2077–2088 (2010)
20. Ma, X., Wu, H.: Power system short-term load forecasting based on cooperative co-evolutionary immune network model. In: Proceedings of 2nd International Conference on Education Technology and Computer, pp. 582–585 (2010)
21. Makridakis, S., Andersen, A., Carbone, R., Fildes, R., Hibon, M., Lewandowski, R., Newton, J., Parzen, E., Winkler, R.: The accuracy of extrapolation (time series) methods: Results of a forecasting competition. J. Forecast. **1**(2), 111–153 (1982)
22. Makridakis, S., Hibon, M.: The m3-competition: results, conclusions and implications. Int. J. Forecast. **16**(4), 451–476 (2000)
23. Maus, A., Sprott, J.C.: Neural network method for determining embedding dimension of a time series. Commun. Nonlinear Sci. Numer. Simul. **16**(8), 3294–3302 (2011)
24. Parras-Gutierrez, E., Garcia-Arenas, M., Rivas, V., del Jesus, M.: Coevolution of lags and rbfns for time series forecasting: L-co-r algorithm. Soft Comput. **16**(6), 919–942 (2012)
25. Pea, D.: Análisis de Series Temporales. Alianza Editorial (2005)
26. Potter, M., De Jong, K.: A cooperative coevolutionary approach to function optimization. In: Proceedings of Parallel Problem Solving from Nature, LNCS, vol. 866, pp. 249–257. Springer, Heidelberg (1994)
27. Rivas, V., Merelo, J., Castillo, P., Arenas, M., Castellano, J.: Evolving rbf neural networks for time-series forecasting with evrbf. Inf. Sci. **165**(3–4), 207–220 (2004)
28. Sheskin, D.: Handbook of parametric and nonparametric statistical procedures. Chapman & Hall/CRC, Boca Raton (2004)
29. Snyder, R.: Recursive estimation of dynamic linear models. J. Roy. Stat. Soc. Ser. B (Methodological) **47**(2), 272–276 (1985)
30. Takens, F.: Dynamical systems and turbulence, Lecture Notes In Mathematics, vol. 898, Chapter Detecting Strange Attractor in Turbulence, pp. 366–381. Springer, New York, NY (1980)
31. Tong, H.: On a threshold model. Pattern Recogn. signal process. NATO ASI Ser. E: Appl. Sc. **29**, 575–586 (1978)
32. Tong, H.: Threshold models in non-linear time series analysis. Springer, Berlin (1983)
33. Whitehead, B., Choate, T.: Cooperative-competitive genetic evolution of radial basis function centers and widths for time series prediction. IEEE Trans. Neural Netw. **7**(4), 869–880 (1996)
34. Winters, P.: Forecasting sales by exponentially weighted moving averages. Manage.Sci. **6**(3), 324–342 (1960)

Tree Automata Mining

Michal R. Przybylek

Abstract This paper [The article is an essentially revised version of conference paper (Przybylek (2013) International Conference on Evolutionary Computation Theory and Applications)] describes a new approach to mine business processes. We define bidirectional tree languages together with their finite models and show how they represent business processes. We offer an algebraic explanation for the phenomenon of an evolutionary metaheuristic "skeletal algorithms", and show how this explanation gives rise to algorithms for recognition of bidirectional tree automata. We use the algorithms in process mining and in discovering mathematical theories.

Keywords Evolutionary algorithms · Process mining · Language recognition · Minimum description length

1 Introduction

Nowadays, there is no longer any question that the quality of a company's business processes has a crucial impact on its sales and profits. The degree of innovation built into these business processes, as well as their flexibility and efficiency, are critically important for the success of the company. The importance of business processes is further revealed when their are considered as the link between business and IT; business applications only become business solutions when the processes are supported efficiently. The essential task of any standard business software is and always will be to provide efficient support of internal and external company processes.—Torsten Scholz

In order to survive in today's global economy more and more enterprises have to redesign their business processes. The competitive market creates the demand for high quality services at lower costs and with shorter cycle times. In such an envi-

This work has been partially supported by Polish National Science Center, project DEC-2011/01/N/ST6/02752.

M.R. Przybylek (✉)
Faculty of Mathematics, Informatics and Mechanics,
University of Warsaw, Warsaw, Poland
e-mail: mrp@mimuw.edu.pl

© Springer International Publishing Switzerland 2016
K. Madani et al. (eds.), *Computational Intelligence*,
Studies in Computational Intelligence 613,
DOI 10.1007/978-3-319-23392-5_8

ronment business processes must be identified, described, understood and analysed
to find inefficiencies which cause financial losses. One way to achieve this is by
modelling. Business modelling is the first step towards defining a software system. It
enables the company to look afresh at how to improve organization and to discover
the processes that can be solved automatically by software that will support the business. However, as it often happens, such a developed model corresponds more to
how people think of the processes and how they wish the processes would look like,
then to the real processes as they take place.

Another way is by extracting information from a set of events gathered during
executions of a process. Process mining [3, 10–20] is a growing technology in the
context of business process analysis. It aims at extracting this information and using
it to build a model. Process mining is also useful to check if the "a priori model"
reflects the actual situation of executions of the processes. In either case, the extracted
knowledge about business processes may be used to reorganize the processes to
reduce they time and cost for the enterprise.

Figure 1 shows a typical event-log gathered during executions of a service process
in an online store. We assume that with every such an event-log there are associated:

Fig. 1 An event log

Case	Actions	Actor	Timestamp
33	Incoming order	Jan Kowalski	09:33:01 06.20.2013
33	Check if products are available	Jan Kowalski	09:37:11 06.20.2013
33	Inform the client	Jan Kowalski	09:57:02 06.20.2013
127	Incoming order	John Doe	09:57:22 06.20.2013
127	Check if products are available	John Doe	09:57:59 06.20.2013
33	Response	Jan Kowalski	10:00:01 06.20.2013
127	Prepare products, Prepare invoice	John Doe	10:00:49 06.20.2013
127	Deliver	John Doe	10:41:08 06.20.2013
33	Order from a contrahent	Jan Kowalski	10:41:55 06.20.2013
33	Available	Jan Kowalski	10:42:00 06.20.2013
127	End	John Doe	10:44:37 06.20.2013
33	Prepare products, Prepare invoice	John Smith	08:07:19 06.24.2013
33	Deliver	John Smith	08:34:02 06.24.2013
33	End	John Smith	08:35:33 06.24.2013
791	Incoming order	Adam Smith	09:37:01 07.20.2013
791	Check if products are available	Adam Smith	09:37:15 07.20.2013
791	Inform the client	Adam Smith	09:51:00 07.20.2013
…	…	…	…

- an identifier referring to the execution (the case) of the process that generated the event
- a unique timestamp indicating the particular moment when the event occurred
- an observable set of actions of the parallel events.

Figure 2 shows a model recognized from this sample.

The aim of this paper is twofold: to extend and revise methods for exploration of business processes developed in [7, 8] to improve their effectiveness in a business environment; and to provide an algebraic explanation of the phenomenon of skeletal algorithms. We show some sample applications of our algorithms: in mining business processes and in rediscovering a mathematical theory.

Fig. 2 Discovered model

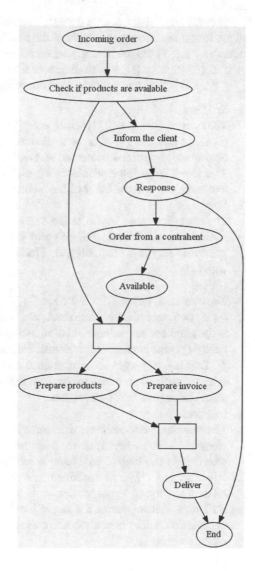

2 Skeletal Algorithms

Skeletal algorithms [7, 8] are a new branch of evolutionary metaheuristics [1, 4–6, 9] focused on data and process mining. The basic idea behind the skeletal algorithm is to express a problem in terms of congruences on a structure, build an initial set of congruences, and improve it by taking limited unions/intersections, until a suitable condition is reached. Skeletal algorithms naturally arise in the context of data/process mining, where the skeleton is the "free" structure on initial data and a congruence corresponds to similarities in the data. In such a context, skeletal algorithms come equipped with fitness functions measuring the complexity of a model.

Skeletal algorithms, search for a solution of a problem in the set of quotients of a given structure called the skeleton of the problem. More formally, let S be a set, and denote by $Eq(S)$ the set of equivalence relations on S. If $i \in S$ is any element, and $A \in Eq(S)$ then by $[i]_A$ we shall denote the abstraction class of i in A—i.e. the set $\{j \in S: jAi\}$. We shall consider the following skeletal operations on $Eq(S)$:

1. Splitting
   The operation $split\colon \{0, 1\}^S \times S \times Eq(S) \rightarrow Eq(S)$ takes a predicate $P\colon S \rightarrow \{0, 1\}$, an element $i \in S$, an equivalence relation $A \in Eq(S)$ and gives the largest equivalence relation R contained in A and satisfying: $\forall_{j \in [i]_A} iRj \Rightarrow P(i) = P(j)$. That is—it splits the equivalence class $[i]_A$ on two classes: one for the elements that satisfy P and the other of the elements that do not.

2. Summing
   The operation $sum\colon S \times S \times Eq(S) \rightarrow Eq(S)$ takes two elements $i, j \in S$, an equivalence relation $A \in Eq(S)$ and gives the smallest equivalence relation R satisfying iRj and containing A. That is—it merges the equivalence class $[i]_A$ with $[j]_A$.

3. Union
   The operation $union\colon S \times Eq(S) \times Eq(S) \rightarrow Eq(S) \times Eq(S)$ takes one element $i \in S$, two equivalence relations $A, B \in Eq(S)$ and gives a pair $\langle R, Q \rangle$, where R is the smallest equivalence relation satisfying $\forall_{j \in [i]_B} iRj$ and containing A, and dually Q is the smallest equivalence relation satisfying $\forall_{j \in [i]_A} iQj$ and containing B. That is—it merges the equivalence class corresponding to an element in one relation, with all elements taken from the equivalence class corresponding to the same element in the other relation.

4. Intersection
   The operation $intersection\colon S \times Eq(S) \times Eq(S) \rightarrow Eq(S) \times Eq(S)$ takes one element $i \in S$, two equivalence relations $A, B \in Eq(S)$ and gives a pair $\langle R, Q \rangle$, where R is the largest equivalence relation satisfying $\forall_{x,y \in [i]_A} xRy \Rightarrow x, y \in [i]_B \vee x, y \notin [i]_B$ and contained in A, and dually Q is the largest equivalence relation satisfying $\forall_{x,y \in [i]_B} xQy \Rightarrow x, y \in [i]_A \vee x, y \notin [i]_A$ and contained in B. That is—it intersects the equivalence class corresponding to an element in one relation, with the equivalence class corresponding to the same element in the other relation.

Furthermore, we assume that there is also a fitness function. There are many things that can be implemented differently in various problems.

2.1 Construction of the Skeleton

As pointed out earlier, the skeleton of a problem should correspond to the "free model" build upon sample data. Observe, that it is really easy to plug in the skeleton some priori knowledge about the solution—we have to construct a congruence relation induced by the priori knowledge and divide by it the "free unrestricted model". Also, this suggests the following optimization strategy—if the skeleton of a problem is too big to efficiently apply the skeletal algorithm, we may divide the skeleton on a family of smaller skeletons, apply to each of them the skeletal algorithm to find quotients of the model, glue back the quotients and apply again the skeletal algorithm to the glued skeleton.

2.2 Construction of the Initial Population

Observe that any equivalence relation on a finite set S may be constructed by successively applying *sum* operations to the identity relation, and given any equivalence relation on S, we may reach the identity relation by successively applying *split* operations. Therefore, every equivalence relation is constructible from *any* equivalence relation with *sum* and *split* operations. If no priori knowledge is available, we may build the initial population by successively applying to the identity relation both *sum* and *split* operations.

2.3 Selection of Operations

For all operations we have to choose one or more elements from the skeleton S, and additionally for a split operation—a splitting predicate $P : S \rightarrow \{0, 1\}$. In most cases these choices have to reflect the structure of the skeleton—i.e. if our models have an algebraic or coalgebraic structure, then to obtain a quotient model, we have to divide the skeleton by an equivalence relation *preserving* this structure, that is, by a congruence. The easiest way to obtain a congruence is to choose operations that map congruences to congruences. Another approach is to allow operations that move out congruences from they class, but then "improve them" to congruences, or just punish them in the intermediate step by the fitness function.

2.4 Choosing Appropriate Fitness Function

Data and process mining problems frequently come equipped with a natural fitness function measuring the total complexity of data given a particular model. One of the crucial conditions that such a function has to satisfy is the ability to easily adjust its value on a model obtained by applying skeletal operations.

2.5 Creation of Next Population

There is a room for various approaches. We have experimented most successful with the following strategy—append k-best congruences from the previous population to the result of operations applied in the former step of the algorithm.

3 Tree Languages and Tree Automata

Let us first recall the definition of an ordinary tree language and automaton [2]. A ranked alphabet is a function *arity*: $\Sigma \rightarrow \mathcal{N}$ from a finite set of symbols Σ to the set of natural numbers $\mathcal{N}$ called arities of the symbols. We shall write σ/k to indicate that the arity of a symbol $\sigma \in \Sigma$ is $k \in \mathcal{N}$, that is $arity(\sigma) = k$. One may think of a ranked alphabet as of an algebraic signature—then a word over a ranked alphabet is a ground term over corresponding signature.

Example 1 (*Propositional Logic*) A ranked alphabet of the propositional logic consists of symbols:

$$\{\bot/0, \top/0, \vee/2, \wedge/2, \neg/1, \Rightarrow/2\}$$

Every propositional sentence like "$\top \vee \neg\bot \Rightarrow \bot$" corresponds to a word over the above alphabet—in this case to: "$\Rightarrow (\vee(\top, \neg(\bot)), \bot)$", or writing in a tree-like fashion:

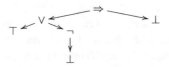

Following [2] we define a finite top-down tree automaton over *arity*: $\Sigma \rightarrow \mathcal{N}$ as a tuple $A = \langle Q, q_s, \Delta \rangle$, where Q is a set of states, $q_s \in Q$ is the initial state, and Δ is the set of rewrite rules, or transitions, of the type:

$$q_0(f(x_1, \ldots c, x_n)) \rightarrow f(q_1(x_1), \ldots c, q_n(x_n))$$

where $f/n \in \Sigma$ and $q_i \in Q$ for $i = 0, \ldots, n$. The rewrite rules are defined on the ranked alphabet $arity \colon \Sigma \to \mathcal{N}$ extended with $q/1$ for $q \in Q$. A word w is recognised by automaton A if $q_s(w) \xrightarrow{\Delta^*} w$, that is, if w may be obtained from $q_s(w)$ by successively applying finitely many rules from Δ.

We shall modify the definition of a tree automaton in two directions. First, it will be more convenient to associate symbols with states of an automaton, rather then with transitions. Second, we extend the definition of a ranked alphabet to allow terms return multiple results; moreover, to fit better the concept of business processes, we identify terms that are equal up to a permutation of their arguments and results.

Definition 1 (*Ranked Alphabet*) A ranked alphabet is a function *biarity* $\colon \Sigma \to \mathcal{N} \times \mathcal{N}^+$. If the ranking function is known from the context, we shall write $\sigma/i/j \in \Sigma$ for a symbol $\sigma \in \Sigma$ having input arity i and output arity j; that is, if $biarity(\sigma) = \langle i, j \rangle$.

A definition of a term is more subtle, so let us first consider some special cases. By a multiset we shall understand a function $\overline{(-)}$ from a set X to the set of positive natural numbers $\mathcal{N}^+$—it assigns to an element $x \in X$ its number of occurrences $\overline{x}$ in the multiset. If X is finite, then we shall write $\{\!\{x_1, \ldots c, x_1, x_2, \ldots c, x_2, \ldots c x_k, \ldots c\}\!\}$, where an element $x_k \in X$ occurs n-times when $\overline{x} = n$, and call the multiset finite. For multisets we use the usual set-theoretic operations $\cup, \cap, /$ defined pointwise—with possible extension or truncation of the domains.

A simple language over a ranked alphabet Σ is the smallest set of pairs, called simple terms, containing $\langle \sigma/0/j, \emptyset \rangle$ for each nullary symbol $\sigma/0/j \in \Sigma$ and closed under the following operation: if $\sigma/i/j \in \Sigma$ and $t_1 = \langle x_1/i_1/j_1, A_1 \rangle, \ldots c, t_k = \langle x_k/i_k/j_k, A_k \rangle$ are simple terms such that $\sum_{s=1}^{k} j_s = i$, then $\langle \sigma/i/j, \{\!\{t_k \colon 1 \le s \le k\}\!\} \rangle$ is a simple term. For convenience we write $\sigma\{\!\{t_1, \ldots c, t_k\}\!\}$ for $\langle \sigma/i/j, \{\!\{t_k \colon 1 \le s \le k\}\!\} \rangle$ and call t_s a subterm of $\sigma\{\!\{t_1, \ldots c, t_k\}\!\}$.

Example 2 (*Ordinary Language*) A word over an ordinary alphabet Σ may be represented as a simple term over the ranked alphabet $biarity(\sigma) = (1, 1)$ for $\sigma \in \Sigma$ and $biarity(\epsilon) = (0, 1)$.

Example 3 (*Ordinary Tree Language*) A word over an ordinary ranked alphabet may be represented as a simple term over the ranked alphabet extended with unary symbols $n/1/1$ for natural numbers $n \in \mathcal{N}$ indicating a position of an argument. A tree-representation of sentence "$\top \vee \neg\bot \Rightarrow \bot$" (compare Example 1) have the form shown on Fig. 3a. Notice, that in all semantics of (any) propositional calculus $A \vee B \equiv B \vee A$, therefore we may use this knowledge on the syntax level and represent sentence "$\top \vee \neg\bot \Rightarrow \bot$" in a more compact form—carrying some extra information about possible models (Fig. 3b).

We extend the notion of a simple term to allow a single term to be a subterm of more than one term. Such extension would be trivial for ordinary terms, but here, thanks to the ability of returning more than one value, it gives us an extra power which is crucial for representing business processes.

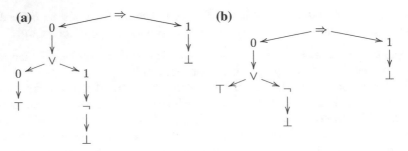

Fig. 3 Tree representations of sentence "$\top \vee \neg\bot \Rightarrow \bot$". **a** Ordinary representation. **b** Representation respecting symmetry of $\vee$

Definition 2 (*Term*) Let Σ be a ranked alphabet. A term over Σ is a finite acyclic coalgebra $\langle S, s_0 \in S, subterm\colon S \to \mathcal{N}^{+S}, name\colon S \to \Sigma \rangle$ satisfying the following compatibility conditions:

$$\forall_{x \in S} \sum_{y \in S} subterm(x)(y) = name(x)_1$$

$$\forall_{y \in S \setminus \{s_0\}} \sum_{x \in S} subterm(x)(y) = name(y)_2$$

where subscripts $_1$ and $_2$ indicates projections on first (i.e. input arity) and second (i.e. output arity) component respectively Two terms $\langle S, s_0, subterm, name \rangle$ and $\langle S', s_0', subterm', name' \rangle$ are equivalent if there exists an isomorphism of the coalgebras, that is, if there exists a bijection $\sigma\colon S \to S'$ such that $\sigma(s_0) = s_0', \mathcal{N}^{+\sigma} \circ subterm \circ \sigma = subterm'$ and $name \circ \sigma = name'$.

We shall not distinguish between equivalent terms.

Example 4 (*Simple Term*) Consider a simple term t over a ranked alphabet Σ. It corresponds to the term $\langle S, s_0 \in S, subterm\colon S \to \mathcal{N}^{+S}, name\colon S \to \Sigma \rangle$, where S is the smallest multiset containing t and closed under subterms, $s_0 = t$, $name(\sigma \{\!\{ t_1, \ldots c, t_k \}\!\}) = \sigma$ and $subterm(\sigma \{\!\{ t_1, \ldots c, t_k \}\!\}) = \{\!\{ t_1, \ldots c, t_k \}\!\}$.

In line with the above example, we shall generally represent a term as a sequence of equations (add multiple variables, please):

$$\sigma_0 \{\!\{ t_{0,1}, \ldots c, t_{0,k_0} \}\!\} \ \textit{in free variables } x_1, \ldots c, x_n$$
$$x_1 = \sigma_1 \{\!\{ t_{1,1}, \ldots c, t_{1,k_0} \}\!\} \ \textit{in free variables } x_2, \ldots c, x_n$$
$$\ldots$$
$$x_n = \sigma_n \{\!\{ t_{n,1}, \ldots c, t_{n,k_n} \}\!\} \ \textit{without free variables}$$

where $t_{i,j}$ are simple terms and x_i are multisets of variables.

Corollary 1 *Terms are tantamount to finite sets of equations of the form* $x = \sigma\{\!\{t_1, \ldots c, t_k\}\!\}$ *over simple terms without cyclic dependencies of free variables.*

Example 5 (*Terms from a Business Process*) Consider a business process:

which starts in the "start" state and ends in the "end" state. The semantics of the process is that one have to preform simultaneously task B and at least one task A and then either finish or repeat the whole process. Some terms t_1, t_2, t_3 generated by this process are:

$$t_1 = \text{start } \{\!\{\text{fork } \{\!\{A \{\!\{x\}\!\}, B \{\!\{x\}\!\}\}\!\}\}\!\}$$
$$x = \text{join } \{\!\{\text{end}\}\!\}$$
$$t_2 = \text{start } \{\!\{\text{fork } \{\!\{A \{\!\{A \{\!\{x\}\!\}\}\!\}, B\{\!\{x\}\!\}\}\!\}\}\!\}$$
$$x = \text{join } \{\!\{\text{end}\}\!\}$$
$$t_3 = \text{start } \{\!\{\text{fork } \{\!\{A \{\!\{A \{\!\{A \{\!\{x\}\!\}\}\!\}\}\!\}, B\{\!\{x\}\!\}\}\!\}\}\!\}$$
$$x = \text{join } \{\!\{\text{fork } \{\!\{A \{\!\{A \{\!\{x\}\!\}\}\!\}, B\{\!\{y\}\!\}\}\!\}\}\!\}$$
$$y = \text{join } \{\!\{\text{end}\}\!\}$$

Generally, every term t generated by this process has to be of the following form:

$$t = \text{start } \{\!\{\text{fork } \{\!\{A^{k_1} \{\!\{x_1\}\!\}, B\{\!\{x_1\}\!\}\}\!\}\}\!\}$$
$$x_1 = \text{join } \{\!\{\text{fork } \{\!\{A^{k_2} \{\!\{x_2\}\!\}, B\{\!\{x_2\}\!\}\}\!\}\}\!\}$$
$$\ldots$$
$$x_{n-1} = \text{join } \{\!\{\text{fork } \{\!\{A^{k_n} \{\!\{x_m\}\!\}, B\{\!\{x_m\}\!\}\}\!\}\}\!\}$$
$$x_n = \text{join } \{\!\{\text{end}\}\!\}$$

The whole business process cannot be represented as a single term. One could write the following set of equations:

$$t = \text{start } \{\!\{x\}\!\}$$
$$x = \text{fork } \{\!\{A \{\!\{y\}\!\}, B\{\!\{z\}\!\}\}\!\}$$
$$y = A \{\!\{y\}\!\} \vee y = z$$
$$z = \text{join } \{\!\{x\}\!\} \vee z = \text{join } \{\!\{\text{end}\}\!\}$$

However, there is no term corresponding to this set—there are cyclic dependencies between variables (for example y depends on y, also x depends on z, z depends on x), and there are disjunctions in the set of equations.

Definition 3 (*Tree Automaton*) A tree automaton over a ranked alphabet Σ is a tuple $A = \langle Q, q_0, \Delta, name \rangle$, where:

- Q is the set of states of the automaton
- $q_0 \in Q$ is the initial state of the automaton
- *name* is a function from set of states Q to $\Sigma \sqcup \{\epsilon/0/1\}$
- Δ is a set of rewrite rules (transitions) of the form:

$$\{x_0, \ldots c, x_k\} \to^{\delta} \{x'_0, \ldots c, x'_l\}$$

with:

$$\sum_{i=0}^{k} name(x_i)_1 = \sum_{i=0}^{l} name(x'_i)_0$$

where $x_0, \ldots c, x_k, x'_0, \ldots c, x'_l \in Q$.

Notice that in the above definition there is a single initial state, but there are no final states—an automaton finishes its run if it is in neither of the states.

Example 6 (*Business Process as Tree Automaton*) We shall use the following graphical representation of a tree automaton: every state is denoted by a circle with the letter associated to the state inside the circle, every rule $\{x_0, \ldots c, x_k\} \overset{\delta}{\longrightarrow} \{x'_0, \ldots c, x'_l\}$ is denoted by a rectangle (optionally with letter δ inside); moreover this rectangle is connected by ingoing arrows from circles denoting states $\{x_0, \ldots c, x_k\}$ and outgoing arrows to circles denoting states $\{x'_0, \ldots c, x'_l\}$:

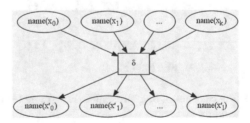

For convenience we shell sometimes omit the intermediating box of a singleton rule $\{x\} \to \{x'\}$ and draw only a single arrow from the node representing x to the node representing x'. The business process from Example 5 defines over a signature $\Sigma = \{start/1/0, fork/2/1, A/1/1, B/1/1, join/1/2, end/0/1\}$ an automaton $\langle start, \Sigma, \Delta, id \rangle$ with rules Δ:

$$\{\, \text{start} \,\} \xrightarrow{\delta_1} \{\, \text{fork} \,\}$$

$$\{\, \text{fork} \,\} \xrightarrow{\delta_2} \{A, B\}$$

$$\{A\} \xrightarrow{\delta_3} \{A\}$$

$$\{A, B\} \xrightarrow{\delta_4} \{\, \text{join} \,\}$$

$$\{\, \text{join} \,\} \xrightarrow{\delta_5} \{\, \text{fork} \,\}$$

$$\{\, \text{join} \,\} \xrightarrow{\delta_6} \{\, \text{end} \,\}$$

$$\{\, \text{end} \,\} \xrightarrow{\delta_7} \{\}$$

which may be represented as:

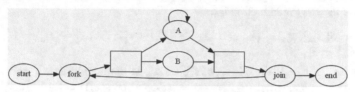

Example 7 (*Term as a Skeletal Tree Automaton*) The automaton corresponding to a term t is constructed in two steps. First we define the following automaton. For every $s \in S$ with $name(s) = \sigma/i/j$ define a multiset:

$$E_s = \{\, \epsilon_{s,1}, \epsilon_{s,2}, \ldots c, \epsilon_{s,j} \,\}$$

and a rule:

$$\{s\} \rightarrow E_s$$

and for every $p \in S$ with $k = subterm(p)(s)$ choose any k-element subset X_p of E_p and put a rule:

$$\bigcup_{p \in S} X_p \rightarrow \{s\}$$

Then, for convenience, we simplify the automaton by cutting at ϵ-states. That is: every pair of rules

$$X \rightarrow \{Y, E\}$$

$$\{E\} \rightarrow Z$$

where E consists only of ϵ-states, is replaced by a single rule:

$$X \rightarrow \{Y, Z\}$$

The next picture illustrates the skeletal automaton constructed from term t_2 from Example 5.

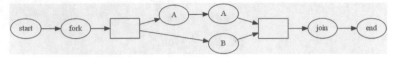

Given a finite multiset X, a rule $\{x_0, \ldots c, x_k\} \xrightarrow{\delta} \{x_0', \ldots c, x_l'\}$ is applicable to X if $\{x_0, \ldots c, x_k\}$ is a multisubset of X. In such a case we shall write $\delta[X]$ for the multiset $(X \backslash \{x_0, \ldots c, x_k\}) \cup \{x_0', \ldots c, x_l'\}$. We say that a term $t = \langle S, s_0, subterm_t, name_t \rangle$ is recognised by an automaton $A = \langle Q, q_0, \Delta_A, name_A \rangle$ if there is a finite sequence $\langle\{q_0\}, \{q_0 \mapsto s_0\}\rangle = T_0, T_1, \ldots c, T_n = \langle\{\}, \{\}\rangle$ with $name(q_0) = name(s_0)$ satisfying for all $0 < m < n$ the induction laws:

- $T_{m+1} = \langle \delta[X_m], \pi_m[x_1 \nmapsto, \ldots c, x_k \nmapsto][x_1' \mapsto r_1', \ldots c, x_l' \mapsto r_l']\rangle$
- $\langle X_m, \pi_m \rangle = T_m$
- a rule $\{x_0, \ldots c, x_k\} \xrightarrow{\delta} \{x_0', \ldots c, x_l'\} \in \Delta_A$ is applicable to X_m and $subterm_t$ $(\pi_m(x_0)) = subterm_t(\pi_m(x_1)) = \cdots = subterm_t(\pi_m(x_k)) = \{r_0, \ldots c, r_l\}$
- if $name_A(x_i') = \epsilon$ then $r_i' = \epsilon\{r_i\}$
- if $name_A(x_i') \neq \epsilon$ then $name_t(r_i) = name_A(x_i')$ and $r_i' = r_i$

Notice that because $X_n = \{\}$, the last applied rule has to be of the form $\{x_0, \ldots c, x_k\}$ $\xrightarrow{\delta} \{\}$ and due to the compatibility condition on rules of a tree automaton:

$$\sum_{i=0}^{k} name_A(x_i)_1 = 0$$

which means that the states $x_0, \ldots c, x_k$ generate only nullary letters. Therefore the corresponding subterms $\{\pi(x_0), \ldots c, \pi(x_k)\}$ of t are nullary.

Example 8 Let us show that term t_2 from Example 5 is recognised by automaton $\langle start, \Sigma, \Delta, id \rangle$ from Example 6. Since $name(t_2) = start = id(start)$ we may put $T_0 = \langle\{ strat \}, strat \mapsto t\rangle$ and consider the following sequence:

- $T_1 = \langle\{ fork \}, fork \mapsto fork\{A\{A\{x\}\}, B\{x\}\}\rangle$ by δ_1
- $T_2 = \langle\{A, B\}, A \mapsto A\{A\{x\}\}, B \mapsto B\{x\}\rangle$ by δ_2
- $T_3 = \langle\{A, B\}, A \mapsto A\{A\{x\}\}, B \mapsto B\{x\}\rangle$ by δ_3
- $T_4 = \langle\{A, B\}, A \mapsto A\{x\}, B \mapsto B\{x\}\rangle$ by δ_3
- $T_5 = \langle\{ join \}, join \mapsto join\{end\}\rangle$ by δ_4
- $T_6 = \langle\{ end \}, end \mapsto end \rangle$ by δ_6
- $T_7 = \langle\{\}, \{\}\rangle$ by δ_7

it is easy to verify that each T_m is constructed according to the induction laws.

4 Skeletal Algorithms in Tree Mining

Given a finite list K of sample terms over a common alphabet Σ, we shall construct the skeletal automaton $skeleton(K) = \langle q_0, S, \Delta, name \rangle$ of K in the following way. For each term $K_i, 0 \le i < length(K)$ let $skeleton(K_i) = \langle q_0^i, S^i, \Delta^i, name^i \rangle$ be the skeletal automaton of K_i constructed like in Example 7, then:

- $S = \{START\} \sqcup \bigcup_i S^i$
- $q_0 = START$
- $\Delta = \{\{\!\{ START \}\!\} \rightarrow \{\!\{ q_0^i \}\!\}: 0 \le i < length(K)\} \sqcup \bigcup_i \Delta^i$
- $name(q) = \begin{cases} START & if\, q = START \\ name^i(q) & if\, q \in S^i \end{cases}$

That is $skeleton(K) = \langle \Sigma, S, l, \delta \rangle$ constructed as a disjoint union of skeletal automatons for t_k enriched with two states $start$ and end. So the skeleton of a sample is just an automaton corresponding to the disjoint union of skeletal automaton corresponding to each of the terms enriched with a single starting state. Such automaton describes the situation, where all actions are different. Our algorithm will try to glue some actions that give the same output (shall search for the best fitting automaton in the set of quotients of the skeletal automaton). The next figure shows the skeletal automaton of the sample t_1, t_2 from Example 7.

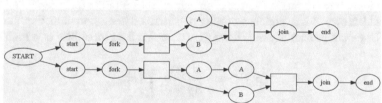

Given a finite list of sample data K, our search space $Eq(K)$ consists of all equivalence relations on the set of states S of the skeletal automaton for K.

4.1 Skeletal Operations

1. Splitting
   For a given congruence A, choose randomly a state $q \in skeleton(K)$ and make use of two types of predicates

   - split by output: $P(p) \Leftrightarrow \underset{q' \in [q]_A \underset{X \xrightarrow{\delta} Y}{}}{\exists} \quad \exists\, p \in X \wedge q' \in Y$
   - split by input: $P(p) \Leftrightarrow \underset{q' \in [q]_A \underset{X \xrightarrow{\delta} Y}{}}{\exists} \quad \exists\, q' \in X \wedge p \in Y$

2. Summing
   For a given congruence A, choose randomly two states p, q such that $name(p) = name(q)$.

3. Union/Intersection
   Given two skeletons A, B choose randomly a state $q \in skeleton(K)$.

Let us note that by choosing states and predicates according to the above description, all skeletal operations preserve congruences on $skeleton(K)$.

4.2 Fitness

The idea behind the fitness function for bidirectional tree automata is the same as for ordinary finite automata analysed in [7]. The additional difficulty comes here from two reasons: a bidirectional tree automaton can be simultaneously in a multiset of states; moreover, two transitions may non-trivially depend on each other. Formally, let us say that two transitions $X \overset{\delta}{\to} Y$ and $X' \overset{\delta'}{\to} Y'$ are depended on each other if $X \cap X' \neq \{\!\{\}\!\}$, and are fully depended if $X = X'$. Unfortunately, extending the Bayesian interpretation to our framework yields a fitness function that is impractical from the computational point of view. For this reason we shall propose a fitness function that agrees with Bayesian interpretation only on some practical class of bidirectional tree automata—directed tree automata. A directed tree automaton is a bidirectional tree automaton whose each pair of rules is either fully depended or not depended. Now if δ is a sequence of rules of a directed tree automaton, then similarly to the Bayesian probability in [7], we may compute the probability of a multiset of states X:

$$p^\delta(X) = \frac{\Gamma(k)}{\Gamma(n+k)} \prod_{i=1}^{k} c_i^{\frac{c_i}{i}}$$

where:

- k is the number of rules $X \overset{\delta_i}{\to} Y$ for some Y of the automaton
- c_i is the total number of i-th rule $X \overset{\delta_i}{\to} Y$ used in δ
- $n = \sum_{i=1}^{k} c_i$ is the total number of rules of the form $X \to Y$ for some Y used in δ

and the total distribution as:

$$p(\delta) = \prod_{X \subseteq S} p^\delta(X)$$

which corresponds to the complexity:

$$p(\delta) = -\sum_{X \subseteq S} log(p^\delta(X))$$

This complexity does not include any information about the exact model of an automaton. Therefore, we have to adjust it by adding "the code" of a model. By using two-parts codes, we may write the fitness function in the following form:

$$fitness(A) = length(skeleton(K)/A) - \sum_{X \subseteq S} log(p^\delta(X))$$

where $length(skeleton(K)/A)$ is the length of the quotient of the skeletal automaton $skeleton(K)$ by congruence A under any reasonable coding, and S is the set of states of the quotient automaton. For sample problems investigated in the next section, we chose this length to be:

$$clog(|S|)|\{\langle \delta, x \rangle : X \overset{\delta}{\to} Y \in \Delta, 0 \leq x < size(X) + size(Y)\}|$$

for constant $1 \leq c \leq 2$.

5 Sample Applications

5.1 Process Mining

We shall start with a business process similar to one investigated in Example 5, but extended with multiple states generating the same action A:

This process starts in state *start* then performs simultaneously at least three tasks A and exactly one task B, and then finishes in *end* state. Figure 4a, b shows automata mined from 4 and 8 random samples. Notice that the first mined automaton correspond to the minimal automaton recognizing any sample, and after seeing 8 samples the initial model is fully recovered.

5.2 Theory Discovery

In this section we show a direct application of the above idea to theory discovery. Given a finite theory over a ranked alphabet, we use true sentences from the theory as sample data. Let us consider the following signature:

$$\Sigma = \{=/2/1, cons/2/1, nil/0/1, +/2/1, 0/0/1, 1/0/1\}$$

A natural number n will be represented as a term:

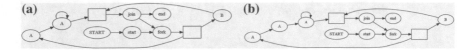

Fig. 4 Discovered models. **a** Model discovered after seeing 4 samples. **b** Model discovered from 10 samples

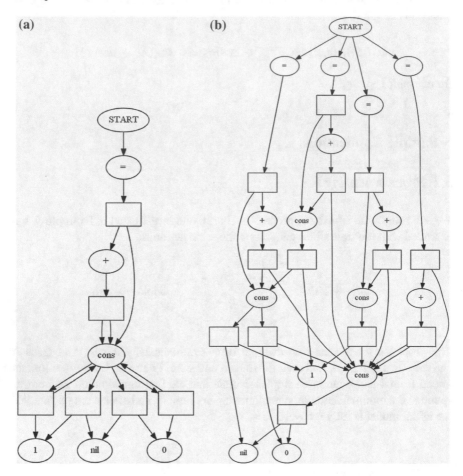

Fig. 5 Natural numbers. **a** Syntax for natural numbers. **b** Automaton mined from 16 samples

$$cons\{b_k \cdots cons\{b_1, cons\{b_0, nil\}\} \cdots \}$$

where b_i is either 0 or 1, $b_k \neq 0$ if $n \neq 0$ and $n = \sum_{i=0}^{k} 2^k b_k$. Our sample data K consists of equations $\langle a + b = c \rangle$, $for 0 \leq a, b \leq 4, c = a + b$ on natural numbers. Notice, that the minimal automaton that recognises K generates the syntax for the equational theory of natural numbers with addition. Figure 5a shows minimal automaton mined from K, and Fig. 5b shows a model discovered from 16 samples, which accurately describes three-bit addition on natural numbers.

6 Conclusions

In this paper we defined bidirectional tree automata, and showed how they can represent business process. We adapted skeletal algorithms introduced in [7] to mine bidirectional tree automata, resolving the problem of mining nodes that corresponds to parallel executions of a process (i.e. AND-nodes). In future works we will be mostly interested in validating the presented algorithms in industrial environment and apply them to real data.

References

1. Bremermann, H.J.; Optimization through evolution and recombination. In: Yovitts, M.C. et al. (eds.) Self-Organizing Systems 1962, p. 93106. Spartan Books, Washington (1962)
2. Comon, H., Dauchet, M., Gilleron, R., Löding, C., Jacquemard, F., Lugiez, D., Tison, S., Tommasi, M.: Tree automata techniques and applications (2007)
3. de Medeiros, A., van Dongen, B., van der Aalst, W., Weijters, A.: Process mining: extending the alpha-algorithm to mine short loops. In BETA Working Paper Series, Eindhoven. Eindhoven University of Technology (2004)
4. Friedberg, R.M.: A learning machines part I. IBM J. Res. Dev. 2 (1956)
5. Friedberg, R.M., Dunham, B., North, J.H.: A learning machines part II. IBM J. Res. Dev. 3 (1959)
6. Holland, J.H.: Adaption in Natural and Artificial Systems. The University of Michigan Press, Ann Arbor (1975)
7. Przybylek, M.R.: Skeletal algorithms in process mining. In: Studies in Computational Intelligence, vol. 465. Springer, Berlin (2013)
8. Przybylek, M.R.: Process mining through tree automata. In: International Conference on Evolutionary Computation Theory and Applications (2013)
9. Rechenberg, I.: Evolutions strategie–optimierung technischer systeme nach prinzipien der biologischen evolution. Ph.D. thesis (1971) [Reprinted by Fromman-Holzboog, 1973]
10. Ren, C., Wen, L., Dong, J., Ding, H., Wang, W., Qiu, M.: A novel approach for process mining based on event types. In: IEEE SCC 2007, pp. 721–722 (2007)
11. Valiant, L.: A theory of the learnable. In: Communications of The ACM, vol. 27 (1984)
12. van der Aalst, W.: Process Mining: Discovery, Conformance and Enhancement of Business Processes. Springer, Berlin (2011)
13. van der Aalst, W., de Medeiros, A.A., Weijters, A.: Process equivalence in the context of genetic mining. BPM Center Report BPM-06-15. www.BPMcenter.org (2006)
14. van der Aalst, W., Pesic, M.S.M.: Beyond process mining: from the past to present and future. BPM Center Report BPM-09-18. www.BPMcenter.org (2009)
15. van der Aalst, W., ter Hofstede, A., Kiepuszewski, B., Barros, A.: Workflow patterns. BPM Center Report BPM-00-02. www.BPMcenter.org (2000)
16. van der Aalst, W., van Dongen, B.: Discovering workflow performance models from timed logs. In: Engineering and Deployment of Cooperative Information Systems, pp. 107–110 (2002)
17. van der Aalst, W., Weijters, A., Maruster, L.: Workflow mining: discovering process models from event logs. In: BPM Center Report BPM-04-06. www.BPMcenter.org (2006)

18. Weijters, A., van der Aalst, W.: Process mining: discovering workflow models from event-based data. In: Proceedings of the 13th Belgium-Netherlands Conference on Artificial Intelligence, pp. 283–290, Maastricht. Springer (2001)
19. Wen, L., Wang, J., Sun, J.: Detecting Implicit Dependencies Between Tasks from Event Logs. Lecture Notes in Computer Science, vol. 3841, pp. 591–603 (2006)
20. Wynn, M., Edmond, D., van der Aalst, W., ter Hofstede, A.: Achieving a general, formal and decidable approach to the or-join in workflow using reset nets. BPM Center Report BPM-04-05. www.BPMcenter.org (2004)

Alternative Topologies for GREEN-PSO

Stephen M. Majercik

Abstract The expense of evaluating the function to be optimized can make it difficult to apply the Particle Swarm Optimization (PSO) algorithm in the real world. Approximating the function is one way to address this issue, but an alternative is conservation of function evaluations. GREEN-PSO (GR-PSO) adopts the latter approach: given a fixed number of function evaluations, GR-PSO conserves them by probabilistically choosing a subset of particles smaller than the entire swarm on each iteration and allowing only those particles to perform function evaluations. Since fewer function evaluations are used on each iteration, the algorithm can use more particles and/or more iterations for a given number of function evaluations. GR-PSO has been shown to be effective using the global topology, performing as well as, or better than, the standard PSO algorithm (S-PSO) [7]. We extend these results by showing that GR-PSO can achieve significantly better performance than S-PSO, in terms of both best function value achieved and rate of error reduction, using three other topologies—ring, von Neumann, and Moore—on a set of six standard benchmark functions, and that the von Neumann and Moore topologies can be more effective topologies for GR-PSO than the global topology.

Keywords Particle swarm optimization · Swarm intelligence

1 Introduction

Swarm intelligence is a natural phenomenon in which complex behavior emerges from the collective activities of a large number of simple individuals who interact with each other and their environment in very limited ways. A number of swarm based optimization techniques have been developed, among them Particle Swarm Optimization (PSO). PSO, loosely based on the phenomenon of birds flocking, was

S.M. Majercik (✉)
Bowdoin College, 8650 College Station, Brunswick, ME 04011, USA
e-mail: smajerci@bowdoin.edu
URL: http://www.bowdoin.edu/~smajerci

© Springer International Publishing Switzerland 2016
K. Madani et al. (eds.), *Computational Intelligence*,
Studies in Computational Intelligence 613,
DOI 10.1007/978-3-319-23392-5_9

155

introduced by Kennedy and Eberhart [5]. In this algorithm, virtual particles "fly" through the solution space in search of high-quality solutions. In the original algorithm, the search trajectory of a particle is influenced by both the best solution it has found so far (the *personal best*) and the best solution the entire swarm has found so far (the *global best*). The algorithm iteratively updates the velocities and positions of the particles in a way that biases the search toward the personal bests and the global best, converging on a (hopefully) global optimum. PSO is one of the most widely used swarm based algorithms and has been successfully applied to many real world problems [10]. Many variants of the PSO algorithm have been proposed; see [12] for an extensive listing.

In the standard PSO algorithm, every particle evaluates the function being optimized (the objective function) on every iteration in order to determine the fitness of the candidate solution at the particle's new position. A typical PSO algorithm uses 20–40 particles and tens of thousands of iterations to find even a suboptimal, but acceptable, solution, and this high number of function evaluations can be difficult to achieve in real world applications if the function the algorithm is trying to optimize—for example, a complex control mechanism—is computationally or financially expensive to evaluate, or requires time consuming simulations and/or human interaction [6].

A common way of addressing this problem is to use *function approximation*, which can take many forms, such as response surface methods, radial basis functions, Kriging (DACE models), Gaussian process regression, support vector machines, and neural networks [6]. Another approximation technique is *fitness inheritance*, in which the objective function value, or fitness, of an individual is approximated based on the fitnesses of one or more other individuals [11].

Instead of using less expensive—but possibly less effective—function approximations, an algorithm could use fewer function evaluations during each iteration. This is the approach adopted by GREEN-PSO (GR-PSO) [7]. On each iteration, GR-PSO permits only a subset of the particles in the swarm to do function evaluations and uses the conserved function evaluations to increase the number of particles in the swarm and/or the number of iterations that are possible, given a fixed number of function evaluations. Majercik [7] showed that GR-PSO could achieve performance comparable to or, in a number of cases, better than that of the standard PSO algorithm, and as good as a type of function approximation that is similar to GR-PSO. It was unclear, however, to what extent the performance of GR-PSO was dependent on the global topology used in their tests. We extend their results by showing that GR-PSO can achieve significantly better performance than S-PSO, in terms of both best function value achieved and rate of error reduction, using three other topologies—ring, von Neumann, and Moore—and that the von Neumann and Moore topologies can be more effective topologies for GR-PSO than the global topology.

In Sect. 2, we describe the basic PSO algorithm and the GR-PSO algorithm. In Sect. 3, we discuss related work. We describe and discuss the results of our experiments in Sect. 4, and we conclude with some ideas for future work in Sect. 5.

2 PSO and GR-PSO

In this section, we describe the standard PSO algorithm and GREEN-PSO.

2.1 Standard PSO

The standard PSO algorithm (S-PSO) uses a swarm of particles to iteratively search a d-dimensional solution space for good solutions. The particles are guided by their own experience and that of the swarm. The number of particles in the swarm is fixed and each particle's current position and velocity, x_i and v_i, respectively, are initialized randomly. Particle i remembers the best solution it has found so far (the *personal best*, or *pbest*), p_i, and the best solution found so far by the particles in particle i's neighborhood (the *neighborhood best*, or *nbest*), g_i. (The original PSO algorithm used the global topology, in which the neighborhood of every particle is the entire swarm and g_i is the *global best*, or *gbest*.) The velocity v_i of particle i is updated during each iteration such that its motion is biased toward both p_i and g_i, and the new velocity is used to update its position x_i. There are a number of basic PSO algorithms. For purposes of comparison, we adopt the PSO algorithm with a *constriction coefficient* χ and *velocity limits* as described in [10]. The velocity and position update equations are:

$$v_i \leftarrow \chi(v_i + U(0, \phi_1) \otimes (p_i - x_i) + U(0, \phi_2) \otimes (g_i - x_i)) \tag{1}$$

$$x_i \leftarrow x_i + v_i \tag{2}$$

where:

- ϕ_1 and ϕ_2, the *acceleration coefficients* that scale the attraction of particle i to p_i and g_i, respectively, are equal and have the value 2.05,
- $U(0, \phi_i)$ is a vector of real random numbers uniformly distributed in $[0, \phi_i]$, which is randomly generated at each iteration for each particle,
- $\otimes$ is component-wise multiplication, and
- χ is the standard constriction coefficient (approximatley 0.7298).

Finally, each component of v_i is kept within a range $[V_{min}, V_{max}]$, where V_{min} and V_{max} are the minimum and maximum values of the search space and are identical for each dimension.

2.2 GREEN-PSO

The PSO algorithm is often motivated by referencing the human problem solving process, in which an individual, confronted with a problem, looks for a solution based partially on her own experience solving that problem in the past and partially on the experience of others who have solved that problem before. Continuing that analogy, the problem solver will need to evaluate her efforts periodically, but it is unclear when she should do that. Some evaluation is necessary to direct her efforts productively, but (1) evaluation may be expensive, in terms of one or more resources, (2) a potentially successful path of inquiry might be prematurely terminated if it is difficult to accurately assess the value of the current solution, and (3) it is not always clear when the quality of the current solution is sufficiently different from that of the most recently evaluated solution to justify a new evaluation. For these reasons, while trying to improve the best solution she has found in the past, she might suspend evaluation of her efforts for a period of time.

In S-PSO, the points at which an evaluation might be done are well-defined. On each iteration, a particle moves and the new position might be a better solution. S-PSO requires that a particle's position be evaluated whenever it moves. The analogy above suggests that it might be better to evaluate a particle's position less frequently. GR-PSO puts this idea into practice in the following way: GR-PSO operates like S-PSO, except that each particle, after calculating a new velocity and updating its position according to that velocity, performs a function evaluation (FE) on its new position with some probability *probFE*, where $0.0 < \text{probFE} < 1.0$. This means that on every iteration, the expected number of particles doing a function evaluation is $(n \times \text{probFE})$, where n is the number of particles in the swarm, so the expected number of iterations is $(\text{FEs}_{\max}/(n \times \text{probFE}))$, where $\text{FEs}_{\max}$ is the total number of function evaluations available. This allows the swarm to use more particles and/or more iterations. For example, in the standard PSO algorithm, 20,000 function eval-

---

Algorithm 1: GR-PSO.

---

1 **Inputs:**
2   n, the size of the swarm
3   f, the function to be optimized (minimized)
4    $\text{FEs}_{\max}$, the maximum number of function evaluations
5   *probFE*, the probability of doing a function evaluation
6 **Outputs:**
7   x^*, the position of the minimum function value found
8    $f(x^*)$, the value of the function at that position
9 **for** $i \leftarrow 1 \ldots n$ **do**
10 | Initialize particle i with a random position and velocity;
11 **while** *number FEs performed* $< \text{FEs}_{\max}$ **do**
12   **for** $i \leftarrow 1 \ldots n$ **do**
13       $p_i \leftarrow$ position of best solution particle i has found so far;
14       $g_i \leftarrow$ position of best solution found by particle i's neighborhood so far;
15       $v_i \leftarrow$ velocity of particle i updated from Eq. 1 using p_i and g_i;
16       $x_i \leftarrow$ position of particle i updated from Eq. 2;
17   **if** *randomDouble* $<$ *probFE* **then**
18 | Calculate $f(x_i)$ and update p_i, g_i, and x^*;

19 **return** x^* and $f(x^*)$

---

uations would allow a swarm of 100 particles to do 200 iterations. If each particle does a function evaluation with a probability of 0.5, we could, for example, increase the number of iterations to 400, or increase the size of the swarm to 200 particles, or some combination of these two strategies. See Algorithm 1 for pseudocode for GR-PSO.

3 Related Work

Fitness inheritance, the approximation technique closest to the GR-PSO approach, approximates the value of the objective function for a particle's current position based on the function values of some set of particles designated as its "parents," thereby avoiding function evaluations. GR-PSO can be viewed as an extreme form of fitness inheritance; a particle that does not do a function evaluation is its own parent, inheriting its own objective function value directly.

The work of [11] is the only work we know of that incorporates fitness inheritance into a PSO algorithm. They tested the effectiveness of twelve fitness inheritance techniques and four fitness approximation techniques in a multi-objective PSO algorithm. Majercik [7] compared GR-PSO using the global topology to the best three of these techniques (as ranked by [11]) and found that the performance of all three was never better than the best GR-PSO algorithms and that the majority of GR-PSO algorithms was better than all three of these techniques, although the differences in performance were, in some cases, quite small.

Akat and Gazi [1] describe a decentralized, asynchronous approach that allows the PSO algorithm to be implemented on multiple processors with very weak requirements on communication between those processors. Particles reside on different processors. At each time step, each particle/processor has access only to some subset of other particles/processors; thus, there may be significant intervals during which a particle p has received no information from particle p'; it may even be the case that, on a given iteration, a particle receives no information from any other particles, in which case its position and velocity remain the same. They report that the performance of their approach was comparable to standard PSO implementations.

GR-PSO is similar to their work in that a particle in GR-PSO that is not doing a function evaluation can be thought of as residing on a processor that no other particle is in communication with (including itself). An important difference, which may account for the difference in performance (GR-PSO performs better than the standard PSO algorithm), is that the communication links between particles/processors are always changing, so that when a particles/processor finds a new personal best, this information will be immediately available only to those other particles/processors that are communicating with that processor at the time. In GR-PSO, however, if a particle does a function evaluation and finds a new personal best, all other particles in its neighborhood have access to that information immediately.

In other work, [2] considered the effect of the *information flow topology* on the performance of three approaches to creating non-reciprocal dynamic neighborhoods, which can be represented as directed graphs. If these digraphs are strongly connected *over time*, i.e. if there is a fixed interval such that the union of the digraphs over every interval of iterations of that length is strongly connected, then information flow in the swarm will be preserved and every particle eventually has access to the information gathered by every other particle. An interesting open question is whether GR-PSO is operating in a similar fashion, creating temporary, smaller neighborhoods, where a neighborhood is the subset of particles that are doing function evaluations. The inhabitants of those neighborhoods are constantly changing, but they are connected over time. It is possible that the probability of doing a function evaluation is regulating the connectedness of these shifting neighborhoods. An investigation into this possibility could shed light on the performance of GR-PSO and the performance of PSO algorithms that use dynamic neighborhoods, in general.

Finally, the results of [7] suggest an intriguing relationship with work of García-Nieto and Alba [4]. They tested a variant of the standard PSO algorithm in which the neighborhood for each particle on each iteration is constructed by choosing k other particles, or "informants," randomly. They tested the algorithm over a range of values for k and found evidence for a quasi-optimal neighborhood size of approximately 6. In a sense, the expected number of particles doing function evaluations in GR-PSO during an iteration can be viewed as the number of informants for every particle in each iteration, since it is these particles that could potentially provide new information. Majercik [7] suggested that there might be an optimal range for the expected number of particles doing function evaluations during an iteration, and that this range might be similar to the optimal range for the number of informants in the work of García-Nieto and Alba. This is still an open question.

4 Experimental Results

We tested GR-PSO on six standard benchmark functions in 30 dimensions: Sphere, Rosenbrock, Ackley, Griewank, Rastrigin, and Penalized Function P8. (See [3] for definitions of these functions.) Sphere and Rosenbrock are uni-modal functions, while Ackley, Griewank, Rastrigin, and Penalized Function P8 are multi-modal functions with many local optima. The optimum (minimum) value for all of these functions is 0.0 and, in all but two cases (the Rosenbrock Function and the Penalized Function P8), this value is obtained at $(0, 0, \ldots, 0)$. For the latter two functions, this value is obtained at $(1, 1, \ldots, 1)$ and $(-1, -1, \ldots, -1)$, respectively. We used asymmetric initialization and randomly shifted the location of the optimum away from the center of the search space in order to avoid the tendency of PSO algorithms to converge to the center [9].

We compared the performance of GR-PSO to that of the standard PSO algorithm described in Sect. 2.1, conducting tests on four standard PSO topologies: global, ring, von Neumann, and Moore. As noted above, the neighborhood of each particle in the global topology is the entire swarm. In the ring topology, the particles can be thought of as being arranged in a ring, and the neighbors of a particle are the two particles on either side of it. In the von Neumann and Moore topologies, the particles can be thought of as being arranged on a toroidal grid. In the von Neumann topology, the neighbors of a particle are the four particles to the north, south, east, and west of it. In the Moore topology, the neighbors are the eight particles around it. We fixed the number of function evaluations at 10,000, and averaged the results of 100 runs for each function.

Tests in [7] indicated that a swarm size of 20 or 50 particles produced good results, given 10,000 FEs. As they noted, although a probFE of less than 1.0 provides additional iterations, a smaller swarm (10 particles) is unable to explore the space effectively, in spite of the additional iterations, while larger swarms (100 and 200 particles) consume too many of those additional iterations, degrading performance. Their tests also indicated that probFE values of 0.05, 0.1, and 0.2 were good choices. For S-PSO, their tests indicated that 20 and 50 particles were good choices for swarm size.

We tested GR-PSO for each of the four topologies on a range of swarm sizes and probFEs and our findings confirmed these choices for swarm size and probFE. Thus, we show results for six GR-PSO algorithms—20 particles with probFEs of 0.05, 0.1, and 0.2, and 50 particles with probFEs of 0.05, 0.1, and 0.2—and two S-PSO algorithms—20 particles and 50 particles—applied to each of the four topologies: global, ring, von Neumann, and Moore.

The results of our tests are presented in Tables 1 and 2. For each of the 192 combinations of function, algorithm, and topology we show the mean and standard deviation of the lowest function value found over 100 runs. For each function and topology, the results of the three best algorithms are shown in bold-face; the best result is italicized as well.

Of the 24 function-topology cases, GR-PSO has the best performance in 19 of those (79%), and in 12 of those cases has the lowest standard deviation as well. S-PSO has the best performance in five of the 24 cases (21%), and, in three of those cases, the lowest standard deviation. Notably, in 18 of the 24 cases (75%), S-PSO has the worst performance of the eight algorithms, the 50-particle S-PSO algorithm being responsible for all of them. This is not surprising, given that the larger swarm reduces the number of iterations to 200, whereas all of the other algorithms have at least 2,000 iterations and as many as 10,000 iterations (the 20 particle GR-PSO algorithm with a probFE of 0.05).

The performance gain of GR-PSO over S-PSO is particularly notable in the case of the Sphere function. The minimum function value found by the best GR-PSO algorithm was five orders of magnitude lower than that found by the best S-PSO algorithm. This performance was achieved by the global topology algorithm using the smaller swarm size and the lowest probFE (20 particles, probFE of 0.05), which is to be expected, since the smooth surface of this function means that any reasonable algorithm will tend to get closer to the minimum, given more iterations, and the smaller swarm size and lowest probFE provide the most iterations of all the GR-PSO algorithms.

Although they did not report the results, [7] indicated that GR-PSO using the ring topology did not yield the same performance gains over S-PSO with the ring

Table 1 Performance of GR-PSO and S-PSO (average over 100 runs)

Function	Algorithm	Mean function value and standard deviation			
		Ring	von Neumann	Moore	Global
Sphere	GR-PSO-20-0.05	**2.05e-02**	*3.82e-06*	*4.56e-08*	*1.03e-08*
		2.90e-02	*4.80e-06*	*9.92e-08*	3.79e-08
	GR-PSO-20-0.1	*9.26e-03*	5.67e-06	1.92e-07	4.20e-08
		9.83e-03	8.45e-06	3.82e-07	1.69e-07
	GR-PSO-20-0.2	**1.17e-02**	**1.49e-04**	**6.97e-07**	**3.08e-07**
		1.41e-02	**1.42e-03**	**1.62e-06**	**1.07e-06**
	GR-PSO-50-0.05	1.08e+02	1.08e+00	5.67e-02	3.71e-05
		1.09e+02	1.06e+00	5.26e-02	4.71e-05
	GR-PSO-50-0.1	8.03e+01	1.22e+00	9.07e-02	1.53e-04
		6.96e+01	9.57e-01	8.02e-02	1.91e-04
	GR-PSO-50-0.2	6.99e+01	1.96e+00	2.42e-01	1.53e-03
		5.42e+01	1.55e+00	2.01e-01	2.40e-03
	S-PSO-20	4.38e-01	2.12e-02	6.27e-03	3.35e-03
		3.20e-01	1.82e-02	7.36e-03	6.97e-03
	S-PSO-50	3.13e+02	4.92e+01	1.79e+01	7.86e-01
		1.76e+02	2.37e+01	9.33e+00	6.43e-01
Rosenbrock	GR-PSO-20-0.05	4.36e+01	*3.32e+01*	3.76e+01	**3.52e+01**
		3.11e+01	*2.10e+01*	2.23e+01	**2.40e+01**
	GR-PSO-20-0.1	**4.22e+01**	**3.35e+01**	3.49e+01	3.65e+01
		2.85e+01	**1.87e+01**	2.36e+01	2.38e+01
	GR-PSO-20-0.2	**4.03e+01**	**3.42e+01**	**3.43e+01**	3.53e+01
		2.70e+01	**1.84e+01**	**2.11e+01**	2.09e+01
	GR-PSO-50-0.05	6.79e+01	3.73e+01	**3.40e+01**	3.59e+01
		3.84e+01	2.06e+01	**1.84e+01**	2.12e+01
	GR-PSO-50-0.1	5.87e+01	4.01e+01	*3.40e+01*	**3.49e+01**
		3.23e+01	2.44e+01	*1.67e+01*	**1.92e+01**
	GR-PSO-50-0.2	5.65e+01	3.97e+01	3.81e+01	3.82e+01
		2.86e+01	2.16e+01	2.17e+01	2.38e+01
	S-PSO-20	*3.90e+01*	3.56e+01	3.48e+01	*3.36e+01*
		2.32e+01	2.03e+01	1.90e+01	*1.91e+01*
	S-PSO-50	8.03e+01	4.32e+01	3.87e+01	4.31e+01
		2.86e+01	2.05e+01	1.97e+01	2.68e+01
Ackley	GR-PSO-20-0.05	*6.43e+00*	6.54e+00	7.67e+00	1.06e+01
		8.38e+00	8.42e+00	7.97e+00	7.24e+00
	GR-PSO-20-0.1	**7.50e+00**	7.10e+00	7.83e+00	9.83e+00
		8.91e+00	8.56e+00	8.40e+00	7.46e+00
	GR-PSO-20-0.2	7.87e+00	6.20e+00	6.14e+00	8.85e+00
		9.05e+00	8.47e+00	7.91e+00	7.74e+00
	GR-PSO-50-0.05	8.81e+00	*5.33e+00*	**5.74e+00**	**7.19e+00**

(continued)

Table 1 (continued)

Function	Algorithm	Mean function value and standard deviation			
		Ring	von Neumann	Moore	Global
		7.05e+00	*8.15e+00*	**8.98e+00**	**8.80e+00**
	GR-PSO-50-0.1	9.21e+00	**5.46e+00**	**5.25e+00**	**7.48e+00**
		7.23e+00	**8.11e+00**	**8.69e+00**	**9.06e+00**
	GR-PSO-50-0.2	8.13e+00	6.04e+00	*4.73e+00*	*6.91e+00*
		6.95e+00	8.27e+00	*8.25e+00*	*8.95e+00*
	S-PSO-20	**6.47e+00**	**5.75e+00**	6.77e+00	9.00e+00
		8.70e+00	**8.47e+00**	8.64e+00	8.23e+00
	S-PSO-50	9.97e+00	7.81e+00	6.11e+00	7.58e+00
		6.78e+00	7.42e+00	7.02e+00	8.91e+00

Key: GR-PSO-n-p = GR-PSO with n particles and probFE = p
S-PSO-n = S-PSO with n particles

Table 2 Performance of GR-PSO and S-PSO (average over 100 runs)

Function	Algorithm	Mean function value and standard deviation			
		Ring	von Neumann	Moore	Global
Griewank	GR-PSO-20-0.05	**8.54e-02**	*1.27e-02*	3.73e-02	4.51e-02
		9.17e-02	*1.67e-02*	7.92e-02	6.01e-02
	GR-PSO-20-0.1	*4.85e-02*	**1.35e-02**	*1.80e-02*	7.83e-02
		5.25e-02	**1.87e-02**	*2.33e-02*	2.89e-01
	GR-PSO-20-0.2	**5.19e-02**	**1.47e-02**	**2.94e-02**	5.29e-02
		6.32e-02	**2.24e-02**	**6.98e-02**	1.50e-01
	GR-PSO-50-0.05	2.24e+00	7.40e-01	1.25e-01	**1.76e-02**
		1.28e+00	2.09e-01	8.75e-02	**2.59e-02**
	GR-PSO-50-0.1	1.97e+00	7.67e-01	2.18e-01	*1.18e-02*
		9.99e-01	1.91e-01	1.54e-01	*1.49e-02*
	GR-PSO-50-0.2	1.76e+00	9.04e-01	3.77e-01	**2.06e-02**
		7.01e-01	1.57e-01	2.08e-01	**2.97e-02**
	S-PSO-20	4.84e-01	6.72e-02	**3.50e-02**	8.81e-02
		2.08e-01	7.74e-02	**5.32e-02**	3.92e-01
	S-PSO-50	3.73e+00	1.48e+00	1.18e+00	6.15e-01
		1.64e+00	2.46e-01	9.96e-02	2.71e-01
Rastrigin	GR-PSO-20-0.05	1.18e+02	9.08e+01	9.83e+01	1.27e+02
		4.78e+01	3.75e+01	5.34e+01	8.01e+01
	GR-PSO-20-0.1	**1.15e+02**	8.38e+01	1.05e+02	1.23e+02
		3.95e+01	3.37e+01	5.31e+01	8.22e+01
	GR-PSO-20-0.2	**1.10e+02**	7.92e+01	8.13e+01	1.14e+02
		3.92e+01	2.72e+01	3.27e+01	6.20e+01
	GR-PSO-50-0.05	1.33e+02	8.00e+01	**6.93e+01**	**8.12e+01**
		4.41e+01	2.65e+01	**2.22e+01**	**3.72e+01**

(continued)

Table 2 (continued)

Function	Algorithm	Mean function value and standard deviation			
		Ring	von Neumann	Moore	Global
	GR-PSO-50-0.1	1.35e+02	**7.54e+01**	**7.09e+01**	8.58e+01
		5.17e+01	**2.34e+01**	**2.30e+01**	3.81e+01
	GR-PSO-50-0.2	1.27e+02	**7.88e+01**	***6.49e+01***	**7.43e+01**
		3.85e+01	**2.06e+01**	***2.04e+01***	**2.74e+01**
	S-PSO-20	***9.81e+01***	**7.35e+01**	7.51e+01	8.52e+01
		3.00e+01	**2.35e+01**	2.20e+01	3.21e+01
	S-PSO-50	1.33e+02	9.11e+01	7.95e+01	***7.41e+01***
		2.99e+01	2.04e+01	1.74e+01	***2.40e+01***
Penalized Fnc P8	GR-PSO-20-0.05	***4.46e+00***	7.09e-01	**8.13e-01**	7.31e-01
		3.20e+00	*1.03e+00*	**1.34e+00**	1.26e+00
	GR-PSO-20-0.1	**5.11e+00**	**8.27e-01**	***4.93e-01***	**6.87e-01**
		3.33e+00	1.09e+00	***7.14e-01***	1.00e+00
	GR-PSO-20-0.2	4.86e+00	**1.08e+00**	**6.18e-01**	8.87e-01
		3.00e+00	**1.61e+00**	**8.72e-01**	1.30e+00
	GR-PSO-50-0.05	7.12e+04	4.77e+00	2.14e+00	***4.96e-01***
		2.54e+05	3.60e+00	1.92e+00	***7.04e-01***
	GR-PSO-50-0.1	2.74e+04	6.95e+00	2.61e+00	**6.50e-01**
		9.32e+04	4.60e+00	1.96e+00	**8.63e-01**
	GR-PSO-50-0.2	1.36e+04	6.58e+00	3.04e+00	9.23e-01
		4.36e+04	4.98e+00	2.01e+00	1.26e+00
	S-PSO-20	5.38e+00	2.45e+00	1.56e+00	1.51e+00
		3.17e+00	1.73e+00	1.37e+00	1.47e+00
	S-PSO-50	9.52e+03	1.12e+01	6.10e+00	3.38e+00
		2.26e+04	6.54e+00	3.64e+00	2.20e+00

Key: GR-PSO-n-p = GR-PSO with n particles and probFE = p
S-PSO-n = S-PSO with n particles

topology, as GR-PSO with the global topology achieved compared to S-PSO with the global topology (although GR-PSO with the global topology outperformed S-PSO with the ring topology in many cases). Our results confirm the latter result, but differ from the first result, in that GR-PSO with the ring topology frequently outperforms S-PSO with that topology.

In 16 of the 24 function-topology cases, GR-PSO algorithms are the three best performing algorithms. The force of this observation is weakened by the fact that more GR-PSO algorithms were tested than S-PSO algorithms, so, in each function-topology case, we applied a Mann Whitney test to compare the performance of the best GR-PSO algorithm to that of the best S-PSO algorithm. In each case, we rank ordered the best function values for the 100 runs of the two algorithms, and applied a 2-tailed Mann-Whitney U-test to compare the ranks. Since the samples were large enough (> 20), the distribution of the U statistic approximates a normal distribution, so we report the Z-score, which is typically used in such cases, as well as the U-score and p-value (see Table 3). The results indicate a statistically significant difference in the distributions of the two groups at the 0.01 level for 16 of the 19 cases in which the best GR-PSO algorithm outperforms the best S-PSO algorithm. Three of the five cases in which the best S-PSO algorithm outperforms the best GR-PSO algorithm are statistically significant, two at the 0.05 level and one at the 0.01 level. Thus, a GR-PSO algorithm was better than either S-PSO algorithm in 16 cases and, statistically, as good as the best S-PSO algorithm in two more cases.

We also compared the performance of the best GR-PSO algorithm and the best S-PSO algorithm for each function, regardless of topology. The performance of GR-PSO was better than that of S-PSO in all cases and the difference was significant in five of the six cases at the 0.01 level: Sphere ($p = 0.0$), Rosenbrock ($p = 0.0099$), Griewank ($p = 0.0$), Rastrigin ($p = 0.0031$), and Penalized Function P8 ($p = 0.0$). The difference in the case of Ackley was not significant at that level ($p = 0.0751$).

None of the 24 GR-PSO algorithm-topology cases is clearly best, but our results do provide some guidance. If we look at the distribution of the three best-performing and the three worst-performing algorithm-topology combinations over all functions (the three best and worst, instead of the best and worst, to help compensate for the small differences among performances), we find that any algorithm with the ring

Table 3 Statistical significance of differences between performance of best GR-PSO algorithm and performance of best S-PSO algorithm

Function	Mann Whitney statistics			
	Ring	von Neumann	Moore	Global
Sphere	U = 3	U = 0	U = 0	U = 0
	Z = 13.44	Z = 13.44	Z = 13.44	Z = 13.44
	p = 0.0	p = 0.0	p = 0.0	p = 0.0
Rosenbrock	U = 6178	U = 5767	U = 6311	U = 5540
	Z = 2.10	Z = 2.85	Z = −1.85	Z = −3.27
	p = 0.0357	p = 0.0044	p = 0.0643	p = 0.0011
Ackley	U = 6801	U = 6950	U = 2879	U = 5208
	Z = −0.95	Z = −0.68	Z = 8.16	Z = 3.88
	p = 0.3421	p = 0.4965	p = 0.0	p = 0.0001
Griewank	U = 128	U = 1600	U = 4977	U = 3160
	Z = 13.21	Z = 10.51	Z = 4.30	Z = 7.64
	p = 0.0	p = 0.0	p = 0.0	p = 0.0
Rastrigin	U = 6098	U = 6972	U = 5318	U = 7117
	Z = −2.24	Z = −0.64	Z = 3.68	Z = 0.37
	p = 0.0251	p = 0.5222	p = 0.0002	p = 0.7114
Penalized Function P8	U = 5859	U = 2508	U = 3033	U = 3285
	Z = 2.68	Z = 8.84	Z = 7.87	Z = 7.41
	p = 0.0074	p = 0.0	p = 0.0	p = 0.0

topology is a poor choice. Algorithms using the ring topology are never responsible for one of the best three, and are responsible for one of the worst three in 16 of the 36 cases, confirming the judgment in [7] that ring is a poor choice of topology for GR-PSO. The global topology is responsible for the other two worst performance cases. Of the 18 best performance cases, five are von Neumann, nine are Moore, and four are global.

We conjecture that the poor performance of the ring topology is due to the fact that the ring topology and GR-PSO both operate to reduce the availability of information in the swarm, GR-PSO by delaying the discovery of better solutions, and the ring topology by reducing the immediate impact of a better solution to four other particles. It may be that these two effects combine to reduce the availability of new information to a degree that compromises the effectiveness of the swarm.

The same numbers for the von Neumann, Moore, and global topologies are shown in the first three columns of Table 4, where the number of best three performances is shown along with the number of worst three performances in parentheses. The worst performing cases from ring are distributed almost entirely among the von Neumann 50-particle algorithms and the global 20-particle algorithms. Moore seems to be a good choice, since it is not responsible for any of the worst performing cases, but its best performing cases are spread among the six algorithms, offering no clear overall recommendation. We get more guidance if we look at these numbers for each of the von Neumann, Moore, and global topologies individually (the last three columns of Table 4). We see that in the Moore topology all of the GR-PSO algorithms are responsible for about the same number of best and worst cases, and each algorithm is responsible for almost the same number of best cases as worst cases (identical in four algorithms). Thus, there is no clear algorithm recommendation for this topology.

Performance in the von Neumann and global cases, however, is not as evenly distributed. The 20-particle algorithms using the von Neumann topology and the 50-particle algorithms using the global topology are clearly better choices. For each probFE, the 20-particle von Neumann algorithm is responsible for at least as many best performance cases, and fewer worst performance cases, as the corresponding

Table 4 Best and worst performing algorithm-topology combinations

GR-PSO Algorithm	von Neumann, Moore, and Global					
	von Neumann	Moore	Global	von Neumann	Moore	Global
GR-PSO-20-0.05	2 (0)	1 (0)	1 (2)	4 (2)	3 (3)	2 (4)
GR-PSO-20-0.1	2 (0)	1 (0)	1 (2)	4 (2)	3 (3)	2 (4)
GR-PSO-20-0.2	0 (0)	1 (0)	0 (2)	5 (1)	4 (2)	2 (4)
GR-PSO-50-0.05	1 (3)	1 (0)	1 (0)	1 (5)	3 (3)	4 (2)
GR-PSO-50-0.1	0 (4)	3 (0)	1 (0)	2 (4)	3 (3)	5 (1)
GR-PSO-50-0.2	0 (4)	2 (0)	0 (1)	2 (4)	2 (4)	3 (3)

GR-PSO-n-p = GR-PSO with n particles and probFE = p
# of best three performances (# of worst three performances)

20-particle Moore algorithm. Furthermore, each of the 20-particle von Neumann algorithms has at least twice as many best performance cases as worst performance cases. The same is true for the 50-particle global topology algorithms and their corresponding Moore algorithms (except that the 50-particle global topology with a probFE of 0.2 has the same number of best performance and worst performance cases. Over all the topologies, however, the 20-particle von Neumann topology is responsible for more best performances than the 50-particle global topology, so our overall recommendation (although tentative since it is based on a small number of tests), is to use a 20-particle swarm with the von Neumann topology and a probFE of 0.05, 0.1, or 0.2.

Since the motivation behind GR-PSO is to achieve good performance using fewer function evaluations, it makes sense to look at how rapidly the GR-PSO algorithms reduce error compared to the S-PSO algorithms. In Fig. 1, we show, for each function, the reduction in error for the first 5,000 FEs achieved by the three 20-particle GR-PSO algorithms with the von Neumann topology, and the 20-particle and 50-particle S-PSO algorithms. We show the 20-particle GR-PSO algorithms because they are always able to reduce error more rapidly than any of the 50-particle algorithms (supporting our recommendation above of the 20-particle algorithms). We note that the error reduction results for the Moore topology are virtually identical to the von Neumann topology results shown; the results for the global topology are similar. Note that the performance of the three GR-PSO algorithms is often so close as to be indistinguishable. Note also that the scale of the y-axis is not the same for all functions since the necessary ranges of values varies greatly.

In all but one function (Penalized Function P8), all three of the GR-PSO algorithms reduce error faster than both of the S-PSO algorithms. By FE 5,000, with the exception of one function (Sphere), the 20-particle S-PSO algorithm has caught up with the GR-PSO algorithms. The difference between the three GR-PSO algorithms and the 50-particle S-PSO algorithm, however, is clear over the entire interval of FEs for every function. While this difference at FE 5,000 is only one order of magnitude or less for the Rosenbrock, Ackley, Griewank, and Rastrigin functions, it is three and four orders of magnitude for the Sphere and Penalized P8 functions, respectively.

Finally, a natural question to ask is whether the effectiveness of GR-PSO depends on the *neighborhood influence model*, which specifies which particles in a particle's neighborhood will affect that particle's trajectory. The tests reported above used the neighborhood-best influence model, in which a particle's trajectory is affected by its own personal best and the neighborhood best. The other major neighborhood influence model is the fully informed particle swarm (FIPS), in which the trajectory of a particle is biased towards the personal bests of all the particles in its neighborhood [8]. We ran the same tests reported above using FIPS-GR-PSO, a version of GR-PSO that uses the FIPS neighborhood influence model. In the version of FIPS that we used, a particle is affected by the personal bests of all the particles in its neighborhood, including its own personal best, equally.

The results of the FIPS tests were unambiguous. In every one of the 24 function-topology cases, one of the S-PSO algorithms was responsible for the best performance and, in 22 of those cases, the other S-PSO algorithm was responsible for the second

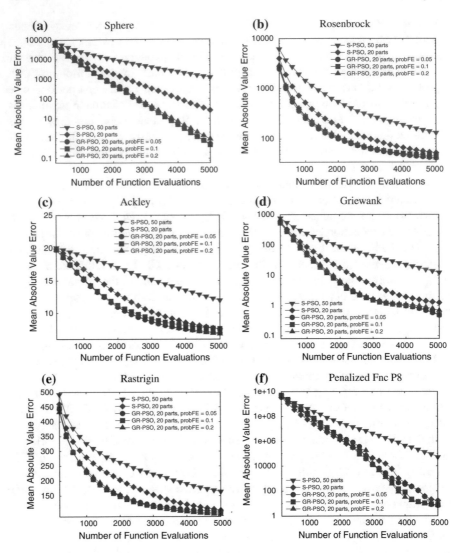

Fig. 1 Mean absolute value error as a function of number of function evaluations (von Neymann topology)

best performance. GR-PSO is clearly not effective in the FIPS setting. If, as we did for the neighborhood-best influence model, we look at the distribution of the three best-performing and three worst-performing algorithm-topology combinations over all functions, we find that the ring topology is responsible for all 18 of the three best performances and is not responsible for any of the three worst. The global topology is responsible for 15 of the three worst performances and the Moore topology is responsible for the remaining three. This is in stark contrast to the results for the neighborhood-best algorithms, in which the ring topology was responsible for the best performance in only one of the 36 cases, and the worst performance in 28 of those cases. So, the ring topology, which was the worst topology in the neighborhood-best setting, is now the best.

We conjecture that the poor performance of FIPS-GR-PSO can be explained by looking at the degree to which the impact of newly discovered personal bests is diluted by the inclusion of all the neighborhood personal bests in the velocity updates of the particles. Since function evaluations are being done so infrequently (in the recommended configuration, between one and four function evaluations in the entire swarm during an iteration), new personal bests are discovered infrequently. Let us assume that one particle has discovered a new personal best by doing an FE, but that all the other particles have not found new personal bests recently. Suppose that this new personal best is also a neighborhood best. In the global topology using the neighborhood-best influence model, this new (potentially better) personal best will be responsible for half of the change in velocity of every other particle. In the FIPS model, however, a particle is affected by all the personal bests in its neighborhood so, in this scenario, the new personal best will be responsible for only $1/n$ of the change in velocity of every other particle, where n is the number of particles in the swarm. So, even though new bests are discovered less frequently in GR-PSO, suggesting that it might be better to strengthen their impact, the FIPS model weakens their impact.

This suggests that FIPS-GR-PSO would perform better when the neighborhood is smaller, since this would decrease the ratio of old to new information, given the same amount of new information. We would also expect that performance would be better when probFE is higher, since this would tend to increase the number of recently discovered personal bests in a neighborhood. Our FIPS-GR-PSO tests strongly confirm the first conjecture, since the ring topology outperformed the other topologies in every case, and, although not as strong, there was evidence supporting the second conjecture (the 20-particle ring topology algorithm with the highest probFE was responsible for six of the eight best performances of the 20-particle ring algorithms).

We tested the conjecture that the age of personal bests matters by modifying FIPS-GR-PSO so that a particle includes another particle in its neighborhood only if that particle has done an FE during the previous three iterations, thereby imposing a maximum age on personal bests. The performance of this version of FIPS-GR-PSO was better in five of the function-algorithm cases (by two orders of magnitude in three cases, and four and seven orders of magnitude in the other two cases). Not surprisingly, smaller neighborhoods and higher probFEs were still better—three of these cases used the ring topology with a probFE of 0.2. This provides some support for our conjecture, but more work is needed to determine whether different maximum ages or a more sophisticated version of this approach would improve performance more broadly. In any case, this suggests the possiblity that the information provided by bests should be used differently depending on how old it is, regardless of which neighborhood influence model is being used.

5 Conclusions and Further Work

We have shown that GR-PSO can outperform the standard PSO algorithm using any one of three topologies—von Neumann, Moore, and global—and that the von Neumann and Moore topologies can be more effective than the global topology. In particular, our results indicate that a 20-particle swarm using the von Neumann topology and a probFE of 0.05, 0.1, or 0.2 is a good overall configuration for GR-PSO. We also tested FIPS-GR-PSO, a version of GR-PSO that uses the FIPS neighborhood influence model, and found that the GR-PSO function evaluation conservation strategy was not effective in the FIPS setting.

The function evaluation conservation technique we tested is very simple; there are a number of possibilities for more sophisticated conservation mechanisms. We are currently exploring a version of GR-PSO in which each particle has its own probFE that is adjusted each time an FE is done. If the FE produces a new personal best, the particle's probFE is increased to 0.5 to make it more likely that any continued improvement will be noticed. If the FE results in a value that is not even as good as the value of the current position, the probFE is reduced to 0.01 in order to discourage further function evaluations in an area of the search space that does not appear promising. If the FE reveals a solution that is better than the current one, but not a new personal best, the probFE is set to an intermediate level of 0.2. In a sense, this mechanism provides a novel way to control exploration and exploitation. Given a lower probability of doing function evaluations, personal and neighborhood bests are discovered less frequently and, since particles are moving based on older information, the balance is tipped away from exploitation toward exploration.

The decision about whether to do a function evaluation does not need to be probabilistic. It might be productive to think about FE usage in terms of a *priority queue*. In the current version of GR-PSO, every particle does an FE with the same probability on each iteration. Instead, each particle might be assigned an FE priority that is periodically updated, and the particle with the highest priority is given the next FE. In the simplest version of this idea, a particle's priority would increase by a small amount every time it does not get an FE and would be greatly reduced whenever it does. This simple mechanism would result in particles taking turns using FEs, but a priority udpate mechanism that incorporated other information, such as the recent history of the particle (e.g. the distance it has moved and the change in its function value over the last k FEs), the recent history of the neighborhood best, and the recent FE usage in the particle's neighborhood, has the potential to distribute FEs in a more effective manner.

Regardless of the mechanism by which function evaluations are conserved, it might be useful for a particle to have the opportunity to revisit past solutions that were not evaluated, in an effort to recover missed high quality solutions. For example, a particle could save some number of positions for which it did not do a function evaluation and, under certain circumstances (e.g. if stagnation is detected in the particle's neighborhood), pick one of those positions randomly and evaluate it.

Finally, a better understanding of the impact of FE conservation on the information dynamics of the swarm would be useful. In any algorithm that seeks to take advantage of swarm behavior, the mechanism that generates and propagates information is critical. GR-PSO operates by reducing the rate at which new information is generated, and the amount of this reduction is the same for all iterations. Perhaps it would be better to vary the amount of this reduction during the optimization process. For example, preliminary experiments sugggest that it is better to reduce the generation of new information earlier in the process. This makes some sense since, as noted above, a reduction in new information encourages exploration, which would be especially valuable earlier in the process. An understanding of the reasons behind this might allow us to improve performance by managing information flow more explicitly.

References

1. Akat, S., Gazi, V.: Decentralized asynchronous particle swarm optimization. In: Swarm Intelligence Symposium, SIS 2008, IEEE, pp. 1–8 (2008a)
2. Akat, S., Gazi, V., Particle swarm optimization with dynamic neighborhood topology: Three neighborhood strategies and preliminary results. In: Swarm Intelligence Symposium, SIS 2008, IEEE, pp. 1–8 (2008b)
3. Bratton, D., Kennedy, J.: Defining a standard for particle swarm optimization. In: Swarm Intelligence Symposium, SIS 2007, IEEE, pp. 120–127 (2007)
4. García-Nieto, J., Alba, E.: Why six informants is optimal in PSO. In: Proceedings of the fourteenth international conference on Genetic and evolutionary computation conference, GECCO '12, pp. 25–32 (2012)
5. Kennedy, J., Eberhart, R.: Particle swarm optimization. In: Proceedings of IEEE, pp. 1942–1948 (1995)
6. Landa-Becerra, R., Santana-Quintero, L.V., Coello Coello, C.A.: Knowledge incorporation in multi-objective evolutionary algorithms. In: Multi-Objective Evolutionary Algorithms for Knowledge Discovery from Databases, pp. 23–46 (2008)
7. Majercik, S.M.: GREEN-PSO: Conserving function evaluations in particle swarm optimization. In: Proceedings of the Fifth International Conference on Evolutionary Computation Theory and Applications, pp. 160–167 (2013)
8. Mendes, R., Kennedy, J., Neves, J.: The fully informed particle swarm: simpler, maybe better. IEEE Trans. Evol. Comput. **8**(3), 204–210 (2004)
9. Monson, C.K., Seppi, K.D.: Exposing origin-seeking bias in PSO. In: GECCO, pp. 241–248 (2005)
10. Poli, R., Kennedy, J., Blackwell, T.: Particle swarm optimization: an overview. Swarm Intell. **1**, 33–57 (2007)
11. Reyes-Sierra, M., Coello Coello, C.A.: A study of techniques to improve the efficiency of a multi-objective particle swarm optimizer. In: Evolutionary Computation in Dynamic and Uncertain Environments, Studies in Computational Intelligence vol. 51, pp. 269–296 (2007)
12. Sedighizadeh, D., Masehian, E.: Particle swarm optimization methods, taxonomy and applications. Int. J. Comput. Theory Eng. **1**(5), 486–502 (2009)

A Hybrid Shuffled Frog-Leaping Algorithm for the University Examination Timetabling Problem

Nuno Leite, Fernando Melício and Agostinho C. Rosa

Abstract The problem of examination timetabling is studied in this work. We propose a hybrid solution heuristic based on the Shuffled Frog-Leaping Algorithm (SFLA) for minimising the conflicts in the students's exams. The hybrid algorithm, named Hybrid SFLA (HSFLA), improves a population of frogs (solutions) by iteratively optimising each memeplex, and then shuffling the memeplexes in order to distribute the best performing frogs by the memeplexes. In each iteration the frogs are improved based on three operators: crossover and mutation operators, and a local search operator based on the Simulated Annealing metaheuristic. For the mutation and local search, we use two well known neighbourhood structures. The performance of the proposed method is evaluated on the 13 instances of the Toronto datasets from the literature. Computational results show that the HSFLA is competitive with state-of-the-art methods, obtaining the best results on average in seven of the 13 instances.

Keywords Examination timetabling · Memetic algorithm · Shuffled Frog-Leaping Algorithm · Simulated annealing · Toronto benchmarks

1 Introduction

Examination timetabling is an important practical problem faced by schools and universities every epoch. This problem, known as the *Examination Timetabling Problem* (ETTP), consists in scheduling students's exams into a limited number of time slots

N. Leite (✉) · F. Melício
Instituto Superior de Engenharia de Lisboa/IPL, R. Conselheiro Emídio Navarro, 1,
1959-007 Lisboa, Portugal
e-mail: nleite@cc.isel.ipl.pt

N. Leite · F. Melício · A.C. Rosa
LaSEEB-System and Robotics Institute, Av. Rovisco Pais 1 TN 6.21,
1049-001 Lisboa, Portugal

A.C. Rosa
Department of Bioengineering/IST, Universidade de Lisboa,
Av. Rovisco Pais, 1, 1049-001 Lisboa, Portugal
e-mail: acrosa@laseeb.org

© Springer International Publishing Switzerland 2016
K. Madani et al. (eds.), *Computational Intelligence*,
Studies in Computational Intelligence 613,
DOI 10.1007/978-3-319-23392-5_10

and rooms subject to a set of constraints. The constraints are divided into *hard* (cannot be violated) and *soft* (can be violated) constraints. One universal hard constraint specifies that cannot exist any student with two or more exams scheduled for the same time period. An universal soft constraint involves spacing out the student's exams according to some measure. Usually, the optimisation goal is then to minimise the violations of this constraint or/and other soft constraints while satisfying the hard ones. This problem is further classified as Uncapacitated ETTP (if the room capacity is infinite) or Capacitated ETTP (if the room capacity is limited).

The ETTP, like other educational timetabling problems (e.g. school and course timetabling), belong to the class of NP-complete problems, which constrains the application of exact solution methods (e.g. Mathematical programming techniques or Dynamic Programming) to problem instances of limited size. Often, real instances found in practice are too large to be solved by exact methods, so several heuristic based methods have recently been proposed with great success. The first works to solve the ETTP were proposed in the 1960s decade. Until now this problem was approached using very different techniques. Carter [14] first classified these approaches into four types: sequential methods, cluster methods, constraint-based methods and generalised search. Later, Petrovic and Burke [22] specified more six types: hybrid evolutionary algorithms, metaheuristics, multi-criteria approaches, case based reasoning techniques, hyper-heuristics and adaptive approaches. A recent and detailed overview of the proposed methods to solve the ETTP can be found in [23].

The Shuffled Frog-Leaping Algorithm (SFLA) is a memetic metaheuristic proposed in 2003 [16, 17]. The SFLA was applied to many areas and problems, namely: TSP [30], Clustering [4], Flow-shop Scheduling [29], multiobjective optimisation [24], ETTP [28], among others. The SFLA was first applied to the ETTP by Wang et al. [28] (in chinese). The authors proposed a Discrete SFLA (DSFLA) where solutions are encoded using a time permutation scheme suited to be manipulated by the DSFLA. As the original SFLA is only suitable for continuous optimisation problems, a specific update operator was defined for the discrete case of ETTP. The algorithm manipulates both feasible and infeasible solutions, being these last ones penalized to avoid further exploring them. The DSFLA is evaluated on four datasets of the Capacitated *Toronto* benchmark data (Toronto variant c in [23]).

In our previous work [20] we presented a novel adaptation of SFLA for solving the ETTP. The proposed algorithm is applied to the complete set of Uncapacitated Toronto benchmark data (Toronto variant b in [23]), and compared with other techniques in the literature. In the present work, we extend and improve the previous algorithm by applying three operators when updating the worst frog: crossover and mutation operators and a local search step based on the Simulated Annealing metaheuristic. The proposed hybrid solution heuristic was evaluated on the Toronto benchmark data achieving competitive results.

The paper is organized as follows. The next section presents the Examination Timetabling problem formulation. Section 3 presents the original SFLA model. Section 4 describes the proposed hybrid heuristic algorithm for solving the ETTP. Section 5 presents simulation results and analysis of the proposed algorithm. Finally, conclusions and future work are presented in Sect. 6.

2 Problem Description and Formulation

We now describe the studied Uncapacitated ETTP problem in more detail as well as the examination timetabling data used. The problem formulation was adapted from the descriptions presented in [2, 5]. The following terms were defined:

- E_i is a set of N examinations $(i = 1, \ldots, N)$.
- T is the number of time slots.

- $C = (c_{ij})_{N \times N}$, *Conflict matrix*, is a symmetric matrix of size N where each element, denoted by c_{ij} $(i, j \in \{1, \ldots, N\})$, represents the number of students attending exams i and j. The diagonal elements c_{ii} denote the total of students enrolled in exam i.
- M is the number of students.
- t_k $(1 \le t_k \le T)$ denotes the assigned time slot for exam k $(k \in \{1, \ldots, N\})$.

The uncapacitated problem studied has one hard constraint where exams that have common students cannot be scheduled in the same time slot. A soft constraint is defined for measuring the proximity cost of conflicting exams, which should be scheduled as far as possible from each other. The optimisation objective is to minimise the sum of proximity costs given as:

$$\text{minimise} \quad \frac{1}{M} \cdot \sum_{i=1}^{N-1} \sum_{j=i+1}^{N} c_{ij} \cdot prox(i, j) \tag{1}$$

where

$$prox(i, j) = \begin{cases} 2^{5-|t_i - t_j|} & if\ 1 \le |t_i - t_j| \le 5 \\ 0 & otherwise, \end{cases} \tag{2}$$

subject to

$$\sum_{i=1}^{N-1} \sum_{j=i+1}^{N} c_{ij} \cdot \lambda(t_i, t_j) = 0 \quad and \quad \lambda(t_i, t_j) = \begin{cases} 1\ if\ t_i = t_j \\ 0\ otherwise \end{cases}. \tag{3}$$

Equation (2) measures the proximity cost of exams i and j which is greater than zero for exams that are five or less time slots apart. Equation (3) represents the hard constraint mentioned above.

The set of benchmarks used to evaluate our algorithm are known as the Toronto datasets, and its specifications are presented in Table 1 [23].

Table 1 Specifications of the 13 Toronto benchmark instances (version I)

Dataset	Students	Exams	Enrolments	Density of the conflict matrix	Time slots
car91	16925	682	56877	0.13	35
car92	18419	543	55522	0.14	32
ear83	1125	190	8109	0.27	24
hec92	2823	81	10632	0.42	18
kfu93	5349	461	25113	0.06	20
lse91	2726	381	10918	0.06	18
pur93	30032	2419	120681	0.03	42
rye92	11483	486	45051	0.07	23
sta83	611	139	5751	0.14	13
tre92	4360	261	14901	0.18	23
uta92	21266	622	58979	0.13	35
ute92	2750	184	11793	0.08	10
yor83	941	181	6034	0.29	21

The density of the conflict matrix in the fifth column is computed as the ratio of the number of non-zero elements of this matrix to the total number of matrix elements. The last column shows the minimum number of time slots of a feasible solution

3 Shuffled Frog-Leaping Algorithm

In the SFLA, a population of F frogs, denoted $U(i)$, $i = 1, \ldots, F$, with identical structure, but different adaptation to the environment, is maintained. The F frogs are divided into m substructures called memeplexes, where they "search for food" (they are optimised, in the algorithm sense) and meanwhile, exchange information (exchange memes) with other frogs, trying to reach the food localisation (global optimum). Each memeplex is comprised of n frogs, so that $F = mn$. After searching locally in their memeplex, the frogs are ranked and shuffled in order to go, eventually, to a different memeplex and exchange their memes with the frogs located there. The main steps of the SFLA are depicted in Fig. 1a. The local search employed corresponds to the so called Frog-Leaping local search Fig. 1b. The ranking comprises sorting the frogs in descending order of performance. The partition of frogs is as follows. The first frog (the frog with the best fitness) in the sorted list is allocated to the memeplex 1; the second frog is allocated to the memeplex 2, and so on, so that the frog m will go to memeplex m; then, the $m + 1$ frog will go to memeplex 1, the $m + 2$ frog will go to memeplex 2, and the process continues in this fashion for the remainder frogs.

In the original SFLA [16], in order to prevent the algorithm getting stuck in a local optima, a submemeplex of size $q < n$ is constructed in each memeplex. The individual frogs in the memeplex are selected to form a submemeplex according to their fitness. The selection strategy is to give higher weights to frogs that have higher performance values and less weight to those with lower performance values [16].

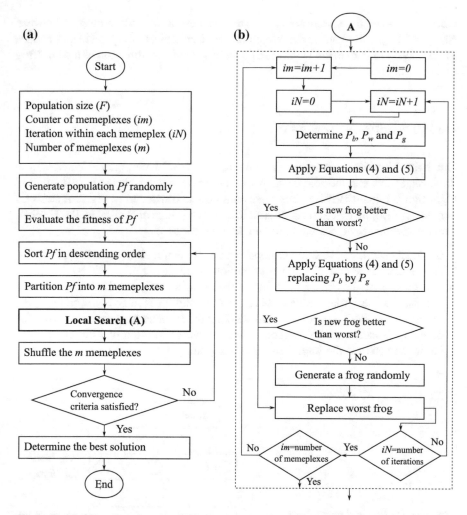

Fig. 1 Shuffled Frog-Leaping Algorithm: **a** Main algorithm steps; **b** Frog-Leaping local search. Illustrations adapted from [4]

In each submemeplex, the P_b and P_w vectors denote, respectively, the best and the worst frog. At the end of each iteration of the Frog-Leaping local search, the worst frog in the submemeplex is updated according to the following rule:

$$S = \begin{cases} \min \{\text{int} \, [rand * (P_b - P_w)] \, , S_{max}\} \, , & \text{for a positive step} \\ \max \{\text{int} \, [rand * (P_b - P_w)] \, , -S_{max}\} \, , & \text{for a negative step} \end{cases} \quad (4)$$

$$U(q) = P_w + S \quad (5)$$

where S denotes the update step size, *rand* represents a random number between $(0, 1)$ and S_{max} is defined as the maximum step size that any frog can take. The idea of this step is to update the worst frog position towards the direction of the best frog in the memeplex.

4 Hybrid SFLA for ETTP

Our hybrid heuristic algorithm for solving the ETTP incorporates features from the basic SFLA and the Simulated Annealing (SA) metaheuristic [19]. The hybrid algorithm is named HSFLA. The algorithm flow is illustrated in Fig. 2. It starts by generating a population of feasible solutions which is then optimised by the HSFLA. The SA metaheuristic has the following known features:

- SA local search can lead to near optimal solutions if a slow annealing process is conducted, at the cost of a longer execution time.
- The quality of the optimised solution depends not only on the SA parameters but also on the initial solution. If the initial solution is not very optimised, the improvement attained could be considerable; on the other way, when we rerun SA on an optimised solution, we could obtain a worse solution or a better solution, but in the last case the improvement is marginal.

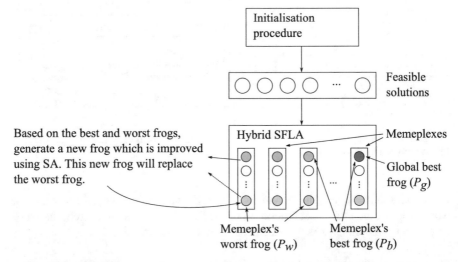

Fig. 2 Flow of the hybrid heuristic algorithm. A set of feasible solutions is obtained using the Saturation Degree graph colouring heuristic. Next, these solutions are organized into memeplexes and optimised using the hybrid SFLA. In the SFLA local search, the memeplex's worst frog is replaced by a new frog which results from the combination of the memeplex's best and worst frogs followed by application of the SA metaheuristic

The HSFLA was designed taking these points into consideration. It works like a multi-start SA optimising different initial solutions. It maintains elitism by keeping the global best frog. After shuffling the memeplexes, the SA is executed again on solutions of a given memeplex, and the process is repeated for all the memeplexes for a given number of time loops (Fig. 1a). In the next sections we describe the main aspects of the algorithm which are the following: (1) the solution representation, (2) the initialisation procedure, (3) the neighbourhood structure, and (4) the SFLA's worst frog improvement and random frog generation.

4.1 Solution Representation

Each individual frog (solution) is represented by an array of dimension equal to the number of time slots, where each position contains an array of exams scheduled at that time slot. Figure 4 shows the graphical representation of three solutions (the t_i's are the time slots and the e_j's are the exams). In our method, only feasible solutions are manipulated as all the operators produce feasible timetables. The fitness of a solution is the value of the proximity cost function that we are minimising, which is a measure of the soft constraint violations (see Eq. (1)).

4.2 Initialisation Procedure

The initial frog population is created using a construction algorithm that is based on the Saturation Degree graph colouring heuristic [12]. To construct each solution, the approach begins with an empty timetable and the most difficult exams to insert (exams with the least number of available periods) are selected next for insertion (in case of ties, one of the eligible exams is selected randomly). The remainder, less complex, exams to be inserted have more feasible periods available, which will facilitate their insertion. In this heuristic only the hard constraints are met when searching for feasible time slots where to schedule the exams.

4.3 Neighbourhood Structure

Local search methods like SA start from an initial solution and explore other candidate solutions in the neighbourhood. The neighbourhood comprises a set of solutions that are reached from the initial one by applying a move. The search progresses by moving to a candidate solution (which may or may not improve the previous solution, depending on the method) and repeating the process until a given stopping criterion is met. In our approach we use two neighbourhoods, *PeriodSwap* and *KempeChain*,

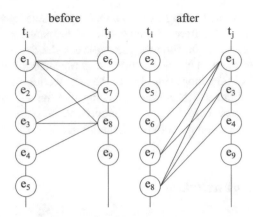

Fig. 3 An example of the Kempe chain heuristic (Adapted from [15].) Moving exam e_1 from period t_i to period t_j while maintaining feasibility implies moving the conflicting exams (e_6, e_7, and e_8) from period t_j to period t_i. In each turn, conflicting exams remaining in t_i (e_3 and e_4) also have to be moved to period t_j. In the worst case, when all exams have conflicts, a swap of the exams between the two time slots occurs

published in the literature. The neighbourhoods are, respectively, denoted N_1 and N_2, and are defined as follows.

Neighbourhood N_1: exchange exams in time slot t_i with exams in time slot t_j, where t_i and t_j are two randomly chosen time slots. This neighbourhood was introduced in [9]. It maintains the solution feasibility since all exams in a time slot are swapped.

Neighbourhood N_2: perturb, in a feasible fashion, an exam included in a *Kempe chain*. Figure 3 illustrates the application of this operator. The Kempe chain neighbourhood was first applied to the ETTP in [26].

Neighbourhoods N_1 and N_2 are applied in the HSFLA's worst frog improvement step, described in detail next.

4.4 Worst Frog Improvement and Random Frog Generation

In the original SFLA, each solution (frog) is a vector, and the worst performance frog within each submememplex is updated towards the direction of the best frog, according to Eqs. (4) and (5) (see Fig. 1b). In our adaptation of the SFLA for the ETTP, we update the worst frog by applying three operators. These are specified in the Algorithm 1 and described in the sequel. In Step 1, we combine the worst and best frogs in order to produce the new candidate frog. The crossover operation is illustrated and explained on Fig. 4. In Step 2, we apply a mutation operator using neighbourhood N_1 to the solution P'_w obtained in the Step 1. In Step 3, we apply the

SA metaheuristic with neighbourhood N_2 to the solution P'_w obtained in the Step 2. The SA implemented is described in Algorithm 2. Each step is executed with a given probability, so in the event that no operator is applied, the new candidate frog P'_w is equal to P_b.

Algorithm 1. Worst frog improvement procedure.

Input: P_b, P_w: Memeplex's best and worst frogs.
Output: P'_w: New candidate frog. It will replace P_w if it is better.

Set $P'_w = P_b$.
1: With probability pc make $P'_w = crossover(P_b, P_w)$, according to Fig. 4.
2: With probability pm make $P'_w = mutation_{N_1}(P'_w)$.
3: With probability pi make $P'_w = SA_{N_2}(P'_w)$.

The crossover operator in Step 1 was adapted from the crossover operator of [3]. As can be observed from Fig. 4, this operator produces feasible solutions, so no special constraint-handling techniques are needed.

Executing the steps of the SFLA (Fig. 1b), the new frog is going to replace the worst frog if it is better than this last one. Otherwise, the procedure is repeated but

(a)

t_1	e_2	e_{14}	e_{10}	e_3	e_{16}
t_2	e_1	e_{11}	e_4		
t_3	e_9	e_{20}	e_5	e_{18}	
t_4	e_6	e_{13}	e_7		
t_5	e_8	e_{12}	e_{15}	e_{17}	e_{19}

(b)

t_1	e_{15}	e_{20}			
t_2	e_9	e_2	e_{12}	e_{10}	e_7
t_3	e_6	e_1	e_{17}	e_{13}	
t_4	e_5	e_{18}	e_4	e_{16}	
t_5	e_8	e_{14}	e_{11}	e_3	e_{19}

(c)

t_1	e_2	~~e_{14}~~	e_{10}	~~e_3~~	e_{16}		
t_2	e_1	e_{11}	e_4	~~e_8~~	e_{14}	~~e_{11}~~	e_3 e_{19}
t_3	e_9	e_{20}	e_5	e_{18}			
t_4	e_6	e_{13}	e_7				
t_5	e_8	e_{12}	e_{15}	e_{17}	~~e_{19}~~		

Fig. 4 Crossover between P_b and P_w. The new frog P'_w in (c) is the result of combining the best frog P_b (a) with the worst frog P_w (b). Initially, make $P'_w = P_b$. Then, insert into P'_w, at a random time slot (shown *dark shaded* in (a) and (c)), exams chosen from a random time slot from solution P_w (shown *dark shaded* in (b)). When inserting these exams (shown *light gray* in (c)), some could be infeasible or already existing in that time slot (respectively, the case of e_8 and e_{11} in (c)). These exams are not inserted. The duplicated exams in the other time slots are removed. We apply the above procedure three times, combining three random time slots from P_w with three time slots from P_b

substituting P_b by the global best frog, P_g. If the new frog doesn't still improve over the worst frog $U(q)$, then a random solution is generated as the new $U(q)$, replacing the worst frog. To generate a random frog, we use the construction method (based on the graph colouring heuristic), described in Sect. 4.2.

Algorithm 2. Simulated Annealing (algorithm for minimisation of f).

Input: T_{max}: Initial temperature. R: Temperature decreasing rate.
          k: Number of iterations at a fixed temperature. T_{min}: Final temperature.
Output: u: optimised solution after the annealing process.

1: Make $T = T_{max}$, $t = 0$, and choose a solution u (at random)
2: Select a neighbour of u, say v
   If $f(v) < f(u)$ **Then** make $u = v$ **Else** make $u = v$ with probability
       $p(fu, fv) = \exp((fu - fv)/(T \cdot fu))$
   Repeat Step 2 k times
3: Make $t = t + 1$. Set $T = g(t)$, where $g(t) = T_{max} \cdot e^{-R \cdot t}$
   If $T < T_{min}$ **Then** *Stop* **Else** go to Step 2

---

5 Experiments and Comparisons

5.1 Problem Instances and Experimental Settings

The performance of the HSFLA was evaluated using the Toronto datasets (see Table 1). The algorithm was programmed in the C++ language. The hardware and software specifications are: Intel Core i7-2630QM, CPU @ 2.00 GHz × 8, with 8 GB RAM; OS: Ubuntu 12.04, 32 bit; Compiler used: GCC v. 4.6.3. The parameters of SFLA are: Population size $F = 50$, Memeplex count $m = 10$, Memeplex and Submemeplex size $n = q = 5$ (no submemeplexes were defined), and Number of time loops (convergence criterion) $L = 3$. The Simulated Annealing parameters are: $T_{max} = 0.1$, $r = 0.00001$, $k = 5$, and $T_{min} = 0.0000001$. For this cooling schedule the number of evaluations done in each SA is 6 907 760. The crossover, mutation and improvement probabilities, respectively, pc, pm, and pi, were set equal to 0.1. The parameter values were chosen empirically, in a way to achieve a reasonable balance between global and local exploration, and also establish a satisfactory compromise between solution quality and execution time. To obtain our computational results, the HSFLA is run five times on each instance with different random seeds.

5.2 Comparative Results and Discussion

Tables 2 and 3 show the best results of the HSFLA on the Toronto datasets as well as a selection of the best results available in the literature. In the last two rows of each table, the *TP* and *TP (11)* indicate, respectively, the total penalty for the 13

Table 2 Computational results of IISFLA and comparison with selection of best algorithms from literature

Data Set	Carter et al. [13]	Burke and Newall [10]	Merlot et al. [21]	Burke and Newall [7]	Burke et al. [5]	Kendall and Hussin [18]
car91	7.10	4.65	5.10	5.00	4.80	5.37
car92	6.20	4.10	4.30	4.30	4.20	4.67
ear83	36.40	37.05	35.10	36.20	35.40	40.18
hec92	10.80	11.54	10.60	11.60	10.80	11.86
kfu93	14.00	13.90	13.50	15.00	13.70	15.84
lse91	10.50	10.82	10.50	11.00	10.40	–
pur93	3.90	–	–	–	4.80	–
rye92	7.30	–	8.40	–	8.90	–
sta83	161.50	168.73	157.30	161.90	159.10	157.38
tre92	9.60	8.35	8.40	8.40	8.30	8.39
uta92	3.50	3.20	3.50	3.40	3.40	–
ute92	25.80	25.83	25.10	27.40	25.70	27.60
yor83	41.70	37.28	37.40	40.80	36.70	–
TP (11)	327.10	325.45	310.80	325.00	312.50	
TP	338.30				326.20	
Data Set	Yang and Petrovic [31]	Burke and Bykov [8]	Burke and Bykov [9]	Caramia et al. [11]	Abdullah et al. [2]	Sabar et al. [25]
car91	4.50	4.42	4.58	6.60	4.42	4.79
car92	3.93	**3.74**	3.81	6.00	3.76	3.90
ear83	33.71	32.76	32.65	**29.30**	32.12	34.69
hec92	10.83	10.15	10.06	**9.20**	9.73	10.66
kfu93	13.82	12.96	12.81	13.80	**12.62**	13.00
lse91	10.35	9.83	9.86	**9.60**	10.03	10.00
pur93	–	–	4.53	**3.70**	–	–
rye92	8.53	–	7.93	**6.80**	–	10.97
sta83	158.35	157.03	157.03	158.20	156.94	157.04
tre92	7.92	7.75	7.72	9.40	7.86	7.87
uta92	3.14	3.06	3.16	3.50	**2.99**	3.10
ute92	25.39	24.82	24.79	**24.40**	24.90	25.94
yor83	36.35	34.84	34.78	36.20	34.95	36.18
TP (11)	308.29	301.36	301.25	306.20	**300.32**	307.17
TP			**313.71**	316.70		

Values in bold represent the best results reported. "–" indicates that the corresponding instance is not tested or a feasible solution cannot be obtained

Table 3 Computational results of HSFLA and comparison with selection of best algorithms from literature (continued)

Data Set	Burke et al. [6]	Abdullah et al. [3]	Turabieh and Abdullah [27]	Demeester et al. [15]	Abdullah and Alzaqebah [1]	HSFLA
car91	4.90	**4.35**	4.81	4.52	4.76	4.59
car92	4.10	3.82	4.11	3.78	3.94	3.86
ear83	33.20	33.76	36.10	32.49	33.61	32.72
hec92	10.30	10.29	10.95	10.03	10.56	10.08
kfu93	13.20	12.86	13.21	12.90	13.44	12.87
lse91	10.40	10.23	10.20	10.04	10.87	9.85
pur93	–	–	–	5.67	–	4.47
rye92	–	–	–	8.05	8.81	8.00
sta83	**156.90**	**156.90**	159.74	157.03	157.09	157.03
tre92	8.30	8.21	8.00	**7.69**	7.94	7.78
uta92	3.30	3.22	3.32	3.13	3.27	3.15
ute92	24.90	25.41	26.17	24.77	25.36	24.76
yor83	36.30	36.35	36.23	**34.64**	35.74	34.85
TP (11)	305.80	305.40	312.84	301.02	306.58	301.54
TP				314.74		314.01

Table 4 Computational results of HSFLA and comparison with the best algorithms from literature

Data Set	HSFLA	(five runs)		Burke and Bykov [8]	Caramia et al. [11]	A bdullah et al. [2] (five runs)	
	f_{min}	f_{ave}	σ	f_{min}	f_{min}	f_{min}	f_{ave}
car91	4.59	**4.62**	0.03	4.42	6.60	4.42	4.81
car92	3.86	3.87	0.01	**3.74**	6.00	3.76	3.95
ear83	32.72	32.80	0.07	32.76	**29.30**	32.12	33.69
hec92	10.08	10.10	0.01	10.15	**9.20**	9.73	10.10
kfu93	12.87	**12.91**	0.03	12.96	13.80	**12.62**	12.97
lse91	9.85	**9.90**	0.06	9.83	**9.60**	10.03	10.34
pur93	4.47	**4.49**	0.03	–	**3.70**	–	–
rye92	8.00	**8.03**	0.03	–	**6.80**	–	–
sta83	157.03	**157.03**	0.00	157.03	158.20	156.94	157.30
tre92	7.78	7.84	0.05	7.75	9.40	7.86	8.20
uta92	3.15	3.18	0.02	3.06	3.50	**2.99**	3.32
ute92	24.76	**24.80**	0.02	24.82	**24.40**	24.90	25.41
yor83	34.85	35.00	0.09	34.84	36.20	34.95	36.27
TP (11)	301.54	**302.05**		301.36	306.20	**300.32**	306.36
TP	314.01	**314.57**			316.70		

Values in bold represent the best results reported. "–" indicates that the corresponding instance is not tested or a feasible solution cannot be obtained

Table 5 Computational results of HSFLA and comparison with the best algorithms from literature (continued)

Data Set	Abdullah et al. [3] (five runs)	Burke et al. [6] (100 runs)	Demeester et al. [15] (10 runs) Burke and Bykov [9] (20 runs)			
	f_{min}	f_{min}	f_{min}	f_{ave}	f_{min}	f_{ave}
car91	**4.35**	4.90	4.52	4.64	4.58	4.68
car92	3.82	4.10	3.78	**3.86**	3.81	3.92
ear83	33.76	33.20	32.49	**32.69**	32.65	32.91
hec92	10.29	10.30	10.03	**10.06**	10.06	10.22
kfu93	12.86	13.20	12.90	13.24	12.81	13.02
lse91	10.23	10.40	10.04	10.21	9.86	10.14
pur93	–	–	5.67	5.75	4.53	4.71
rye92	–	–	8.05	8.20	7.93	8.06
sta83	**156.90**	**156.90**	157.03	157.05	157.03	157.05
tre92	8.21	8.30	**7.69**	**7.79**	7.72	7.89
uta92	3.22	3.30	3.13	**3.17**	3.16	3.26
ute92	25.41	24.90	24.77	24.88	24.79	24.82
yor83	36.35	36.30	**34.64**	**34.83**	34.78	35.16
TP (11)	305.40	305.80	301.02	302.42	301.25	303.07
TP			314.74	316.37	**313.71**	315.84

instances and the total penalty except the `pur93` and `rye92` instances. Tables 4 and 5 compare HSFLA with the top seven best algorithms. For the HSFLA we present the lowest penalty value f_{min}, the average penalty value f_{ave}, and the standard deviation σ over five independent runs. For the reference algorithms we present the best and average (where available) results and the number of runs. The authors analysed in Tables 4 and 5 mention computation times that are within several minutes—1 h, to several hours (12 h maximum). Demeester et al. [15] mention 12 h of running time for all the instances. Table 6 compares the execution times of the HSFLA and Demeester et al.'s algorithm. For the largest instance, `pur93`, the stopping criterion was the completion of a single run of the SA metaheuristic.

The best results obtained by HSFLA are comparable with the ones produced by state-of-the-art algorithms, and it is able to produce some of the best average results. We also observe that the HSFLA obtains the lowest sum of average cost on the *TP* and *TP (11)* quantities, for the Toronto datasets. For the larger instances, the HSFLA registers long execution times, however good solutions are obtained soon after the first SA searches. Further studies should focus on the HSFLA parameters optimisation in order to shorten the execution time while not degrading the performance significantly.

Table 6 Minimum and average fitness and standard deviation comparison

Data set	HSFLA (5 runs)				Demeester et al. [15] (10 runs)			
	Execution time (h)	f_{min}	f_{ave}	σ	Stopping criterion (h)	f_{min}	f_{ave}	σ
car91	27	4.59	**4.62**	0.03	4	4.68	4.75	0.05
					12	4.52	4.64	0.05
car92	14	3.86	3.87	0.01	4	3.84	3.94	0.05
					12	3.78	**3.86**	0.06
ear83	6	32.72	32.80	0.07	2	32.82	33.02	0.16
					12	32.49	**32.69**	0.13
hec92	1	10.08	10.10	0.01	1	10.09	10.20	0.13
					12	10.03	**10.06**	0.03
kfu93	17	12.87	**12.91**	0.03	2	13.06	13.45	0.31
					12	12.90	13.24	0.20
lse91	10	9.85	**9.90**	0.06	2	10.06	10.38	0.19
					12	10.04	10.21	0.13
pur93	15	4.47	**4.49**	0.03	4	6.45	6.57	0.07
					12	5.67	5.75	0.05
rye92	17	8.00	**8.03**	0.03	4	8.18	8.31	0.10
					12	8.05	8.20	0.12
sta83	4	157.03	**157.03**	0.00	1	157.03	157.05	0.01
tre92	8	7.78	7.84	0.05	2	7.73	7.91	0.06
					12	**7.69**	**7.79**	0.07
uta92	30	3.15	3.18	0.02	2	3.32	3.37	0.03
					12	3.13	**3.17**	0.03
ute92	3	24.76	**24.80**	0.02	2	24.83	24.99	0.24
					12	24.77	24.88	0.17
yor83	5	34.85	35.00	0.09	2	34.79	35.06	0.25
					12	**34.64**	**34.83**	0.14
		Total		σ_{ave}		Total		σ_{ave}
	157	314.01	**314.57**	**0.0346**	32	316.88	319.00	0.13
					145	314.74	316.37	0.0916

For the pur93 instance the HSLFA was stopped after executing one local search with SA

6 Conclusions and Future Research Directions

We presented a hybrid solution heuristic that combines features from the SFLA and the SA metaheuristic. The experimental evaluation of the HSFLA shows that it is competitive with state-of-the-art methods. In the set of the 13 instances of the Toronto benchmark data it attains the lowest sum of average cost with a low standard deviation. In seven out of the 13 instances our approach gets better results on average when compared with the methods from the literature. The algorithm main disadvantage is the time taken on the larger instances, which is too high. On the smaller instances, however, the method is time competitive with one of the best methods analysed (work of Demeester et al.). In the simulations done we've used the same parameters without special fine tuning.

Further studies should address the parameter optimisation of the SA metaheuristic or other competitive metaheuristic (e.g. Tabu search, Great Deluge Algorithm) could be used. Further analysis on the remainder parameters of HSFLA should be carried out in order to get optimal algorithm performance.

As future research, we intend to apply our solution method to the instances of the 1st Track (Examination Timetabling) of the 2nd International Timetabling Competition (ITC2007), which contain more hard and soft constraints.

Acknowledgments Nuno Leite wishes to thank FCT, Ministério da Ciência e Tecnologia, his Research Fellowship SFRH/PROTEC/67953/2010. This work was supported by the FCT Project PEst-OE/EEI/LA0009/2013.

References

1. Abdullah, S., Alzaqebah, M.: A hybrid self-adaptive bees algorithm for examination timetabling problems. Appl. Soft Comput. **13**(8), 3608–3620 (2013)
2. Abdullah, S., Turabieh, H.: A hybridization of electromagnetic-like mechanism and great deluge for examination timetabling problems. In: Blesa, C.J., Blum, C., Gaspero, L.D., Roli, A., Sampels, M. (eds.) Hybrid Metaheuristics. Lecture Notes in Computer Science, pp. 60–72. Springer, Berlin (2009)
3. Abdullah, S., Turabieh, H., McCollum, B.: A tabu-based memetic approach for examination timetabling problems. In: Yu, J., Greco, S., Lingras, P., Wang, G. (eds.) RSKT. Lecture Notes in Computer Science, pp. 574–581. Springer, Berlin (2010)
4. Amiri, B., Fathian, M., Maroosi, A.: Application of shuffled frog-leaping algorithm on clustering. Int. J. Adv. Manuf. Technol. **45**(1–2), 199–209 (2009)
5. Burke, E., Bykov, Y., Newall, J., Petrovic, S.: A time-predefined local search approach to exam timetabling problems. IIE Trans. **36**(6), 509–528 (2004)
6. Burke, E., Eckersley, A., McCollum, B., Petrovic, S., Qu, R.: Hybrid variable neighbourhood approaches to university exam timetabling. Eur. J. Oper. Res. **206**(1), 46–53 (2010)
7. Burke, E., Newall, J.: Solving examination timetabling problems through adaption of heuristic orderings. Ann. Oper. Res. **129**(1–4), 107–134 (2004)
8. Burke, E.K., Bykov, Y.: Solving exam timetabling problems with the flex-deluge algorithm. In: Proceedings of the 6th International Conference on the Practice and Theory of Automated Timetabling, pp. 370–372 (2006). ISBN:80–210–3726–1
9. Burke, E.K., Bykov, Y. (2008) A late acceptance strategy in Hill-Climbing for exam timetabling problems. In: PATAT'08 Proceedings of the 7th International Conference on the Practice and Theory of Automated Timetabling (2008)
10. Burke, E.K.: Enhancing timetable solutions with local search methods. In: Burke, E.K., Causmaecker, P.D. (eds.) PATAT. Lecture Notes in Computer Science, pp. 195–206. Springer, Belin (2002)
11. Caramia, M., Dell'Olmo, P., Italiano, G.F.: Novel local-search-based approaches to university examination timetabling. INFORMS J. Comput. **20**(1), 86–99 (2008)

12. Carter, M., Laporte, G.: Recent developments in practical examination timetabling. In: Burke, E., Ross, P. (eds.) Practice and Theory of Automated Timetabling. Lecture Notes in Computer Science, vol. 1153, pp. 1–21. Springer, Berlin (1996)
13. Carter, M., Laporte, G., Lee, S.Y.: Examination timetabling: algorithmic strategies and applications. J. Oper. Res. Soc. **47**(3), 373–383 (1996)
14. Carter, M.W.: A survey of practical applications of examination timetabling algorithms. Oper. Res. **34**(2), 193–202 (1986)
15. Demeester, P., Bilgin, B., Causmaecker, P.D., Berghe, G.V.: A hyperheuristic approach to examination timetabling problems: benchmarks and a new problem from practice. J. Sched. **15**(1), 83–103 (2012)
16. Eusuff, M., Lansey, K., Pasha, F.: Shuffled frog-leaping algorithm: a memetic meta-heuristic for discrete optimization. Eng. Optim. **38**(2), 129–154 (2006)
17. Eusuff, M.M., Lansey, K.E.: Optimization of water distribution network design using the shuffled frog leaping algorithm. J. Water Res. Plan. Manag. **129**(3), 210–225 (2003)
18. Kendall, G., Hussin, N.: An investigation of a tabu-search-based hyper-heuristic for examination timetabling. In: Kendall, G., Burke, E., Petrovic, S., Gendreau, M. (eds.) Multidisciplinary Scheduling: Theory and Applications, pp. 309–328. Springer, New York (2005)
19. Kirkpatrick, S., Gelatt, C.D., Vecchi, M.P.: Optimization by simulated annealing. Science **220**, 671–680 (1983)
20. Leite, N., Melício, F., Rosa, A.: Solving the examination timetabling problem with the shuffled frog-leaping algorithm. In: Proceedings of the 5th International Joint Conference on Computational Intelligence, pp. 175–180 (2013)
21. Merlot, L., Boland, N., Hughes, B., Stuckey, P.: A hybrid algorithm for the examination timetabling problem. In: Burke, E., Causmaecker, P. (eds.) Practice and Theory of Automated Timetabling IV. Lecture Notes in Computer Science, vol. 2740, pp. 207–231. Springer, Berlin (2003)
22. Petrovic, S., Burke, E.: University timetabling. Handbook of Scheduling: Algorithms, Models, and Performance Analysis, chapter 45. Chapman Hall/CRC Press, London (2004)
23. Qu, R., Burke, E., McCollum, B., Merlot, L.T.G., Lee, S.Y.: A survey of search methodologies and automated system development for examination timetabling. J. Sched. **12**, 55–89 (2009)
24. Rahimi-Vahed, A., Mirzaei, A.H.: A hybrid multi-objective shuffled frog-leaping algorithm for a mixed-model assembly line sequencing problem. Comput. Ind. Eng. **53**(4), 642–666 (2007)
25. Sabar, N.R., Ayob, M., Kendall, G.: Solving examination timetabling problems using honey-bee mating optimization (etp-hbmo). In: Blazewicz, J., Drozdowski, M., Kendall, G., McCollum, B. (eds.) Proceedings of the 4th Multidisciplinary International Scheduling Conference: Theory and Applications (MISTA 2009), 10–12 August 2009, Ireland, Dublin, pp. 399–408 (2009)
26. Thompson, J.M., Dowsland, K.A.: A robust simulated annealing based examination timetabling system. Comput. OR **25**(7–8), 637–648 (1998)
27. Turabieh, H.: A hybrid fish swarm optimisation algorithm for solving examination timetabling problems. LION. Lecture Notes in Computer Science, pp. 539–551. Springer, Berlin (2011)
28. Wang, Y.-M., Pan, Q.-K., Ji, J.-Z.: Discrete shuffled frog leaping algorithm for examination timetabling problem. Comput. Eng. Appl. **45**(36), 40 (2009)
29. Xu, Y., Wang, L., Liu, M., Wang, S.-Y.: An effective shuffled frog-leaping algorithm for hybrid flow-shop scheduling with multiprocessor tasks. Int. J. Adv. Manuf. Technol. 1–9 (2013)
30. Xue-hui, L., Ye, Y., Xia, L.: Solving TSP with shuffled frog-leaping algorithm. In: Eighth International Conference on Intelligent Systems Design and Applications, 2008, ISDA'08, vol. 3, pp. 228–232 (2008)
31. Yang, Y., Petrovic, S.: A novel similarity measure for heuristic selection in examination timetabling. In: Burke, E., Trick, M. (eds.) Practice and Theory of Automated Timetabling V. Lecture Notes in Computer Science, vol. 3616, pp. 247–269. Springer, Berlin (2005)

Part II
Fuzzy Computation Theory and Applications

Model-Based Fuzzy System for Multimodal Image Segmentation

Joanna Czajkowska

Abstract In this paper, a new model-based fuzzy system for multimodal 3-D image segmentation in MR series is introduced. The presented fuzzy system calculates affinity values for fuzzy connectedness segmentation procedure, which is the main stage of the processing. The fuzzy rules, generated for the system simulating a radiological analysis, are structured on the basis of Gaussian mixture model of analyzed image regions. For the model parameters estimation, different MR modalities, acquired during a single examination, are used. The segmentation abilities of a prototype system have been tested on two medical databases. The first one consists of 27 examinations with bone tumors, which are visualized with two different MR sequences. The second one is the database of brain tumors with ground truth description obtained from the MICCAI 2012 Challenge on Multimodal Brain Tumor Segmentation.

Keywords Model based analysis · Fuzzy interference system · Fuzzy connectedness analysis · Gaussian mixture model

1 Introduction

According to [1, 2], a bone tumor is an abnormal growth of cells within a single bone, spreading to another one, muscles or soft tissue in their surroundings. They are usually found in children and young adults and their early diagnosis can be crucial for the applied treatment. The diversity of bone tumors in children still features

This work was funded by the German Research Foundation (DFG) as part of the research training group GRK 1564 *Imaging New Modalities*..

J. Czajkowska (✉)
Media Systems Group, Institute for Vision and Graphics, University of Siegen,
Hoelderlinstr. 3, Siegen, Germany
e-mail: joanna.czajkowska@uni-siegen.de

J. Czajkowska
Department of Biomedical Engineering, Silesian University of Technology,
ul. Roosevelta 40, Gliwice, Poland

© Springer International Publishing Switzerland 2016
K. Madani et al. (eds.), *Computational Intelligence*,
Studies in Computational Intelligence 613,
DOI 10.1007/978-3-319-23392-5_11

191

many diagnostic and therapeutic problems. Determination of their nature requires experience and close collaboration of specialists from various areas. Despite the significant progress of imaging techniques, many cases are diagnosed too late.

The bone tumor diagnostics is mostly based on Magnetic Resonance (MR) imaging, where during a single examination, series in different MR modalities are acquired. However, different tumor types vary in their appearance even in the same modality. Largely, only the comparative radiological analysis which takes several acquired series into consideration enables a reliable diagnosis.

Due to the fact that bone tumors are quite rare, the problem of their segmentation is not often discussed in literature. Varied intensity levels in MR sequences of different tumors cause the described segmentation methods [3–5] to be dedicated to only one tumor type. A wide range of currently available imaging techniques differentiates the segmentation procedure to dynamic MR based [4, 6] as well as static MR based [3, 5] techniques. All the procedures presented in the mentioned papers combine the information coming from different MR modalities. The segmentation algorithm proposed in [3] is based on fuzzy connectedness analysis developed by [7, 8] and is commonly used in different medical applications [9, 10]. The fuzzy connectedness principles have been tested in dozens of studies in past 15 years.

The brain tumors, which are detailed, examined and described in literature [11] still remain one of the most difficult tumors to be diagnosed and treated. They are histologically very weakly delimited from healthy tissue [12]. Necrosis and extended edema are also frequently visible there. These make the proper delineation of active tumor a laborious and difficult task [11, 12].

Most of the approaches to brain tumor analysis are studies on automatic segmentation algorithms. However, because of the variety of their shapes and locations, the semi-automatic methods are also common. The existing approaches use region growing and fuzzy connectedness based methods [10] as well as active contours and its geodesic modification [13].

Different fuzzy logic based techniques are available which are dedicated to medical applications [14–16]. The fully automated fuzzy topology method used for brain image analysis is proposed in [17]. Brain analysis investigating its morphological changes based on a combination of Bayesian classification with Gaussian Mixture Model (GMM) and fuzzy active surface is presented in [14].

The differences in grey intensity levels, which build bone tumor areas depending on their location in the human body, make reliable automatic segmentation and direct application of mentioned procedures impossible. Conversely, this paper presents a novel segmentation algorithm which is insensitive to tumor type and location in the body. This novel algorithm combines GMM and fuzzy inference systems in the fuzzy connectedness procedure. The developed 3-D segmentation method is based on previously segmented tumor and surrounding tissue regions on a single slice. Additionally, it adopts the fuzzy inference system parameters to enable the analysis of the whole study.

As already mentioned, radiological diagnostics of tumors relies on comparative analysis of different MR images which are acquired during one examination. Keeping this in mind, in the presented methodology a parallel analysis of two different MR series is applied.

In the following section, a short introduction to the fuzzy connectedness-based segmentation procedure is given. In Sect. 3 the membership functions, structured based on GMM, are described. Section 4 introduces the used fuzzy inference system and Sect. 5 presents the developed algorithm. Discussion concerning performed experiments and achieved segmentation results is given in Sect. 6. Then, the last section (Sect. 7) concludes the work and presents some plans for the future.

2 Fuzzy Connectedness Based Segmentation

The idea of fuzzy connectedness analysis in image processing and image segmentation has been given in works [7, 8]. Their methodology operates on multidimensional, multifeature sets of data by connecting and ordering them. The points classified into an object are strongly connected while other relations have relatively lower values when it comes to points outside the object. In medical image segmentation a multifeature data set often consists of grey intensity levels of pixels (or voxels—in volumetric data) in acquired CT, MR, US etc. studies. The image fusion methods, applied afterwards, make it possible to analyze all of them simultaneously. In the presented study only MR data of bone and brain tumors are collected; however, the multifeature data set is constructed with different MR modalities, namely T2-weighted, T1-weighted, T1-weighted contrast enhanced, etc.

Then, the segmentation procedure takes local similarities of the analyzed voxels into consideration while exploring their position $\underline{e} = (e_x, e_y, e_z)$ and reading their gray intensity levels $I^d(\underline{e})$, where $d = \{1, \ldots, D\}$ is the dimensionality of feature space.

Fuzzy connectedness of two image points is estimated on the basis of their fuzzy relation—fuzzy affinity κ

$$\kappa = \{((\underline{e}, \underline{d}), \mu_\kappa(\underline{e}, \underline{d})) : (\underline{e}, \underline{d}) \in C \times C\}, \tag{1}$$

where $\mu_\kappa \in [0, 1]$ is the fuzzy affinity membership function of spels (spatial elements) $\underline{e}$ and $\underline{d}$. The reflexive: $\mu_\kappa(\underline{e}, \underline{e}) = 1$ and symmetric: $\mu_\kappa(\underline{e}, \underline{d}) = \mu_\kappa(\underline{d}, \underline{e})$ fuzzy affinity is mostly given as

$$\mu_\kappa(\underline{e}, \underline{d}) = \mu_\alpha(\underline{e}, \underline{d}) \cdot g(\mu_\omega(\underline{e}, \underline{d}), \mu_\psi(\underline{e}, \underline{d})), \tag{2}$$

where μ_α is the functional form of adjacency relation α, while μ_ω and μ_ψ are intensity-based and intensity gradient-based components of the affinity, respectively. There are different forms of (2) discussed in [10], from which the most popular in medical applications is the weighted Gaussian variant

$$\mu_\kappa(\underline{e}, \underline{d}) = \mu_\alpha \cdot (w_1 H_1(\underline{e}, \underline{d}) + w_2 H_2(\underline{e}, \underline{d})), \tag{3}$$

with parameters w_1 and w_2 denoting positive constants which fulfill

$$w_1 + w_2 = 1. \tag{4}$$

Components H_1 and H_2 are defined as:

$$H_1(\underline{e}, \underline{d}) = \exp\left(-\tfrac{1}{2\sigma_1^2}\left(\tfrac{I(\underline{e})+I(\underline{d})}{2} - \lambda_1\right)^2\right),$$
$$H_2(\underline{e}, \underline{d}) = \exp\left(-\tfrac{1}{2\sigma_2^2}\left(|I(\underline{e}) - I(\underline{d})| - \lambda_2\right)^2\right). \tag{5}$$

Pairs λ_1, σ_1 and λ_2, σ_2 are the expected parameters of the segmented object, describing its gray intensity and gradient.

To determine the relations of spels $\underline{e}$ and $\underline{d}$ the concept of digital path has been introduced [8]. A nonempty path p_{ed} from $\underline{e} = \underline{e}^{(1)}$ to $\underline{d} = \underline{e}^{(m)}$ is any sequence of elements $\langle \underline{e}^{(1)}, \underline{e}^{(2)}, \ldots, \underline{e}^{(m)} \rangle$, such that for any $i \in [1, m-1]$ pair $\langle \underline{e}^i, \underline{e}^{(i+1)} \rangle$ a link exists. The strength of a path is given by the strength of its weakest link (with the smallest affinity). The strength of the "strongest" path between two image points (spels) $\underline{e}$ and $\underline{d}$ describes their connectedness.

Finally, the fuzzy κ-connectedness relation K between two image points $\underline{e}$ and $\underline{d}$ is given as follows

$$\mu_K(\underline{e}, \underline{d}) = \max_{p_{ed} \in P_{ed}} [\mu_N(p_{ed})], \tag{6}$$

where

$$\mu_N(p_{ed}) = \min_i \{\mu_\kappa(\underline{e}^{(i)}, \underline{e}^{(i+1)})\}. \tag{7}$$

Fuzzy affinity scene ℓ_o with respect to object's starting point $\underline{o}$ is then given by

$$\ell_o(\underline{e}) = \mu_K(\underline{o}, \underline{e}). \tag{8}$$

Then, the segmented fuzzy object $O(\underline{o})$ containing starting point $\underline{o}$ is obtained using the thresholding procedure on scene ℓ_o. The problem of threshold selection is solved by introducing the second object, treated as a background region with its own seed point $\underline{b}$. Therefore, spel $\underline{e}$ belongs to object $O(\underline{o})$ if $\mu_K(\underline{o}, \underline{e}) > \mu_K(\underline{b}, \underline{e})$. The already described approach is called the Relative Fuzzy Connectedness method and it is discussed in detail in [8, 10]. To solve the shortest path problem the authors use the dynamic programming approach, which in [18] has been replaced by the Dijkstra's Algorithm. In the later FC applications [19] also the single seed points, belonging to the object as well as the background, have been replaced by the seed points sets.

In cases of clearly visible tissues and sharp enough edges, the already described FC-based segmentation method yields very good results because based on the selected seed point areas, the required intensity and intensity gradient parameters are easy to estimate. The pathologies, like soft tissue tumors, with a more complex

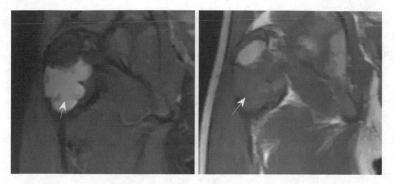

Fig. 1 A bone cyst which is visualized on a single slice coming from two different MR coronal series: *left*—STIR, *right*—FSE T1-weighted

structure can be segmented by applying a clustering based FC analysis, presented in [20]. The analyzed image data are first clustered and the calculated clusters parameters are then utilized in the FC step.

The idea introduced in this paper is to adopt the fuzzy connectedness approach to multifeature tumor analysis. The presented methodology is based on two different MR modalities. Two exemplary images of a bone cyst on coronal Short Time Inversion Recovery (STIR) and Fast Spin-Echo (FSE) T1-weighted series are shown in Fig. 1. In the radiological diagnosis the comparative analysis of both series is utilized. Hence, first, different 3-D multifeature clustering procedures [21, 22] have been applied. However, none of the implemented multifeature algorithms has yielded satisfactory final segmentation results. Therefore, a fuzzy inference system which simulates experts reasoning, described in the following sections, has been developed.

Let us assume that there are two reference regions selected on a single slice: tumor area and the background, respectively. The areas are then transferred into both analyzed modalities. Let the already given regions constitute the expert knowledge. Due to the fact that the grey intensity levels constructing the tumor areas vary, depending on the analyzed lesion and its location in the patients body, the fuzzy rules and dictionary are adaptively created for each single tumor case. The varying grey intensity levels building tumor as well as background areas on both MR sequences are described using the Gaussian Mixture Model. The generated models are then used both in a fuzzy system dictionary and the rule base. The crisp value at the output of the developed fuzzy inference system is the fuzzy affinity value $\mu_K(\underline{o}, \underline{e})$.

3 Gaussian Mixture Model

Gaussian Mixture Model (GMM) is a semi-parametric technique which enables estimating a probability density function with a mixture distribution [21, 23], defined as a weighted sum of its components.

Let be a set of N vectors $\underline{x}_n = [x_n^1, x_n^2, \ldots, x_n^D]^T$, $n \in \{1, \ldots, N\}$, where D is the dimensionality of feature space. Consider a mixture model with $K(K > 1)$ components in R^n for $n \geq 1$. The probability density function of vector $\underline{x}$ in the mixture is given as

$$p(\underline{x}) = \sum_{k=1}^{K} \pi_k p_k(\underline{x}), \qquad (9)$$

where $p_k(\underline{x})$ is the density of kth component and $\pi_k \in [0, 1]$ are the mixing proportions coefficients fulfilling

$$\sum_{k=1}^{K} \pi_k = 1. \qquad (10)$$

In the Gaussian Mixture Model each group of the data is assumed to be generated by a normal probability distribution

$$p_k(\underline{x}) = \frac{1}{(2\pi)^{\frac{D}{2}} det(\Sigma_k)^{\frac{1}{2}}} \exp \left\{ -\frac{1}{2} (\underline{x} - \lambda_k)^T \Sigma_k^{-1} (\underline{x} - \lambda_k) \right\}, \qquad (11)$$

where λ_k and Σ_k are the parameters of D-dimensional normal probability distribution $N(\lambda_k, \Sigma_k)$, mean values vector and covariance matrix, respectively.

The maximum likelihood estimator of parameter $\Theta = \{\Theta_1, \Theta_2, \ldots, \Theta_K\}$, where $\Theta_k = \{\lambda_k, \Sigma_k\}$, of a parametric probability distribution is found using the Expectation-Maximization (EM) algorithm [23]. Since the EM procedure is dedicated to incomplete data sets analysis, it iteratively alternates between finding the greatest lower bound to the likelihood function, making guesses about the complete data and then maximizing this bound by finding the Θ that maximizes $p(\underline{x}|\Theta)$ over Θ.

The EM algorithm requires starting points and a pre-selected number of clusters. The required parameters are estimated applying the unsupervised cascade clustering procedure and Kernelized CS cluster validity measure, discussed in detail in [24].

In the proposed segmentation procedure the adaptively generated tumor and background model constitute the basis for fuzzy inference system. The obtained components parameters are used in the fuzzy dictionary, which defines the membership functions of fuzzy rules. The combination of different components which build particular image regions is the basis for fuzzy rules generation.

4 Fuzzy Inference System

Different existing fuzzy reasoning systems are described in literature [14, 15, 25] and also applied in medical tasks. The basic structure of such systems consists of three components: dictionary defining the membership functions, base of fuzzy IF-THEN rules and reasoning mechanism.

Historically, the first fuzzy control system, based on Zadeh's formulations from 1973, was introduced by Mamdani in 1974 [26]. In Mamdani's system, the input numbers are translated into linguistic terms, and the fuzzy rules map them onto linguistic terms on output. Then, the output linguistic terms are translated back into the numbers. The procedures of translations are known as fuzzification and defuzzification, respectively. A typical fuzzy rule in such a system is constructed as follows:

$$IF\ input1\ is\ A_i^1\ AND\ input2\ is\ A_i^2$$
$$THEN\ output\ is\ B_i \tag{12}$$

It tries to formulate the expert knowledge by some linguistic rules. An exemplary rule dedicated to the task of tumor segmentation can be simply described as:

$$\begin{aligned}&\textbf{"IF}\ \textit{the intensity level of the area}\\&\textit{in}\ T2 - \textit{weighted series is}\ \textbf{very high}\\&\textbf{AND}\ \textit{the intensity level of the area}\\&\textit{in}\ T1 - \textit{weighted series is}\ \textbf{very low}\\&\textbf{THEN}\ \textit{the analyzed region}\\&\textbf{might be a tumour"}\end{aligned} \tag{13}$$

There are different combinations of grey intensity levels suggestive of a tumor, defined by the experts, and consequently different linguistic rules connected with them. Simultaneously, a set of linguistic rules exists which defines the healthy tissues. The developed fuzzy inference system attempts to describe the majority of them.

Let the fuzzy sets in the fuzzy premises of ith rule be given as A_i^1 and A_i^2, respectively and the fuzzy set in the conclusion of ith rule is denoted as B_i. In the exemplary radiologist reasoning rule, the fuzzy sets A_i^1 and A_i^2 are given as "high" and "low" and B_i as "might be tumor".

The fuzzy control algorithm, developed by Mamdani, is based on two concepts: fuzzy implication and compositional rule of inference [27]. Assume two fuzzy sets: A of the universe of discourse $\mathbb{X}$ and B of $\mathbb{Y}$ defined by their fuzzy membership functions μ_A and μ_B. The membership function of a fuzzy implication S: "IF A then B" is then defined as

$$\mu_s(x, y) = \min[\mu_A(x); \mu_B(y)], \quad x \in \mathbb{X}, \ y \in \mathbb{Y}. \tag{14}$$

For such given implication S, fuzzy set B' of the universe of discourse $\mathbb{Y}$ inferred by a given fuzzy set A' of $\mathbb{X}$, has a membership function estimated as

$$\mu_{B'}(y) = \max_x \min[\mu_{A'}(x); \mu_s(x, y)], \quad x \in \mathbb{X}, \ y \in \mathbb{Y}. \tag{15}$$

In fuzzy systems found in research applications, there are different rules which describe one phenomenon. When the rules conditions are matched, a set of actions will be activated. Each rule, with the antecedent non-zero matching degree, contributes an output with the activation value equal to it. The final system output, which takes all the activated rules into consideration, is constructed using an aggregation operation. Its most common implementation is operator max; however, different aggregation operators are found in real applications, like algebraic sum or the bounded product.

Coming back to the tumor analysis, the already described fuzzy system is used in the fuzzy connectedness analysis in order to estimate the fuzzy affinity value of the spels connection, instead of the functions given by (4) and (5).

5 Algorithm

Before the segmentation procedure begins, thanks to the positioning information provided by the DICOM header, the positions of voxels belonging to the two analyzed MR series are matched.

The segmentation procedure starts on the basis of exemplary regions, which have been selected by an expert. The automated part of analysis begins with the adaptive 3-D filtering method [28]. The there required parameters are adaptively estimated based on the assumptions given in [29]. The goal of this analysis step is firstly the reduction of noise and thereby an increase in the signal-to-noise or contrast-to-noise ratios, while maintaining the edge lines. Secondly, as a result of smoothing of the object areas, the number of groups for the clustering procedure, which is the next step, decreases and the analysis is not sensitive to outliers.

The main part of the performed analysis constitutes of four steps, discussed in previous sections. The combination of these four steps is shown on the block diagram in Fig. 5.

First, based on the reference expert selections, personalized GMMs of tumor and background areas are generated, separately. The detailed discussion of the algorithm is given in Sect. 3. Let the estimated GMMs of tumor area be given as $G_t^{1,2}$ and of the background as $G_b^{1,2}$, respectively. Both GMM pairs are the sets of mixture components parameters θ and voxels C classified into each of K_t or K_b groups

$$
\begin{aligned}
G_t^i &= (\theta_{1_t}^i, C_{1_t}^i), (\theta_{2_t}^i, C_{2_t}^i), \ldots, (\theta_{K_t}^i, C_{K_t}^i), \\
G_b^i &= (\theta_{1_b}^i, C_{1_b}^i), (\theta_{2_b}^i, C_{2_b}^i), \ldots, (\theta_{K_b}^i, C_{K_b}^i),
\end{aligned}
\tag{16}
$$

where index $i = \{1, 2\}$ refers to two simultaneously analyzed MR sequences.

Based on them, the input membership functions are defined which describe the intensity levels of tumor as well as background areas. An exemplary set of membership functions, obtained for a bone cyst in STIR and T1-weighted sequences (Fig. 1), is visualized in Fig. 3. The membership functions defined for the tumor area are marked with the black solid lines and the membership functions defined for the background are given by the grey dashed lines.

Since the fuzzy connectedness analysis is used in the segmentation step, the system must be able to calculate the affinity value of two adjacent spels based on the inputs.

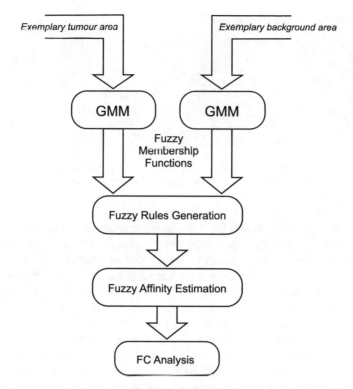

Fig. 2 The flow chart of 3-D tumors segmentation procedure

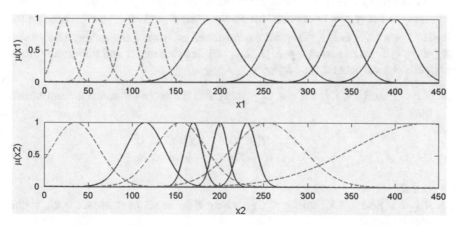

Fig. 3 An exemplary set of membership functions generated for a bone cyst in STIR (*top*) and T1-weighted (*down*) sequences. The membership functions defined for the tumor area are marked with the *black solid lines* and the membership functions defined for the background are given by the *grey dashed lines*

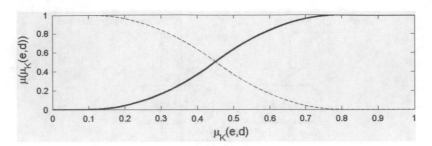

Fig. 4 An exemplary set of membership functions on the output of the fuzzy inference system. The *black solid line* defines the tumour and the *grey dashed line* represents the background

The functions given by (4) and (5) take the mean intensity value $0.5(I(\underline{e}) + I(\underline{d}))$ and gradient $|I(\underline{e}) - I(\underline{d})|$ into consideration. Because the attempts of modeling gradient values characterizing tumor or healthy tissues have not provided any useful information, the inputs to the fuzzy system are defined as $x_i = 0.5(I^i(\underline{e}) + I^i(\underline{d}))$.

Two membership functions in the conclusions of rules are shown in Fig. 4. The output of the system is the affinity value of two adjacent spels. The membership function visualized using the black solid line defines the "high" affinity and the dashed line defines the "low" affinity of spels connection. Moreover, to reduce the computation time associated with the relative FC analysis, the meaning of "big" affinity value is "it might be tumor". Based on the graph of the functions in Fig. 4, the threshold defining the tumor area can be set to 0.45.

Let the fuzzy sets in premises referring to $G_t^{1,2}$ and $G_b^{1,2}$ are given as $A_{k_t}^i$ and $A_{l_b}^i$, and the fuzzy sets in conclusions as B_h—"high" and B_l—"low". Then, on the basis of sets $G_t^{1,2}$ and $G_b^{1,2}$ and all the positions of voxel $\underline{c}$ the unique fuzzy rules R_t^l, $l \in \{1, 2, \ldots L\}$ and R_b^p, $p \in \{1, 2, \ldots P\}$ defining tumor and non-tumor areas, respectively, are generated as follows:

1: **if** $\underline{c} \in T$ **and** $\underline{c} \in C_{k_t}^1$ **and** $\underline{c} \in C_{k_t}^2$, where T is the set of reference tumour voxels
   then
2:
$$R_t^l : \begin{array}{l} IF\ x1\ is\ A_{k_t}^1\ AND\ x2\ is\ A_{l_t}^2 \\ THEN\ \mu_K(\underline{e}, \underline{d})\ is\ B_h \end{array} \quad (17)$$

3: **end if**
4: **if** $\underline{c} \in B$ **and** $\underline{c} \in C_{k_b}^1$ **and** $\underline{c} \in C_{k_b}^2$, where T is the set of reference background
   voxels **then**
5:
$$R_b^p : \begin{array}{l} IF\ x1\ is\ A_{k_b}^1\ AND\ x2\ is\ A_{l_b}^2 \\ THEN\ \mu_K(\underline{e}, \underline{d})\ is\ B_l \end{array} \quad (18)$$

6: **end if**

For each pair of the adjacent voxels, the output linguistic value is then translated into their fuzzy affinity. In the defuzzification step, the center of gravity method is employed. Using the precomputed affinity tables, the multiseeded FC algorithm described in [20] is implemented.

To reduce the false positive regions, in the case when the tumor is connected with the healthy tissues having similar characteristics, a convex hull-based postprocessing technique is applied. Starting from the reference slice, the there calculated tumor convex hull is then mapped into the adjacent slices. The comparison of areas of tumor like regions, covered and uncovered by the convex hull, provides the information concerning the final segmentation results.

6 Experiments and Results

To evaluate the ability of developed methodology, first the database consisting of 27 examinations of 18 patients studies has been used. The therein contained cases have included 5 types of bone tumors: *chondromas, Ewing's sarcomas, osteosarcomas, bone cysts and chondrosarcomas*. In total, 413 pairs of slices have been analyzed. An individual pair have consisted of T1-weighted, T1-weighted contrast enhanced and fat saturated, T2-weighted or STIR sequences in different MR projections: axial, sagittal and coronal. The FC threshold values have been set to 0.45 and 0.5.

All the achieved results have been discussed with an expert, who jugged them on each slice in each examination. As a result, the obtained image regions have been divided into three classes: true positive (TP)—the coherent areas containing a correctly indicated tumor, false positive (FP)—a coherent region containing healthy tissues incorrectly classified as tumorous, false negative (FN)—a coherent region containing tumor areas incorrectly classified as healthy tissue.

The accuracy of presented segmentation procedure has been estimated based on the following similarity coefficient

$$DV = \frac{FP + FN}{TP + FN}, \tag{19}$$

yielding the value equal 0 when the segmentation results are fully correct. The estimated DV value for the bone tumors database has been equal to 0.12, which is sufficient for computer assisted diagnosis systems.

Exemplary results for 3 different types of bone tumors are shown in Figs. 5, 6 and 7.

The original fuzzy connectedness algorithm (FC1), described in [10], as well as its modification (FC2), developed in [20], have been used to compare the obtained results. The results are categorized into two groups: the segmentation results in the homogeneous and in-homogeneous image series. The numerical results (DV values) are summarized in the Table 1, where the last column (FIS) provides the results achieved using the proposed methodology. The first row of Table 1 shows that the

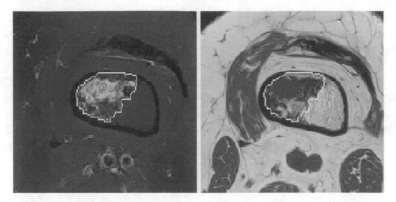

Fig. 5 Final segmentation results of knee Enchondroma visualized on a single slices of axial MR series: *left*—T2 Blade FS, *right*—T1 TSE

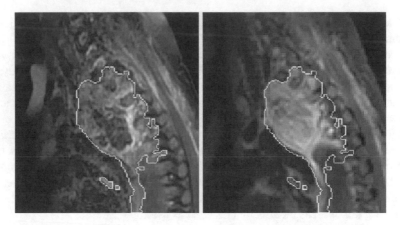

Fig. 6 Final segmentation results of spines Ewing's sarcoma visualized on a single slices of sagittal MR series: *left*—T1+C SE FS, *right*—STIR

segmentation results obtained for homogeneous image data are comparable with other methods. The second row proves a superiority of the proposed method over another approach, which is described in literature, whose results are insufficient for computer assisted diagnosis systems and non-acceptable by a radiologist.

To prove the segmentation abilities of the presented method, the database of brain tumors has been considered. Brain tumor image data used in this work have been obtained from the MICCAI 2012 Challenge on Multimodal Brain Tumor Segmentation organized by B. Menze, A. Jakab, S. Bauer, M. Reyes, M. Prastawa, and K. Van Leemput. The challenge database contains fully anonymized images from the following institutions: ETH Zurich, University of Bern, University of Debrecen, and University of Utah.

The analysis has been performed on two MR modalities: T1-weighted contrast enhanced and T2-weighted series. The initial segmentation procedure, performed

Fig. 7 Final segmentation
results of tibias
Osteosarcoma visualized on
a single slices of coronal MR
series: *left*—T2 FRFSE FS,
right—T1 FSE

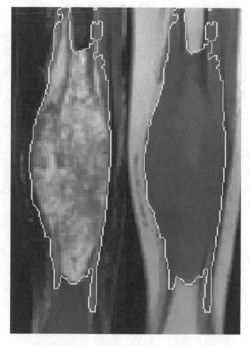

Table 1 Accuracy of
different segmentation
procedures—DV coefficient.

	FC1	FC2	FIS
Homogeneous Series (22) DV	0.16	0.13	**0.13**
In-homogeneous Series (5) DV	0.6	0.5	**0.11**

manually for bone tumors, is in this case based on ground truth data. The slice with maximal area of active tumor, provided by the ground truth, has been used for training step. To reduce the influence of intensity values of voxels which are located on the borders of the tumor, the active tumor as well as edema and healthy tissues masks have been eroded.

The obtained segmentation results have been compared with the expert delineation masks. To evaluate the performance of the developed model based fuzzy system, three similarity measures have been used, namely sensitivity:

$$S = \frac{TP}{TP + FN}, \tag{20}$$

specificity:

$$P = \frac{TN}{TN + FP}, \tag{21}$$

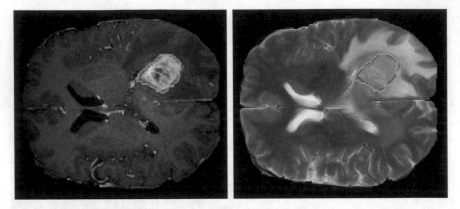

Fig. 8 Final segmentation results of active brain tumor area visualized on a single slices of axial MR series: *left*—T1+C, *right*—T2. The *red line* delineates the ground truth provided by the expert. The *green line* is the segmentation result

and the Dice Similarity Coefficient (DSC) [30]:

$$DSC = \frac{2 \cdot TP}{2 \cdot TP + FP + FN}. \tag{22}$$

Due to the expert delineations available in the database, the TP, TN, FN and FP values are given voxel-wise. The calculated mean values of the similarity measures for the brain tumor database are equal to $S = 0.82$, $P = 0.98$, and $DSC = 0.71$, respectively. Form the work [30] the results with the DSC value greater than 0.7 are considered to be very good (Fig. 8).

7 Conclusions

This paper introduces a new model-based fuzzy system for multimodal image segmentation. The proposed algorithm is insensitive to tumor location and type. It combines Gaussian Mixture Model and fuzzy inference system in the fuzzy connectedness analysis. The proposed procedure has been tested on two different medical databases. The first one consists of 27 examinations of 18 patients with different bone tumor types located in various parts of the body. Each single examination contains two different MR series. The second database provided for the Challenge on Multimodal Brain Tumor Segmentation contains the tumors in two different modalities with their manual expert delineation. The obtained segmentation results encourage to develop this method further. The presented system provides a basis for development of an adaptive learning algorithm by training based on the currently analyzed and verified cases. The problem still remaining to be solved is the normalization of MR sequences so that they can be compared. The plans for future work is to expand the database with new tumor cases and involving the analysis of new features, like texture. The detailed radiological consultation will enable developing fuzzy IF-THEN rules based and reasoning mechanism. In order to improve the segmentation results a fuzzy rules interpolation technique is also planned to be introduced.

Acknowledgments The authors would like to thank the medical staff of the Helimed Diagnostic Imaging Centre, Katowice, for providing the images.

This work was funded by the German Research Foundation (DFG) as part of the research training group GRK 1564 *Imaging New Modalities*.

References

1. Davies, A.M., Sundaram, M., James, S.L.J.: Imaging of Bone Tumors and Tumor-like Lesions, Techniques and Applications. Medical Radiology, Diagnostic Imaging. Springer, Berlin (2009)
2. Husband, J.E., Reznek, R.H.: Imaging in Oncology. Taylor & Francis, London (2004)
3. Ma, J., Li, M., Zhao, Y.: Segmentation of multimodality osteosarcoma MRI with vectorial fuzzy-connectedness theory. Fuzzy Systems and Knowledge Discovery. Lecture Notes in Computer Science, vol. 36(14), pp. 1027–1030. Springer, Berlin (2005)
4. Zhao, Y., Hong, F., Li, M.: Segmentation of osteosarcoma based on analysis of blood-perfusion epi series. In: International Conference on Communications, Circuits and Systems, ICCCAS 2004, vol. 2. IEEE (2004)
5. Pan, J., Li, M.: Segmentation of MR osteosarcoma images. In: International Conference on Computational Intelligence and Multimedia Applications (ICCIMA03). IEEE (2003)
6. Zhao, Y., Hong, F., Li, M.: Multimodality MRI information fusion for osteosarcoma segmentation. In: IEEE EMBS Asian-Pacific Conference on Biomedical Engineering, pp. 166–167 (2003)
7. Rosenfeld, A.: Fuzzy digital topology. Inf. Control **40**(1), 76–87 (1979)
8. Udupa, J.K., Samarasekera, S.: Fuzzy connectedness and object definition: theory, algorithms, and applications in image segmentation. Graph. Models Image Process. **58**(3), 246–261 (1996)
9. Pednekar, A., Kakadiaris, I.A., Kurkure, U.: Adaptive fuzzy connectedness-based medical image segmentation. In: Proceedings of the Indian Conference on Computer Vision, Graphics, and Image Processing (2008)
10. Udupa, J.K., Saha, P.K., Lotufo, R.A.: Relative fuzzy connectedness and object definition: theory, algorithms, and applications in image segmentation. IEEE Trans. Pattern Anal. Mach. Intell. **24**(11), 1485–1500 (2002)
11. Brant, W.E., Helms, C.A.: Fundamentals of Diagnostic Radiology, vol. I. MediPage, Warszawa (2007) Polish translation
12. Kawa, J., Szwarc, P., Bobek-Billewicz, B., Pitka, E.: Multiseries MR data in brain tumours segmentation. In: Pitka, E., Kawa, J., (eds.) Information Technologies in Biomedicine. Volume 69 of Advances in Intelligent and Soft Computing, pp. 53–64. Springer, Berlin (2010)
13. Caselles, V., Kimmel, R., Sapiro, G.: Geodesic active contours. Int. J. Comput. Vis. **22**(1), 61–79 (1997)
14. Yamaguchi, K., Fujimoto, Y., Kobashi, S., Wakata, Y., Ishikura, R., Kuramoto, K., Imawaki, S., Hirota, S., Hata, Y.: Automated fuzzy logic based skull stripping in neonatal and infantile MR images. In: 2010 IEEE International Conference on Fuzzy Systems (FUZZ), pp. 1–7 (2010)
15. Hata, Y., Kobashi, S., Hirano, S., Kitagaki, H., Mori, E.: Automated segmentation of human brain mr images aided by fuzzy information granulation and fuzzy inference. IEEE Trans. Syst. Man Cybern. Part C: Appl. Rev. **30**(3), 381–395 (2000)
16. Tolias, Y., Panas, S.: On applying spatial constraints in fuzzy image clustering using a fuzzy rule-based system. Signal Process. Lett. IEEE **5**(10), 245–247 (1998)

17. Mari, M., Dellepiane, S.: A segmentation method based on fuzzy topology and clustering. In: Proceedings of the 13th International Conference on Pattern Recognition, 1996, vol. 2, pp. 565–569 (1996)
18. Carvalho, B.M., Gau, C.J., Herman, G.T., Kong, T.Y.: Algorithms for fuzzy segmentation. Pattern Analysis & Applications 2, 73–81 (1999)
19. Saha, P.K., Udupa, J.K.: Fuzzy connected object delineation: axiomatic path strength definition and the case of multiple seeds. Comput. Vis. Image Underst. 83(3), 275–295 (2001)
20. Badura, P., Kawa, J., Czajkowska, J., Rudzki, M., Pietka, E.: Fuzzy connectedness in segmentation of medical images. In: International Conference of Fuzzy Computation Theory and Applications, pp. 486–492, October (2011)
21. McLachlan, G., Peel, D.: Finite Mixture Model. Wiley Series in Probability and Statistics (2000)
22. Heo, G., Gader, P.: An extension of global fuzzy c-means using Kernel methods. In: IEEE International Conference on Fuzzy Systems, July (2010)
23. Dempster, A.P., Laird, N.M., Rubin, D.B.: Maximum likelihood from incomplete data via the EM algorithm. J. R. Stat. Soc. Ser. B (Methodological) 39(1), 1–38 (1977)
24. Czajkowska, J., Bugdol, M., Pietka, E.: Kernelized fuzzy c-means method and Gaussian mixture model in unsupervised cascade clustering. In: International Conference of Information Technologies in Biomedicine, Lecture Notes in Bioinformatics, Gliwice, Poland, pp. 58–66, June (2012)
25. Siler, W., Buckley, J.J.: Fuzzy Expert Systems and Fuzzy Reasoning. Wiley, Hoboken (2005)
26. Mamdani, E.: Application of fuzzy algorithms for control of simple dynamic plant. Proc. Inst. Electr. Eng. 121(12), 1585–1588 (1974)
27. Kickert, W.J.M., Mamdani, E.H.: Analysis of a fuzzy logic controller. Fuzzy Sets Syst. 1(1), 29–44 (1978)
28. Perona, P., Shiota, T., Malik, J.: Anisotropic diffusion. Geometry-Driven Diffusion in Computer Vision, pp. 73–92. Kluwer Academic Publishers, Dordrecht (1994)
29. Positano, V., Santarelli, M. F., Landin, L., Benassi, A.: Nonlinear anisotropic filtering as a tool for SNR enhancement in cardiovascular MRI. In: Computers in Cardiology, pp. 707–710. IEEE (2000)
30. Dice, L.R.: Measures of the amount of ecologic association between species. Ecology 26(3), 297–302 (1945)

Multiple Fuzzy Roles: Analysis of Their Evolution in a Fuzzy Agent-Based Collaborative Design Platform

Alain-Jérôme Fougères and Egon Ostrosi

Abstract Design for configurations is a highly collaborative and distributed process. The use of fuzzy agents, that implement the collaborative and distributed design by means of fuzzy logic, is highly recommended due to the fuzzy nature of the collaboration, distribution, interaction and design problems. In this paper, we propose a fuzzy agent model, where fuzzy agents grouped in communities interact and perform multiple fuzzy design roles to converge towards solutions of product configuration. Analysis of both interactions and multiple fuzzy roles of fuzzy agents during product configuration in a collaborative design platform is proposed. The modelling of fuzzy agents and its illustration for a collaborative design platform are presented. The results of analysis have shown the important influence of fuzzy solution agents in the organization of the agent based collaborative design for configurations platform. The more the fuzzy agents share their knowledge, the more their fuzzy roles are complete in every domain of design for configurations. The degree of interactions between fuzzy agents in the design for configurations process has an impact on the emergence of increased activity of some fuzzy agents. The fuzzy function agents, influenced by many fuzzy requirement agents, are the most active in the design process. The simulation shows that this observation can be extended to the fuzzy solution agents. The most active fuzzy solution agents are those which create the best consensual solution. Simulations show that the consensus can be found principally by increasing the degree of interactions.

Keywords Fuzzy agents · Fuzzy roles · Fuzzy agent-based system · CAD

A.-J. Fougères (✉) · E. Ostrosi
Research Laboratory IRTES-M3M, 90010 Belfort, France
e-mail: alain-jerome.fougeres@utbm.fr

E. Ostrosi
e-mail: egon.ostrosi@utbm.fr

A.-J. Fougères
ESTA, School of Business and Engineering, 90004 Belfort, France
e-mail: ajfougeres@esta-belfort.fr

© Springer International Publishing Switzerland 2016
K. Madani et al. (eds.), *Computational Intelligence*,
Studies in Computational Intelligence 613,
DOI 10.1007/978-3-319-23392-5_12

1 Introduction

Design for configurations is the process which generates a set of product configurations based on a configuration model. A product configuration is characterized by a set of solutions, which are designed to satisfy product functions, which in their turn, are supposed to meet customer requirements. This set of solutions should also satisfy the specific process domain constraints. Configuration starts with requirements in the domain of requirements. A customization requirement is manifested by the customer's choice of customizable requirement. The customer perceived value of each requirement indicates the degree of customer satisfaction in the requirement domain. Simultaneously, in the process domain, a constraint is manifested by the expert's choice of process constraint [2]. The expert perceived value of each process constraint indicates the degree of expert satisfaction in the process domain. Therefore, to satisfy customer requirements and process constraints, the mapping from requirements to the solutions as well as the mapping from process constraints to the solutions is applied. It yields a set of consensual solutions from both domains: requirements and process constraints. The consensual solutions problem is how to achieve the maximum consensus degree from a group of distributed experts for the alternative solutions, satisfying customer requirements [30]. Thus the concept of consensus is a problem of the overlapping of experts' and customers' perspectives influencing the design of configurable products simultaneously. Discerning the consensus nucleus can create common ground for moving towards an acceptable configuration [12]. This set of consensual solutions can be distributed in modules to form configurations [27]. Optimal configurations can be generated using some limits of acceptability for objective function values. It enables the early release of possible set of configurations [28].

Following up these phases, configurable product design must be able to deal with various unstable and imprecise requirements coming from the customers, on the one hand, and some distinct form of uncertainty such as imprecision, randomness, fuzziness, ambiguity, and incompleteness, on the other [2]. Uncertainty is thus an integral part of the design for configurations [1, 2].

Fuzzy logic offers a framework for representing uncertainty [35]. In order to capture the uncertainty aspects of design for configurations, the fuzzy sets approach can be used [2]. Design for configurations is a highly collaborative and distributed process. The properties of collaborative and distributed design for configurations are discussed in [28]. It is shown that designs for configurations are fuzzy information and knowledge-based processes. They are fuzzy interaction-based processes and their organizations are heterogeneous, dynamic, and adaptive. Designs for configurations are also fuzzy evolving systems [20]. Therefore, the use of agents, that implement the collaborative and distributed design by means of fuzzy logic, is highly recommended due to the fuzzy nature of the collaboration, distribution, interaction and design problems [12, 28]. Fuzzy agents interact between themselves to adjust their actions using their fuzzy knowledge [12]. They interpret the fuzzy information they receive or perceive. Their evolution is fuzzy [17], when they are designed to interpret fuzzy information and to adopt a fuzzy behavior [12, 28]. Fuzzy agents are also well adapted to model and to design the heterogeneity and the evolving of some organizations [9].

Thus, fuzzy agent modelling based design for configurations is an open-ended question. Indeed, fuzzy agents are currently not sufficiently formalized to support the holistic view of collaborative and distributed designs for configurations with a certain level of uncertainty. In many models of collaborative and distributed agent-based systems, an agent or a group of agents are modelled to perform only one role. Some models allow agents to change their role within their community or the defined organization. In this paper, we propose a model, where a fuzzy agent can perform several roles at any time in their community or the defined organization. The fuzzy agents can perform their roles with varying degrees. This hypothesis of fuzzy agents' model relies on the practice of collaborative and distributed design. Usually, each actor is expert in a main discipline. Furthermore, the actors involved in product design are experienced in solution design. Thus, solution design is a shared domain.

Therefore, this paper proposes to analyze both the evolution of agents' fuzzy roles and the change of their distribution in different communities of an organization, within a collaborative and distributed design for configurations platform. These analyses continue the work we have already done on the interactions between cognitive agents [8, 10, 11], or rather reactive agents [12, 28, 29].

The remainder of the paper is organized in five sections. In the second section, a fuzzy agent model is proposed. In the third section, the proposed fuzzy agent model is illustrated by a design for configurations case study. In this case study, firstly, a fuzzy product configuration model is presented. Then, secondly, an agentification of this model is developed, and thirdly, an analysis of fuzzy interactions and fuzzy roles agents is presented. In the fifth section, the conclusion shows some perspectives and interest in the proposed approach.

2 Fuzzy Agent Modeling

There are at present many definitions of the agent paradigm [6, 9, 15, 19, 23, 34] and several propositions of typologies [26, 31], but new types of agents are continuing to emerge [32]. Thus, fuzzy agents emerged as a tool to model fuzzy behavior problems [10], where agents can decide to act according to a fuzzy-logic rule base [5, 14]. Fuzzy agents are also used in fuzzy reasoning situations, where agents interpret a situation, solve a problem or decide with fuzzy knowledge [3, 4, 13, 16]. Implementations of fuzzy agents are also proposed to solve distributed fuzzy problems [25], or to improve the processing of the fuzziness of information, fuzziness of knowledge and fuzziness of interactions, in collaborative design processes [12, 28]. This section presents a model where agents are completely fuzzy: their knowledge and their behavior are fuzzy, their interactions are fuzzy, their roles in the agent-based system are fuzzy, and their organization in the agent-based system is also fuzzy.

2.1 Fuzzy Agent Model

An agent-based system is fuzzy if agents that make it up are fuzzy, which means that:

- *Their Knowledge and their Behaviors are Fuzzy.* Knowledge of an agent is defined by fuzzy values. Behavior of an agent depends on the fuzzy evaluation of its fuzzy perceptions, its fuzzy decisions, and its fuzzy actions.
- *Their Interactions are fuzzy.* Relationships between agents (affinities) are weighted by a fuzzy value. Interactions provide a relative interest to fuzzy agents based on roles that they perform at a given time.
- *Their Roles are Fuzzy.* At a given time, it is possible to determine what roles a fuzzy agent performs based on fuzzy values of its roles and a threshold value setting the minimum value an agent should invest in these roles.
- *Their Organization is fuzzy.* The distribution of roles performed by fuzzy agents is continually evolving. This defines self-organizing agents which is the result both of their fuzzy interactions and the continuing evolution of their roles.

Agents developed in our different collaborative platform could perform reflex actions, routine actions, and actions in new situations (creative or cooperative) [7, 8]. Recently, we integrated fuzziness characteristics in our agent model [12, 28] (Fig. 1).

A fuzzy agent-based system is described by the following tuple (1):

$$\tilde{M}_\alpha = < \tilde{A}, \tilde{I}, \tilde{P}, \tilde{O} > \tag{1}$$

where $\tilde{A}$, $\tilde{I}$, $\tilde{P}$, and $\tilde{O}$, are respectively a fuzzy set of agents, a fuzzy set of interactions between fuzzy agents, a fuzzy set of roles that fuzzy agents can perform, a fuzzy set of organizations (or communities) defined for fuzzy agents of $\tilde{A}$.

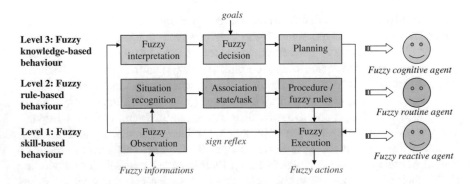

Fig. 1 Behavior of fuzzy agents, based on Rasmussen's model

A fuzzy agent $\tilde{\alpha}_i \in \tilde{A}$ is described by the following tuple (2):

$$\tilde{\alpha}_i = < \Phi_{\tilde{\Pi}(\tilde{\alpha}_i)}, \Phi_{\tilde{\Delta}(\tilde{\alpha}_i)}, \Phi_{\tilde{\Gamma}(\tilde{\alpha}_i)}, \tilde{K}_{\tilde{\alpha}_i} > \tag{2}$$

where $\Phi_{\tilde{\Pi}(\tilde{\alpha}_i)}$, $\Phi_{\tilde{\Delta}(\tilde{\alpha}_i)}$ and $\Phi_{\tilde{\Gamma}(\tilde{\alpha}_i)}$ are respectively functions of observation, decision and action [9]. The set of fuzzy knowledge $\tilde{K}_{\tilde{\alpha}_i}$ includes decision rules, values of domain, acquaintances, and dynamic knowledge (observed events, internal states).

2.2 Fuzzy Interaction, Fuzzy Organization, and Fuzzy Role

In agent-based systems, as in human organizations, actions, interactions and communications, are closely linked and interdependent [15]. Interaction is an exchange between agents and their environment. This exchange depends on the intrinsic properties of the world in which agents are active. Perception of agents may be passive when receiving messages/signals, or active, when it is the result of voluntary actions. Communication is an exchange between the agents themselves, using a language.

A fuzzy interaction $\tilde{\iota}_{s,r} \in \tilde{I}$ between two fuzzy agents is defined by (3):

$$\tilde{\iota}_{s,r} = < \tilde{\alpha}_s, \tilde{\alpha}_r, \tilde{P}_{\tilde{\alpha}_s}, \tilde{\gamma}_i > \tag{3}$$

where $\tilde{\alpha}_s$ is the fuzzy agent source of the interaction, $\tilde{\alpha}_r$ is the fuzzy agent destination, $\tilde{P}_{\tilde{\alpha}_s}$ is the fuzzy set of roles performed by $\tilde{\alpha}_s$, and $\tilde{\gamma}_i$ is a fuzzy act of cooperation. Interactions are fuzzy: the destination agent also always evaluates an interaction (fuzzy value) to determine the interest this interaction can take for it.

Problems due to the partial view of agents (local goals, interleaving activities, etc.) require the development of strong coordination mechanisms [18]. The organization shall allow an agent-based system to behave as a coherent whole, to solve a problem unequivocally. It controls and coordinates the interaction between agents of the system, thus structuring their activities with the goal of convergence. Ferber et al. [7] distinguish between "organizational structure" and "organization", corresponding to the process of designing the structure. Wooldridge [34] proposed a more practical definition: "a collection of roles that stand in certain relationships to one another and that take part in systematic institutionalized patterns of interactions with other roles".

From the numerous definitions of agent organization [6, 7, 15, 17, 21, 33, 34], we extracted the following properties, before interpreting them in the fuzzy field (Fig. 2a):

- $P1$. An organization is partitioned into groups or communities of agents.
- $P2$. A community is comprised of agents sharing a goal and characteristics.
- $P3$. An agent can belong to several communities.
- $P4$. An agent performs one or several roles within the community(ies) to which it belongs.

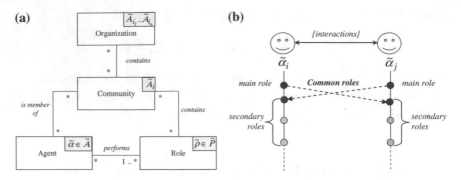

Fig. 2 Fuzzy agents: **a** organization model, **b** interactions and induced common roles

- *P5.* A role is an abstract representation of a function performed by agents within one or several communities.
- *P6.* An agent interacts with the agents of its community or other communities to perform its roles.
- *P7.* An agent that interacts with another agent then participates in the same role as the latter (Fig. 2b).

In a collaborative structure different roles are performed by agents. Modelling the notion of roles for the agent paradigm can take many forms (Fig. 3):

- In many models of distributed agent-based systems, agents perform only one role in their community or the defined organization: the role for which they are designed. Sometimes several agents can perform the same role.

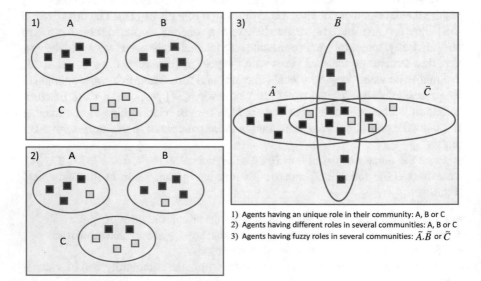

1) Agents having an unique role in their community: A, B or C
2) Agents having different roles in several communities: A, B or C
3) Agents having fuzzy roles in several communities: $\tilde{A}, \tilde{B}$ or $\tilde{C}$

Fig. 3 Distributions of agents in communities based on roles they perform at a given time

- Some models allow agents to change their role within their community or the defined organization. At any given time, an agent will perform one role. Agents change roles at times determined by the context of problem solving or group activity. In this case, the role change corresponds to a context switch.
- A more innovative model where agents can perform several roles at any time in their community or the defined organization. In this case, the agents perform their roles with varying degrees, which means that a role may be singled out and others are active. In this case, fuzzy set theory is well suited to modelling and designing such roles. This is the solution that we will develop in this paper.

During our experiments on collaborative and distributed design, we observed that designers were more widely involved in terms of their unique area of expertise [11, 22, 24]. This is observable in sequences of creativity, where designers perform several roles in the same sequence with greater or lesser degrees. We model this property with the theory of fuzzy sets. We also proposed that the roles of agents are considered fuzzy. An agent in this organization can have several fuzzy roles at a given time. In that case, the fuzzy set of roles performed by a fuzzy agent $\tilde{\alpha}_i$ is defined by (4):

$$\tilde{P}(\tilde{\alpha}_i) = \left\{ \mu_{\tilde{\rho}_1}(\tilde{\alpha}_i), \mu_{\tilde{\rho}_2}(\tilde{\alpha}_i), \ldots, \mu_{\tilde{\rho}_q}(\tilde{\alpha}_i) \right\} \tag{4}$$

During cooperative activities, a fuzzy agent performs roles according to its knowledge and its fuzzy interactions. A fuzzy agent interacts by sending messages within its initial community (performing its main role), or within other communities (performing other roles). A fuzzy agent $\tilde{\alpha}_i$ by interacting with a fuzzy agent $\tilde{\alpha}_j$ of another community then participates in the same role as $\tilde{\alpha}_j$ (5):

$$\forall \tilde{\alpha}_i \in \tilde{A} \supset [\exists x : \tilde{\rho}_x \in \tilde{P} \wedge \alpha_j \in \tilde{A}_x, \Phi_{\tilde{P}}(\tilde{\alpha}_j, \tilde{\rho}_x) \wedge \tilde{\lambda}_{i,j}(\tilde{\alpha}_i, \tilde{\alpha}_j, \tau, \tilde{\eta}) \supset \Phi_{\tilde{P}}(\tilde{\alpha}_i, \tilde{\rho}_x)] \tag{5}$$

3 Product Configuration Approach

To analyze roles of fuzzy agents within a collaborative design platform, a "*chair configurable product*" is chosen because of both the simplicity and accessibility of this illustration. A chair is made up of a few elements, but it can be configured in multiple ways satisfying both customer's requirements and different experts' process views.

3.1 Fuzzy Product Configuration Model

The configurable product design is a mapping process between product requirement view, functional view, physical solution view, process view and fuzziness of collaborative design process. We proposed a fuzzy approach for searching configu-

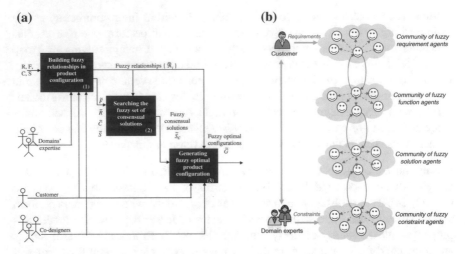

Fig. 4 **a** Product configuration approach, and **b** agent-based architecture of *FAPIC* platform

ration structures [2, 28, 29]. This approach is carried out into three phases (Fig. 4a): (1) *Fuzzy relationships in engineering design*: the results are the different engineering design models, from requirements to solutions, necessary for the configuration of a product (the fuzzy sets $\tilde{R}, \tilde{F}, \tilde{C}, \tilde{S}$); (2) *Searching the fuzzy set of consensual solutions*: the result is a fuzzy set of consensual solutions (the fuzzy set $\tilde{S}_c$); and (3) *Fuzzy optimal solution agents based product configuration*: the results are optimal solutions (the fuzzy set $\tilde{G}$).

3.2 Agentification of the Configuration Approach

Requirements, functions, constraints and solutions are fuzzy agents, with a degree of membership in each community defined for configuration ($\tilde{R}, \tilde{F}, \tilde{C}, \tilde{S}$). Cooperative interaction can occur between fuzzy agents in communities of functions and solutions (*intra-communities interaction*), or between fuzzy agents of different communities (*inter-communities interaction*). A fuzzy interaction is defined by (3) and the degree of interest of a fuzzy interaction $\mu_{\tilde{\alpha}_j}(\tilde{\iota}_{i,j})$ for a fuzzy agent $\tilde{\alpha}_j$ is defined by (6):

$$\mu_{\tilde{\alpha}_j}(\tilde{\iota}_{i,j}) = \min\left(\mu_{\tilde{\alpha}_j}(\tilde{\alpha}_i), \mu_{\tilde{\rho}_r}(\tilde{\alpha}_j), \mu_{\tilde{\rho}_r}(\tilde{\alpha}_i)\right) \tag{6}$$

where $\mu_{\tilde{\alpha}_j}(\tilde{\alpha}_i)$ is the degree of affinity between $\tilde{\alpha}_j$ and $\tilde{\alpha}_i$, $\mu_{\tilde{\rho}_r}(\tilde{\alpha}_j)$ and $\mu_{\tilde{\rho}_r}(\tilde{\alpha}_i)$ are the membership functions of $\tilde{\alpha}_j$ and $\tilde{\alpha}_i$ in performing the role $\tilde{\rho}_r$.

A fuzzy agent-based platform called *FAPIC* (Fuzzy Agents for Product Integrated Configuration) was developed for product configuration (Fig. 4b). In *FAPIC*, fuzzy agents are organized in four communities (7):

$$\tilde{A}_r \subseteq \tilde{A}, \ \tilde{A}_f \subseteq \tilde{A}, \ \tilde{A}_c \subseteq \tilde{A}, \ \tilde{A}_s \subseteq \tilde{A} \tag{7}$$

Each community has a clear objective, which determines the main role that fuzzy agents perform in their communities [28]. This means that each fuzzy agent belongs to a community of reference in which it plays its main role (8):

$$\forall \tilde{\alpha} \in \tilde{A} \supset [\exists x \in \{r, f, c, s\}, \ \tilde{\alpha} \in \tilde{A}_x \wedge \ \Phi_{\tilde{p}}(\tilde{\alpha}, \tilde{\rho}_x)] \tag{8}$$

4 Illustration for a Chair Configuration

4.1 Presentation of the Case Study

This section gives a detailed illustration for the three phases of the proposed approach (Fig. 4a).

In the first phase (*Fuzzy agents based systems building*) communities of fuzzy agents are built. In this case study, 11 fuzzy requirement agents, 4 fuzzy function agents, 20 fuzzy solution agents, and 16 fuzzy constraint agents, are built (cf. Appendix II). Then, interactions between fuzzy agents of all communities are built.

The second phase (*Searching fuzzy set of consensual solution*) comprises six steps:

- *Step 1: Definition of Fuzzy Set of Requirements.* The fuzzy set of requirements for a particular customer is defined. The fuzzy requirement agents observe this fuzzy set and take the corresponding fuzzy values.
- *Step 2: Emergence of Fuzzy Product Functions.* It spells out functions that the configuration product will support. The fuzzy set of product function agents are computed using the fuzzy relationship between requirement agents and product function agents.
- *Step 3: Emergence of Fuzzy Set of Solutions.* The fuzzy set of solutions is computed from interaction between the set of active function agents and solution agents.
- *Steps 4 and 5: Definition and Integration of Fuzzy Set of Constraints.* The fuzzy constraints agents observe what the constraints of a particular process view are and they decide to take the corresponding fuzzy values.
- *Step 6: Emergence of Consensual Fuzzy set of Solutions.* Fuzzy constraint agents interact with fuzzy solution agents to converge towards a consensual fuzzy set of solutions.

In the third phase (*Fuzzy optimal solution for configuration*), the consensual solution agents are structured into modules, through their interactions, using their affinities from the fuzzy solution agents' structure. The fuzzy optimal solution agents represent

Table 1 Optimal configuration: local point of view of fuzzy solution agents

Agent	Optimal configuration	Value	Agent	Optimal configuration	Value
$\tilde{s}_1$	$\tilde{s}_1 - \tilde{s}_6 - \tilde{s}_{16}$	**2.25**	$\tilde{s}_{11}$	–	0
$\tilde{s}_2$	$\tilde{s}_2 - \tilde{s}_6 - \tilde{s}_{16}$	2.1	$\tilde{s}_{12}$	–	0
$\tilde{s}_3$	$\tilde{s}_3 - \tilde{s}_7 - \tilde{s}_{17}$	1.95	$\tilde{s}_{13}$	–	0
$\tilde{s}_4$	$\tilde{s}_4 - \tilde{s}_6 - \tilde{s}_{16}$	1.5	$\tilde{s}_{14}$	–	0
$\tilde{s}_5$	$\tilde{s}_5 - \tilde{s}_6 - \tilde{s}_{16}$	1.8	$\tilde{s}_{15}$	–	0
$\tilde{s}_6$	$\tilde{s}_1 - \tilde{s}_6 - \tilde{s}_{16}$	**1.4**	$\tilde{s}_{16}$	$\tilde{s}_1 - \tilde{s}_9 - \tilde{s}_{16}$	**1.7**
$\tilde{s}_7$	$\tilde{s}_1 - \tilde{s}_7 - \tilde{s}_{19}$	1.2	$\tilde{s}_{17}$	$\tilde{s}_1 - \tilde{s}_9 - \tilde{s}_{17}$	1.45
$\tilde{s}_8$	$\tilde{s}_2 - \tilde{s}_8 - \tilde{s}_{17}$	1.15	$\tilde{s}_{18}$	$\tilde{s}_1 - \tilde{s}_7 - \tilde{s}_{18}$	1.4
$\tilde{s}_9$	$\tilde{s}_1 - \tilde{s}_9 - \tilde{s}_{19}$	1.0	$\tilde{s}_{19}$	$\tilde{s}_1 - \tilde{s}_9 - \tilde{s}_{19}$	1.5
$\tilde{s}_{10}$	$\tilde{s}_1 - \tilde{s}_{10} - \tilde{s}_{16}$	1.2	$\tilde{s}_{20}$	$\tilde{s}_1 - \tilde{s}_6 - \tilde{s}_{20}$	1.2

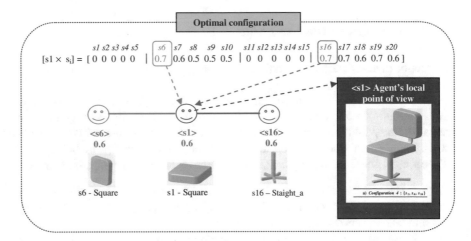

Fig. 5 Configuration: local point of view of agent $\tilde{s}_1$

a network of fuzzy solution agents which maximize the objective function. Results of this phase are given in Table 1. For instance, considering the fuzzy solution agent $\tilde{s}_1$ as solution for the class Cl_1 (Seat, *Appendix II*), its optimal network is formed by the solution agents $[\tilde{s}_1\text{-}\tilde{s}_6\text{-}\tilde{s}_{16}]$, with a value of objective function equal to 2.25 (Fig. 5).

4.2 Analysis of Fuzzy Agents Roles

In *FAPIC*, the set of fuzzy roles $\tilde{P} = \{\tilde{\rho}_r, \tilde{\rho}_f, \tilde{\rho}_c, \tilde{\rho}_s\}$ is defined. Then, the fuzzy set of roles an agent $\tilde{\alpha}_i$ performs is defined by (9):

$$\tilde{P}(\tilde{\alpha}_i) = \left\{ \mu_{\tilde{\rho}_r}(\tilde{\alpha}_i), \mu_{\tilde{\rho}_f}(\tilde{\alpha}_i), \mu_{\tilde{\rho}_c}(\tilde{\alpha}_i), \mu_{\tilde{\rho}_s}(\tilde{\alpha}_i) \right\} \tag{9}$$

Let us consider Phase 2 of the configuration process and the fuzzy agents $\tilde{r}_1$, $\tilde{f}_1$, $\tilde{c}_{11}$ and $\tilde{s}_1$ (traced agents) (Fig. 6). The fuzzy values of roles performed by an agent $\tilde{\alpha}_i$ are calculated by the formula (10):

$$(n_e/n_a)/((n_e/n_a) + 1) \tag{10}$$

where n_e is the number of exchanges between $\tilde{\alpha}_i$ and agents of the community corresponding to the target role and n_a is the number of agents in the community corresponding to the target role.

The following steps are illustrated in Fig. 6:

- *Step 1.* $\tilde{r}_1$ interacts with the 10 other members of the requirements community $\tilde{R}$. At this time $\tilde{r}_1$ performs one role: $\ll$ Definition of requirements $\gg$.
- *Step 2.* $\tilde{r}_1$ interacts with $\tilde{f}_1$, and participates in the role of $\ll$ the definition of functions $\gg$; then $\tilde{f}_1$ interacts with the 3 other members of the fuzzy functions community $\tilde{F}$. At this time $\tilde{f}_1$ performs two roles: $\ll$ Integration of requirements $\gg$ and $\ll$ Definition of functions $\gg$.
- *Step 3.* $\tilde{f}_1$ interacts with $\tilde{s}_1$, and participates in the role of $\ll$ Definition of solutions $\gg$. Then $\tilde{s}_1$ interacts with the 19 other members of the fuzzy solutions community $\tilde{S}$. At this time, $\tilde{s}_1$ performs two roles: $\ll$ Integration of functions $\gg$ and $\ll$ Definition of solutions $\gg$.
- *Step 4.* $\tilde{c}_{11}$ interacts with the 15 other members of the constraints community $\tilde{C}$. At this time, $\tilde{c}_{11}$ performs one role: $\ll$ Definition of constraints $\gg$.
- *Step 5.* $\tilde{c}_{11}$ interacts with $\tilde{s}_1$, and participates in the role of $\ll$ definition of solutions $\gg$; then $\tilde{s}_1$ interacts with the 19 other members of the fuzzy solutions community $\tilde{S}$. At this time, $\tilde{s}_1$ performs two roles: $\ll$ Integration of constraints $\gg$ and $\ll$ Definition of solutions $\gg$.
- *Step 6.* $\tilde{s}_1$ interacts again with the 19 other members of the solutions community $\tilde{S}$. At this time, $\tilde{s}_1$ performs the role: $\ll$ Definition of consensus solutions $\gg$.

The six tables presented in Fig. 6 show the change step by step of the fuzzy values of agents' roles during Phase 2. These tables indicate for each step of the Phase 2 and each of the four tracks fuzzy agents $\left(\tilde{r}_1, \tilde{f}_1, \tilde{c}_{11} \text{ and } \tilde{s}_1 \right)$: (1) the number of exchanges between these fuzzy agents and other fuzzy agents of *FAPIC* (inter or intra-community interactions: $\tilde{R}/\tilde{R}, \tilde{R}/\tilde{F}, \tilde{F}/\tilde{F}, \tilde{F}/\tilde{S}, \tilde{C}/\tilde{C}, \tilde{C}/\tilde{S}, \tilde{S}/\tilde{S}$), and (2) the fuzzy values of the different fuzzy roles performed by the fuzzy agents (a vector of fuzzy roles corresponding to $\tilde{P} = \{ \tilde{\rho}_r, \tilde{\rho}_f, \tilde{\rho}_c, \tilde{\rho}_s \}$).

Finally, after a full configuration, we obtain for fuzzy agents $\tilde{r}_1, \tilde{f}_1, \tilde{c}_{11}, \tilde{s}$ (our track agents), the number of inter/intra-communities exchanges and the fuzzy values of roles given in the first table of the Fig. 7 (Fig. 7a).

(a)	**(b)**

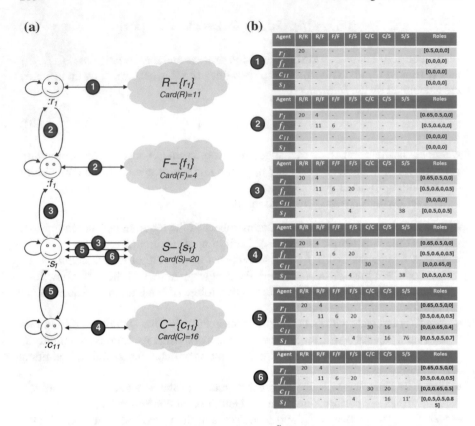

Fig. 6 **a** Illustration of interactions of fuzzy agents $\tilde{r}_1$, $\tilde{f}_1$, $\tilde{c}_{11}$, $\tilde{s}$ during Phase 2 of configuration; and **b** evolution of their fuzzy roles

The three tables presented in the figure below (Fig. 7b, c, d) show the evolving roles for three different and frequent scenarios:

- The first scenario corresponds to the change in requirements made by the customer when the results are not fully satisfactory. The consequences of this change are: (a) the roles of requirements and functions are enhanced for fuzzy agents $\tilde{r}_1$ and $\tilde{f}_1$, and (b) the roles of functions and solutions are reinforced for fuzzy agent $\tilde{s}_1$.
- The second scenario is the change of constraints by one of the expert domains (here the domain of production) after obtaining the results of the configuration and that it does not fully comply. In this case, we find that: (a) the role of constraint is enhanced for the fuzzy agent $\tilde{c}_{11}$, and (b) the roles of constraints and solutions are reinforced for the fuzzy agent $\tilde{s}_1$.

(a)

Agent	R/R	R/F	F/F	F/S	C/C	C/S	S/S	Roles
r_1	220	44	-	-	-	-	-	[1,0.9,0,0]
f_1	-	11	66	220	-	-	-	[0.5,1,0,0.9]
c_{11}	-	-	-	-	1512	560	-	[0,0,1,0.9]
s_1	-	-	-	80	-	560	12800	[0,0.9,0.9,1]

(c)

Agent	R/R	R/F	F/F	F/S	C/C	C/S	S/S	Roles
r_1	20	4	-	-	-	-	-	[0.65,0.5,0,0]
f_1	-	11	6	20	-	-	-	[0.5,0.6,0,0.5]
c_{11}	-	-	-	-	60	40	-	[0,0,0.79,0.67]
s_1	-	-	-	4	-	32	152	[0,0.5,0.67,0.88]

(b)

Agent	R/R	R/F	F/F	F/S	C/C	C/S	S/S	Roles
r_1	40	8	-	-	-	-	-	[0.78,0.67,0,0]
f_1	-	22	12	40	-	-	-	[0.67,0.75,0,0.67]
c_{11}	-	-	-	-	30	20	-	[0,0,0.65,0.44]
s_1	-	-	-	8	-	16	152	[0,0.67,0.5,0.88]

(d)

Agent	R/R	R/F	F/F	F/S	C/C	C/S	S/S	Roles
r_1	40	8	-	-	-	-	-	[0.78,0.87,0,0]
f_1	-	22	12	40	-	-	-	[0.67,0.75,0,0.67]
c_{11}	-	-	-	-	30	20	-	[0,0,0.65,0.5]
s_1	-	-	-	8	-	16	114	[0,0.67,0.5,0.85]

Fig. 7 Fuzzy values of roles: **a** at the end of the process, and **b, c, d** for three basic scenarios

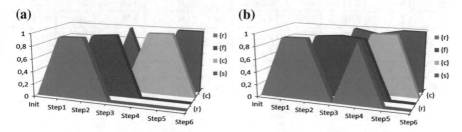

(a) **(b)**

Fig. 8 Activity of fuzzy agents during the 6 steps of Phase 2: **a** without change, **b** with change of requirements in step 3 (cf. Fig. 7d)

- The third scenario is the change in requirements by the customer before he has received the results of the configuration (for instance, he realizes that he has ill-defined his need, and he does not expect the outcome of configuration to change it). Then the results of this change are: (a) the roles of requirements and functions are enhanced for the fuzzy agents $\tilde{r}_1$ and $\tilde{f}_1$ in the same way as in the first scenario, and (b) only the role of function is enhanced for the fuzzy agent $\tilde{s}_1$ (i.e., the step 6 of phase 2 is not duplicated in this case).

Figure 8 enables the distribution and volume of activity of each community of fuzzy agents during the 6 steps of phase 2 to be visualized. This figure presents two cases: (a) without the intervention of customers or expert of the domains (Fig. 8a), and (b) with the intervention of one of the actors (Fig. 8b)—here the intervention of the customer, according to the third scenario presented above. In the latter case, the increased activity of fuzzy agents (requirements, functions and solutions agents) and their respective roles are clearly visible.

Let us now examine the impact of interactions on the fuzzy configuration. To do this, we will analyze the roles of fuzzy function agents during Phase 2 of configuration. The number of interactions for a fuzzy function agent with fuzzy agents of other communities is equal to $[\tilde{R} \rightarrow 11, \tilde{F} \rightarrow 6, \tilde{C} \rightarrow 0, \tilde{S} \rightarrow 20]$: a total of 37 interactions per fuzzy function agent during this phase. Without weighting of fuzzy

Table 2 Fuzzy values of fuzzy function agent roles

Agent	R	F	C	S	Agent	R	F	C	S
$\tilde{f}_1$	**0.89**	**0.86**	**0**	**0.36**	$\tilde{f}_3$	0.36	0.53	0	0.18
$\tilde{f}_2$	0.73	0.58	0	0.18	$\tilde{f}_4$	0.71	0.48	0	0.18

Table 3 Values of fuzzy solution agent roles

Agent	R	F	C	S	Agent	R	F	C	S
$\tilde{s}_1$	**0**	**0.23**	**0.48**	**0.79**	$\tilde{s}_{11}$	**0**	**0.2**	**0.31**	**0.71**
$\tilde{s}_2$	0	0.23	0.52	0.66	$\tilde{s}_{12}$	0	0.2	0.33	0.63
$\tilde{s}_3$	0	0.18	0.49	0.74	$\tilde{s}_{13}$	0	0.2	0.32	0.53
$\tilde{s}_4$	0	0.13	0.44	0.68	$\tilde{s}_{14}$	0	0.15	0.26	0.56
$\tilde{s}_5$	0	0.15	0.4	0.54	$\tilde{s}_{15}$	0	0.13	0.24	0.59
$\tilde{s}_6$	**0**	**0.18**	**0.29**	**0.79**	$\tilde{s}_{16}$	**0**	**0.38**	**0.28**	**0.74**
$\tilde{s}_7$	0	0.23	0.32	0.69	$\tilde{s}_{17}$	0	0.4	0.34	0.68
$\tilde{s}_8$	0	0.23	0.29	0.58	$\tilde{s}_{18}$	0	0.35	0.35	0.68
$\tilde{s}_9$	0	0.13	0.28	0.59	$\tilde{s}_{19}$	0	0.38	0.33	0.68
$\tilde{s}_{10}$	0	0.15	0.25	0.51	$\tilde{s}_{20}$	0	0.3	0.26	0.49

interactions, we obtain for the fuzzy function agents the following set of degrees of membership of fuzzy roles: [0.5, 0.6, 0, 0.5]. With weighting of fuzzy interactions, we obtain the following set of degrees of membership of fuzzy roles: [0.89, 0.86, 0, 0.36].

The following table (Table 2) shows the results for each of the four fuzzy function agents. The preponderance of the activity of the fuzzy function agent $\tilde{f}_1$ (function: *Support the lower-body weight of a person in a sitting position*) is visible in Table 2. We will now analyze the roles of fuzzy solution agents during Phase 2 of configuration. The number of interactions for a fuzzy solution agent with fuzzy agents of other communities is equal to [$\tilde{R} \to$ 0, $\tilde{F} \to$ 4, $\tilde{C} \to$ 16, $\tilde{S} \to$ 114]: a total of 134 interactions per fuzzy solution agent during this phase. Without weighting of fuzzy interactions, we obtain for the fuzzy solution agents the following set of degrees of membership of fuzzy roles: [0, 0.5, 0.5, 0.85]. With weighting of fuzzy interactions, we obtain the following set of degrees of membership of fuzzy roles: [0, 0.23, 0.48, 0.79].

The following table (Table 3) shows the results for each of the 20 fuzzy solution agents. The increased activity of fuzzy solution agents $\tilde{s}_1, \tilde{s}_6, \tilde{s}_{16}$ is visible during Phase 2. These agents will provide the best and consensual solution to the end of phase 3; what was seen in Table 1.

This analysis shows that organizations in *FAPIC* platform are fuzzy evolving systems. Indeed, dynamic adaptive organizations emerge from the fuzzy interaction of heterogeneous fuzzy agents and their fuzzy roles. The analysis of the behavior of fuzzy agents during design collaborations has shown that the distribution of roles performed by fuzzy agents is continually changing. Fuzzy agents are characterized by fuzzy organizations. The last one is the result of the evolution of agents' fuzzy roles due to their fuzzy interactions.

5 Discussion and Conclusions

This paper has presented the analysis of the evolution of multiple fuzzy roles of four communities of fuzzy agents in a collaborative design for configurations platform. Fuzzy agents have been modeled to have fuzzy knowledge and fuzzy interactions. Fuzzy agents are modeled to play multiple fuzzy roles. In addition, the resulting organizations are also fuzzy.

Fuzzy agents have been developed and used for product configuration because of their similarity to the actors' behavior and reasoning. In the proposed agent-based *FAPIC* platform (Fuzzy Agents for Product Integrated Configuration), requirements, functions, solutions, process constraints are fuzzy agents grouped in four fuzzy communities characterized by a main fuzzy role and other secondary roles. The aim of the application described in this paper was to analyze the fuzzy behavior of fuzzy agents, particularly the analysis of the evolution of fuzzy roles and their fuzzy interactions.

Tradeoff between actor's interventions and fuzzy agents has been considered to be an important issue. This has been extended to the tradeoff between customer intervention and fuzzy agents. The results of analysis of these tradeoffs have shown the important influence of fuzzy solution agents in the organization of the agent-based collaborative design platform. The fuzzy role of fuzzy solution agents is strongly influenced by the variations and changing of requirements and process constraints.

Another finding is the influence of sharing of knowledge between the communities of agents. The more the fuzzy agents share their knowledge, the more their fuzzy roles are complete in every domain of design for configurations. The simulation shows that fuzzy requirement agents perform well their own main role, but they do not play at all the role of fuzzy constraint agents. The same observation can be done for the role played by fuzzy constraint agents in the domain of fuzzy requirement agents. This is due to the lack of knowledge sharing between these two communities of fuzzy agents.

The influence of the degree of interactions in the design for configurations process should be outlined. The fuzzy function agents, influenced by many fuzzy requirement agents, are the most active in the design process. The simulation shows that this observation can be extended to the fuzzy solution agents. The most active fuzzy solution agents are those which create the best consensual solution. It shows that the consensus can be found principally by increasing the degree of interactions.

Appendix

I: Notation Used in the Fuzzy Agent Model

$\tilde{A} = \{\tilde{\alpha}_i\}$ is the finite fuzzy set of fuzzy agents

$\tilde{I} = \{\tilde{\iota}_i\}$ is the finite fuzzy set of interactions defined for all fuzzy agents

$\tilde{P} = \{\tilde{\rho}_i\}$ is the finite fuzzy set of roles to be performed by all fuzzy agents

$\tilde{O} = \{\tilde{o}_i\}$ is the finite fuzzy set of organizations of all fuzzy agents into communities

$\tilde{\Sigma} = \{\tilde{\sigma}_i\}$ is the finite fuzzy set of states defined in agent-based system

$\tilde{\Sigma}_{\tilde{\alpha}_i} \subseteq \tilde{\Sigma}$ is the finite fuzzy set of states of fuzzy agent $\tilde{\alpha}_i$

$\tilde{\Pi} = \{\tilde{\pi}_i\}$ is the finite fuzzy set of perceptions in agent-based system

$\tilde{\Pi}_{\tilde{\alpha}_i} \subseteq \tilde{\Pi}$ is the finite fuzzy set of perceptions of fuzzy agent $\tilde{\alpha}_i$

$\tilde{\Delta} = \left\{\tilde{\delta}_i\right\}$ is the finite fuzzy set of fuzzy decisions, with $\tilde{\Delta}_{\tilde{\alpha}_i} = < \tilde{E}_{\tilde{\alpha}_i}, \tilde{X}_{\tilde{\alpha}_i}, \tilde{\Gamma}_{\tilde{\alpha}_i} >$

$\tilde{\Gamma} = \{\tilde{\gamma}_i\}$ is the finite fuzzy set of actions

$\tilde{\Gamma}_{\tilde{\alpha}_i} \subseteq \tilde{\Gamma}$ is the finite fuzzy set of actions that fuzzy agent $\tilde{\alpha}_i$ can process

$\tilde{\Lambda}_{\tilde{\alpha}_i} \subseteq \tilde{\Gamma}$ is the specific finite fuzzy set of communication acts that fuzzy agent $\tilde{\alpha}_i$ can process; $\tilde{\lambda}_{s,r} = < \tilde{\lambda}, \tilde{\alpha}_s, \tilde{\alpha}_r, \tilde{P}_{\tilde{\alpha}_s}, \tau, \tilde{\eta} >$ is a fuzzy communication between $\tilde{\alpha}_s$ and $\tilde{\alpha}_r$

$\tilde{K} = \{\tilde{\kappa}_i\}$ is the finite fuzzy set of fuzzy knowledge in agent-based system

$\tilde{K}_{\tilde{\alpha}_i} \subseteq \tilde{K}$ is the finite fuzzy set of fuzzy knowledge of fuzzy agent $\tilde{\alpha}_i$, with $\tilde{K}_{\tilde{\alpha}_i} = \tilde{P}_{\tilde{\alpha}_i} \cup \tilde{\Sigma}_{\tilde{\alpha}_i} \cup \tilde{\Sigma}_{\tilde{M}_{\tilde{\alpha}_i}}$

$\tilde{E} = \{\tilde{\varepsilon}_i\}$ is the finite fuzzy set of fuzzy events observed in agent-based system

$\tilde{E}_{\tilde{\alpha}_i} \subseteq \tilde{E}$ is the finite fuzzy set of fuzzy events that fuzzy agent $\tilde{\alpha}_i$ can observe

$\tilde{X} = \{\tilde{\chi}_i\}$ is the finite fuzzy set of conditions in agent-based system

$\tilde{X}_{\tilde{\alpha}_i} \in \tilde{X}$ is the finite fuzzy set of conditions associated to internal states of fuzzy agent $\tilde{\alpha}_i$

$\tilde{B} = \left\{\tilde{\beta}_i\right\}$ is the finite fuzzy set of speech acts

$\tilde{H} = \{\tilde{\eta}_i\}$ is the finite fuzzy set of messages

$\tilde{T} = \{\tilde{\tau}_i\}$ is the finite set of types of messages

$\tilde{M}_\alpha = < \tilde{A}, \tilde{I}, \tilde{P}, \tilde{O} >$ is the tuple defining an agent-based system

$\Phi_{\tilde{\Pi}(\tilde{\alpha}_i)} : \tilde{\Sigma} \times \tilde{\Sigma}_{\tilde{M}_{\tilde{\alpha}_i}} \to \tilde{\Pi}_{\tilde{\alpha}_i}$ is the function of observations of fuzzy agent $\tilde{\alpha}_i$

$\Phi_{\tilde{\Delta}(\tilde{\alpha}_i)} : \tilde{\Pi}_{\tilde{\alpha}_i} \times \tilde{\Sigma}_{\tilde{\alpha}_i} \to \tilde{P}_{\tilde{\alpha}_i}$ is the function of decisions of fuzzy agent $\tilde{\alpha}_i$

$\Phi_{\tilde{\Gamma}(\tilde{\alpha}_i)} : \tilde{\Delta}_{\tilde{\alpha}_i} \times \tilde{\Sigma} \to \tilde{\Gamma}_{\tilde{\alpha}_i}$ is the function of actions of fuzzy agent $\tilde{\alpha}_i$

II: Characteristics Defined for the Case Study

Domains	Fuzzy Agents	Description	Fuzzy Agents	Description
Requirements	$\tilde{r}_1$	Size	$\tilde{r}_7$	Classic
	$\tilde{r}_2$	Weight	$\tilde{r}_8$	Comfortable
	$\tilde{r}_3$	Price	$\tilde{r}_9$	Practical
	$\tilde{r}_4$	Office	$\tilde{r}_{10}$	Durable
	$\tilde{r}_5$	Bar	$\tilde{r}_{11}$	Stable
	$\tilde{r}_6$	Classroom		
Functions	$\tilde{f}_1$	Support the lower-body weight of a person	$\tilde{f}_3$	Support the arms of a person in a sitting position
	$\tilde{f}_2$	Support the back of a person in a sitting position	$\tilde{f}_4$	Offer movement space for the legs of a person in a sitting position
Solutions	$\tilde{s}_1$	Cl_1: Seat	$\tilde{s}_{11}$	Cl_3: Armrest
	$\tilde{s}_2$		$\tilde{s}_{12}$	
	$\tilde{s}_3$		$\tilde{s}_{13}$	
	$\tilde{s}_4$		$\tilde{s}_{14}$	
	$\tilde{s}_5$		$\tilde{s}_{15}$	
	$\tilde{s}_6$	Cl_2: Back	$\tilde{s}_{16}$	Cl_4: Stand
	$\tilde{s}_7$		$\tilde{s}_{17}$	
	$\tilde{s}_8$		$\tilde{s}_{18}$	
	$\tilde{s}_9$		$\tilde{s}_{19}$	
	$\tilde{s}_{10}$		$\tilde{s}_{20}$	
Constraints: example for the view "Production"	$\tilde{c}_{11}$	Aim at simple shapes	$\tilde{c}_{41}$	Provide adequate support surfaces
	$\tilde{c}_{21}$	Avoid differences in cross-section	$\tilde{c}_{51}$	Avoid unnecessary machining
	$\tilde{c}_{31}$	Avoid large curvatures	$\tilde{c}_{61}$	Avoid excessively thin sections

References

1. Agard, B., Barajas, M.: The use of fuzzy logic in product family development: literature review and opportunities. J. Intell. Manuf. **23**(5), 1445–1462 (2012)
2. Deciu, E.R., Ostrosi, E., Ferney, M., Gheorghe, M.: Configurable product design using multiple fuzzy models. J. Eng. Des. **16**(2–3), 209–235 (2005)
3. Doctor, F., Hagras, H., Callaghan, V.: An intelligent fuzzy agent approach for realising ambient intelligence in intelligent inhabited environments. IEEE Trans. SMC, Part A: Syst. Hum. **35**(1), 55–65 (2005)
4. Duman, H., Hagras, H., Callaghan, V.: Intelligent association exploration and exploitation of fuzzy agents in ambient intelligent environments. J. Uncertain Syst. **2**(2), 133–143 (2008)
5. Epstein, J.-G., Möhring M., Troitzsch K.G.: Fuzzy-logical rules in a multi-agent system. In: Proceedings of SimSocVI Workshop, Groningen, Netherlands, 19–21 September 2003
6. Ferber, J.: Multi-agent Systems. An Introduction to Distributed Artificial Intelligence. Addison Wesley, London (1999)
7. Ferber J., Stratulat T., Tranier J.: Towards an integral approach of organizations in multi-agent systems: the MASQ approach. In: Multi-agent Systems: Semantics and Dynamics of Organizational Models, Virginia Dignum (Ed), IGI (2009)
8. Fougères A.-J.: Agents to cooperate in distributed design process. In: IEEE International Conference on Systems, Man and Cybernetics, (SMC'04), The Hague, vol. 3, pp. 2629–2634 (2004)
9. Fougères, A.-J.: Modelling and simulation of complex systems: an approach based on multi-level agents. Int. J. Comput. Sci. Issues **8**(6), 8–17 (2011)
10. Fougères, A.-J.: A modelling approach based on fuzzy agents. Int. J. Comput. Sci. Issues **9**(6), 19–28 (2012)
11. Fougères, A.-J., Choulier, D., Ostrosi, E.: ADEA–a multi agent system for design activity analysis. In: Proceedings of the 19th ISPE International Conference on Concurrent Engineering (CE'2012), Trier, Germany, 3–7 September 2012, vol. 1, pp. 485–496 (2012)
12. Fougères, A.-J., Ostrosi, E.: Fuzzy agent-based approach for consensual design synthesis in product integrated configuration. Integr. Comput.-Aided Eng. **20**(3), 259–274 (2013)
13. Ghasem-Aghaee, N., Ören, T.I.: Cognitive complexity and dynamic personality in agent simulation. Comput. Hum. Behav. **23**, 2983–2997 (2007)
14. Skarmeta, A.F.G., Barberá, H.M., Alonso M.S.: A fuzzy agents architecture for autonomous mobile robots. In: Proceedings of IFSA'99, Taiwan (1999)
15. Jennings, N.R.: On agent-based software engineering. AI **117**, 277–296 (2000)
16. Fard, M.K., Zaeri, A., Aghaee, N.G., Bakhsh, M.A.N., Mardukhi, F.: Fuzzy emotional COCOMO II software cost estimation (FECSCE) using multi-agent systems. Appl. Soft Comput. **11**(2), 2260–2270 (2011)
17. Kota, R., Gibbins, N., Jennings, N.R.: Self-organising agent organisations. In: Proceedings of the 8th International Conference on Autonomous Agents and Multiagent Systems, vol. 2, pp. 797–804 (2009)
18. Kubera, Y., Mathieu, P., Picault, S.: IODA: an interaction-oriented approach for multi-agent based simulations. AAMAS **23**(3), 303–343 (2011)
19. Leitão, P.: Agent-based distributed manufacturing control: a state-of-the-art survey. Eng. Appl. Artif. Intell. **22**(7), 979–991 (2009)
20. Lughofer, E.: Evolving Fuzzy Systems—Methodologies. Advanced Concepts and Applications. Springer, Berlin (2011)
21. Di Marzo, S.G., Gleizes, M.-P., Karageorgos, A.: Self-organization in multi-agent systems. Knowl. Eng. Rev. **20**(2), 165–189 (2005)
22. Micaëlli, J.-P., Fougères, A.-J.: L'Évaluation creative. UTBM Press, Belfort (2007)
23. Monostori, L., Vancza, J., Kumara, S.R.T.: Agent-based systems for manufacturing. Ann. CIRP **55**(2), 697–720 (2006)
24. Movahed-Khah, R., Ostrosi, E., Garro, O.: Analysis of interaction dynamics in collaborative and distributed design process. Int. J. Comput. Ind. **61**(2), 2–14 (2010)

25. Munoz-Hernandez, S., Gomez Perez, J.M.: Solving Collaborative Fuzzy Agents Problems with CLP(FD). Lecture Notes in Computer Science, vol. 3350/2005, pp. 187–202 (2005)
26. Nwana, H.S.: Software agents: an overview. Knowl. Eng. Rev. **11**(2), 205–244 (1996)
27. Ostrosi, E., Bi, S.T.: Generalised design for optimal product configuration. Int. J. Adv. Manuf. Technol. **49**(1–4), 13–25 (2010)
28. Ostrosi, E., Fougères, A.-J., Ferney, M.: Fuzzy agents for product configuration in collaborative and distributed design process. Appl. Soft Comput. **8**(12), 2091–2105 (2012)
29. Ostrosi, E., Fougères, A.-J., Ferney, M., Klein, D.: A fuzzy configuration multi-agent approach for product family modelling in conceptual design. J. Intell. Manuf. **23**(6), 2565–2586 (2012)
30. Ostrosi, E., Haxhiaj, L., Fukuda, S.: Fuzzy modelling of consensus during design conflict resolution. Res. Eng. Des. **23**(1), 53–70 (2012)
31. Shen, W., Norrie, D.H., Barthès, J.-P.: Multi-Agent Systems for Concurrent Intelligent Design and Manufacturing. Taylor and Francis, London (2001)
32. Tweedale, J., Ichalkaranje, N.: Innovations in multi-agent systems. J. Netw. Comput. Appl. **30**(3), 1089–1115 (2007)
33. van Aart, C.: Organizational Principles for Multiagent Architectures. Birkhauser Verlag, Basel (2005)
34. Wooldridge, M.: Agent-based software engineering. IEE Proc. Softw. Eng. **144**(1), 26–37 (1997)
35. Zadeh, L.A.: Fuzzy sets. Information and control **8**, 338–353 (1965)

Multi-distance and Fuzzy Similarity Based Fuzzy TOPSIS

Mikael Collan, Mario Fedrizzi and Pasi Luukka

Abstract This article introduces a new extension to fuzzy TOPSIS. In the extension we have used as a basis a fuzzy similarity based fuzzy TOPSIS that uses an additional component of multi-distance in forming a closeness coefficient. Ordered weighted averaging is also used in the aggregation process over fuzzy similarity values. For the ordered weighted averaging operator weight generation we use the O'Hagan's method, to find optimal weights. Several different, predefined orness values are tested and an overall ranking is computed, based on the rankings resulting from multiple different orness values. The presented method is numerically applied to a research and development project selection problem.

Keywords Fuzzy similarity · Fuzzy TOPSIS · Multi-distances · OWA · O'Hagan's method

1 Introduction

This paper investigates and presents a new extension of the fuzzy similarity based fuzzy Technique for Order Performance by Similarity to Ideal Solution (fuzzy TOP-SIS). Fuzzy TOPSIS was originally introduced by Chen in [1] and later extended to include trapezoidal fuzzy numbers in [2]. In these contributions a vertex based fuzzy distance method was used as a measure of distance from ("similarity to") the ideal solutions. A similarity measure based version of fuzzy TOPSIS was introduced

M. Collan · P. Luukka
School of Business and Management, Lappeenranta University of Technology,
Lappeenranta, Finland
e-mail: mikael.collan@lut.fi

P. Luukka
e-mail: pasi.luukka@lut.fi

M. Fedrizzi (✉)
Department of Industrial Engineering, University of Trento,
Via Mesiano 77, 38123 Trento, Italy
e-mail: mario.fedrizzi@unitn.it

© Springer International Publishing Switzerland 2016
K. Madani et al. (eds.), *Computational Intelligence*,
Studies in Computational Intelligence 613,
DOI 10.1007/978-3-319-23392-5_13

in [3], where the similarity to the ideal solutions is calculated by using a fuzzy similarity measure. This strain of research was continued by [4], where two different fuzzy similarity measures were considered and by [5], where four fuzzy similarity measure based fuzzy TOPSIS variants and a way of holistic overall ranking of projects was presented.

Fuzzy TOPSIS uses fuzzy numbers as inputs and is thus able to incorporate inaccurate and imprecise information in the analysis (not having to simplify reality by using crisp numbers). The main difference in using the fuzzy similarity measures and the (crisp) distance measures in the TOPSIS environment with fuzzy numbers is that fuzzy similarity measures can take into consideration more of the information that is stored in the fuzzy number, e.g., with regards to the perimeter and the area of the fuzzy number. The crisp distance measures essentially defuzzify the fuzzy number in order to calculate a distance between the resulting crisp number and the ideal solution. Using a defuzzified crisp distance based measure may cause a loss of relevant information. The fuzzy similarity measure used here is introduced in Hejazi et al. [6] and can take into account the perimeter and the area of fuzzy numbers. This similarity measure was previously studied in the context of fuzzy similarity based TOPSIS method in [4, 5].

The new contribution of this paper concentrates on the application of multi-distances in creating additional information for project ranking by similarity coefficients, after they have been analyzed with fuzzy similarity measure based fuzzy TOPSIS. Multi-distances are used in analyzing the "level" of similarity between analyzed criteria. High level of similarity means a low multi-distance and can be interpreted as homogeneity or consistency of, e.g., performance or expectations. Such information may be valuable in the analysis and offers an additional differentiator between objects. Multi-distances were examined by Martin and Mayor [7], and presented as a generalization of the notion of distance. Martin and Mayor proposed the construction of multi-distances by means of OWA functions in [8]. The OWA based multi-distances functions [9], used here, combine the distance values of all pairs of elements in the collection into OWA-based multi-distances. Using the multi-distance in the aggregation will add a step of pairwise distance measurement of similarities between criteria (values) in the procedure. Use of multi-distances with fuzzy TOPSIS is, to the best of our knowledge a new approach.

The remainder of the paper is organized as follows. In Sect. 2 the fuzzy similarity relation between fuzzy numbers, the OWA operator, multi-distances, and total ordering of fuzzy numbers are introduced. Section 3 is devoted to the description of the new approach to fuzzy TOPSIS based on fuzzy similarity and multi-distances. A numerical example is introduced in Sect. 4 and some conclusions in Sect. 5 close the paper.

2 Preliminaries

In this section some preliminary mathematical concepts, used in the MCDM method, are shortly introduced. They include: fuzzy similarity measures, the OWA-operator, and an often-used method to generate the weights for the OWA operator the O'Hagan's method. Multi-distances are defined following the work of Martin and Mayor [7] and the relationship between the OWA-operator and multi-distances is presented. Additionally, a way to find a total ordering for fuzzy numbers is shortly introduced.

2.1 Fuzzy Similarity of Fuzzy Numbers

By focusing on uncertain objects like in fuzzy sets or fuzzy numbers, the notion of a fuzzy subset generalizes that of the classical subset, where the concept of similarity can be considered as a many-valued generalization of the classical notion of equivalence [10]. As an equivalence relation is a familiar way to classify similar objects, fuzzy similarity is an equivalence relation that can be used to classify multi-valued objects [4]. The concept of a similarity measure is of high importance in this work, and it is defined as follows:

Definition 1 For any fuzzy subset $F \neq \emptyset$ of $\mathbb{R}^n$, and for any elements $A, B \in F$ the function of a similarity measure is defined as [11]:

$$s(A, B) : F \times F \to [0, 1]$$

The defined similarity measures s satisfy the following properties for any x, y, $z \in F$,

- $s(x, x) = s(y, y), \forall x, y \in F$ (*Reflexivity*)
- $s(x, y) \leq s(y, y), \forall x, y \in F$ (*Minimality*)
- $s(x, y) = s(y, x)$ (*Symmetry*)
- If $s(x, y) = s(x, z)$ it implies that $s(x, y) = s(x, z) = s(y, z)$ (*Transitivity*)

Since fuzzy numbers can be considered to be a certain type of restricted fuzzy sets, the similarity measures for generalized fuzzy numbers come from similarity measures for fuzzy sets.

Represented by Chen [12], a generalized trapezoidal fuzzy number's notation is $\tilde{A} = (a, b, c, d; w)$, where a, b, c and d are real values and $0 < w \leq 1$. The membership function $\mu_{\tilde{A}}$ satisfies the following conditions [12]:

1. $\mu_{\tilde{A}}$ is a continuous mapping from the universe of discourse X to the closed interval in $[0, 1]$
2. $\mu_{\tilde{A}} = 0$, where $-\infty < x \leq a$
3. $\mu_{\tilde{A}}$ is monotonically increasing in $[a, b]$
4. $\mu_{\tilde{A}} = w$, where $b \leq x \leq c$
5. $\mu_{\tilde{A}}$ is monotonically decreasing in $[c, d]$
6. $\mu_{\tilde{A}} = 0$, where $d \leq x < \infty$

Due to the fit and the applicability of similarity measures in the context of decision-making, various similarity measures have been proposed for the calculation the degree of similarity between fuzzy numbers of [12]. In this work, a recently introduced similarity measure by Hejazi et al. in [6] is used. The similarity measure takes into consideration the perimeter and the area of fuzzy numbers. The similarity measure is denoted $s(M, N)$, and involves fuzzy numbers $M = (m_1, m_2, m_3, m_4; \omega_m)$ and $N = (n_1, n_2, n_3, n_4; \omega_n)$ with $0 \leq m_1 \leq m_2 \leq m_3 \leq m_4 \leq 1, 0 \leq n_1 \leq n_2 \leq n_3 \leq n_4 \leq 1$, and $M(x_i)$ and $N(x_i)$ their corresponding membership functions with $i \in \{1, 2, 3, 4\}$ for generalized trapezoidal fuzzy numbers, where ω_m and ω_n are their corresponding heights. The definition is as follows:

$$s(M, N) = (1 - \frac{\sum_{i=1}^{4} |m_i - n_i|}{4})$$
$$\times \frac{min(p(m), p(n))}{max(p(m), p(n))}$$
$$\times \frac{min(a(m), a(n)) + min(\omega_m, \omega_n)}{max(a(m), a(n)) + max(\omega_m, \omega_n)} \qquad (1)$$

where the values $p(m)$ and $p(n)$ represent the perimeters of the trapezoidal fuzzy numbers M and N, and are defined as:

$$p(m) = \sqrt{(m_1 - m_2)^2 + \omega_m^2} + \sqrt{(m_3 - m_4)^2 + \omega_m^2}$$
$$+ (m_3 - m_2) + (m_4 - m_1)$$

and

$$p(n) = \sqrt{(n_1 - n_2)^2 + \omega_n^2} + \sqrt{(n_3 - n_4)^2 + \omega_n^2}$$
$$+ (n_3 - n_2) + (n_4 - n_1)$$

The values $a(m)$ and $a(n)$ represent the areas of the trapezoidal fuzzy numbers M and N, they are defined as:

$$a(m) = \frac{1}{2}\omega_m(m_3 - m_2 + m_4 - m_1),$$

and

$$a(m) = \frac{1}{2}\omega_n(n_3 - n_2 + n_4 - n_1).$$

Notice that the result of the above similarity measure $s(M, N)$ belongs to the unit interval [0, 1] and the larger the value of the similarity measure is, the stronger is the similarity between the fuzzy numbers M and N.

2.2 The OWA Operator

In 1988 Yager introduced an aggregation operator, called ordered weighted averaging operator (OWA) [13] and formalized it as follows:

An ordered weighted averaging (OWA) operator of dimension m is a mapping $R^m \to R$ that has associated weighting vector $W = [w_1, w_2, \ldots, w_m]$ of dimension m with

$$\sum_{i=1}^{m} w_i = 1, w_i \in [0, 1] \text{ and } 1 \le i \le m$$

such that:

$$OWA(a_1, a_2, \ldots, a_m) = \sum_{i=1}^{m} w_i b_i \qquad (2)$$

where b_i is the ith largest element of the collection of objects $a_1, a_2, \ldots, a_m$. The function value $OWA(a_1, a_2, \ldots, a_m)$ determines the aggregated values of arguments $a_1, a_2, \ldots, a_m$. One of the measures related to the OWA is the so called "orness" measure. For a given weighting vector $W = [w_1, w_2, \ldots, w_m]^T$ the measure of orness of the OWA aggregation operator for W is given as

$$orness(W) = \frac{1}{m-1} \sum_{i=1}^{m} (m - i) w_i. \qquad (3)$$

It can be observed that the weighting vector has an important role in the OWA operator; the weighting vector determines how large a weight, each aggregated object receives. The distribution of weights depends on the selected value of orness that can be selected from between [0, 1]. If orness is 0 then the first ordered object gets all weight and the rest of the objects get a weight of zero. If the orness value is 1, then the weight is evenly distributed among all objects and the weighting is actually the same as a normal non-weighted average. In 1988 O'Hagan [14] introduced a technique for computing the weights used in OWA. The procedure assumes a predefined degree of orness—the weights are obtained by maximizing the entropy $-\sum_{i=1}^{m} w_i \, ln(w_i)$. The solution is based on the constrained optimization problem

$$\text{maximize} \quad -\sum_{i=1}^{m} w_i ln(w_i)$$

$$\text{subject to} \quad \alpha = \frac{1}{m-1} \sum_{i=1}^{m} (m-1) w_i$$

$$\sum_{i=1}^{m} w_i = 1 \text{ and } w_i \geq 0.$$

The above constrained optimization problem can be solved by using different methods. Here an analytical solution introduced by [15] is used. Below this weighting scheme is presented:

a. if $m = 2$, implies that $w_1 = \alpha$ and $w_2 = 1 - \alpha$
b. if $\alpha = 0$ or $\alpha = 1$ implies that the corresponding weighting vectors are $w = (0, \dots 0, 1)$ or $w = (1, 0, \dots, 0)$ respectively.
c. if $m \geq 3$ and $0 \leq \alpha \leq 1$ then, we have,

$$w_i = \left(w_1^{m-i} \cdot w_m^{i-1} \right)^{\frac{1}{m-1}}$$

$$w_m = \frac{((m-1)\cdot\alpha - m)\cdot w_1 + 1}{(m-1)\cdot\alpha + 1 - m\cdot w_1}$$

$$w_1[(m-1)\cdot\alpha + 1 - m\cdot w_1]^m = ((m-1)\cdot\alpha)^{m-1} \cdot [((m-1)\cdot\alpha - m)\cdot w_1 + 1]$$

For $m \geq 3$, the weights are computed by initially obtaining the first weight, followed by the last weight of the weighting vector, before other weights are computed.

2.3 Multi-distances

A multi-distance is a representation of the notion of multi-argument distances. The set X is a union of all m-dimensional lists of elements of X, multi-distance is defined as a function $D : X \rightarrow [0, \infty)$ on a non empty set X provided that the following properties are satisfied for all m and $x_1, x_2, \dots, x_m, y \in X$

c1. $D(x_1, x_2, \dots, x_m) = 0$ if and only if $x_i = x_j$ for all $i, j = 1, 2, \dots, m$
c2. $D(x_1, x_2, \dots, x_m) = D(x_{\sigma(1)}, x_{\sigma(2)}, \dots, x_{\sigma(m)})$ for any permutation σ of $i, j = 1, 2, \dots, m$
c3. $D(x_1, x_2, \dots, x_m) \leq D(x_1, y) + D(x_2, y) + \cdots + D(x_m, y)$.

We say that D is a strong multi-distance if it satisfies $c1, c2$, and

$c3^*$ $D(\mathbf{x}_1, \mathbf{x}_2, \dots, \mathbf{x}_m) \leq D(\mathbf{x}_1, \mathbf{y}) + D(\mathbf{x}_2, \mathbf{y}) + \cdots + D(\mathbf{x}_m, \mathbf{y})$. for all $\mathbf{x}_1, \mathbf{x}_2, \dots, \mathbf{x}_m, \mathbf{y} \in X$

In application contexts, the estimation of distances between more than two elements of the set X can be constructed using multi-distances by means of the OWA operator as suggested by Martin and Mayor [7].

$$D_w(x_1, x_2, \ldots, x_m) = OWA_w(d(x_1, x_2), d(x_2, x_3), \ldots,$$
$$d(x_{m-1}, x_m)) \tag{4}$$

In this case, elements $x_1, x_2, \ldots, x_m$ are obtained from the similarity measure (1), and the distance applied is $d(x, y) = |x - y|$.

2.4 Total Ordering of Fuzzy Numbers

Set inclusion of fuzzy sets is only a partial order, where all fuzzy sets are not comparable. Kaufmann and Gupta [16] propose that when trying to find a *total order* or *linear order* for fuzzy numbers, where all fuzzy numbers and fuzzy intervals are comparable, we have to first check that it is possible to find a linear order by giving different emphases to different properties of fuzzy sets. To reach a *total order* or a *linear order* of fuzzy numbers, an importance order must be given to three criteria. If the first criterion does not give a unique linear order, then the second criterion should be used. One continues in this way as long as it is needed. The description of the three different criteria used in the Kaufmann and Gupta ordering method is given below:

1st *The removal*: Let us consider an ordinary number $k \in \mathbb{R}$ and a fuzzy number A. The left side removal of A with respect to k, denoted by $R_l(A, k)$, is defined as the area bounded by k and the left side of the fuzzy number A. Similarly, the right side removal, $R_r(A, k)$ is defined. The removal of the fuzzy number A with respect to k is defined as the mean of $R_l(A, k)$ and $R_r(A, k)$. Thus,

$$R(A, k) = \frac{1}{2}(R_l(A, k) + R_r(A, k)). \tag{5}$$

The position of k can be located anywhere on the x-axis including $k = 0$. By definition, the areas are positive quantities, but here they are evaluated by integration taking into account the position (negative, zero, or positive) of the variable x; therefore, $R(A, k)$ can be positive, negative or null.

The first criterion, used in ordering is the removal with respect to k. However, two different fuzzy numbers can have the same removal with respect to the same k. In fact, this criterion decomposes a set of fuzzy numbers into classes having the same removal number. In other words, fuzzy numbers are ranked into classes that can then be ordered according to the removal number; if there is only one fuzzy number per each class, then we have a linear ordering of the fuzzy numbers.

The removal number $R(A, k)$ defined in this criterion, relocated to $k = 0$ is equivalent to an "ordinary representative" of the fuzzy number. In the case of a triangular fuzzy number this ordinary representative is given by:

$$\widehat{A} = \frac{a_1 + 2a_2 + a_3}{4}, \tag{6}$$

where $A = (a_1, a_2, a_3)$.

If after using the removal criteria there are classes with multiple fuzzy numbers, one has to go forward and use the second criteria for ordering the fuzzy numbers within the "multiple number" classes.

2nd *The mode*: In each class of (multiple) fuzzy numbers, one should look for the mode of each fuzzy number in the class; these modes will generate sub-classes. If the fuzzy numbers under consideration have a non-unique mode, one takes the mean position of the modal values. It must be noted that this is only one way of obtaining sub-classes, and one may need the following third divergence criterion for further sub-classification.

$$Mode(A) = \{x \in U | A(x) = 1\} \tag{7}$$

If there are still classes (or rather sub classes) with multiple fuzzy numbers, one will resort to the third ordering criterion.

3rd *The divergence*: The consideration of the divergence around the mode in each sub-class leads to the sub-sub-classes, and this criterion may be sufficient to obtain the final linear ordering of fuzzy numbers (Fig. 1).

$$Divergence(A) = \sup(supp(A)) - \inf(supp(A)) \tag{8}$$

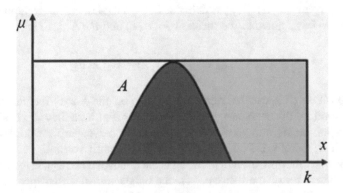

Fig. 1 Removal number

Summarizing the method: when one orders fuzzy numbers to size order, one proceeds as follows. Apply the above presented criteria in the exact given order, such that if the unique linear order is not obtained then move to the next criterion.

3 Fuzzy Similarity Based Fuzzy TOPSIS with Multi-distances

A fuzzy extension to the Technique for Order Performance by Similarity to Ideal Solution (TOPSIS) was presented by Chen [1] and it has been extended to solve problems involving trapezoidal fuzzy numbers and applied, e.g., to solving supplier selection problems [2]. It is a Multiple Criteria Decision Making (MCDM) method [2, 17]useful in ranking objects, based on the similarity of the object characteristics to the characteristics of an ideal object (ideal solution). The method is based on the idea that objects are ranked higher, the shorter their distance is from the *Fuzzy Positive Ideal Solution (FPIS)* and the further away they are from the *Fuzzy Negative Ideal Solution (FNIS)*. One advantage of having extended the TOPSIS method to the fuzzy environment is that a linguistic assessment can be properly used, instead of being constrained to using only numerical values; linguistic variables can be mapped to corresponding fuzzy numbers [1, 2].

Solution to the project selection problem, when using the $TOPSIS$ approach, can be presented by considering a situation of a finite set of projects $P = \{P_i | i = 1, 2, \ldots, m\}$, which need to be evaluated by a committee of decision-makers $D = \{D_l | l = 1, 2, \ldots, k\}$, by considering a finite set of given criteria $C = \{C_j | j = 1, 2, \ldots, n\}$.

Let us consider a decision matrix representing a set of performance ratings of each alternative project $P_i, i = 1, 2, \ldots, m$ with respect to each criterion $C_j, j = 1, 2, \ldots, n$, as follows [18]:

$$
\mathbf{X} = \begin{bmatrix} x_{11} & x_{12} & \cdots & x_{1n} \\ x_{21} & x_{22} & \cdots & x_{2n} \\ \cdots & \cdots & \cdots & \cdots \\ x_{m1} & x_{m2} & \cdots & x_{mn} \end{bmatrix}
$$

Let us also assume the weight w_j of the jth criterion C_j, such that the weight vector is represented as follows:

$$
\mathbf{W} = \begin{bmatrix} w_1, & w_2, & \ldots, & w_n \end{bmatrix}
$$

where m rows represent m possible alternatives, n columns represent n relevant criteria, and x_{ij} represent the performance rating of the ith project P_i with respect to the jth criterion C_j. The above fuzzy ratings for each decision-maker $D_l, l = 1, 2, \ldots, k$ are represented by positive trapezoidal fuzzy numbers

$\hat{R}_l = (a_l, b_l, c_l, d_l), l = 1, 2, \ldots, k$ with the respective membership function $\mu_{\hat{R}_l}(x)$. As the rating $\hat{R}_l = (a_l, b_l, c_l, d_l)$ is for the lth decision-maker, the aggregated fuzzy number that can stand for all decision-makers' rating is:

$$\hat{R} = (a, b, c, d)$$

with:
$a = \min_l\{a_l\}$, $b = \frac{1}{k}\sum_{l=1}^{k} b_l$, $c = \frac{1}{k}\sum_{l=1}^{k} c_l$, $d = \max_l\{d_l\}$. The fuzzy rating and importance weight of the lth decision-maker can respectively be represented by $x_{ijl} = (a_{ijl}, b_{ijl}, c_{ijl}, d_{ijl})$ and $\hat{w} = (w_{jl1}, w_{jl2}, w_{jl3}, w_{jl4})$ with $i = 1, 2, \ldots, m$; $j = 1, 2, \ldots, n$. Then, the aggregated fuzzy ratings x_{ij} of alternatives, with respect to each criterion are:

$$x_{ij} = (a_{ij}, b_{ij}, c_{ij}, d_{ij}),$$

calculated as: $a_{ij} = \min_l\{a_{ijl}\}$, $b_{ij} = \frac{1}{k}\sum_{l=1}^{k} b_{ijl}$, $c_{ij} = \frac{1}{k}\sum_{l=1}^{k} c_{ijl}$, $d_{ij} = \max_l\{d_{ijl}\}$. The aggregated fuzzy weight $\hat{w}_j$ of each criterion can be calculated as:

$$\hat{w}_j = (w_{j1}, w_{j2}, w_{j3}, w_{j4})$$

with $w_{j1} = \min_l\{w_{jl1}\}$, $w_{j2} = \frac{1}{k}\sum_{l=1}^{k} w_{jl2}$, $w_{j3} = \frac{1}{k}\sum_{l=1}^{k} w_{jl3}$, $w_{j4} = \max_l\{w_{jl4}\}$. After aggregation the decision matrix and the weight vector are of the following form $X = \{x_{ij}\}_{m \times n}$ and $W = \{w_j\}_{1 \times n}$, where $i = 1, 2, \ldots, m$ and $j = 1, 2, \ldots, n$.

These matrices' elements are given by positive trapezoidal fuzzy numbers as: $x_{ij} = (a_{ij}, b_{ij}, c_{ij}, d_{ij})$ and $w_j = (w_{j1}, w_{j2}, w_{j3}, w_{j4})$.

A linear scale transformation is used to transform the various criteria scales into comparable scales, in order to avoid overly complex mathematical operations in a decision process. The set of criteria can be divided into *benefit criteria B*, where the larger the rating, the greater the preference and to *cost criteria C*, where the smaller the rating, the greater the preference. A normalization method designed to preserve the property, in which the elements are normalized trapezoidal fuzzy numbers, is used. The normalized value of x_{ij} is r_{ij}, and the normalized fuzzy decision matrix is then represented as:

$$R = [r_{ij}]_{m \times n} \qquad (9)$$

with

$$r_{ij} = \left(\frac{a_{ij}}{d_j^+}, \frac{b_{ij}}{d_j^+}, \frac{c_{ij}}{d_j^+}, \frac{d_{ij}}{d_j^+}\right), j \in B$$

$$r_{ij} = \left(\frac{a_j^-}{d_{ij}}, \frac{a_j^-}{c_{ij}}, \frac{a_j^-}{b_{ij}}, \frac{a_j^-}{a_{ij}}\right), j \in C$$

where $d_j^+ = \max_i\{d_{ij}\}$, $j \in B$ and $a_j^- = \min_i\{a_{ij}\}$, $j \in C$. The weighted normalized value of r_{ij} is called v_{ij}, and by considering the importance of each criterion, the weighted normalized fuzzy decision matrix is represented as:

$$V = [v_{ij}]_{m \times n} \tag{10}$$

where $v_{ij} = r_{ij} \cdot w_j$. For all i, j, the elements v_{ij} are now normalized positive trapezoidal fuzzy numbers.

Next, the ideal solutions must be determined and taken from the given criteria, which are linguistically expressed; they are commonly referred to as *Fuzzy Positive Ideal Solution (FPIS)* and *Fuzzy Negative Ideal Solution (FNIS)*. By considering a finite set of given criteria $C = \{C_j | j = 1, 2, \ldots, n\}$, the ways to select the *FPIS(P^+)* and the *FNIS(P^-)* come from the weighted normalized decision matrix $V = (v_{ij})_{m \times n}$, where the obtained weighted normalized values v_{ij} are fuzzy numbers expressed as:

$$v_{ij} = (v_{ij1}, v_{ij2}, v_{ij3}, v_{ij4})$$

The fuzzy positive-ideal solution P^+ and the fuzzy negative-ideal solution P^-, respectively are:

$$P^+ = [v_1^+, v_2^+, \ldots, v_n^+] \tag{11}$$

$$P^- = [v_1^-, v_2^-, \ldots, v_n^-] \tag{12}$$

A way for choosing the *FPIS* (P^+) and the *FNIS* (P^-) has been explained in [3], and is given as follows:

Every element of P^+ is the maximum for all i weighted normalized value, and every element of P^- is the minimum for all i weighted normalized value:

$$v_j^+ = (\max_i v_{ij1}, \max_i v_{ij2}, \max_i v_{ij3}, \max_i v_{ij4}) \tag{13}$$

$$v_j^- = (\min_i v_{ij1}, \min_i v_{ij2}, \min_i v_{ij3}, \min_i v_{ij4}) \tag{14}$$

This approach is also used here. The similarity measure between each project and the ideal solutions P^+ and P^- will be needed later, when calculating the closeness coefficients to determine the ranking order of all possible alternative projects.

The similarities s_i^+ from the positive and negative ideal solution are calculated as:

$$s_i^+ = \{s_{i1}(v_{i1}, v_1^+), s_{i2}(v_{i2}, v_2^+), \ldots, s_{in}(v_{in}, v_n^+)\} \tag{15}$$

$$s_i^- = \{s_{i1}(v_{i1}, v_1^-), s_{i2}(v_{i2}, v_2^-), \ldots, s_{in}(v_{in}, v_n^+)\} \tag{16}$$

where for similarity we used the similarity measure from equation (1).

These similarity vectors are then aggregated using OWA, as follows:

$$S_{iw}^+ = OWA_w(s_{i1}^+, s_{i2}^+, \ldots, s_{in}^+) \tag{17}$$

$$S_{iw}^- = OWA_w(s_{i1}^-, s_{i2}^-, \ldots, s_{in}^-) \tag{18}$$

Besides this we also aggregate s_i^+ vector by using multi-distance as

$$D_{iw}^+(s_{i1}^+, s_{i2}^+, \ldots, s_{in}^+) = OWA_w(d(s_{i1}^+, s_{i2}^+), d(s_{i2}^+, s_{i3}^+),$$
$$\ldots, d(s_{i(n-1)}^+, s_{in}^+)) \tag{19}$$

In the closeness coefficient we now want to take both into account, the similarities from the positive and the negative ideal solution, but also the multi-distance. This is now done by modifying the closeness coefficient in form given in Eq. (20). The closeness coefficients of the alternative project P_i with respect to the positive ideal solution by using the distance matrix (CC_i) are defined as:

$$CC_i = \frac{S_{iw}^- + D_{iw}^+}{D_{iw}^+ + S_{iw}^+ + S_{iw}^-}, i = 1, 2, \ldots, m \tag{20}$$

Next we rank the projects by closeness coefficients, now using an ascending order. For all $i = 1, 2, \ldots, m$ and $j = 1, 2, \ldots, n$. Different steps for the given $TOPSIS$ algorithm can be presented as follows:

Step 1: Form a decision-makers' committee, and identify the evaluation criteria.
Step 2: Choose appropriate linguistic variables for the importance weight of the criteria and the linguistic ratings for alternative projects.
Step 3: Aggregate the weight of criteria to get the aggregated fuzzy weight $\hat{w}_j$ of the criterion C_j, and join decision-makers' ratings to get an aggregated fuzzy rating x_{ij} of the project P_i in consideration of the criterion C_j.
Step 4: Construct a fuzzy decision matrix and a normalized fuzzy decision matrix.
Step 5: Construct a weighted normalized fuzzy decision matrix.
Step 6: Determine a fuzzy positive (and negative) ideal solution $FPIS$ (and $FNIS$).
Step 7: Construct a similarity matrix by calculating the similarity measure of each project from the $FPIS$ (and $FNIS$).
Step 8: Calculate aggregated similarity values for each project with regards to the $FPIS$ and the $FNIS$ by using OWA.

Step 9: Calculate a multi-distance value for each project with regards to the *FPIS*.

Step 10: Calculate a closeness coefficient for each project, in order to determine the projects' ranking within the set of projects.

For steps 8 − 10: use multiple orness values for each alternative to get multiple ranking results.

Step 11: Rank the set of alternatives for each orness value and calculate the minimum, the mean, and the maximum ranking of each alternative, to form a triangular fuzzy ranking score for each alternative by using the three values.

Step 12: Make an overall ranking of the alternatives using the method by Kaufmann and Gupta [16], based on the fuzzy ranking score.

4 Numerical Example

This numerical example uses the same data that is also used in [19]. A pharmaceutical company can select a certain number of projects to invest in from among 20 R&D projects. Criteria that are used in the example come from costs, revenues, budget constraints, and real option values (ROV), calculated for each project by using the

Table 1 Evaluation of R&D projects

Project	C_1	C_2	C_3	C_4
P_1	(53, 62, 68, 78)	(43, 50, 55, 63)	(115, 128, 128, 141)	0.06
P_2	(83, 98, 108, 123)	(85, 100, 110, 125)	(126, 140, 140, 154)	0.0594
P_3	(157, 185, 204, 231)	(170, 200, 220, 250)	(170, 189, 189, 208)	18
P_4	(204, 240, 268, 300)	(170, 200, 220, 250)	(164, 182, 182, 200)	0.54
P_5	(259, 305, 336, 381)	(510, 600, 660, 750)	(209, 232, 232, 255)	3.10
P_6	(85, 100, 110, 125)	(85, 100, 110, 125)	(185, 206, 206, 227)	5
P_7	(259, 305, 336, 381)	(510, 600, 660, 750)	(209, 232, 232, 255)	3.10
P_8	(94, 110, 121, 138)	(85, 100, 110, 125)	(177, 197, 197, 217)	1.58
P_9	(140, 165, 182, 206)	(153, 180, 198, 225)	(238, 264, 264, 290)	17.15
P_{10}	(190, 223, 245, 279)	(323, 380, 418, 475)	(257, 285, 285, 314)	1.65
P_{11}	(60, 70, 77, 88)	(68, 80, 88, 100)	(148, 164, 164, 180)	10.03
P_{12}	(91, 107, 118, 134)	(85, 100, 110, 125)	(144, 160, 160, 176)	2.39
P_{13}	(247, 290, 319, 363)	(34, 40, 44, 50)	(297, 330, 330, 363)	0
P_{14}	(370, 435, 479, 544)	(595, 700, 770, 875)	(338, 375, 375, 413)	278.25
P_{15}	(166, 195, 215, 244)	(425, 500, 550, 625)	(279, 310, 310, 341)	320.25
P_{16}	(221, 260, 286, 325)	(255, 300, 330, 375)	(315, 350, 350, 385)	39.66
P_{17}	(235, 277, 305, 346)	(298, 350, 385, 438)	(311, 346, 346, 381)	72.48
P_{18}	(281, 330, 363, 413)	(468, 550, 605, 688)	(331, 368, 368, 405)	231
P_{19}	(344, 405, 446, 506)	(680, 800, 880, 1000)	(365, 406, 406, 447)	414.75
P_{20}	(451, 530, 583, 663)	(978, 1150, 1265, 1438)	(394, 438, 438, 482)	651

Table 2 Project closeness coefficient values and rankings, for $\alpha = 0.1, 0.5, 1$

Project	$CC_{(\alpha=0.1)}$	Rank	$CC_{(\alpha=0.5)}$	Rank	$CC_{(\alpha=1)}$	Rank
P_1	0.83	19	0.710	8	0.709	8
P_2	0.813	13	0.728	11	0.728	11
P_3	0.79	10	0.741	15	0.741	15
P_4	0.80	11	0.751	19	0.751	19
P_5	0.61	2	0.614	2	0.615	2
P_6	0.821	17	0.74	14	0.739	14
P_7	0.819	15	0.717	10	0.716	10
P_8	0.824	18	0.744	17	0.743	17
P_9	0.809	12	0.751	18	0.75	18
P_{10}	0.68	6	0.681	7	0.682	7
P_{11}	0.81	14	0.716	9	0.715	9
P_{12}	0.82	16	0.737	13	0.736	13
P_{13}	0.85	20	0.798	20	0.797	20
P_{14}	0.64	4	0.674	6	0.674	6
P_{15}	0.58	1	0.612	1	0.613	1
P_{16}	0.77	9	0.742	16	0.743	16
P_{17}	0.75	8	0.729	12	0.729	12
P_{18}	0.64	3	0.672	5	0.673	5
P_{19}	0.68	5	0.661	4	0.662	4
P_{20}	0.71	7	0.652	3	0.651	3

pay-off method for real option valuation [20]; the values of these four criteria are represented by trapezoidal fuzzy numbers. The first and the third criteria are cost criteria and the second and the fourth, benefit criteria.

In Table 1 one can see evaluations of the different criteria by using trapezoidal fuzzy numbers. The fourth (ROV) criterion is carried out in computations as a fuzzy number of form $A = (a_1, a_2, a_3, a_4)$, where $a_1 = a_2 = a_3 = a_4$.

For different α values, the Table 2 shows the computed closeness coefficients and the rankings for each of the twenty projects for three different orness values α. In computation of basic statistics (see Table 3) and in creation of fuzzy numbers (see Table 4) a larger set of values of α was used. There α values were $\alpha = 0.1, 0.2, \ldots, 0.9, 1$.

Table 3 The minimum, mean, and the maximum rankings

Project	Minimum	Mean	Maximum
P_1	8	10.3	19
P_2	11	11.9	14
P_3	10	13.7	15
P_4	11	17.1	19
P_5	2	2	2
P_6	14	15.2	18
P_7	10	11	15
P_8	17	17.5	19
P_9	12	17.1	18
P_{10}	6	6.8	7
P_{11}	9	10	14
P_{12}	13	14.1	17
P_{13}	20	20	20
P_{14}	4	5.4	6
P_{15}	1	1	1
P_{16}	9	13.6	16
P_{17}	8	10.5	12
P_{18}	3	4.4	5
P_{19}	4	4.3	5
P_{20}	3	4.1	7

In Table 3, the minimum, the mean, and the maximum rankings from our experimental setup are summarized. These are then used in the formation of a triangular fuzzy number ranking for each project.

A total ordering is found for the fuzzy numbers presented in Table 3 by using the method introduced by Kaufmann and Gupta [16]. For this purpose removal number, dispersion, and modal value are calculated in a way presented above—Table 4 presents the resulting overall ranking. According to the result the top five projects are 15, 5, 18, 19, and 20.

Table 4 Overall rankings of the R&D projects using removal number, dispersion, and modal value

Project	Rank	Removal no	Div	Mode
P_{15}	1	1	0	1
P_5	2	2	0	2
P_{18}	3	4.2	2	4.4
P_{19}	4	4.4	1	4.3
P_{20}	5	4.55	4	4.1
P_{14}	6	5.2	2	5.4
P_{10}	7	6.65	1	6.8
P_{17}	8	10.25	4	10.5
P_{11}	9	10.75	5	10
P_7	10	11.75	5	11
P_1	11	11.9	11	10.3
P_2	12	12.2	3	11.9
P_{16}	13	13.05	7	13.6
P_3	14	13.1	5	13.7
P_{12}	15	14.55	4	14.1
P_6	16	15.6	4	15.2
P_9	17	16.05	6	17.1
P_4	18	16.05	8	17.1
P_8	19	17.75	2	17.5
P_{13}	20	20	0	20

5 Conclusions

A new multiple-criteria decision making approach was presented; it is an extension
for the fuzzy similarity based fuzzy TOPSIS. OWA was used for aggregating similarity to fuzzy negative and positive ideal solutions for each criterion and multi-distance
was used in collecting information about the "similarity of these similarities" that
can be understood as a measure of homogeneity or consistency of a given project.
This has allowed the inclusion of more relevant information than is possible when
using a simple defuzzification procedure. The method was applied to a R&D project
selection problem. The results are dependent on the proper selection of the orness
parameter, α, when the weights are generated for the OWA operator. This weight generation was done by using O'Hagan's method that finds the weights as an optimal
solution for a predefined (given) orness value (α). We examined the effect that the
pre-selection of orness has by testing the ranking with a number of orness values. We
presented a way to take in to consideration the "created" information with different
orness values by forming fuzzy numbers from ten rankings of the projects created
by using ten different orness values. A measure of homogeneity of similarity of the
different criteria of each project to the fuzzy positive ideal solution was calculated by

using multi-distances. This was done to include information about the consistency of the level of goodness of projects (by the selected criteria). This information was included in the closeness coefficient that was used in the ranking of the projects. The final ranking thus includes information about the goodness of each project (as ranked by TOPSIS) and about the "stability" of the level of goodness of each of the criteria of each project. The top five projects from the numerical example were found to be 15, 5, 18, 19, and 20. Notable from the results is that projects 15 and 5 were always top 2 choices, but project 20 varied between rankings 3 to 7 so that with lower values of orness ranking was lower and after orness value 0.6 it was always the third best choice. Forming a fuzzy number from different rankings allows one to include different points of view and creating an intelligent overall ranking. Using multiple orness values in forming the final ranking is relevant in situations, where there is uncertainty or imprecision with regards to the correct parameter selection. It is clear that if there is absolutely no uncertainty involved in the orness parameter selection that then one should use the certain parameter alone in creating the ranking. Furthermore, more relevant information is carried along in the analysis, until the ranking stage, enabling the ranking to take more things into consideration and thus being based on a more holistic view of the problem. Interesting future research directions include research into aggregation operators in general and into how different types of multi-distances may be used in connection with decision-making problems in general.

References

1. Chen, C.T.: Extensions of the TOPSIS for group decision-making under fuzzy environment. Fuzzy Sets Syst. **114**, 1–9 (2000)
2. Chen, C.-T., Lin, C.-T., Huang, S.-F.: A fuzzy approach for supplier evaluation and selection in supply chain management. Int. J. Prod. Econ. **102**(2), 289–301 (2006)
3. Luukka, P.: Fuzzy similarity in multicriteria decision-making problem applied to supplier evaluation and selection in supply chain management. Advances in Artificial Intelligence, pp. 1–9. Hindawi Publishing Corporation (2011)
4. Niyigena, L., Luukka, P., Collan, M.: Supplier evaluation with fuzzy similarity based fuzzy TOPSIS with new fuzzy similarity measure. In: 13th IEEE International Symposium on Computational Intelligence and Informatics. Budapest, November, 2012 (2013)
5. Collan, M., Luukka, P.: Evaluating R&D projects as investments by using an overall ranking from four new fuzzy similarity measure based TOPSIS variants. IEEE Trans. Fuzzy Syst. **26**(6), 1–11 (2013)
6. Hejazi, S.R., Doostparast, A., Hosseini, S.M.: An improved fuzzy risk analysis based on a new similarity measures of generalized fuzzy numbers. Expert Syst. Appl. **38**(8), 1–7 (2011)
7. Martin, J., Mayor, G.: Some properties of multi-argument distances and Fermat multidistance. In: Proceedings of IPMU 2010, pp. 703-711 (2010)
8. Martin, J., Mayor, G., Valero, O.: Multi-argument distances. Fuzzy Sets Syst. **167**, 92–100 (2011)
9. Brunelli, M., Fedrizzi, M., Fedrizzi, M., Molinari, F.: On some connections between multidistances and valued m-ary adjacency relations, Advances on computational intelligence. In: Proceedings of the 14th International Conference on Information Processing and Management of Uncertainty in Knowledge-Based Systems, IPMU 2012, vol. 297, Part I, pp. 201–207. Catania, Italy, 9–13 July 2012

10. Zadeh, L.: Similarity relations and fuzzy orderings. Inf. Sci. **3**, 177–200 (1971)
11. Shepard, R.N.: Toward a universal law of generalization for psychological science. Science **237**, 1317–1323 (1987)
12. Chen, S.H.: Operations on fuzzy numbers with function principal. Tamkang J. Manag. Sci. **6**(1), 13–25 (1985)
13. Yager, R.R.: On ordered weighted averaging operators in multicriteria decision making. IEEE Trans. Syst. Man Cybern. **18**, 183–190 (1988)
14. O'Hagan, M.: Aggregating template or rule antecedents in real time expert systems with fuzzy set logic. In: Proceedings of 22nd Annual IEEE Asilomar Conference on Signals, Systems, Computers, pp. 681–689. Pacific Grove, California (1988)
15. Fullér, R., Majlander, P.: An analytical approach for obtaining maximal entropy OWA operator weights. Fuzzy Sets Syst. **124**, 53–57 (2001)
16. Kaufmann, M., Gupta, M.: Fuzzy Mathematical Models in Engineering and Management Science. Elsevier, New York (1988)
17. Socorro, M., García-Cascales, M., Lamata, T.: Solving a decision problem with linguistic information. Pattern Recognit. Lett. **28**(16), 2284–2294 (2007). Elsevier
18. Cui, Z.X., Yoo, H.K., Choi, J.Y., Youn, H.Y., Multi-criteria group decision making with fuzzy logic and entropy based weighting. In: Proceedings of the 5th ICUIMC'11, pp. 1–7. Feb 2011
19. Hassanzadeh, F., Collan, M., Modarres, M.: A practical approach to R&D portfolio selection using fuzzy set theory. IEEE Trans. Fuzzy Syst. **20**, 615–622 (2012)
20. Collan, M., Fullér, R., Mezei, J.: A fuzzy pay-off method for real option valuation. J. Appl. Math. Decis. Sci. http://ideas.repec.org/p/pra/mprapa/13601.html (2009). Accessed 20 Mar 2012

Methodology of Virtual Wood Piece Quality Evaluation

Jeremy Jover, Vincent Bombardier and Andre Thomas

Abstract This paper presents a way to evaluate the quality of virtual wood products according to their tomographic image. The main objective is to anticipate a sawmill divergent process in order to enhance the production plan. From a virtual representation of the product, singularity features are extracted and their impact on the product virtual quality is assessed thanks to the Choquet integrals. Next, the visual quality is evaluated by merging singularity impacts and singularity number criterion using suitable operators. Three operators are compared to the mean operator which is the commonly used one when there is little knowledge on the decision process. Finally the measure is express in the Sawmill expert language using linguistic variables which give the possibility degree that the product belongs to each quality. This degree could be understood as the risk to attribute the concerned quality. It is finally used to determine which quality is to attribute to product in order to satisfy customer needs and maximize sawyers benefit by a linear programming algorithm.

Keywords Virtual product · Quality · RX computed tomography · Information fusion · Divergent bill of material

1 Introduction

Divergent processes have always presented a difficulty for the traceability implementation. The raw material cutting leads two problems:

- The link between material and production information is difficult to establish and to maintain all along the product cycle life.

J. Jover · V. Bombardier (✉) · A. Thomas
Centre de Recherche en Automatique de Nancy (CRAN), Université de Lorraine, CNRS,
Boulevard des Aiguillettes B.P. 239, 54 506 Vandœuvre-lès-Nancy, France
e-mail: vincent.bombardier@univ-lorraine.fr

J. Jover
e-mail: jover.jeremy@gmail.com

A. Thomas
e-mail: andre.thomas@univ-lorraine.fr

© Springer International Publishing Switzerland 2016
K. Madani et al. (eds.), *Computational Intelligence*,
Studies in Computational Intelligence 613,
DOI 10.1007/978-3-319-23392-5_14

- Foresee the finish products features, which are very useful for traceability and production management, is full of uncertainty.

Some solutions are proposed to overcome this problem especially in the food industry. These solutions are based on marking and documenting batches [7]. However a part of the root information is lost and a unique identification is not still possible (what is ideally expected).

The wood industry is also concerned by these divergent process problems. From a tree, products satisfying the end customer's needs must be produced. Moreover the product origin traces have to be conserved for traceability reasons [13]. The wood, being a heterogeneous material, increases complexity. Structurally, the wood is composed of aligned fibers following a longitudinal axe. It not reacts in the same way following the different axes (longitudinal, radial, and tangential). The wood color is not homogeneous too: the growing rings alternation, the singularity presence or the fungal attacks (blue stain) create heterogeneity on visual and mechanical points of view.

In sawmills, the optimization, in order to have the right products, is an important and complicated task. Sawyers have to saw products which have characteristics needed by customers, from a raw material which internal characteristics are unknown. Dimensionally, it is easy to foresee and have the right product dimension (apply the cutting pattern), but other features as the color or the mechanical resistance are more complicated to estimate and to characterize before the log is sawed due to their subjectivity character and the wood heterogeneity.

Our researches are concerned by the information loss reduction in the wood industry. We have proposed a solution to mark and maintain the origin information of the trees [11]. In this study, we propose a way to determine the wood product characteristics before sawing operation in order to satisfy the customer needs and the optimal determination of the production element (net requirement for each product quality class). The proposed approach aims to automate the product qualification process (quality product estimation) usually done by an operator. The global process is described Fig. 1.

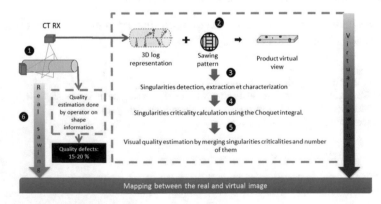

Fig. 1 Extraction and exploitation process of virtual products

In this article, we shortly present the wood quality notion, the sawing optimization process and the virtual sawing concept allowing to extract the virtual product. Then we explain how we characterize singularities and calculate their impact on product with the Choquet integral. Impact calculation is improved by using learning process to calculate the Choquet capacities. To finish, we describe a way to estimate the wood products quality by using the impact singularities. We demonstrate the feasibility with distinguish wood pieces.

2 Production Foresee in Sawmill

2.1 The Visual Quality of Wood Products

Concerning the needs of the first transformation customers, there are three kinds of quality: the dimensional quality, the mechanical quality and the visual (aesthetic) quality. The dimensional quality is easy to characterize (dimension piece precision). The mechanical quality is more skillful to evaluate. The clear wood has a mechanical resistance which can be reduced by the singularities presence (knots, crack, rot . . .). Techniques based on the vibrations give results as explained in [10].

The last one is the visual quality. It is the most complicated to evaluate because the visual quality is a subjective decision. The visual quality is defined in the standard NF EN 1611. This standard defines different classes (five) of qualities based on the singularity feature measurements (size, numbers, type . . .). But the evaluation is done by a Human Expert which has to estimate the quality within a short time (according to sawmill high production rate). In this short time the expert cannot evaluate singularity features as precisely as the standard defines them. So the standard definition is not adapted to the evaluation. More over the wood is "intrinsically fuzzy" [3]. Boundaries between clear wood and singularities are not so easy to determine and impact the characteristic measures. A big part of the price is based on this quality, so its determination is important for customer and sawyers.

2.2 Raw Material Optimization

Sawyers optimize wood by estimating which cutting plan allows to have the best material yield and the customer requirement. That is why it is essential to foresee product features which would be cut in the log.

The Expert (present at the optimization post) estimates product features (dimension, mechanical resistance, and visual quality) according to the external log features and his experiment. He is able to determine approximately which defects are present in the wood (according to the external log features) and decides which cutting plan is the most appropriate to obtain the customer requirement. So final product aspect and

Fig. 2 Example of a virtual
sawing extracted side

quality are more or less well determined. We easily understand that all singularities are not visible on the surface and singularities which are visible give only incomplete information on their shapes in wood. Lot of researches have proposed solutions taking into account the external shapes of logs [15]. But these optimizations are only based on the log dimensional features and do not take into account the visual and mechanical characteristics. This paper proposed a way (virtual sawing) to address this issue.

2.3 Virtual Sawing

The use of non-destructive control techniques [4], in particular X ray computed tomography, allows to have a 3D representation of the log (internal and external) to be cut. [1] use volumetric information to improve part log quality determination and their sorting. In our proposition, we investigate this step and the global process is described in the Fig. 1.

The log representation is virtually sawed with ad-hoc software according to a cutting plan. This leads to obtain a numerical view of all product faces which should be obtained. The Fig. 2 shows one face for one product. The obtained image represents the face of a product according to the density data. Some information cannot be obtained (the color) and the distinction between detected object is not so easy. All of these add imperfection, imprecision and uncertainty and make the quality determination harder.

2.4 Problematic

Our aim is to propose a process to estimate the wood product quality according to the face picture extracted by the virtual sawing stage. We consider that the singularity features are computed in a similar way as [3]. So we propose a way to determine quality products from these measurements. As obtained information is uncertain, incomplete and imprecise, we use methods allowing taking into account this imperfection especially the Choquet Integral and fuzzy fusion operators. In this paper, we decide to estimate the singularity impact on the product and then to determine the piece quality.

3 Singularity Impact Evaluation on Virtual Product

3.1 Singularity Criterion Measurement

In our study, we evaluate the visual quality of the wood. So the criteria have to reflect singularity impact on the visual quality. [1] defines forty criteria to evaluate quality. From these forty criteria, only around twenty concern the final product and only a dozen the visual quality. To evaluate singularity impact on density data, we only use the four of them which are measurable on a grey scale image.

The first criterion μ_t, described by the Eq. 1, reflects the singularity size. Bigger a singularity is, more the visual quality is down grading. Moreover the expert judgment stipulates a singularity higher than 50 mm is considered as highly critical.

$$\begin{cases} If \ l \leq 50 \, mm, \quad U_t = 1 - T_s/l \\ If \ l > 50 \, mm \ and \ T_s \leq 50 \, mm, \quad U_t = 1 - T_s/50 \\ If \ l > 50 \, mm \ and \ T_s > 50 \, mm, \quad U_t = 0 \end{cases} \quad (1)$$

The second criterion u_x, described by the Eq. 2, reflects the position of the singularity on the product length. More a singularity is close to the product end, more this singularity lost importance and the quality becomes higher.

$$u_x = \frac{|L/2 - X_s|}{L/2} \quad (2)$$

The third criterion μ_y, given by the Eq. 3, reflects the position of the singularity on the product width. More a singularity is close to the product edge, more this singularity lost important and the quality becomes higher.

$$u_y = \frac{|l/2 - Y_s|}{l/2} \quad (3)$$

The last criterion μ_c, (Eq. 4), reflect contrast between the singularity and the product background. More the contrast is weak, less this singularity is visible and more the quality increases.

$$u_c = 1 - \frac{|I_s - I_p|}{I/p} \quad (4)$$

In the following part, we proposed a method using Choquet integral according to the fact that the singularity characteristic measurements are full of imperfection and imprecision (see Sect. 2).

3.2 Impact Calculation Using the Choquet Integrals

The Choquet integrals were proposed by Gustave Choquet in 1954 [5]. Their use
in the multi criteria decision making domain appears in the nineties in different
context (car industry, strategical placement …) and similar classification problems are
usually process with the Choquet integrals [8]. They allow taking into consideration
importance of each criterion and the interactions existing between each of them.

Let $\{X\}:\{x1,\ldots,xn\}$ be a set of normalized criteria, consider a capacity $\mu:P(X)\rightarrow$
$[0,1]$ on this set, verifying Eq. 5:

$$\begin{cases} \mu(\emptyset) = 0 \\ \mu(X) = 1 \\ \mu(A) < \mu(B), \ \forall A \in B \ and \ B \in X \end{cases} \tag{5}$$

The capacity defines all weights and interactions. Then Choquet integral is defined
by Eq. 6:

$$C_\mu(u_1,\ldots u_n) = \sum_{i=1}^{n} (u_{\sigma k} - u_{\sigma k-1})\mu(A_{\sigma k}) \tag{6}$$

where σ is the index permutation satisfying Eq. 7:

$$0 = \mu_{\sigma 0} \leq \mu_{\sigma 1} \leq \cdots \leq \mu_{\sigma n-1} \leq \mu_{\sigma n}$$
$$\mu_{\sigma 1} = Min(u_i) \ and \ \mu_{\sigma n} = Max(u_i) \tag{7}$$

and $A\sigma k = \{g\sigma k, \ldots, g\sigma n\}$ the features non used in previous step.

In our case, the $C\mu(u)$ corresponds to the measure of the singularity impact on the
product when the Choquet integral is apply on the criteria μi. More the value tends
to 1, less the singularity is important (our own standards). The Choquet integral is
useful when the knowledge and the learning batches are low. The greatest challenge
is the definition of the capacities [8]. To do so, some approaches were developed to
learn the capacities.

3.3 Learning Process of the Capacities

In order to have a better definition of the capacities used in the Choquet integral,
we decide to use a learning process. Different approaches can be used to identify
capacities [8]:

- The Least Square approach (LS): based on the expert knowledge on each element.
  The expert attributes a target impact value to each element (expected value) in the
  learning lot and system searches capacities values that minimize the difference
  square between the computed value and the expected value (Eq. 8).

$$Min\ F_{LS}(\mu) := \sum_{x \in O} [C_\mu(u(x)) - y(x)]^2 \tag{8}$$

- The Linear Programming approach (LP): proposed by Marichal and Roubens in [12], it is based on the expert knowledge on the global ranking of the batch elements (Eq. 9). The approach looks for the value which satisfy as closely as possible the ranking establish by the expert.

$$Max\ F_{LP}(\varepsilon) := \varepsilon$$

$$subj.\ to \begin{cases} \Sigma_{T \subseteq S} m_v(T \cup i) \geq \forall i, \forall S \\ \Sigma_{T \subseteq N} m_v(T) = 1 \\ C_v(u(A)) - C_v(u(B)) \geq \delta_c \\ \vdots \end{cases} \tag{9}$$

In [8], authors explain that the least square approach is appropriate when it is possible to attribute precisely the desired value. They explain too that the linear programming is better for cases which it is easy to give a pre-order between the learning lot elements. This is our case for the evaluation of the singularity impact because it is hard for the expert to give a score for each singularity (due to number and variation of the cases). The expert decides of a pre-order between elements composing the learning batch (with a δ_C corresponding to the minimum margin to respect the ranking). This constraint, noted E, can be translate by equation (10):

$$C_\mu(a) > C_\mu(b) > \cdots > C_\mu(k)$$
$$with\ C_\mu(u(i)) \leq C_\mu(u(i+1)) + \delta_c \tag{10}$$

Some conditions can be imposed, over the element pre-order, on the criterion importance and/or interaction. The expert can express the criterion importance against one another. By the used of the Shapley indexes ϕ (which indicates the global importance of each criterion), the expert expresses the equality between two criteria. The value δ_ϕ is the maximal distance between two Shapley values to consider two criteria are equal. This constraint, noted S, can express for a couple of criteria A and B as:

$$-\delta_\phi \leq \phi_v(\mu_A) - \phi_v(\mu_B) \leq \delta_\phi \tag{11}$$

More over the Expert can express constraints on the interaction between the criteria. The interaction between two criteria can be easily expressed by the expert because the phenomenon is understandable. But the interaction between more than two criteria is harder to understand and express. The last condition (apply on the interaction indices) is only expressed on interaction between a pair of criteria. This constraint, noted I, can be:

- negative (redundancy): $I_v(\{A, B\}) < 0 - \delta_I$
- positive (synergies): $I_v(\{A, B\}) > 0 + \delta_I$
- null (no interaction): $I_v(\{A, B\}) = 0 \pm \delta_I$

δI is the minimum threshold value in absolute value to consider the interaction as significant.

Using the software R and the Kappalab R package, we compute the LP approach in order to determine the capacities and influence of the calculated values on the Choquet integral results.

3.4 Result of the Learning Process LP

The learning batch described in the Table 1 is composed of singularities which are commonly found in the wood.

Expert constraints are described below:

- About the elements' batch (E): The singularities are ranked as they are stored in the Table 1 from the best to the worst with $\delta_C = 0.05$.
- About the importance criterion (S): criteria $[\mu_t, \mu_c]$ have the same importance, criteria $[\mu_x, \mu_y]$ too and criteria $[\mu_t, \mu_c]$ are more important than $[\mu_x, \mu_y]$ ($\delta_\mu = 0.1$):

$$\phi_v(\mu_T) = \phi_v(\mu_C) > \phi_v(\mu_X) = \phi_v(\mu_Y) \tag{12}$$

- About the influence among criterion (I): criteria $[\mu_t, \mu_c]$ are in synergy and $[\mu_x, \mu_y]$ too ($\delta_I = 0.05$):

$$I_v(\{T, C\}) > 0 \, and \, I_v(\{X, Y\}) > 0 \tag{13}$$

The results of the Choquet value, the Shapley values and the interaction indices are respectively presented in the Tables 2, 3 and 4. In Table 2, the second column

Table 1 Learning batch

Singularity	a	b	c	d	e	f
T	0.75	0.41	0.39	0.69	0.42	0.75
X	0.72	0.09	0.27	0.36	0.75	0.81
Y	0.2	0	0.24	0.8	0.32	0.52
C	0.93	0.98	0.9	0.37	0.3	0.09

Table 2 Results for the LP approach for the different constraints

Singularity	a	b	c	d	e	f
Ø	0.65	0.37	0.45	0.55	0.46	0.49
E	0.878	0.815	0.753	0.45	0.387	0.325
E+S	0.878	0.815	0.753	0.569	0.422	0.359
E+S+I	0.873	0.799	0.738	0.593	0.532	0.471

Table 3 Shapley indices for the different constraints

Shapley value	T	X	Y	C
E	0.152	0.117	0.069	0.662
E+S	0.397	0.074	0.045	0.484
E+S+I	0.322	0.169	0.098	0.411

Table 4 Interaction indices for the different constraints (symmetric matrix)

Constraints E

Criteria	T	X	Y	C
T	NA	0.006	−0.138	−0.119
X		NA	0.131	0.137
Y			NA	0.007

Constraints E + S

Criteria	T	X	Y	C
T	NA	−0.133	−0.090	0.113
X		NA	0.179	−0.133
Y			NA	−0.090

Constraints E + S + I

Criteria	T	X	Y	C
T	NA	−0.219	0.032	0.050
X		NA	0.050	−0.285
Y			NA	−0.226

corresponds to values obtained without any importance (singleton capacity equal to 0.25) and constraints (other capacity equal to the sum of the singleton capacities) between criteria.

Imposed Constraints are respected at each step (E, E + S, E + S + I). We can see in the Table 2 that the order of the singularity is the same as the expert ranking.

The first constraint (based on the elements ranking) accords lot of importance to the contrast (Shapley value $\phi_v(\mu_c = 0.762)$) and few on the other criteria. Moreover interaction between the criteria [T C] is negative (that is not corresponding to the expert choice). The Fig. 2 shows the variation of the Choquet integrals value function of [X Y]. The criterion Y has little influence except up to 0.75. This translates an

Fig. 3 Influence of the X and Y criteria modification on the Choquet value with the weights obtained under the three different constrain and no variation of the T and C criteria (u = (0.75, Δ(X), Δ(Y), 0.09)). **a** Under E constrains, **b** under E + S constrains, **c** under E + S + I constrains

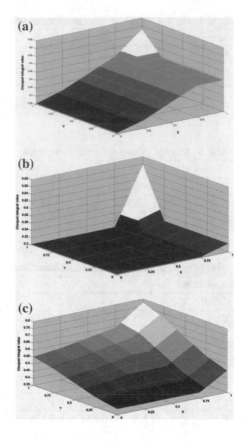

expert view: the singularity position on the width upgrade the singularity only when it is very close to the side.

The addition of the constraint S (on Shapley indices) offers a positive interaction between [T, C] but does not with [X, Y]. Moreover, the importance of the criteria [X,Y] is so small that they have few influences on the Choquet value. The Fig. 3b shows the Choquet value variation function of X and Y variation for a singularity (u = (0.75, Δ(X), Δ(Y), 0.09)). The criteria have no action when theirs values are below 0.75. This comportment means that singularity position is only important when a singularity is close to the end and the side of the product.

The addition of the last constraint gives capacities which allow the respect of the constraints given by the expert. Moreover all criteria have impact on the calculation result. The Fig. 3c translates the [X, Y] impact. Thereafter we use this weight for the product quality estimation.

4 Virtual Product Quality Evaluation

Once singularity impacts are determined, an estimation of the quality of a virtual product is evaluated by merging these impacts and the criterion related with the singularity number. In this part, we describe this criterion and the fusion operation.

4.1 Singularity Number Criterion

In the quality evaluation, the number of singularities is important. A product with a lot of singularities is more down grading because the clear wood homogeneity is broken.

To evaluate the number, we used the criterion R_{nb} defined by (12). This criterion represents the expert vision: more there are singularities, more the product is down grading. Moreover when the number reaches towards a particular value, the criterion reaches towards 0. We choose to use an exponential function. Following the particular number of singularities fixed by the expert, the k coefficient can be changed. In our case, we determine that up to 20 singularities, the value starts to become constant (k = 1.1).

$$u_{nb} = k^{-NB_s} \ with \ k = 1.1 \tag{14}$$

4.2 Quality Determination by Data Fusion

In order to determine the quality product, we merge singularity impacts and the singularity number criterion. There are three kinds of merging operators [2]: Severe, Indulgent and compromise operators.

In the quality evaluation, Expert never evaluates products on the best singularities. So, indulgent operators cannot be used. The two other kinds of operators translate different visions from the expert in quality evaluation. We propose to compare different operators which appear to be well adapted to our use and Expert quality evaluation.

The first operator which can be used is the operator defined by Perez-Orama in [14] and describe by 15. This operator (PO) is a compromise operator when the minimum value is under 0.5; otherwise it is an indulgent operator. This characteristic can be interesting to isolate product with few singularities and evaluate the worst product.

$$F(a, b) = \min(1, \frac{\min(a, b)}{1 - \min(a, b)}) \tag{15}$$

The Hamacher operator, described in 16, is a severe operator. That means this operator gives result under the worst singularity. This can be useful to evaluate quality for product where visual quality is very important (joinery, cabinet ...) because only products with high quality are highlighted.

$$F(a, b) = \frac{ab}{a + b - ab} \tag{16}$$

The Ordered Weight Average (OWA) adapted to our situation too, described in 17, is a compromised operator. Usually the product quality is based on a part of the worst singularities (represented by α which represent the percent of product fusion). The OWA allows to attribute weight only on this part of the singularities impact C_i.

$$F(u_1, \ldots, u_n) = \sum_{i=1}^{n} w_i C_i$$
$$and \ C_1 \geq C_2 \geq \cdots \geq C_{n-1} \geq C_n$$
$$with \begin{cases} w_i = 0, \forall i \in [1, \lfloor \alpha n \rfloor] \\ w_i = \frac{1}{\alpha * NB_s}, \forall i \in [\lceil \alpha n \rceil, n] \end{cases} \tag{17}$$

The results provided by these three operators will be compared to the arithmetic mean (used as a benchmark) which is a classical operator when the aggregation comportment is unknown.

4.3 Sample Set Presentation

We will consider the batch described in the Table 5 taken into sawmill. The first column indicates the piece number, the second the criterion R_{nb} (function of the number of singularities), the third, the impact of all the singularities present on the product and the last corresponds to the product aesthetics class given by the sawmill Expert. The aim is to compare some fusion operators with the expert vision so as to find the closest.

Table 5 List of the piece used to compare fusion operators

Product	R_{nb}	Cu								
1	0.91	0.888								
2	0.91	0.439								
3	0.42	0.888	0.866	0.826	0.776	0.748	0.746	0.674	0.601	0.520
4	0.75	0.814	0.740	0.259						
5	0.75	0.888	0.372	0.332						
6	0.51	0.814	0.740	0.694	0.601	0.565	0.432	0.312		
7	0.51	0.725	0.667	0.587	0.479	0.423	0.372	0.332		

- Piece 1: only one singularity with few impact (impact value close to 1), used for cabinet work.
- Piece 2: only one singularity with high impact, used for joinery work.
- Piece 3: lot of singularity with little impact, function of the use, the quality can be high or not depending on the singularities number. In our case expert classes the product for joinery.
- Pieces 4 and 5: the same number of singularities but one with more singularities with few impact (4) used for joinery work and the other (5) used in industrial carpentry.
- Pieces 6 and 7: more singularities than pieces 4 and 5. Used respectively in industrial carpentry and traditional carpentry.

To compare the operators, two features are studied: the products ranking (cf. Table 6) and the distance between them (cf. Fig. 4). Operators have to be compared on the distance and the groups of products they make. If the ranking is good but groups are totally different from the Expert choice (two products by in the same aesthetic class on the Expert judgment have to be close one to the other) the operator is less efficient than an operator which wrong ranks but keeps the right groups of products.

The Hamacher operator, as it is the only pessimist operator, gives the lowest results. This operator is very efficient to highlight product which have good features. When there are a lot of singularities, this operator reaches towards highly downgrade product (up to 2 singularities, maximum value is 0.5). Three groups of product are made: (a, b) on the top as the expert, (e, d) and (c, f, g) on the low. These seconds' two groups mix quality product express by the Expert. So this operator is useful to evaluate high quality products.

The Perez-Orama operator gives high importance to product with all singularity impacts up to 0.5 and downgrades the others. It assumes that products with less than 8 singularities with an impact up to 0.5, have a quality equal to 1. This operator is particularly useful for a first ranking and extracts products previously described (less than 8 singularities with impact up to 0.5). The operator places on opposite ends a and d and mades two groups, (b, c) and (e, f, g). This classification is close

Table 6 Piece ranking for each fusion operator

Rang	CE	QE	Hama	PO	OWA ($\alpha = 0.2$)	OWA ($\alpha = 0.8$)	Means
1	a	0	a (0.82)	a (1)	a (0.89)	a (0.90)	a (0.90)
2	b	1	b (0.42)	b (0.43)	c (0.47)	b (0.67)	c (0.71)
3	c		d (0.05)	c (0.43)	b (0.44)	c (0.66)	b (0.67)
4	d		e (0.04)	e (0.31)	g (0.37)	d (0.64)	d (0.64)
5	e	2	c (0.00)	g (0.31)	f (0.35)	e (0.59)	e (0.59)
6	f		f (0.00)	e (0.33)	f (0.55)	f (0.29)	f (0.58)
7	g	3	g (0.00)	d (0.26)	g (0.48)	d (0.23)	g (0.51)

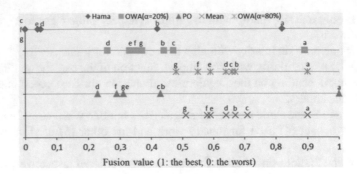

Fig. 4 Result of the fusion operation for each piece

to the expert choice (except for *d*). This operator is efficient to classifie low quality products.

α factor in the OWA operator may change value function of the singularity numbers to be evaluated. We propose to compare α = 0.2 and α = 0.8. In the α = 0.2 case, the ranking gives, as the PO operator, four classes ((a), (c, b),(e, f, g) and (d)) and the same observation as the previous operator. In α = 0.8 case, the ranking is the same as the expert. Moreover, product groups do by this operator is the same as the expert classification ((a), (b, c, d), (e, f), (g)). Classes are close to each other but allowed to classify product as the Expert estimation. The Expert who chooses this ranking should have a decision process following this view.

The last is the mean operator. It gives ranking different to the expert ranking but the product groups are respected. This operator may be used to group products with the same features without respecting the ranking. This behavior is interesting for carpentry products for which the ranking is not important.

In the following sections, we use OWA (α = 0.8) results which are the closest results to the Expert view.

5 Quality Class Determination

The aim of this section is to propose a way to determine the best quality to be attributed to virtual products in relation with the production data (storage, needs ...). There are two last steps in our methodology, the first to express the subjectivity and the hesitation which are present in the quality determination and the second one which determines the profitable quality to be attributed to virtual product so as to foresee production and generate product BOM.

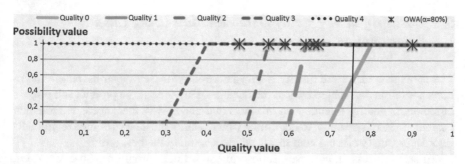

Fig. 5 Membership functions of the visual quality classes based on OWA($\alpha = 0.8$) values

5.1 Quality Measure Expression in the Expert Vocabulary

In order to express the quality measure in the expert vocabulary, we propose to use fuzzy linguistic variables. Its definition is based on constraints coming from the expert definition of the aesthetic quality:

- There are five classes of quality from 0 (the best) to 4 (the worst)
- Lower qualities are included in the best qualities; this means that a product with quality 1 can satisfy needs for the same product with lower quality (2, 3 or 4) but not 0 quality needs.
- Each membership function has core, which means for each quality that there is at least one singularity
- All products have a quality, which means for each quality measure, there is a non-null quality possibility measure.
- We consider that distributions are empirically trapezoidal.

All these constraints allow to determine membership functions for quality classes as describes in the Fig. 5. To define the membership function cores we, use the OWA ($\alpha = 0.8$) value by considering the Expert ranking.

This definition of these linguistic variables gives the possibility degree for each virtual quality. For example, a product with an aesthetic quality value equal to 0.75 (black vertical line) has a vector of possibilities $Q_v = \{0.5\ 1\ 1\ 1\ 1\}$. This measure could be understood by different ways [6], in our case these values correspond to the membership possibility to a quality classes or to the risk taken by the sawyer to attribute a quality class to a product (with the example: 0.75 to quality 0 (a little part of risk), 1 to quality 1, 2, 3, 4 (no risk).

5.2 Most Profitable Quality Determination Using Linear Programming

The previously expressed quality is very interesting for the sawyer because it maintains hesitation and subjectivity in the decision. However as the product sell price is based on its quality, clients need products that match with their needs without any hesitation. So it is essential to determine which quality must be attributed to products in order to satisfy all customer needs while generating the best profit.

To perform this choice, we propose to use a linear program whose variables, constraints and goal are described below, to know if sawyer has to sell product:

- In the best quality with the upper possibility degree, taking no risk of product return
- In the best quality non-null (upgrade the product) and taking the risk of a possible product returns but avoiding storage and maximizing sell price.
- In lower quality (downgrade product) avoiding a too longer storage period and lost link to it.

The risk is evaluated in the goal function by taking directly the possibility degree for a quality.

Variables et constraints
Indices
i :   *Sawed product index*
j :   *Quality index ($j \in [1, 5]$).*

Variable
x_{ij} :   *Function variable $\{0, 1\}$;*

Sales constants
Qij :   *Quality possibility (same as the risk take to attribute the quality)*
$Cpij$:   *Sale cost for product i in quality j*
P_{ij} :   *Matrix of the needs for each product reference (1 when needs is present ,0 then*

Storage constants
Cs_{ij} :   *Storage cost for the product i*
T_i :   *Storage time*

Log constant
Cg :   *Log cost*

Constraints
$x_{ij} = [0, 1]; P_{ij} = [0, 1]; Q_{ij} = [0,1]; j \in \{0,1\}$
$\forall i, \sum_{i} x_{ij} = 1$
Goal function

Table 7 Result of the optimization to determine the profitable quality

Product	Section	F	$Q_{i,0}$	$Q_{i,1}$	$Q_{i,2}$	$Q_{i,3}$	$Q_{i,4}$	Q
2	0.25/0.25	0.7033	0.03	1	1	1	1	1
7	0.25/0.25	0.6331	0	0.66	1	1	1	1
5	0.25/0.25	0.269	0	0.54	1	1	1	3

Table 8 Production data used to determine the profitable quality

Section	Quality	Product need	Cell price	Storage cost	Delay (days)
0.25/0.25	0	6	100	20	12
	1	4	75	20	3
	2	0	60	20	90
	3	10	50	20	7

$$Pour\ i = 1 \ldots n,\ Max\left(\sum_i \left(\left(C_{p_{ij}} * P_{ij} * Q_{ij} - C_{s_{ij}} * T_{ij}\right) * x_{ij}\right)\right) - C_g = GD$$

We implement this linear programing on a product batch (Table 7 is an extract) whose qualities have to be determined in relation with the production data presented in the Table 8. As we can see, there are three cases to attribute the product quality. The first one is to attribute the best quality with a possibility degree ($Q_{i,j}$) equal to 1 (see product 2). In this case, we insure to satisfy the client need without risk of product come back and a high cell price. The second cases is to attribute the high quality possible with a possibility degree included in]0,1[(see product 7). In this case, we take the risk that the customer can be unsatisfied and to have a product return but to sell it with the higher price. The last case is to attribute a lower quality than the quality used in the first case. In this case, the product is downgrading, the sawyer takes no risk and sells it with a lower price, but avoids a too high storage imposed by a log storage period (see product 5).

6 Conclusion

In this article, we present a way to determine the product quality in the wood industry. We decided to base the product quality evaluation on the singularities impact. As the information used to determine the singularity impact and the quality product are uncertain, imprecise, imperfect, we have to use operators which take into account of them.

The singularity impact is evaluated on criteria (size, position and contrast) which are linked by interaction. Moreover, the poorness of the sample and the knowledge on the process decision, lead us to use the Choquet integral to determine impact. By the use of learning process, we have determined the capacities in order to satisfy the Expert vision.

The quality measure is done by merging the singularity impact and the number of singularities. The use of different operators allows us to cover the majority of cases concerning the product quality determination. The comparison with the expert ranking and classification allows to conclude OWA operator with $\alpha = 0.8$ reflects as close as his choice.

Then quality measures are express in the expert vocabulary by the use of a linguistic variable which transcribes expert decision subjectivity. Finally the uncertainty is removed by determining the most profitable quality function of production data like the storage cost and the sell price.

Acknowledgments The authors gratefully acknowledge the financial support of the CPER 2007-2013 "Structuration du Pôle de Compétitivité Fibres Grand'Est" (Competitiveness Fibre Cluster), through local (Conseil Général des Vosges), regional (Région Lorraine), national (DRRT and FNADT) and European (FEDER) funds.

References

1. Almecija, B., Bombardier, V., Charpentier, P.: Modeling quality knowledge to design log sorting system by X rays tomography. In: Information Control Problems in Manufacturing, vol. 14, pp. 1190–1195 (2012)
2. Bloch, I.: Information combination operators for data fusion: a comparative review with classification. IEEE Trans.Syst. Man Cybern. Part A Syst. Hum. **26**(1), 52–67 (1996)
3. Bombardier, V., Mazaud, C., Lhoste, P., Vogrig, R.: Contribution of fuzzy reasoning method to knowledge integration in a defect recognition system. Comput. Ind. **58**(4), 355–366 (2007)
4. Bucur, V.: Techniques for high resolution imaging of wood structure: a review. Measur. Sci. Technol. **14**(12), 91–98 (2003)
5. Choquet, G.: Theory of capacities. Annales de l'institut Fourier **5**, 87 (1953)
6. Dubois, D., Prade, H.: The three semantics of fuzzy sets. Fuzzy Sets Syst. **90**(2), 141–150 (1997)
7. Dupuy, C.: Analyse et conception d'outils pour la traçabilité de produits agroalimentaire afin d'optimiser la dispersion des lots de fabrication. Ph.D. thesis, Institue National des Sciences Appliquées de lyon (2004)
8. Grabisch, M., Kojadinovic, I., Meyer, P.: A review of methods for capacity identification in Choquet integral based multi-attribute utility theory: applications of the kappalab R package. Eur. J. Oper. Res. **186**(2), 766–785 (2008)
9. Grabisch, M., Labreuche, C.: A decade of application of the Choquet and Sugeno integrals in multicriteria decision aid. 4OR. Q. J. Oper. Res. **6**(1), 1–44 (2008)
10. Guillot, J., Lanvin, J., Sandoz, J.: Classement structure du sapin/épicéa par réseaux neuronaux à partir des mesures sylvatest. In: Sciences et industries du bois, pp. 377–384. Colloque (1996)
11. Jover, J., Thomas, A., Bombardier, V.: Marquage du bois dans la masse: intérêts et perspectives. In: 9e Congr'es International de Génie Industriel, CIGI 2011, St Sauveur, Canada (2011)
12. Marichal, J.-L., Roubens, M.: Determination of weights of interacting criteria from a reference set. Eur. J. Oper. Res. **124**(3), 641–650 (2000)
13. PEFC1: Pefc st 1003:2010: Sustainable Forest Management (2010)
14. Perez Oramas, P.: Contribution une méthodologie d'intégration de connaissances pour le traitement d'images. Application la détection de contours par régles linguistiques floues. Ph.D. thesis, université Henry Poincaré (2000)
15. Todoroki, C., Rönnqvist, M.: Dynamic control of timber production at a sawmill with log sawing optimization. Scand. J. For. Res. **17**(1), 79–89 (2002)

Fuzzy Discretization Process from Small Datasets

José M. Cadenas, M. Carmen Garrido and Raquel Martínez

Abstract A classification problem involves selecting a training dataset with class labels, developing an accurate description or a model for each class using the attributes available in the data, and then evaluating the prediction quality of the induced model. In this paper, we focus on supervised classification and models which have been obtained from datasets with few examples in relation with the number of attributes. Specifically, we propose a fuzzy discretization method of numerical attributes from datasets with few examples. The discretization of numerical attributes can be a crucial step since there are classifiers that cannot deal with numerical attributes, and there are other classifiers that exhibit better performance when these attributes are discretized. Also we show the benefits of the fuzzy discretization method from dataset with few examples by means of several experiments. The experiments have been validated by means of statistical tests.

Keywords Data mining · Fuzzy discretization · Numerical attributes · Bagging

1 Introduction

In the real world, there are many situations where organizations have to work from datasets with few examples. The extraction of valuable information from these data sources requires purposeful application of rigorous analysis techniques such as Data Mining or Machine learning, [17].

Data mining methods are often employed to understand the patterns present in the data and derive predictive models with the purpose of predicting future behavior.

J.M. Cadenas · M.C. Garrido · R. Martínez (✉)
Faculty of Informatics, University of Murcia, Campus of Espinardo,
30100 Murcia, Spain
e-mail: jcadenas@um.es

M.C. Garrido
e-mail: carmengarrido@um.es

R. Martínez
e-mail: raquel.m.e@um.es

© Springer International Publishing Switzerland 2016
K. Madani et al. (eds.), *Computational Intelligence*,
Studies in Computational Intelligence 613,
DOI 10.1007/978-3-319-23392-5_15

263

A classification method aims to induce a model with the purpose of predicting cate-
gorical class labels for new samples given a training set of samples each with a class
label. A typical implementation of the classification problem involves selecting a
training dataset with class labels, developing an accurate description or a model for
each class using the attributes available in the data, and then evaluating the prediction
quality of the induced model.

This paper focus on supervised classification and models which have been
obtained from datasets with few examples in relation with the number of attributes.
From the computational learning viewpoint, these datasets are very important in
machine learning problems, because the information contained can be more difficult
to extract.

While an exact relationship between the probability of misclassification, the number of training examples, the number of attributes and the true parameters of the
class-conditional densities is difficult to establish, some guidelines have been suggested regarding the ratio of the sample size to dimensionality, [14]. As a general
rule, a minimum number of $(10 \cdot |A| \cdot |C|)$ training examples is required for a $|A|$-
dimensionality classification problem of $|C|$ classes, [14]. Following this rule, when
datasets have few examples (hereafter called small size datasets), we should take into
account in the learning algorithm some measure in order to build better models to get
good accuracy in classification. We focus on discretization of numerical attributes
in small size datasets. The discretization of numerical attributes is a crucial step in
machine learning problems since there are classifiers that cannot deal with numerical attributes, and there are other classifiers that exhibit better performance when
these attributes are discretized, since discretization reduces the number of numerical
attribute values, enabling faster and more accurate learning [1].

More specifically we propose the introduction of a measure in the discretization method presented in [8], OFP_CLASS, that allows it to get good accuracy in
classification when it deals with small size datasets. This measure is based on the
use of bagging in order to improve the stability and accuracy of the discretization
process. Bagging is a special case of the model averaging approach. We have called
the resulting method BAGOFP_CLASS.

This paper is organized as follows. First, in Sect. 2, we describe some of the
different methods reported in literature to discretize numerical attributes highlight-
ing the fulfilling degree of the rule. Next, in Sect. 3, the BAGOFP_CLASS is
described. Then, in Sect. 4, some experimental results illustrating the performance of
BAGOFP_CLASS method is presented. Finally, in Sect. 5 remarks and conclusions
are presented.

2 Discretization Process and Methods

The term "cut point" refers to a real value within the range of continuous values
that divides the range into two intervals, one interval is less than or equal to the cut
point and the other interval is greater than the cut point. Cut point is also known as

split point. A typical discretization process broadly consists of four steps: (1) sorting the continuous values of the feature to be discretized, (2) evaluating a cut point for splitting or adjacent intervals for merging, (3) according to some criterion, splitting or merging intervals of continuous value, and (4) finally stopping at some point.

After sorting, the next step in the discretization process is to find the best cut point to split a range of continuous values or the best pair of adjacent intervals to merge. There are numerous evaluation functions found in the literature such as entropy measures and statistical measures.

Generally, the discretization methods can be categorised as: (1) Supervised or Unsupervised; (2) Static or Dynamic; (3) Local or Global; (4) Top-down or Bottom-up; (5) Direct or Incremental.

There are some classification algorithms which can only take nominal data as inputs and some of them need to discretize numerical data into nominal data before the learning process. Therefore, discretization is needed as a pre-processing step to partition each numerical attribute into a finite set of adjacent distinct intervals/items. A good discretization algorithm should not only characterize the original data to produce a concise summarization, but also help the classification performance.

Discretization algorithms can be categorized from different viewpoints depending on the measure that is being focused on. If the class label is used to build partitions, the discretization methods can be categorized into two ways, namely unsupervised and supervised. With the focus on the kind of logic, discretization algorithms can be categorized into fuzzy discretization and crisp discretization. On the one hand, when the fuzzy logic is used, we can create fuzzy partitions where a value can belong to more than one partition. On the other hand, when the classical logic is used, we create crisp partitions where a value can only belong to one partition. Furthermore, discretization methods can be classified taking into account the kind of measure that it is used or the way that the partitions are created.

In this section, we focus on how several discretization methods consider the management of small size datasets explicitly. From this viewpoint, there are methods that build partitions taking into account the possible problems that datasets with few examples can induce in the discretization process and later in loss of accuracy. However, there are methods which are not specific for small size datasets but get good precision for this kind of datasets.

- For instance, in [2] a method, called ε-procedure, that constructs crisp partitions on the range of an attribute taking numerical values is proposed. These partitions can be seen as refinements of the ones given by the expert or the ones given by a standard discretization method. Moreover, the method can be seen as "similar" to the fuzzy discretization methods since the ε-procedure takes into account the neighborhood of the thresholds given by the crisp discretization methods.

  Another method that does not specify anything about managing small size datasets is proposed in [18]. The discretization quality is improved by increasing the certainty degree of a decision table in terms of deterministic attribute relationship, which is revealed by the positive domain ratio in rough set theory. Furthermore, they take into account both the decrement of uncertainty level and increment of certainty degree to induce a Coupled Discretization algorithm. This algorithm selects the best cut point according to the importance function composed of the information entropy and positive domain ratio in each run. The algorithm builds crisp partitions because it is focused on classical logic.

- In [16] a novel soft decision tree method that uses soft of fuzzy discretization instead of traditional crisp cuts is proposed. They take into account the dataset size and use a resampling based technique to generate soft discretization points. They use a fuzzy decision tree and an ordinary bootstrap as a method for resampling. Bootstrap allows them to get global partitions and a better estimation toward the entire population.

Typical datasets which have few examples are the microarrays datasets. In [15] this kind of data is used. That paper presents a novel classification approach that integrates fuzzy class association rules and support vector machines. Also, a fuzzy discretization technique based on fuzzy c-means clustering algorithm is employed to transform the training set, particularly quantitative attributes, to a format appropriate for association rule mining. A hill-climbing procedure is adapted for automatic thresholds adjustment and fuzzy class association rules are mined accordingly.

It is notable that some of the above methods use some kind of repetitions to manage small size datasets. This approach is also used in other types of machine learning algorithms to manage small size datasets.

Bagging was proposed by Breiman in [5], and is based on bootstrapping and aggregating concepts, so it incorporates the benefits of both approaches. Bagging uses the same training set multiple times, and has been shown to outperform a single classifier trained from the training set. In bagging, the training set is randomly sampled k times with replacement, producing k training sets with sizes equal to the original training set. Since the original set is sampled with replacement, some training instances are repeated in the new training sets, and some are not present at all. Bagging has several advantages. First, because different classifiers make different errors, combining multiple classifiers generally leads to superior performance when compared to a single classifier, and thus it is more noise tolerant. Second, bagging can be computationally efficient in training because it can be implemented in a parallel or distributed way. Finally, bagging is able to maintain the class distribution of the training set.

In the next section, we describe the BAGOFP_CLASS method which also obtains good results when discretizing numerical attributes of small size datasets.

3 BAGOFP_CLASS Method

The proposed BAGOFP_CLASS method can be classified a supervised, dynamic, local, top-down and incremental. Following the methodology of OFP_CLASS [8], BAGOFP_CLASS is composed of two phases. The former uses a fuzzy decision tree in order to find possible cut points to create the partitions. The latter takes as input the cut points of the first phase to optimize and to build the final fuzzy partitions by means of a genetic algorithm. It should be noted that if the second phase is not carry out, with the cut points it could be possible to construct intervals and crisp partitions would be obtained for the attributes instead of fuzzy partitions.

For the design of these two phases, we use bagging. The use of bagging is motivated by the fact that bagging can give substantial gains in accuracy when the classification method used is unstable. If perturbing the learning set can cause significant changes in the classifier constructed, the bagging can improve accuracy [5]. Instability was studied in [6] where it was pointed out that neural nets, classification and regression trees, and subset selection in linear regression were unstable. The result, both experimental and theoretical, is that bagging can push a good but unstable procedure a significant step towards optimality.

In addition, instability of decision trees can be increased by using small size datasets because in a decision tree when a node is split, the attribute with higher information gain is selected. When the number of examples is small relative to the number of the attributes, the probability that redundant attributes exist is increased. In this way, there will be several attributes that have the same information gain and the selection of one of them is random. With a certain probability, the unselected attributes are not partitioned and, therefore, these attributes might not be part of the final partition generated by the decision tree. When a bagging process is introduced, the probability that these attributes can be part of the final partition is increased. In this situation, the method will work with different bagging of the dataset. By repeating the process, the selection of other attributes is allowed.

Therefore, since the first phase of the algorithm is based on decision trees, can be improved with the use of bagging obtaining a better partitioning from the set of decision trees generated and enabling that a greater number of attributes to be part of the partition when the information gain of various attributes is the same. The second phase of the algorithm is responsible for selecting the most relevant attributes for classification. On the other hand, the genetic algorithm in the second phase of the OFP_CLASS method, uses a fitness function which tries to find the best set of partitions to divide the examples with respect to the class attribute. Again, the second phase of OFP_CLASS method can be improved with a bagging process with the motivation of reducing variability in the partitioning results via averaging. The fitness function is modified in order to obtain a better value for those partitions (individuals) with better average performance when we using a bagging of the dataset. Also with bagging the genetic algorithm studies different region of the searching space and it can get more global fuzzy partitions.

Therefore, we will introduce bagging into the two phases of BAGOFP_CLASS method: (1) in the construction of the decision tree which generates the cut points of the first phase, and (2) in the fitness function that will be used in the genetic algorithm of the second phase.

The fuzzy partitions created with BAGOFP_CLASS have the same characteristics as the fuzzy partitions created with the OFP_CLASS method. This means that the domain of each numerical attribute is partitioned in trapezoidal fuzzy sets and the partitions are:

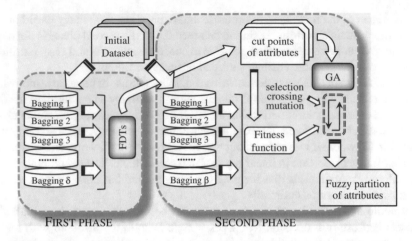

Fig. 1 BAGOFP_CLASS method

- Completeness (no point in the domain is outside the fuzzy partition), and
- Strong (it verifies that $\forall x \in \Omega_i$, $\sum_{f=1}^{F_i} \mu_{B_f}(x) = 1$ where $B_1, .., B_{F_i}$ are the F_i fuzzy sets for the partition of the i numerical attribute with Ω_i domain and $\mu_{B_f}(x)$ are its functions membership).

The domain of each i numerical attribute is partitioned in trapezoidal fuzzy sets and the membership functions $B_1, B_2, \ldots, B_{F_i}$ are calculated as following:

$$\mu_{B_1}(x) = \begin{cases} 1 & b_{11} \leq x \leq b_{12} \\ \frac{(b_{13}-x)}{(b_{13}-b_{12})} & b_{12} \leq x \leq b_{13} \\ 0 & b_{13} \leq x \end{cases} ; \quad \mu_{B_2}(x) = \begin{cases} 0 & x \leq b_{12} \\ \frac{(x-b_{12})}{(b_{13}-b_{12})} & b_{12} \leq x \leq b_{13} \\ 1 & b_{13} \leq x \leq b_{23} \\ \frac{(b_{24}-x)}{(b_{24}-b_{23})} & b_{23} \leq x \leq b_{24} \\ 0 & b_{24} \leq x \end{cases} ;$$

$$\cdots ; \quad \mu_{B_{F_i}}(x) = \begin{cases} 0 & x \leq b_{(F_i-1)3} \\ \frac{(x-b_{(F_i-1)3})}{(b_{(F_i-1)4}-b_{(F_i-1)3})} & b_{(F_i-1)3} \leq x \leq b_{(F_i-1)4} \\ 1 & b_{F_i3} \leq x \end{cases}$$

In the next subsection, we detail the two phases of the BAGOFP_CLASS method. Figure 1 shows, in an illustrative way, the process of BAGOFP_CLASS method.

3.1 Searching the Cut Points

The first phase of BAGOFP_CLASS uses a fuzzy decision tree as base method to search the possible cut points to build the fuzzy partitions, but the input dataset to this fuzzy decision tree is each dataset obtained by performing a bagging process over the original dataset.

The kind of values, with which this fuzzy decision tree can work, are nominal attributes, numerical discretized attributes by means of a fuzzy partition, non-discretized numerical attributes described with crisp values, interval and fuzzy values and furthermore it allows the existence of missing values in all of them. The fuzzy decision tree behavior changes depending on the kind of attribute that is analyzed by the tree in order to find the best of them to divide a node.

On the one hand, when the attribute is numerical and does not have partitions available, the process to follow is similar to the decision tree C4.5 process. The thresholds selected in each node of the tree for these attributes will be the cut points that delimit the intervals. On the other hand, for the other kind of attributes, the methodology followed is the same as the fuzzy decision tree presented in [8]. During the rest of the process the method behavior is the same for both discretize attributes and non-discretized attributes. The procedure of building the fuzzy decision tree has a priority tail which is used to arrange tree nodes depending on the number of examples each one has. The reason for incorporating the priority tail is that the nodes with more examples are analized first because these nodes will have more information than those with less examples.

Comparing the new method with the OFP_CLASS method, the main difference between them is that for datasets which do not have $10 \cdot |A| \cdot |C|$ training examples, the cut points obtained by OFP_CLASS are not rich enough in information to provide a good partition, because these cut points are too specific and partitions obtained are not global. In order to prevent getting cut points so specific and more attributed can be discretized, the input to the fuzzy decision tree in BAGOFP_CLASS is a bagging obtained from the whole dataset.

It must be remembered that in the bagging process, it is possible that a cut point may appear several times. In this case the cut point will be included once.

Algorithm 1 describes all the process to get all possible cut points.

All the cut points obtained after the first phase are introduced as input in the second phase in order to build optimal fuzzy partitions.

3.2 Building Fuzzy Partitions

In the second phase of the method, we are going to use a genetic algorithm to get the fuzzy sets that make up the partitioning of non-discretized attributes. We have decided to use a genetic algorithm, because these algorithms are very powerful and robust, as in most cases they can successfully deal with an infinity of problems from very diverse areas and specifically in Data Mining. These algorithms are normally used in problems without specialized techniques or even in those problems where a technique does exist, but is combined with a genetic algorithm to obtain hybrid algorithms that improve results [9].

In this section we briefly describe the several elements which compose the algorithm. We focus on those that characterizing the BAGOFP_CLASS method. For

---

Algorithm 1. Getting Cut points.

FindingCutPoints(*in* : E; *out* : *Cut points*)
begin
  1. Initialize the set of cut points to each numerical attribute k to empty: $CPS_k = \emptyset$
  2. Initialize bagging size δ.
  3. For j=1 to δ

      3.1 Obtain the dataset BA_j consisting of $|E|$ examples selected from E randomly and with replacement.
      3.2 Obtain a fuzzy decision tree from BA_j:
         i. Start at the root node, which is placed in the initially empty priority tail. Initially, in the root node the set of examples BA_j with an initial weight are found.
        ii. Extract the first node from the priority tail.
        iii. Select the best attribute to split this node using information gain, which is explained in detail in [8], as the criterion.
        iv. Having selected the attribute to expand node, all the descendants generated are introduced in the tail. If the selected attribute is nominal, a descendant is generated to each possible value of that attribute. If the selected attribute (SA) is numerical it is necessary to obtain the corresponding *cut point* and two descendants are generated. In this case, CPS_{SA} is updated as follows: $CPS_{SA} = CPS_{SA} \bigcup \{cut\ point\}$
        v. Go back to step two to continue constructing the tree until no nodes remain in the priority tail or until another stopping condition occurs, such as reaching nodes with a minimum number of examples allowed by the algorithm.

  4. Return CPS_k sets for each numerical attribute k.
end

---

remaining elements originating from the OFP_CLASS method a more detailed description can be found in [8].

The genetic algorithm takes as input the cut points which have been obtained in the first phase, but it is important to mention that the genetic algorithm will decide what cut points are more important to construct the fuzzy partitions, so it is possible that many cut points are not used to obtain the optimal fuzzy partitions. If the first phase gets F cut points for the attribute i, the genetic algorithm can make up $F_i + 1$ fuzzy partitions for the attribute i at the most. However, if the genetic algorithm considers that the attribute i will not have a lot of relevance in the dataset, this attribute will not be partitioned.

The different elements which compose this genetic algorithm are as follows:

Encoding. An individual has a real coding and its size will be the sum of the number of cut points that the fuzzy decision tree will have provided for each attribute in the first phase. Each gene represents the quantity to be added to and subtracted from each attribute's split point to form the fuzzy partition. Also, each gene is associated with a boolean value which indicates whether this gene or cut point has to be taken into account or not, in other words, if this gene or cut point is active or not. We must consider that if a gene is not active the domain of the adjacent gene can change. The domain of each gene is an interval defined by $[0, min(\frac{p_r - p_{r-1}}{2}, \frac{p_{r+1} - p_r}{2})]$ where p_r is the r-th cut point of attribute i represented by this gene except in the first (p_1) and last (p_u) cut point of each attribute whose domains are, respectively: $[0, min(p_1, \frac{p_2 - p_1}{2}]$

and $[0, min(\frac{p_u - p_{u-1}}{2}, 1 - p_u]$. When $F_i = 2$, the domain of the single cut point is defined by $[0, min(p_1, 1 - p_1]$.

Initialization. Firstly, it is determined if each gene is active or not. We must consider that at least one gene from each individual must be active because if all genes were inactive, any attribute would be discretized. Once the boolean value of each gene of the individual has been initialized, the domain of each gene is calculated, considering which cut points are active and which are not. After calculating the domain of each gene, each gene is randomly initialized generating a value within its domain.

Fitness Function. The fitness function of each individual is defined according to the information gain defined in [3]. In this case, in the same way as in the first phase of the method BAGOFP_CLASS, to calculate the fitness for each individual a bagging process is applied. In this case, the fitness of the individual is an average fitness using different bagging of the input dataset. In this way, the algorithm obtains a more robust fitness for each individual because the average fitness gives a more general vision over different data subsets and is a more reliable measure. Algorithm 2 implements the fitness function.

In the algorithm $\mu_{if}(\cdot)$ is the membership function corresponding to fuzzy set f of attribute i. This fitness function, based on the information gain, indicates how dependent the attributes are with regard to class, i.e., how discriminatory each attribute's partitions are. If the fitness obtained for each individual is close to zero, it indicates that the attributes are totally independent of the classes, which means that the fuzzy sets obtained do not discriminate classes. On the other hand, as the fitness value moves further away from zero, it indicates that the partitions obtained are more than acceptable and may discriminate classes with good accuracy.

Selection. Individual selection is by means of tournament, taking subsets with size 2. It must be taken into account that the best individual is always selected due to the elitism that the method carries out.

Crossover. The crossover operator is applied with a certain probability, crossing two individuals through a single point, which may be any one of the positions on the vector. Not all crossings are valid, since one of the restrictions imposed on an individual is that the individual must not have all its genes inactive. When crossing two individuals and this situation occurs, the crossing is invalid, and individuals remain in the population without interbreeding. If the crossing is valid, the domain for each gene is updated in the individuals generated. As in selection, the method applies elitism in this operator, so the best individual is not crossed.

Mutation. Mutation is carried out according to a certain probability at interval $[0.01, 0.2]$.

Since each gene has a boolean associated value which indicates whether the gene is active or not and the gene value represents the amount to add and subtract the cut point, the mutation operator is hybrid. In BAGOFP_CLASS method when a gene has to mutate by chance, there are two options:

Algorithm 2. Fitness Function.

Fitness($in : E, out : FinalFitness$)
begin
  1. Initialize bagging size β
  2. Initialize $ValueFitness = 0$
  3. For j=1 to β

      3.1 Obtain the dataset BA_j consisting of $|E|$ examples selected from E randomly and with replacement.

      3.2 For each attribute $i = 1, \ldots, |A|$:
         i. For each set $f = 1, \ldots, F_i$ of attribute i

           For each class $k = 1, \ldots, |C|$ calculate $P_{ifk} = \dfrac{\sum_{e \in BA_{jk}} \mu_{if}(e)}{\sum_{e \in BA_j} \mu_{if}(e)}$

         ii. For each class $k = 1, \ldots, |C|$ calculate $P_{ik} = \sum_{f=1}^{F_i} P_{ifk}$
         iii. For each $f = 1, \ldots, F_i$ calculate $P_{if} = \sum_{k=1}^{|C|} P_{ifk}$
         iv. For each $f = 1, \ldots, F_i$ calculate the information gain of attribute i and set f
           $I_{if} = \sum_{k=1}^{|C|} P_{ifk} \cdot \log_2 \frac{P_{ifk}}{P_{ik} \cdot P_{if}}$
         v. For each $f = 1, \ldots, F_i$ calculate the entropy $H_{if} = -\sum_{k=1}^{|C|} P_{ifk} \cdot \log_2 P_{ifk}$
         vi. Calculate I and H of attribute i: $I_i = \sum_{f=1}^{F_i} I_{if}$, $H_i = \sum_{f=1}^{F_i} H_{if}$

      3.3 Calculate the fitness as : $ValueFitness = ValueFitness + \dfrac{\sum_{i=1}^{|A|} I_i}{\sum_{i=1}^{|A|} H_i}$

  4. $FinalFitness = ValueFitness/\beta$
  5. Return $FinalFitness$
end

---

1. to activate or deactivate the gene and
2. to modify the amount to add and subtract to the cut point.

The first option allows us to explore new search spaces. When the second option is applied the slopes of the partition are modified. This last option is included in the method in order to adjust the slopes as much as possible and in this way to get better accuracy.

The method applies the first option 50 % of the time and the second option 50 % of the time.

When a gene is activated or deactivated, first, the boolean value associated to the gene is modified and then a check is made to make sure there are still active genes. If there are still active genes, their domains and the domains of adjacent genes must be updated. If all the individual genes are deactivated, the mutation process is not performed.

When the mutation have to modify the value to add and subtract, a randomly calculated value within its domain is generated.

Stopping. The stopping condition is determined by the number of generations.

The genetic algorithm should find the best possible solution in order to achieve a more efficient classification.

In the next section we show some computational experiments where we are going to compare using several datasets the accuracy in classification of discretizing with OFP_CLASS method and with BAGOFP_CLASS method. In these experiments, we can see that the behavior of the BAGOFP_CLASS method with datasets with a few examples is better than the previous method.

4 Experimental Study

The aim of this experimental study is to analyse the performance of using the BAGOFP_CLASS method with datasets with a few examples. In this section, we show the details of the experimental framework. We present the datasets employed and we describe the experimental setup, the performance measure and the statistical test employed to analyze the results. Finally, we present and analyze the results obtained.

4.1 The Datasets and the Experimental Setup

The proposed approach is going to evaluate by means of experiments on various datasets selected from UCI machine learning repository [11]. In addition, three high dimensional datasets (ADE, PRO, SRB) available in [10] have been added. These datasets used to test the proposed approach are summarized in Table 1.

Table 1 shows the number of examples ($|E|$), the number of attributes ($|A|$) and the number of classes ($|C|$) for each dataset. All the various dataset attributes are numerical except AUS dataset having six numerical attributes and eight nominal attributes. In addition, the last column "rule" of the Table 1 shows whether the dataset verifies the condition $|E| \geq 10 \cdot |A| \cdot |C|$. "Abbr" indicates the abbreviation of the dataset used in the experiments.

In order to evaluate the partitions obtained both in the OFP_CLASS method and the BAGOFP_CLASS method, we use an ensemble classifier called Fuzzy Random Forest (FRF) [7]. This ensemble needs as input data a fuzzy partition for the numerical attributes. FRF ensemble was originally presented in [4], and then extended in [7] to handle imprecise and uncertain data. The ensemble is composed of a set of fuzzy decision trees (FDTs).

The experimental parameters are as follows:

- Parameters of the FRF ensemble.

  - Ensemble Size: 500 FDTs
  - Random selection of attributes from the set of available attributes: $\sqrt{|A|}$

- Parameters for GA in both methods (OFP_CLASS and BAGOFP_CLASS).

Table 1 Datasets

Dataset	Abbr	$|E|$	$|A|$	$|C|$	Rule	Dataset	Abbr	$|E|$	$|A|$	$|C|$	Rule
Australian cr.	AUS	690	14	2	Yes	Glass	GLA	214	9	7	No
Breast C.W.	BCW	699	9	2	Yes	Ionosphere	ION	351	34	2	No
Statlog heart	HEA	270	13	2	Yes	Prostate	PRO	102	6033	2	No
Iris plant	IRP	150	4	3	Yes	Sonar	SON	208	60	2	No
Vehicle	VEH	946	18	4	Yes	Spectf	SPE	267	44	2	No
						Srbct	SRB	63	2308	4	No
Adenocarc.	ADE	76	9868	2	No	W.D.B. cancer	WDB	569	31	2	No
Apendicitis	APE	106	7	2	No	Wine	WIN	178	13	3	No

    o Individual Number: 100
    o Generation Number: 250
    o Crossover probability: 0.8
    o Mutation probability: 0.1

- Parameters for BAGOFP_CLASS method.

    o In first phase—δ bagging size: 20
    o In second phase—β bagging size: 30

As we have commented in Sects. 2 and 3, the use of bagging is justified by the fact that bagging can give substantial gains in accuracy when the classification method used is unstable (in our case, fuzzy decision trees). If perturbing the learning set can cause significant changes in the classifier constructed, the bagging can improve accuracy.

For the different experiments, we have used the values of bagging (both for the first phase to the second) than indicated above. To obtain these values, we made some tests. In them, we change the values of δ and β of the following way: we set the value of $\beta = 1$, and we vary δ from 1 to 20; next, we set $\delta = 20$, we vary β from 2 to 30. In the first changes of δ, the accuracy (of the FRF classifier) presents the biggest changes. Next, the behavior is stabilizing. Values of $\delta = 20$ and $\beta = 30$ have an acceptable performance with the datasets used in this work.

4.2 Estimation of the Classification Performance and Validating the Experimental Results

To analyze the results obtained in the study, the following performance measure has been employed: to compare the results in the experiments the accuracy (number of successful hits relative to the total number of classifications) is used. More specifically, the accuracy medium obtained as the average accuracy of a 3×5-fold cross-validation is used.

To complete the experimental study, we perform an analysis of them in each subsection using statistical techniques. Following the methodology proposed by García et al. in [12] we use a non-parametric test. We use the Wilcoxon signed-rank test to compare two methods. This test is a non-parametric statistical procedure for performing pairwise comparison between two methods. This is analogous with the paired t-test in non-parametric statistical procedures; therefore, it is a pairwise test that aims to detect significant differences between two sample means, that is, the behavior of two methods. In order to carry out the statistical analysis R packet [13] has been used.

Table 2 Results obtained by FRF after classifying the datasets that verify $|E| \geq 10 \cdot |A| \cdot |C|$ using the discretization obtained by OFP_CLASS and BAGOFP_CLASS

	OFP_CLASS		BAGOFP_CLASS		
	Training	Test	Training	Test	p-value
AUS	$99.85_{0.15}$	$86.42_{2.94}$	$100_{0.00}$	$86.81_{3.12}$	0.1768
BCW	$99.74_{0.10}$	$96.13_{1.27}$	$98.19_{0.22}$	$95.90_{1.55}$	0.7772
HEA	$97.84_{0.54}$	$79.01_{5.34}$	$98.92_{0.41}$	$79.26_{6.56}$	0.5092
IRP	$97.78_{0.92}$	$96.67_{3.26}$	$98.98_{0.62}$	$96.22_{3.16}$	0.6578
VEH	$93.60_{0.47}$	$71.16_{3.48}$	$99.99_{0.02}$	$71.55_{4.80}$	0.4325
Aver.	$97.76_{0.44}$	$85.88_{3.26}$	$99.21_{0.26}$	$85.95_{3.84}$	0.5805

4.3 Evaluation of the Classification Performance

The experiments are designed to evaluate the performance of the proposed approach. We compare the OFP_CLASS method [8] with the proposed method BAGOFP_CLASS using several datasets.

Using Small Size Dataset

First, we compare the two methods using datasets which verify the condition $|E| \geq 10 \cdot |A| \cdot |C|$. The accuracy results are shown in Table 2.

In Table 2, training and test are the percentages of classification average accuracy (mean and standard deviation) for training and test datasets, respectively. Moreover, the obtained p-values, when comparing the results of the test phase using a Wilcoxon signed-rank test are shown.

In Table 2 it is observed that the results obtained by the new method seem to be similar to the results obtained by OFP_CLASS. Analyzing the different p-values with $\alpha = 0.05$, we can conclude:

- There are no significant differences between the datasets.
- Globally, analyzing all the results obtained for different datasets, we can conclude that there are no significant differences between evaluating with the discretization of the method OFP_CLASS and evaluating with the discretization of the new method BAGOFP_CLASS.

Using Non-small Size Dataset

Second, we compare the two methods using datasets that do not verify the condition. The accuracy results are shown in Table 3.

In Table 3, training and test are the percentages of classification average accuracy (mean and standard deviation) for training and test datasets, respectively. Moreover, the obtained p-values, when comparing the results of the test phase using a Wilcoxon signed-rank test are shown.

Table 3 Results obtained by FRF after classifying the datasets that do not verify $|E| \geq 10 \cdot |A| \cdot |C|$ using the discretization obtained by OFP_CLASS and BAGOFP_CLASS

	OFP_CLASS		BAGOFP_CLASS		
	Training	Test	Training	Test	p-value
ADE	$84.32_{2.05}$	$81.94_{9.96}$	$100_{0.00}$	$85.06_{9.83}$	0.01102
APP	$90.33_{1.80}$	$88.33_{7.95}$	$94.02_{0.86}$	$90.25_{5.72}$	0.00529
GLA	$85.82_{1.03}$	$69.80_{5.60}$	$100_{0.00}$	$73.38_{6.68}$	0.01199
ION	$95.89_{0.41}$	$93.07_{2.38}$	$100_{0.00}$	$94.11_{2.05}$	0.00424
PRO	$94.93_{1.36}$	$89.84_{6.98}$	$100_{0.00}$	$93.78_{5.36}$	0.00464
SON	$93.59_{1.59}$	$80.31_{5.11}$	$100_{0.00}$	$83.68_{4.72}$	0.00105
SPE	$79.81_{0.97}$	$79.41_{4.12}$	$100_{0.00}$	$81.78_{4.58}$	0.00108
SRB	$94.58_{1.72}$	$78.46_{13.54}$	$100_{0.00}$	$96.79_{3.77}$	0.00109
WDB	$93.99_{0.41}$	$93.79_{2.23}$	$100_{0.00}$	$95.49_{1.66}$	0.00692
WIN	$94.33_{1.28}$	$93.82_{3.54}$	$100_{0.00}$	$97.57_{1.76}$	0.00253
Aver.	$90.76_{1.26}$	$84.88_{6.14}$	$99.40_{0.09}$	$89.19_{4.61}$	2.2e-16

In Table 3 it is observed that the results obtained by the new method seem to be better than the results obtained by OFP_CLASS. Analyzing the different p-values with $\alpha = 0.05$, we can conclude:

- For all datasets, the analysis indicates that there are significant differences between the two methods, where BAGOFP_CLASS is the best method.
- Globally, analyzing all the results obtained for different datasets, we can conclude that there are significant differences between evaluating with the discretization of the method OFP_CLASS and evaluating with the discretization of the new method. The method which gets the best average accuracy is BAGOFP_CLASS.

4.3.1 Comparing Results

Figure 2 shows, in an illustrative way, the results obtained using OFP_CLASS and BAGOFP_CLASS methods with datasets containing enough and few examples.

In general, with the datasets used (which either verify or do not verify $|E| \geq 10 \cdot |A| \cdot |C|$) we can conclude that the proposed method is useful. When the datasets do not verify the condition, the fundamental difference of the new method of discretization is the partitioning of the attributes. The most important attributes in the classification are partitioned possibly into more parts and more precisely.

From the computational viewpoint, the introduction of bagging and therefore an iterative process in both phases increases the runtime of the method. However, as discussed in Sect. 2, bagging can be computationally efficient because it can be implemented in a parallel or distributed way.

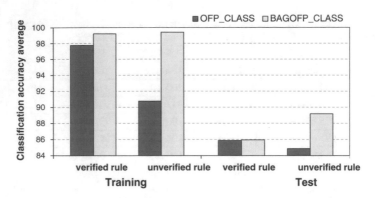

Fig. 2 Comparing results of datasets verifying and do not verifying $|E| \geq 10 \cdot |A| \cdot |C|$

5 Conclusions

In this study we have presented an improvement on a discretization method, in order to get better partitions and better accuracy in the classification task with small size datasets.

The discretization method is divided into two phases. On the one hand, in the first phase the method uses a fuzzy decision tree to search possible cut points to create partitions. On the other hand, in the second phase a genetic algorithm is used to construct the fuzzy partitions taking as input data, the cut points obtained in the first phase.

The way to improve the discretization method is using a bagging process into the two phases. The capacity of bagging to improve the accuracy of unstable classifiers is exploited in the first phase of the method based on fuzzy decision trees. In addition, the use of the bagging procedure allows the discretization of more attributes when datasets have a few training examples compared with the number of attributes, in other word, when the rule $|E| \geq 10 \cdot |A| \cdot |C|$ is not verified.

Also we have presented several experiments where the new proposed method using bagging gets better accuracy in classification with this kind of datasets. These conclusions have been validated by applying statistical techniques to analyze the behavior of different methods in the experiments.

Improve the discretization of numerical attributes in small size datasets is important like previous step to carry out feature selection in microarrays data which is a topic of current interest and we want to carry out the feature selection using a fuzzy ensemble which needs a partition of the numerical attributes.

Acknowledgments Supported by the project TIN2011-27696-C02-02 of the Ministry of Economy and Competitiveness of Spain. Thanks also to "Fundación Séneca - Agencia de Ciencia y Tecnología de la Región de Murcia" (Spain) for the support given to Raquel Martínez by the scholarship program FPI.

References

1. Antonelli, M., Ducange, P., Lazzerini, B., Marcelloni, F.: Learning knowledge bases of multi-objective evolutionary fuzzy systems by simultaneously optimizing accuracy, complexity and partition integrity. Soft Comput. **15**, 2335–2354 (2011)
2. Armengol, E. García-Cerdana, A.: Refining discretizations of continuous-valued attributes. In: The 9th International Conference on Modeling Decisions for Artificial Intelligence, pp. 258–269 (2012)
3. Au, W.H., Chan, K.C., Wong, A.: A fuzzy approach to partitioning continuous attributes for classification. IEEE Trans. Knowl. Data Eng. **18**(5), 715–719 (2006)
4. Bonissone, P.P., Cadenas, J.M., Garrido, M.C., Díaz-Valladares, R.A.: A fuzzy random forest. Int. J. Approx. Reason. **51**(7), 729–747 (2010)
5. Breiman, L.: Bagging predictors. Mach. Learn. **24**(2), 123–140 (1996a)
6. Breiman, L.: Heuristics of instability and stabilization in model selection. Ann. Stat. **24**(6), 2350–2383 (1996b)
7. Cadenas, J.M., Garrido, M.C., Martínez, R., Bonissone, P.P.: Extending information processing in a fuzzy random forest ensemble. Soft Comput. **16**(5), 845–861 (2012a)
8. Cadenas, J.M., Garrido, M.C., Martínez, R., Bonissone, P.P.: Ofp_class: a hybrid method to generate optimized fuzzy partitions for classification. Soft Comput. **16**, 667–682 (2012b)
9. Cox, E.: Fuzzy Modeling and Genetic Algorithms for Data Mining and Exploration. Morgan Kaufmann Publishers, New York (2005)
10. Diaz-Uriarte, R., de Andrés, S.A.: Gene selection and classification of microarray data using random forest. BMC Bioinform. **7**(3) (2006)
11. Frank, A., Asuncion, A.: UCI Machine Learning Repository. School of Information and Computer Sciences, University of California, Irvine (2010)
12. García, S., Fernández, A., Luengo, J., Herrera, F.: A study statistical techniques and performance measures for genetics-based machine learning: accuracy and interpretability. Soft Comput. **13**(10), 959–977 (2009)
13. Ihaka, R., Gentleman, R.R.: A language for data analysis and graphics. J. Comput. Graph. Stat. **5**(3), 299–314 (1996)
14. Jain, A.K.: Statistical pattern recognition: a review. IEEE Transa. Pattern Anal. Mach. Intell. **22**, 4–37 (2000)
15. Kianmehr, K., Alshalalfa, M., Alhajj, R.: Fuzzy clustering-based discretization for gene expression classification. Knowl. Inf. Syst. **24**, 441–465 (2010)
16. Qureshi, T., Zighed, D.A.: A soft discretization technique for fuzzy decision trees using resampling. Intelligent Data Engineering and Automated Learning—IDEAL 2009. Lecture Notes in Computer Science, vol. 5788, pp. 586–593 (2009)
17. Unler, A., Murat, A.: A discrete particle swarm optimization method for feature selection in binary classification problems. Eur. J. Oper. Res. **206**, 528–539 (2010)
18. Wang, C., Wang, M., She, Z., Cao, L.: CD: a coupled discretization algorithm. Advances in Knowledge Discovery and Data Mining. Lecture Notes in Computer Science, vol. 7302, pp. 407–418 (2012)

A Framework for Modelling Real-World Knowledge Capable of Obtaining Answers to Fuzzy and Flexible Searches

Víctor Pablos-Ceruelo and Susana Munoz-Hernandez

Abstract The Internet has become a place where massive amounts of information and data are being generated every day. This information is most of the times stored in a non-structured way, but the times it is structured in databases it cannot be retrieved by using easy fuzzy queries: we need human intervention to determine how the non-fuzzy information stored needs to be combined and processed to answer a fuzzy query. We present a web interface for posing fuzzy and flexible queries and a framework. Our framework allows to represent non-fuzzy concepts, fuzzy concepts and relations between them, giving the programmer the capability to model any real-world knowledge. It is this representation in the framework's language what it uses to (1) determine how to answer the query without any human intervention and (2) provide the search engine with the information it needs to present the user a friendly and easy to use query form. We expect this work contributes to the development of more human-oriented fuzzy search engines.

Keywords Search engine · Fuzzy logic · Framework

This work is partially supported by research projects DESAFIOS10 (TIN2009-14599-C03-00) funded by Ministerio Ciencia e Innovación of Spain, PROMETIDOS (P2009/TIC-1465) funded by Comunidad Autónoma de Madrid and Research Staff Training Program (BES-2008-008320) funded by the Spanish Ministry of Science and Innovation. It is partially supported too by the Universidad Politécnica de Madrid entities Departamento de Lenguajes, Sistemas Informáticos e Ingeniería de Software and Facultad de Informática.

V. Pablos-Ceruelo · S. Munoz-Hernandez (✉)
The Babel Research Group, Facultad de Informática,
Universidad Politécnica de Madrid, Madrid, Spain
e-mail: susana@babel.ls.fi.upm.es
url: http://babel.ls.fi.upm.es

V. Pablos-Ceruelo
e-mail: vpablos@babel.ls.fi.upm.es

© Springer International Publishing Switzerland 2016
K. Madani et al. (eds.), *Computational Intelligence*,
Studies in Computational Intelligence 613,
DOI 10.1007/978-3-319-23392-5_16

1 Introduction

Most of the real-world information is stored in non-fuzzy databases, but most of the queries that we (human beings) wanna pose to a search engine are fuzzy. One example of this is the databases containing the distance of some houses to the center and the user query "I want a house close to the center". Assuming that it is nonsense to teach every search engine user how to translate the (almost always) fuzzy query the user has in mind into a query that a machine can understand and answer, the problem to be solved has two very different parts: recognition of the query and execution of the recognized query.

The recognition of the query has basically two parts: syntactic and semantic recognition. The first one has to be with the lexicographic form of the set of words that compose the query and tries to find a query similar to the user's one but more commonly used. The objective with this operation is to pre-cache the answers for the most common queries and return them in less time, although sometimes it serves to remove typos in the user queries. An example of this is replacing "cars", "racs", "arcs" or "casr" by "car". The detection of words similar to one in the query is called fuzzy matching and the decision to propose one of them as the "good one" is based on statistics of usage of words and groups of words. The search engines usually ask the user if he/she wants to change the typed word(s) by this one(s).

The semantic recognition is work still in progress and it is sometimes called "natural language processing". In the past search engines were tools used to retrieve the web pages containing the words typed in the query, but today they tend to include capabilities to understand the user query. An example is computing 4 plus 5 when the query is "$4 + 5$" or presenting a currency converter when we write "euro dollar". This is still far away from our proposal: retrieving web pages containing "fast red cars" instead of the ones containing the words "fast", "red" and "car".

The execution of the recognized query is the second part. Suppose a query like "I want a restaurant close to the center". If we assume that the computer is able to "understand" the query then it will look for a set of restaurants in the database satisfying it and return them as answer, but the database does not contain any information about "close to the center", just the "distance of a restaurant to the center". It needs a mapping between the "distance" and the meaning of "close", and this knowledge must be stored somewhere.

One of the most successful programming languages for representing knowledge in computer science is Prolog, whose main advantage with respect to the other ones is being a more declarative programming language.[1] Prolog is based on logic. It is usual to identify logic with bi-valued logic and assume that the only available values are "yes" and "no" (or "true" and "false"), but logic is much more than bi-valued logic. In fact we use fuzzy logic (FL), a subset of logic that allow us to represent not only if

---

[1] We say that it is a more declarative programming language because it removes the necessity to specify the flow control in most cases, but the programmer still needs to know if the interpreter or compiler implements depth or breadth-first search strategy and left-to-right or any other literal selection rule.

Fig. 1 Restaurants database
and close fuzzification
function

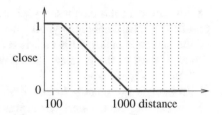

an individual belongs or not to a set, but the grade in which it belongs. Supposing the database contents, the definition for "close" in Fig. 1 and the question "Is restaurant X close to the center?" with FL we can deduce that Il tempietto is "definitely" close to the center, Tapasbar is "almost" close, Ni Hao is "hardly" close and Kenzo is "not" close to the center. We highlight the words "definitely", "almost", "hardly" and "not" because the usual answers for the query are "1", "0.9", "0.1" and "0" for the individuals Il tempietto, Tapasbar, Ni Hao and Kenzo and the humanization of the crisp values is done in a subsequent step by defuzzification.

Name	Distance	Price avg.	Food type
Il_tempietto	100	30	Italian
Tapasbar	300	20	Spanish
Ni Hao	900	10	Chinese
Kenzo	1200	40	Japanese

The simplicity of the previous example introduces a question that the curious reader might have in mind: "Does adding a column "close" of type float to the database and computing its value for each row solves the problem?". The answer is yes, but only if our query is not modifiable: It does not help if we can change our question to "I want a very close to the center restaurant" or to "I want a not very close to the center restaurant". Adding a column for each possible question results into a storage problem, and in some sense it is unnecessary: all this values can be computed from the distance value.

Getting fuzzy answers for fuzzy queries from non-fuzzy information stored in non-fuzzy databases has been studied in some works, for example in [4], the SQLf language. The Ph.D. thesis of Leonid Tineo [20] and the work of Dubois and Prade [6] are good revisions, although maybe a little bit outdated. Most of the works mentioned in this papers focus in improving the efficiency of the existing procedures, in including new syntactic constructions or in allowing to introduce the conversion between the non-fuzzy values needed to execute the query and the fuzzy values in the query, for which they use a syntax rather similar to SQL (reflected into the name of the one mentioned before). The advantages of using a syntax similar to SQL are many (removal of the necessity to retrieve all the entries in the database, SQL programmers can learn the new syntax easily, ...) but there is an important disadvantage: the user needs to teach the search engine how to obtain the fuzzy results from the non-fuzzy values stored in the database to get answers to his/her queries and this includes that he/she must know the syntax and semantics of the language and the structure of the database tables. This task is the one we try to remove by including in the representation of the problem the knowledge needed to link the fuzzy knowledge with the non-fuzzy one.

To include the links between fuzzy and non-fuzzy concepts we could use any of the existing frameworks for representing fuzzy knowledge. Leaving apart the theoretical frameworks, as [22], we know about the Prolog-Elf system [8], the FRIL Prolog system [1], the F-Prolog language [9], the FuzzyDL reasoner [2], the Fuzzy Logic Programming Environment for Research (FLOPER) [15], the Fuzzy Prolog system [7, 21], or Rfuzzy [17]. All of them implement in some way the fuzzy set theory introduced by Lotfi Zadeh in 1965 [23], and all of them let you implement the connectors needed to retrieve the non-fuzzy information stored in databases, but we needed more meta-information than the one they provide.

Retrieving the information needed to ask the query is part of the problem but, as introduced before, it is needed to determine what the query is asking for before answering it. Instead of providing a free-text search field and recognize the query we do it in the other way: we did an in-depth study on which are all the questions that we can answer from the knowledge stored in our system and we created a general query form that allows to introduce any of this questions. This is why in Sect. 3 we do not only present the semantics of our syntactic constructions, but the information that helps us to instantiate the general query form for each domain.

To our knowledge, the works similar to ours are [3, 5, 19]. While the last two seem to be theoretical descriptions with no implementation associated the first one does not appear to be a search engine project. They provide a natural language interface that answers queries of the types (1) does X (some individual) have some fuzzy property, for example "Is it true that IBM is productive?", and (2) do an amount of elements have some fuzzy property, for example "Do most companies in central Portugal have sales_profitability?".

The paper is structured as follows: the syntax needed to understand it goes first (Sect. 2), the description of our framework after (Sect. 3) and conclusions and current work in last place (Sect. 4), as usual.

2 Syntax

We will use a signature Σ of function symbols and a set of variables V to "build" the *term universe* $\mathrm{TU}_{\Sigma,V}$ (whose elements are the *terms*). It is the minimal set such that each variable is a term and terms are closed under Σ-operations. In particular, constant symbols are terms. Similarly, we use a signature Π of predicate symbols to define the *term base* $\mathrm{TB}_{\Pi,\Sigma,V}$ (whose elements are called *atoms*). Atoms are predicates whose arguments are elements of $\mathrm{TU}_{\Sigma,V}$. Atoms and terms are called *ground* if they do not contain variables. As usual, the *Herbrand universe* **HU** is the set of all ground terms, and the *Herbrand base* **HB** is the set of all atoms with arguments from the Herbrand universe. A substitution σ or ξ is (as usual) a mapping from variables from V to terms from $TU_{\Sigma,V}$ and can be represented in suffix $((Term)\sigma)$ or in prefix notation $(\sigma(Term))$.

To capture different interdependencies between predicates, we will make use of a signature Ω of *many-valued connectives* formed by *conjunctions* $\&_1, \&_2, \ldots, \&_k$, *disjunctions* $\vee_1, \vee_2, \ldots, \vee_l$, *implications* $\leftarrow_1, \leftarrow_2, \ldots, \leftarrow_m$, *aggregations* $@_1, @_2, \ldots, @_n$ and tuples of real numbers in the interval $[0, 1]$ represented by $(\mathbf{p}, \mathbf{v})$.

While Ω denotes the set of connective symbols, $\hat{\Omega}$ denotes the set of their respective associated truth functions. Instances of connective symbols and truth functions are denoted by $\&_i$ and $\hat{\&}_i$ for conjunctors, $\vee_i$ and $\hat{\vee}_i$ for disjunctors, $\leftarrow_i$ and $\hat{\leftarrow}_i$ for implicators, $@_i$ and $\hat{@}_i$ for aggregators and $(\mathbf{p}, \mathbf{v})$ and $(\mathbf{p}, \mathbf{v})$ for the tuples.

Truth functions for the connectives are then defined as $\hat{\&} : [0, 1]^2 \rightarrow [0, 1]$ monotone[2] and non-decreasing in both coordinates, $\hat{\vee} : [0, 1]^2 \rightarrow [0, 1]$ monotone in both coordinates, $\hat{\leftarrow} : [0, 1]^2 \rightarrow [0, 1]$ non-increasing in the first and non-decreasing in the second coordinate, $\hat{@} : [0, 1]^n \rightarrow [0, 1]$ as a function that verifies $\hat{@}(0, \ldots, 0) = 0$ and $\hat{@}(1, \ldots, 1) = 1$ and $(\mathbf{p}, \mathbf{v}) \in \Omega^{(0)}$ are functions of arity 0 (constants) that coincide with the connectives.

Immediate examples for connectives that come to mind for conjunctors are: in Łukasiewicz logic ($\hat{F}(x, y) = max(0, x + y - 1)$), in Gödel logic ($\hat{F}(x, y) = min(x, y)$), in product logic ($\hat{F}(x, y) = x \cdot y$), for disjunctors: in Łukasiewicz logic ($\hat{F}(x, y) = min(1, x + y)$), in Gödel logic ($\hat{F}(x, y) = max(x, y)$), in product logic ($\hat{F}(x, y) = x \cdot y$), for implicators: in Łukasiewicz logic ($\hat{F}(x, y) = min(1, 1 - x + y)$), in Gödel logic ($\hat{F}(x, y) = y\, \text{if} x > y\, \text{else} 1$), in product logic ($\hat{F}(x, y) = x \cdot y$) and for aggregation operators[3]: arithmetic mean, weighted sum or a monotone function learned from data.

3 The Framework in Detail

As stated in the introduction, the framework we present provides (1) the syntax needed to model any knowledge domain and (2) an enough expressive syntactical structure for representing any query we can answer with the information stored in the system. We can view it as the sum of three parts: (1) a configuration file (CF) that defines the fuzzy and non-fuzzy concepts of our domain and the relations between them, (2) a framework that understands the CF and provides (2.1) the search capabilities and (2.2) the metainformation that the web application needs to present the user the tools he/she needs to pose the query and (3) a web application that (3.1) reads the metainformation, (3.2) determines the framework capabilities, (3.3)

---

[2] As usually, a n-ary function $\hat{F}$ is called *monotonic in the* i define-th argument ($i \leq n$), if $x \leq x'$ implies $\hat{F}(x_1, \ldots, x_{i-1}, x, x_{i+1}, \ldots, x_n) \leq \hat{F}(x_1, \ldots, x_{i-1}, x', x_{i+1}, \ldots, x_n)$ and a function is called *monotonic* if it is monotonic in all arguments.

[3] Note that the above definition of aggregation operators subsumes all kinds of minimum, maximum or mean operators.

generates an easy to use human-oriented interface for posing queries to the search engine and (3.4) shows the answers found by the framework to the user.

The syntactical structure we use to query the search engine has been defined after studying multiple user queries. It comprises all of them (sometimes with small modifications) while trying to be as expressive as possible and has the form

$$
\left\{
\begin{array}{l}
I'm\ looking\ for\ a/an\ \boxed{\text{individual}} \\
\boxed{\text{not}}\ \boxed{\text{quantifier}}\quad \boxed{\text{fuzzy-pred}} \\
whose\ \boxed{\text{non-fuzzy-pred}}\ \boxed{\text{comp-op}}\ \boxed{\text{value}}
\end{array}
\right\}\ \boxed{\text{AND}}
\tag{1}
$$

where *individual* is the element we are looking for (car, skirt, restaurant, …), *quantifier* is a quantifier (quite, rather, very, …), *fuzzy-pred* is a fuzzy predicate (cheap, large, close to the center, …), *non-fuzzy-pred* is a non-fuzzy predicate (price, size, distance to the center, …) and *comp-op* is a comparison operand (is equal to, is different from, is bigger than, is lower than, is bigger than or equal to, is lower than or equal to and is similar to). The elements in boxes can be modified and the brackets symbolize choosing between a fuzzy predicate query or a comparison between non-fuzzy values (which can be a fuzzy comparison). The "AND" serves to add more lines to the query, to combine multiple conditions. Some examples of use are "I'm looking for a restaurant not very near the city center" (Eq. 2), "I'm looking for a restaurant whose food type is mediterranean" (Eq. 3) and "I'm looking for a restaurant whose food type is similar to mediterranean and near the city center" (Eq. 4). In the examples the empty boxes mean that we do not choose any of the available elements.

$$
\begin{array}{c}
I'm\ looking\ for\ a/an\ \boxed{\text{restaurant}} \\
\boxed{\text{not}}\ \boxed{\text{very}}\ \boxed{\text{near the city center}}\ \square
\end{array}
\tag{2}
$$

$$
\begin{array}{c}
I'm\ looking\ for\ a/an\ \boxed{\text{restaurant}} \\
whose\ \boxed{\text{food type}}\ \boxed{\text{is}}\ \boxed{\text{mediterranean}}\ \square
\end{array}
\tag{3}
$$

$$
\begin{array}{c}
I'm\ looking\ for\ a/an\ \boxed{\text{restaurant}} \\
whose\ \boxed{\text{food type}}\ \boxed{\text{is similar to}} \\
\boxed{\text{mediterranean}}\ \boxed{\text{AND}} \\
\square\square\ \boxed{\text{near the city center}}
\end{array}
\tag{4}
$$

The syntax that we provide to model any knowledge domain is highly coupled to the information that we need to retrieve for providing the values for "individual", "not", "quantifier", "fuzzy-pred", "non-fuzzy-pred", "comp-op" and "value", and to present the answers in a human-readable way. This is why when we provide its semantics we do it in two ways: by providing the formal ones and by providing what the web interface understands from them. We present first a brief but, for our purposes, complete introduction to the multi-adjoint semantics with priorities that we use to give formal semantics to our syntactical constructions. For a more complete description we recommend reading the papers cited below.

The structure used to give semantics to our programs is the multi-adjoint algebra, presented in [10–14, 16]. The interest in using this structure is that we can obtain the credibility for the rules that we write from real-world data, although this time we do not focus in that advantage. We simply highlight this fact so the reader knows why this structure and not some other one.

This structure provides us with the basis, but for our purposes we need that the maximum operator used to decide between multiple rules results the valid one chooses the value of the less generic rule instead of just the higher value. This is why we take as point of departure the work [18]. Definitions needed to understand the formal semantics are given in advance, as usually.

In [18] the meaning of a fuzzy logic program gets conditioned by the combination of a truth value and a priority value. So, the usual truth value $v \in [0, 1]$ is converted into $(p, v) \in \Omega^{(0)}$, a tuple of real numbers between 0 and 1 where $p \in [0, 1]$ denotes the (accumulated) priority. The usual representation (p, v) is sometimes changed into (pv) to highlight that the variable is only one and it can take the value $\perp$. The set of all possible values is symbolized by **KT** and the ordering between its elements is defined as follows:

Definition 1 $(\preccurlyeq_{\mathbf{KT}})$

$$\perp \preccurlyeq_{\mathbf{KT}} \perp \preccurlyeq_{\mathbf{KT}} (p, v)$$
$$(p_1, v_1) \preccurlyeq_{\mathbf{KT}} (p_2, v_2) \leftrightarrow (p_1 < p_2) \ or \ (p_1 = p_2 \ and \ v_1 \leq v_2) \quad (5)$$

where $<$ is defined as usually (v_i and p_j are just real numbers between 0 and 1).

Definition 2 (*Multi-Adjoint Logic Program*) A multi-adjoint logic program is a set of clauses of the form

$$A \xleftarrow{(p, v), \&_i} @_j (B_1, \ldots, B_n) \ \text{if} \ \text{COND} \quad (6)$$

where $(p, v) \in \mathbf{KT}$, $\&_i$ is a conjunctor, $@_j$ an aggregator (unnecessary if $k \in [1..1]$), A and B_k, $k \in [1..n]$, are atoms and $COND$ is a first-order formula (basically a bi-valued condition) formed by the predicates in $\text{TB}_{\Pi, \Sigma, V}$, the predicates $=, \neq, \geq, \leq, >$ and $<$ restricted to terms from $\text{TU}_{\Sigma, V}$, the symbol true and the conjunction $\wedge$ and disjunction $\vee$ in their usual meaning.

Definition 3 (*Valuation, Interpretation*) A *valuation* or *instantiation* $\sigma : V \rightarrow \mathbf{HU}$ is an assignment of ground terms to variables and uniquely constitutes a mapping $\hat{\sigma} : \text{TB}_{\Pi, \Sigma, V} \rightarrow \mathbf{HB}$ that is defined in the obvious way.

A *fuzzy Herbrand interpretation* (or short, *interpretation*) of a fuzzy logic program is a mapping $I : \mathbf{HB} \rightarrow \mathbf{KT}$ that assigns an element in our lattice to ground atoms.[4]

It is possible to extend uniquely the mapping I defined on **HB** to the set of all ground formulas of the language by using the unique homomorphic extension. This extension is denoted $\hat{I}$ and the set of all interpretations of the formulas in a program **P** is denoted $\mathcal{I}_{\mathbf{P}}$.

Definition 4 (*The operator* ∘) The application of some conjunctor $\bar{\&}$ (resp. implicator $\bar{\leftarrow}$, aggregator $\bar{@}$) to elements $(\mathsf{p}, \mathsf{v}) \in \mathbf{KT}\setminus\{\bot\}$ refers to the application of the truth function $\hat{\&}$ (resp. $\hat{\leftarrow}$, $\hat{@}$) to the second elements of the tuples while $\circ_{\&}$ (resp. $\circ_{\leftarrow}$, $\circ_{\&}$) is the one applied to the first ones. The operator ∘ is defined by

$$x \circ_{\&} y = \frac{x + y}{2} \quad \text{and} \quad z \circ_{\leftarrow} y = 2 * z - y.$$

Definition 5 (*Satisfaction, Model*) Let **P** be a multi-adjoint logic program, $I \in \mathcal{I}_{\mathbf{P}}$ an interpretation and $A \in \mathbf{HB}$ a ground atom. We say that a clause $Cl_i \in \mathbf{P}$ of the form shown in Eq. 6 is satisfied by I or I *is a model of the clause* Cl_i ($I \Vdash Cl_i$) if and only if (iff) for all ground atoms $A \in \mathbf{HB}$ and for all instantiations σ for which $B\sigma \in \mathbf{HB}$ (note that σ can be the empty substitution) it is true that

$$\hat{I}(A) \succcurlyeq_{\mathbf{KT}} (\mathsf{p}, \mathsf{v}) \, \bar{\&}_i \, \bar{@}_i(\hat{I}(B_1\sigma), \ldots, \hat{I}(B_n\sigma)) \tag{7}$$

whenever *COND* is satisfied (true). Finally, we say that I *is a model of the program* **P** and write $I \Vdash \mathbf{P}$ iff $I \Vdash Cl_i$ for all clauses in our multi-adjoint logic program **P**.

Now that we have introduced the basics of our formal semantics we introduce the syntax, semantics and what the web interface interprets from them.

The first and most important syntactic structure is the one used to define the individuals we can play with, as "restaurants" or "houses" in the previous examples. Since the database tables storing the information of an individual can be more than one we decided to allow the programmer to use the Prolog facilities for mixing all the information into a predicate and we depart from this predicate. This means that if we have two tables for storing the information of a restaurant, one for the "food type" (*ft*) and another for the "distance to the city center" (*dttcc*) we can write the lines in Eqs. 8–12 to obtain all the information about a restaurant. If instead of that we have all the information of a restaurant in just one table we can make use of the code in Eqs. 13 and 14.

---

[4]The *domain* of an interpretation is the set of all atoms in the Herbrand Base (interpretations are total functions), although for readability reasons we present interpretations as sets of pairs $(A, (\mathsf{p}, \mathsf{v}))$ where $A \in \mathbf{HB}$ and $(\mathsf{p}, \mathsf{v}) \in \mathbf{KT}\setminus\{\bot\}$ (we omit those atoms whose interpretation is the truth value $\bot$).

$$sql\_persistent\_location(rft, \ db('SQL', \ user, \ pass, \ 'host' : port)). \quad (8)$$

$$: -sql\_persistent(rft(integer, \ string), \ rft(id, \ ft), \ rft). \quad (9)$$

$$sql\_persistent\_location(rdttcc, \ db('SQL', \ user, \ pass, \ 'host' : port)). \quad (10)$$

$$: -sql\_persistent(rdttcc(integer, \ integer), \ rdttcc(id, \ dttcc), \ rdttcc). \quad (11)$$

$$restaurant(id, \ ft, dttcc) : -rft(id, \ ft), \ rdttcc(id, \ dttcc). \quad (12)$$

$$sql\_persistent\_location(restaurant, db('SQL', \ user, \ pass, \ 'host' : port)). \quad (13)$$

$$: -sql\_persistent(restaurant(integer, \ string, \ integer, \ integer), \\ restaurant(id, \ ft, \ yso, \ dttcc), restaurant). \quad (14)$$

Once we have all the information accessible we use the syntactical structure in Eq. 15 to define our virtual database table (vdbt), where pT is the name of the vdbt (the individual or subject of our searches), pA is the arity of the predicate or the vdbt, pN is the name assigned to a column of the vdbt pT and pT' is a basic type,[5] one of $\{boolean\_type, enum\_type, integer\_type, float\_type, string\_type\}$. We provide an example in Eq. 16 to clarify, in which the restaurant vdbt has five columns (or the predicate has five arguments), the first for the unique identifier given to each restaurant (its name), the second for the food type served there, the third for the number of years since its opening, the fourth for the restaurant's price average and the last one for the distance to the city center from that restaurant.

$$define\_database(pT/pA, [(pN, pT')]) \quad (15)$$
$$define\_database((id, \ string\_type), \quad (food\_type, \ enum\_type), \\ (years\_since\_opening, \ integer\_type), \\ (price\_average, \ integer\_type), \\ (distance\_to\_the\_city\_center, \ integer\_type)]). \quad (16)$$

This syntactical construction has no formal semantics because it is just for defining the input data, but it provides a lot of information to the web interface and setters/getters that can be used in the programs. We go first for the setters/getters. For each column defined for a vdbt the framework builds for us a setter/getter to store/access the information in/of each cell in the database. The cell selected gets fully determined by the predicate name (the one given to the column) and its first argument. For example, by writing the line in Eq. 16 the framework defines for us the

---

[5]Please note that the types in our framework are not the same as the types used in Eqs. 8–14. Nevertheless, our types are subsets of this ones. We justify in the paragraph below this one why we need this fine-grained type control.

predicates $id(R, Id)$, $food\_type(R, FoodType)$, $years\_since\_opening(R, Years)$, $price\_average(R, Price)$ and $distance\_to\_the\_city\_center(R, Dttcc)$. Each one serves to set/obtain the value in/from the database cell corresponding to the row of restaurant R and the column with the same name as the predicate used (id, $food\_type$, $years\_since\_opening$, $price\_average$ and $distance\_to\_the$ $\_city\_center$). With respect to the web interface, the framework notifies to it that we have a new value for the field "individual" (the value in pT, restaurant in the example), a list of values for *non-fuzzy-pred* (id, food type, years since opening, price average and distance to the city center) and their types (string_type, enum_type, integer_type, integer_type, integer_type). In addition to this explicit information the web interface itself is capable of deriving from the type of each column the values that it can show in *comp-op*. We show them in the table below. It is even able to detect in some cases that it must ask the framework for the values of some field, as in the case of the selection for the field *comp-op* the value "is similar to".

Type	Values for *comp-op*
string_type	"is equal to" and "is different from"
enum_type	"is equal to", "is different from" and "is similar to"
interger_type	"is equal to", "is different from", "is bigger than", "is lower than", "is bigger than or equal to" and "is lower than or equal to"

The second syntactical construction is the one used to define similarity between the individuals of *enum_type*. It is shown in Eq. 17, where pT and pN mean the same as in Eq. 15, $V1$ and $V2$ are possible values for the column pN of the vdbt pT, column that must be of type *enum_type*, and TV is the truth value (a float number between 0 and 1) we assign to the similarity between $V1$ and $V2$. We show an example in Eq. 20, in which we say that the food type mediterranean is 0.7 similar to the spanish food.[6] The syntactical constructions in Eqs. 18 and 19 are optional tails for the syntactical construction in Eq. 17. Since they can appear too as tails of the constructions in Eqs. 17, 23, 27, 31, 32 and 36, we dedicate some paragraphs (just after this one) to explain how the semantics of the constructions change when they are used. With respect to the semantics of Eq. 17, we show them in Eq. 21. For the variables in common we take the values written using the new syntax, while for p, v, $\&_i$ and $COND$ we have by default[7] the values 0.8, 1, *product* and *true*. We show in Eq. 22 the translation of the example in Eq. 20 for the reader to see how the

---

[6]Be careful, we are not saying that the spanish food is 0.7 similar to the mediterranean one. You need to add another clause with that information if you wanna say that too.

[7]The meaning of this "by default" is explained too in the paragraphs after this one.

translation is done in practice. This construction does not provide any information to the web interface.

$$similarity\_between(pT, \; pN(V1), \; pN(V2), \; TV) \tag{17}$$

$$with\_credibility(credOp, \; credVal) \tag{18}$$

$$only\_for\_user\,'UserName' \tag{19}$$

$$similarity\_between(restaurant, \; food\_type(mediterranean),$$
$$food\_type(spanish), 0.7) \tag{20}$$

$$similarity(pT(pN(V1, V2))) \xleftarrow{(p,\,v),\,\&_i} TV \; if \; COND \tag{21}$$

$$similarity(restaurant(food\_type(mediterranean, spanish))) \xleftarrow{(0.8,\,1),\,prod}$$
$$0.7 \; if \; true \tag{22}$$

As outlined in the previous paragraph, the constructions in Eqs. 18 and 19 can be used as tails for the constructions in Eqs. 17, 23, 27, 31, 32 and 36. There is another construction that can be used as tail, the one in Eq. 24, but only for the constructions in Eqs. 23, 27, 31, 32 and 36. This three constructions are meant to change slightly the semantics of the original constructions when they appear as their tails, which is done by modifying at least one of the values given "by default" to the variables p, v, $\&_i$ and $COND$. We explain each case separately.

The tail in Eq. 18 serves to (re)define the credibility of a clause, together with the operator needed to combine it with its truth value. In its syntactic definition in Eq. 18 $credVal$ is the credibility, a number of float type, and $credOp$ is the operator, any conjunctor having the properties defined in Sect. 2. When we use it the values for v and $\&_i$ (usually 1 and $product$) are changed by the values given to the variables $credVal$ and $credOp$.

The tail in Eq. 19 is aimed at defining personalized rules, rules that only apply when the name of the user logged in and the user name in the rule are the same one. In the construction $Username$ is the name of any user, a string. When we use it the value of $COND$ is replaced[8] by $COND \wedge currentUser(Me) \wedge Me = 'UserName'$[9] and the value for p gets increased by 0.1. While the first change is to ensure that the rule is only used when the logged user is the selected user, the second one is to ensure that, when the logged user is the selected user, this rule (considered to be more specialized for the selected user) is chosen before another rule not having this specialization.

The tail in Eq. 24 (not applicable to the construction in Eq. 17) serves to limit the individuals for which we wanna use the fuzzy clause or rule. In the construction pN and pT mean the same as in Eq. 15, $cond$ can take the values is_equal_to,

---

[8]Please note that we not remove the original condition, so we can combine conditions introduced by the semantics of a clause with the conditions introduced by one or more tails.

[9]We use indistinctively ',' and $\wedge$ because the first one is the Prolog symbol for conjunction.

is_different_from, *is_bigger_than*, *is_lower_than*, *is_bigger_than_or_equal_to* and *is_lower_than_or_equal_to* and *value* can be of type *integer_type*, *enum_type* or *string_type*. The only restrictions are that the type of *value* must be the same as the one given to to the column *pN* of *pT* and that if they are of type *enum_type* or *string_type* the only available values for *cond* are *is_equal_to* and *is_different_from*. When we use this tail construction the value of *COND* is changed by $COND \wedge (pN(Individual)\, cond\, value)$, where *Individual* is basically a vdbt row (of type *pT*), and the value for p gets increased by 0.05.

The first tail construction, the one in Eq. 18, is aimed at changing the clause credibility. This is why it only changes the credibility value and the credibility operator in the "by default" semantics (of the clause in which it appears as tail). On the contrary the tails constructions in Eqs. 19 and 24 have as purpose increasing the specialization of the clause. The first one defines that the user prefers the results of this clause to the results of any other clause and the second one defines that, for the subset of individuals of our clause domain delimited by the condition, we prefer the results provided by this clause to the results provided by any other clause. This justifies in part the increasing of the value of p by 0.1 when the clause contains the tail in Eq. 19 and by 0.05 when it is the one in Eq. 24. The missing part, the cause of defining different values for each, comes from a design decision: when choosing between the results of a personalized clause and the ones of a clause defined for a subset of individuals we prefer the first ones. Furthermore, the use of one of the tails' constructions does not disallow the use of the other ones, so we can have personalized rules for a subset of individuals of the clause's domain. And with a defined credibility.

The third construction (shown in Eq. 23) is the one used to define the result of a fuzzy predicate (*fPredName*) when this one is applied to an individual in the selected vdbt (*pT*). It serves to define the rare situation in which for all the individuals in the vdbt we have the same result and, when the construction in Eq. 24 appears as its tail, for subsets of the set of individuals in the vdbt. In Eq. 23 the variables *pT* and *TV* mean the same as in Eqs. 15 and 17 and *fPredName* is the fuzzy predicate we are defining. Equation 25 is an example of use in which we say that the restaurant with id Zalacain is cheap with a truth value of 0.1. The formal semantics for this construction are shown in Eq. 26, where *Individual* is a variable representing the vdbt individual for which the clause will be computed (a restaurant in the example). The default values for p, v, $\&_i$ and *COND* are the values 0.8, 1, *product* and *true*. From the point of view of the interface, the inclusion of a new fuzzy predicate is taken into account and a new predicate appears in the list of predicates from which we can choose one for the field *fuzzy-pred* (see Eq. 1).

$$fPredName(pT) :\sim value(TV) \qquad (23)$$

$$if(pN(pT)\ cond\ value). \qquad (24)$$

$$cheap(restaurant) :\sim value(0.1)$$

$$if(id(restaurant)\ is\_equal\_to\ zalacain). \qquad (25)$$

$$fPredName(Individual) \xleftarrow{(p,\ v),\ \&_i} TV\ if\ COND \qquad (26)$$

The fourth construction serves to define fuzzifications, the computation of fuzzy values for fuzzy predicates from the non-fuzzy value that the individual has in some column in the database. The syntax is presented in Eq. 27, where pN and pT mean the same as in Eq. 15, *fPredName* is the name of the fuzzy predicate that we are defining (the fuzzification), $[(valIn, valOut)]$ is a list of pairs of values such that $valIn$ belongs to the domain of the fuzzification and $valOut$ to its image.[10] An example in which we compute how much traditional is a restaurant from the number of years since its opening is presented in Eq. 28. The formal semantics for this construction are shown in Eq. 29, but only for one sequence of two contiguous points[11] $(valIn1, valOut1)(valIn2, valOut2)$ in Eq. 27. The default values for p, v, &$_i$ and *COND* are the values 0.6, 1, *product* and the *COND'* in Eq. 30. This value for *COND*, *COND'*, serves to limit the domain of the generated clause. Since we generate one clause for each piece of the piecewise function we use *COND'* to ensure that we use the clause designated for the piece our input value belongs to. The web interface assumes that fuzzification functions are fuzzy predicates, so it includes them in the list of available predicates for the field *fp* (see Eq. 1) when they are not there yet.

$$f\,PredName(pT) :\sim function(pN(pT), [(valIn, valOut)]) \quad (27)$$

$$traditional(restaurant) :\sim function(years\_since\_opening(restaurant),$$
$$[(0,0), (5,0.1), (10,0.4), (15,1), (100,1)]). \quad (28)$$

$$f\,PredName(Individual) \xleftarrow{\text{(p, v), }\&_i} pN(Individual) * \frac{(valOut\_2 - valOut\_1)}{(valIn\_2 - valIn\_1)}$$
$$\text{if } COND \quad (29)$$

$$COND' = (valIn\_1 < pN(Individual) \le valIn\_2) \quad (30)$$

The fifth syntactical construction is for defining rules and has two forms, one used when the body depends on more than one subgoal, shown in Eq. 31, and one used when it depends in just one subgoal, shown in Eq. 32. In Eq. 31 *aggr* is the aggregator used to combine the truth values of the subgoals in *complexBody*, which is just a conjunction of names of fuzzy predicates (and the vdbt they are associated to, represented by pT), while in Eq. 32 *simplexBody* is just the name of a fuzzy predicate. In both of them pT means the same as in Eq. 15 and *fPredName* the same as in Eq. 27. We show an example in Eq. 35. The formal semantics for this constructions are respectively shown in Eqs. 33 and 34 and the default values for p, v, &$_i$ and *COND* are the values 0.4, 1, *product* and *true*. With respect to what the web interface receives from this syntactic structure, it considers fuzzy rules as fuzzy predicates, and it always includes fuzzy predicates in the list of available predicates for the field *fp* (see Eq. 1) when they are not there yet.

---

[10] $[(valIn, valOut)]$ is basically a piecewise function definition, where each two contiguous points represent a piece.

[11] This "only for one sequence of two contiguous points" means that we generate one clause of the form in Eq. 29 for each piece defined by two contiguous points.

$$fPredName(pT) :\sim rule(aggr, complexBody) \tag{31}$$

$$fPredName(pT) :\sim rule(simpleBody) \tag{32}$$

$$fPredName(Individual) \xleftarrow{\text{(p, v), \&}_i} aggr(complexBody) \text{ if } COND \tag{33}$$

$$fPredName(Individual) \xleftarrow{\text{(p, v), \&}_i} simplexBody \text{ if } COND \tag{34}$$

$$tempting\_restaurant(restaurant) :\sim rule(min, (near\_the\_city\_center(restaurant),$$
$$cheap(restaurant))) \tag{35}$$

The sixth syntactical construction is the one used to define default values for fuzzy computations. Its main goal is to avoid that the inference process stops when a needed value is missing and it is really useful when a database can have null values. The syntactic form is presented in Eq. 36, where pT means the same as in Eq. 15 and $fPredName$ the same as in Eq. 27. We provide two examples in Eqs. 37 and 38 in which we say that, in absence of information, we consider that a restaurant will not be close to the city center (this is what the zero value means) and that, in absence of information, a restaurant is considered to be medium cheap.[12] The formal semantics for this constructions are shown in Eq. 39 and the default values for p, v, $\&_i$ and $COND$ are the values 0, 1, *product* and *true*. With respect to what the web interface receives from this syntactic structure, the syntactic construction for defining default values is translated as a fuzzy predicate and the web interface always includes fuzzy predicates in the list of available predicates for the field fp (see Eq. 1) when they are not there yet.

$$fPredName(pT) :\sim defaults\_to(TV) \tag{36}$$

$$near\_the\_city\_center(restaurant) :\sim defaults\_to(0). \tag{37}$$

$$cheap(restaurant) :\sim defaults\_to(0.5). \tag{38}$$

$$fPredName(Individual) \xleftarrow{\text{(p, v), \&}_i} TV \text{ if } COND \tag{39}$$

The six constructions defined before and their semantics orchestrate the intended meaning we wanna give to our programs. We summarize in the table below the values given to the variables p, v, $\&_i$ and $COND$ when no tail is attached to the no-tail constructions and explain now the reasons for choosing the values that appear there. For the variables v and $\&_i$ we assign by default the values 1 and *product*. We have chosen this values due to the fact that multiplying by one a number (the rule's truth value) results always in the same number: it does not affect the clause's result. Their value is only changed when the construction in Eq. 18 is used as tail, which means that the programmer wants to change the default credibility of the clause. The variable $COND$ has as goal avoiding the clause from obtaining results when the

---

[12]We include two examples here so if one builds a program by taking all the examples in the contribution the rule in Eq. 35 does not fail to obtain answers because it has not enough information to infer results.

condition is not satisfied. The only construction that needs by default this behaviour is the fuzzification, as explained before. For the other constructions we assign by default the value *true* and modify it when any of the constructions in Eqs. 19 and 24 appears as tails, which affects too to the value of the variable p. The value given to the variable p is used to decide when more than one rule returns results which ones are the valid ones. The only restriction in the selection of its default value is that its range of values is [0, 1]. Since the tails' constructions in Eqs. 19 and 24 can add to it a maximum of 0.15 our default value for p must be always below or equal to 0.85 to ensure that we satisfy the restriction.

Construction	p	v	$\&_i$	COND
Similarity between individuals	0.8	1	Product	True
Fuzzy value	0.8	1	Product	True
Fuzzification function	0.6	1	Product	$COND'$
Fuzzy rule	0.4	1	Product	True
Default fuzzy value	0	1	Product	True

4 Conclusions

We present a framework for modelling the real world knowledge and a web interface for posing fuzzy and flexible queries. As introduced before, the first one has a syntax (and its semantics) with which we can capture the relations between the fuzzy and non-fuzzy knowledge of any domain (inclusive the linking of information from databases with real-world fuzzy concepts) and feed the search engine with the information it needs to provide a friendly and easy to use user interface. The search engine main advantage over the existing ones just derives from this: we avoid the necessity to learn a complex syntax to just pose (fuzzy) queries. This, joint with the possibility to include Prolog code (for complex tasks) makes our framework a very powerful tool for representing the real world and answering questions about it. A link to a beta version of our flexible search engine (with example programs, the possibility to upload new ones, etc.) is available at our web page.

Our current research focus on deriving similarity relations from the modelization of a problem in our framework's language. In this way we could, for example, derive from the RGB composition of two colors their similarity relation.

References

1. Baldwin, J.F., Martin, T.P., Pilsworth, B.W.: Fril—Fuzzy and Evidential Reasoning in Artificial Intelligence. Wiley, New York (1995)
2. Bobillo, F., Straccia, U.: FuzzyDL: an expressive fuzzy description logic reasoner. In: 2008 International Conference on Fuzzy Systems (FUZZ-08), pp. 923–930. IEEE Computer Society (2008)
3. Bordogna, G. Pasi, G.: A fuzzy query language with a linguistic hierarchical aggregator. In: Proceedings of the 1994 ACM Symposium on Applied computing, SAC'94, pp. 184–187. ACM, New York (1994)
4. Bosc, P., Pivert, O.: SQLF: a relational database language for fuzzy querying. IEEE Trans. Fuzzy Syst. 3(1), 1–17 (1995)
5. Bosc, P., Pivert, O.: On a strengthening connective for flexible database querying. In: 2011 IEEE International Conference on Fuzzy Systems (FUZZ), pp. 1233–1238 (2011)
6. Dubois, D., Prade, H.: Using fuzzy sets in flexible querying: why and how? In: Andreasen, T., Christiansen, H., Larsen, H.L. (eds.) Flexible Query Answering Systems, pp. 45–60. Kluwer Academic Publishers, Norwell (1997)
7. Guadarrama, S., Muñoz-Hernández, S., Vaucheret, C.: Fuzzy prolog: a new approach using soft constraints propagation. Fuzzy Sets Syst. (FSS) 144(1), 127–150 (2004). Possibilistic Logic and Related Issues
8. Ishizuka, M., Kanai, N.: Prolog-ELF incorporating fuzzy logic. In: Proceedings of the 9th International Joint Conference on Artificial Intelligence, IJCAI'85, pp. 701–703. Morgan Kaufmann Publishers Inc, San Francisco (1985)
9. Li, D., Liu, D.: A Fuzzy Prolog Database System. Wiley, New York (1990)
10. Medina, J., Ojeda-Aciego, M., Vojtáš, P.: A completeness theorem for multi-adjoint logic programming. In: FUZZ, pp. 1031–1034. IEEE (2001)
11. Medina, J., Ojeda-Aciego, M., Vojtáš, P.: Multi-adjoint logic programming with continuous semantics. In: Eiter, T., Faber, W., Truszczynski, M. (eds.) LPNMR. Lecture Notes in Computer Science, vol. 2173, pp. 351–364. Springer, Berlin (2001)
12. Medina, J., Ojeda-Aciego, M., Vojtáš, P.: A procedural semantics for multi-adjoint logic programming. In: Brazdil, P., Jorge, A. (eds.) EPIA. Lecture Notes in Computer Science, vol. 2258, pp. 290–297. Springer, Berlin (2001)
13. Medina, J., Ojeda-Aciego, M., Vojtáš, P.: A multi-adjoint approach to similarity-based unification. Electron. Notes Theor. Comput. Sci. 66(5):70–85 (2002). UNCL'2002, Unification in Non-Classical Logics (ICALP 2002 Satellite Workshop)
14. Medina, J., Ojeda-Aciego, M., Vojtáš, P.: Similarity-based unification: a multi-adjoint approach. Fuzzy Sets Syst. 146(1), 43–62 (2004)
15. Morcillo, P.J., Moreno, G.: Floper, a fuzzy logic programming environment for research. In: Fundación Universidad de Oviedo (ed.) Proceedings of VIII Jornadas sobre Programación y Lenguajes (PROLE'08), pp. 259–263. Gijón, October 2008
16. Moreno, J.M., Ojeda-Aciego, M.: On first-order multi-adjoint logic programming. In: 11th Spanish Congress on Fuzzy Logic and Technology (2002)
17. Muñoz-Hernández, S., Pablos-Ceruelo, V., Strass, H.: RFuzzy: syntax, semantics and implementation details of a simple and expressive fuzzy tool over prolog. Inf. Sci. 181(10), 1951–1970 (2011). Special Issue on Information Engineering Applications Based on Lattices
18. Pablos-Ceruelo, V., Muñoz-Hernández, S.: Introducing priorities in rfuzzy: syntax and semantics. In: Proceedings of the 11th International Conference on Mathematical Methods in Science and Engineering, CMMSE 2011, vol. 3, pp. 918–929, Benidorm (Alicante), June 2011
19. Ribeiro, R.A., Moreira, A.M.: Fuzzy query interface for a business database. Int. J. Hum.-Comput. Stud. 58(4), 363–391 (2003)
20. Rodriguez, L.J.T.: A contribution to database flexible querying: fuzzy quantified queries evaluation, P.hD. thesis, November 2005

21. Vaucheret, C., Guadarrama, S., Muñoz-Hernández, S.: Fuzzy prolog: a simple general implementation using CLP(R). In: Baaz, M., Voronkov, A. (eds.) LPAR. Lecture Notes in Artificial Intelligence, vol. 2514, pp. 450–464. Springer, Berlin (2002)
22. Vojtáš, P.: Fuzzy logic programming. Fuzzy Sets Syst. **124**(3), 361–370 (2001)
23. Zadeh, L.A.: Fuzzy sets. Inf. Control **8**(3), 338–353 (1965)

On the Deduction Problem in Gödel and Product Logics

Dušan Guller

Abstract We investigate the deduction problem in Gödel and Product logics, both equipped with Gödel negation, in the countable case. Our approach is based on translation of a formula to an equivalent satisfiable finite order clausal theory, consisting of order clauses. An order clause is a finite set of order literals of the form $\varepsilon_1 \diamond \varepsilon_2$ where $\diamond$ is a connective either $=$ or $\prec$. $=$ and $\prec$ are interpreted by th xe equality and standard strict linear order on $[0, 1]$, respectively. We generalise the well-known hyperresolution principle to the standard first-order Gödel logic and devise a calculus operating over order clausal theories. A variant of the *DPLL* procedure in the propositional Product logic exploiting trichotomy and operating over order clausal theories, will be proposed. Both the calculi are refutation sound and complete for the countable case.

Keywords Gödel logic · Product logic · Resolution · *DPLL* procedure · Many-valued logics · Automated deduction

1 Introduction

Exploration of the deduction problems (satisfiability, *SAT*, validity, *VAL*, logical entailment) belongs to all important aims in research on many-valued logics. In addition to investigation of the "classical" many-valued deduction calculi, a perspective from automated deduction has received attractivity during the last two decades. Concerning the three fundamental first-order fuzzy logics, the set of logically valid formulae is Π_2-complete for Łukasiewicz logic, Π_2-hard for Product logic, and Σ_1-complete for Gödel logic, as with classical first-order logic. Among these fuzzy

This work is partially supported by VEGA Grant 1/0592/14 and Slovak Literary Fund.

D. Guller (✉)
Department of Applied Informatics, Comenius University,
Mlynská dolina, 842 48 Bratislava, Slovakia
e-mail: guller@fmph.uniba.sk

© Springer International Publishing Switzerland 2016
K. Madani et al. (eds.), *Computational Intelligence*,
Studies in Computational Intelligence 613,
DOI 10.1007/978-3-319-23392-5_17

logics, only Gödel logic is recursively axiomatisable. Hence, it is necessary to provide a proof method suitable for automated deduction, as one has done for classical logic. In contrast to classical logic, we cannot make shifts of quantifiers arbitrarily and translate a formula to an equivalent (satisfiable) prenex form. In Sect. 2, we solve the deduction problem of a formula from a countable theory in Gödel logic. We generalise the well-known hyperresolution principle to the standard first-order Gödel logic. Our approach is based on the translation of a formula of Gödel logic to an equivalent satisfiable finite order clausal theory, consisting of order clauses. We introduce a notion of quantified atom: a formula a is a quantified atom iff $a = Qx\, p(t_0, \ldots, t_\tau)$ where Q is a quantifier ($\forall$, $\exists$); $p(t_0, \ldots, t_\tau)$ is an atom; x is a variable occurring in $p(t_0, \ldots, t_\tau)$; for all $i \leq \tau$, either $t_i = x$ or x does not occur in t_i. Then an order clause is a finite set of order literals of the form $\varepsilon_1 \diamond \varepsilon_2$ where ε_i is either an atom or a quantified atom, and $\diamond$ is a connective either $=$ or $\prec$. $=$ and $\prec$ are interpreted by the equality and strict linear order on $[0, 1]$, respectively. For an input theory of Gödel logic, the proposed translation produces a so-called admissible order clausal theory. On the basis of the hyperresolution principle, a calculus operating over order clausal theories, is devised. The calculus is proved to be refutation sound and complete for the countable case. Product logic [1, 2] is one of the fundamental fuzzy logics, based on the product t-norm. It has been discovered much later than Gödel and Łukasiewicz logics, known before the beginning of research on fuzzy theory. In Sect. 3, we investigate the deduction problem of a formula from a countable theory in the propositional Product logic. Our approach is based on translation of a formula to an equivalent satisfiable finite order clausal theory, consisting of order clauses. An order clause is a finite set of order literals of the form $\varepsilon_1 \diamond \varepsilon_2$ where ε_i is either a conjunction of propositional atoms or the propositional constant 0 (false) or 1 (true), and $\diamond$ is a connective either $=$ or $\prec$. Trichotomy over order literals, either $\varepsilon_1 \prec \varepsilon_2$ or $\varepsilon_1 = \varepsilon_2$ or $\varepsilon_2 \prec \varepsilon_1$, naturally invokes proposing a variant of the *DPLL* procedure with a trichotomy branching rule, as an algorithm for deciding the satisfiability of a finite order clausal theory. The *DPLL* procedure with its basic rules is proved to be refutation sound and complete in the countable case.

2 An Order Hyperresolution Calculus

2.1 First-Order Gödel Logic

We shall use the standard notions and notation of the first-order Gödel logic and set theory.[1] By $\mathcal{L}$ we denote a first-order language. We assume nullary predicate symbols $0, 1 \in Pred_\mathcal{L}$, $ar_\mathcal{L}(0) = ar_\mathcal{L}(1) = 0$; 0 denotes the false and 1 the true in $\mathcal{L}$. By $Form_\mathcal{L}$ we designate the set of all formulae of $\mathcal{L}$ built up from $Atom_\mathcal{L}$ and $Var_\mathcal{L}$ using the connectives: $\neg$, negation, $\wedge$, conjunction, $\vee$, disjunction, $\rightarrow$, implication, and the quantifiers: $\forall$, the universal quantifier, $\exists$, the existential one.

---

[1]cf. http://ii.fmph.uniba.sk/~guller/sci14.pdf, Sect. 2.1.

In addition, we introduce new binary connectives $=$, equality, and $\prec$, strict order. We denote $Con = \{\neg, \wedge, \vee, \rightarrow, =, \prec\}$. By $OrdForm_{\mathcal{L}}$ we designate the set of all so-called order formulae of $\mathcal{L}$ built up from $Atom_{\mathcal{L}}$ and $Var_{\mathcal{L}}$ using the connectives in Con and the quantifiers: $\forall, \exists$.[2] In the paper, we shall assume that $\mathcal{L}$ is a countable first-order language; hence, all the above mentioned sets of symbols and expressions are countable. By $varseq(\phi)$, $vars(varseq(\phi)) \subseteq Var_{\mathcal{L}}$, we denote the sequence of all variables of $\mathcal{L}$ occurring in ϕ which is built up via the left-right preorder traversal of ϕ.

Gödel logic is interpreted by the standard $\boldsymbol{G}$-algebra augmented by binary operators $=$ and $\prec$ for $=$ and $\prec$, respectively.

$$\boldsymbol{G} = ([0, 1], \leq, \vee, \wedge, \Rightarrow, \overline{}, =, \prec, 0, 1)$$

where $\vee \mid \wedge$ denotes the supremum | infimum operator on $[0, 1]$;

$$a \Rightarrow b = \begin{cases} 1 & \text{if } a \leq b, \\ b & \text{else}; \end{cases} \qquad\qquad \overline{a} = \begin{cases} 1 & \text{if } a = 0, \\ 0 & \text{else}; \end{cases}$$

$$a = b = \begin{cases} 1 & \text{if } a = b, \\ 0 & \text{else}; \end{cases} \qquad\qquad a \prec b = \begin{cases} 1 & \text{if } a < b, \\ 0 & \text{else}. \end{cases}$$

We recall that $\boldsymbol{G}$ is a complete linearly ordered lattice algebra; $\vee \mid \wedge$ is commutative, associative, idempotent, monotone; $0 \mid 1$ is its neutral element; the residuum operator $\Rightarrow$ of $\wedge$ satisfies the condition of residuation:

$$\text{for all } a, b, c \in \boldsymbol{G}, a \wedge b \leq c \Longleftrightarrow a \leq b \Rightarrow c; \tag{1}$$

Gödel negation $\overline{}$ satisfies the condition:

$$\text{for all } a \in \boldsymbol{G}, \overline{a} = a \Rightarrow 0; \tag{2}$$

the following properties, which will be exploited later, hold[3]:
for all a, b, c $\in \boldsymbol{G}$,

$$a \vee b \wedge c = (a \vee b) \wedge (a \vee c), \quad \text{(distributivity of } \vee \text{ over} \wedge) \tag{3}$$
$$a \wedge (b \vee c) = a \wedge b \vee a \wedge c, \quad \text{(distributivity of } \wedge \text{ over} \vee) \tag{4}$$

$$a \Rightarrow (b \vee c) = a \Rightarrow b \vee a \Rightarrow c, \tag{5}$$
$$a \Rightarrow b \wedge c = (a \Rightarrow b) \wedge (a \Rightarrow c), \tag{6}$$

$$(a \vee b) \Rightarrow c = (a \Rightarrow c) \wedge (b \Rightarrow c), \tag{7}$$
$$a \wedge b \Rightarrow c = a \Rightarrow c \vee b \Rightarrow c, \tag{8}$$

---

[2]We assume a decreasing connective and quantifier precedence: $\forall, \exists, \neg, \wedge, \rightarrow, =, \prec, \vee$.
[3]We assume a decreasing operator precedence: $\overline{}, \wedge, \Rightarrow, =, \prec, \vee$.

$$a \Rightarrow (b \Rightarrow c) = a \wedge b \Rightarrow c, \tag{9}$$

$$((a \Rightarrow b) \Rightarrow b) \Rightarrow b = a \Rightarrow b, \tag{10}$$

$$(a \Rightarrow b) \Rightarrow c = ((a \Rightarrow b) \Rightarrow b) \wedge (b \Rightarrow c) \vee c, \tag{11}$$

$$(a \Rightarrow b) \Rightarrow 0 = ((a \Rightarrow 0) \Rightarrow 0) \wedge (b \Rightarrow 0). \tag{12}$$

2.2 Translation to Clausal Form

At first, we introduce a notion of quantified atom. Let $a \in Form_{\mathcal{L}}$. a is a quantified atom of $\mathcal{L}$ iff $a = Qx\, p(t_0, \ldots, t_\tau)$ where $p(t_0, \ldots, t_\tau) \in Atom_{\mathcal{L}}$, $x \in vars(p(t_0, \ldots, t_\tau))$, either $t_i = x$ or $x \notin vars(t_i)$. $QAtom_{\mathcal{L}} \subseteq Form_{\mathcal{L}}$ denotes the set of all quantified atoms of $\mathcal{L}$. Let $Qx\, p(t_0, \ldots, t_\tau) \in QAtom_{\mathcal{L}}$ and $p(t'_0, \ldots, t'_\tau) \in Atom_{\mathcal{L}}$. Let $I = \{i \mid i \leq \tau, x \notin vars(t_i)\}$ and $r_1, \ldots, r_k$, $r_i \leq \tau$, $k \leq \tau$, for all $1 \leq i < i' \leq k$, $r_i < r_{i'}$, be a sequence such that $\{r_i \mid 1 \leq i \leq k\} = I$. We denote

$$freetermseq(Qx\, p(t_0, \ldots, t_\tau)) = t_{r_1}, \ldots, t_{r_k},$$
$$freetermseq(p(t'_0, \ldots, t'_\tau)) = t'_0, \ldots, t'_\tau.$$

We further introduce conjunctive normal form (*CNF*) in Gödel logic. Let $l, \phi \in Form_{\mathcal{L}}$. l is a literal of $\mathcal{L}$ iff either $l = a$ or $l = a \rightarrow b$ or $l = (a \rightarrow b) \rightarrow b$ or $l = a \rightarrow c$ or $l = c \rightarrow a$, $a \in Atom_{\mathcal{L}} - \{0, 1\}$, $b \in Atom_{\mathcal{L}} - \{1\}$, $c \in QAtom_{\mathcal{L}}$. The set of all literals of $\mathcal{L}$ is designated as $Lit_{\mathcal{L}} \subseteq Form_{\mathcal{L}}$. ϕ is a conjunctive | disjunctive normal form of $\mathcal{L}$, in symbols *CNF* | *DNF*, iff either $\phi = 0$ or $\phi = 1$ or $\phi = \bigwedge_{i \leq n} \bigvee_{j \leq m_i} l^i_j \mid \phi = \bigvee_{i \leq n} \bigwedge_{j \leq m_i} l^i_j$, $l^i_j \in Lit_{\mathcal{L}}$.

We finally introduce order clauses in Gödel logic. Let $l \in OrdForm_{\mathcal{L}}$. l is an order literal of $\mathcal{L}$ iff $l = \varepsilon_1 \diamond \varepsilon_2$, $\varepsilon_i \in Atom_{\mathcal{L}} \cup QAtom_{\mathcal{L}}$, $\diamond \in \{=, \prec\}$. The set of all order literals of $\mathcal{L}$ is designated as $OrdLit_{\mathcal{L}} \subseteq OrdForm_{\mathcal{L}}$. An order clause of $\mathcal{L}$ is a finite set of order literals of $\mathcal{L}$. An order clause $\{l_1, \ldots, l_n\}$ is written in the form $l_1 \vee \cdots \vee l_n$. The order clause $\emptyset$ is called the empty order clause and denoted as $\square$. An order clause $\{l\}$ is called a unit order clause and denoted as l. We designate the set of all order clauses of $\mathcal{L}$ as $OrdCl_{\mathcal{L}}$. Let $l, l_0, \ldots, l_n \in OrdLit_{\mathcal{L}}$ and $C, C' \in OrdCl_{\mathcal{L}}$. We define the size of C as $|C| = \sum_{l \in C} |l|$. By $l \vee C$ we denote $\{l\} \cup C$ where $l \notin C$. Analogously, by $l_0 \vee \cdots \vee l_n \vee C$ we denote $\{l_0\} \cup \cdots \cup \{l_n\} \cup C$ where, for all $i, i' \leq n$, $i \neq i'$, $l_i \notin C$ and $l_i \neq l_{i'}$. By $C \vee C'$ we denote $C \cup C'$. C is a subclause of C', in symbols $C \sqsubseteq C'$, iff $C \subseteq C'$. An order clausal theory of $\mathcal{L}$ is a set of order clauses of $\mathcal{L}$. A unit order clausal theory is a set of unit order clauses.

Let $\phi, \phi' \in OrdForm_{\mathcal{L}}$, $T, T' \subseteq OrdForm_{\mathcal{L}}$, $S, S' \subseteq OrdCl_{\mathcal{L}}$, $\mathcal{I}$ be an interpretation for $\mathcal{L}$, $e \in \mathcal{S}_{\mathcal{I}}$. C is true in $\mathcal{I}$ with respect to e, written as $\mathcal{I} \models_e C$, iff there exists $l^* \in C$ such that $\mathcal{I} \models_e l^*$. $\mathcal{I}$ is a model of C, in symbols $\mathcal{I} \models C$, iff, for all $e \in \mathcal{S}_{\mathcal{I}}$, $\mathcal{I} \models_e C$. $\mathcal{I}$ is a model of S, in symbols $\mathcal{I} \models S$, iff, for all $C \in S$, $\mathcal{I} \models C$. $\phi' \mid T' \mid C' \mid S'$ is a logical consequence of $\phi \mid T \mid C \mid S$, in symbols $\phi \mid T \mid C \mid S \models \phi' \mid T' \mid C' \mid S'$, iff, for every model $\mathcal{I}$ of $\phi \mid T \mid C \mid S$ for $\mathcal{L}$, $\mathcal{I} \models \phi' \mid T' \mid C' \mid S'$. $\phi \mid T \mid C \mid S$ is satisfiable iff there exists a model of $\phi \mid T \mid C \mid S$ for $\mathcal{L}$. $\phi \mid T \mid C \mid S$ is equisatisfiable to $\phi' \mid T' \mid C' \mid S'$ iff $\phi \mid T \mid C \mid S$ is satisfiable if and only if $\phi' \mid T' \mid C' \mid S'$ is satisfiable. Let $S \subseteq_{\mathcal{F}} OrdCl_{\mathcal{L}}$. We define the size of S as $|S| = \sum_{C \in S} |C|$. l is a simplified order literal of $\mathcal{L}$ iff $l = \varepsilon_1 \diamond \varepsilon_2$, $\{\varepsilon_1, \varepsilon_2\} \nsubseteq QAtom_{\mathcal{L}}$. The set of all simplified order literals of $\mathcal{L}$ is designated as $SimOrdLit_{\mathcal{L}} \subseteq OrdLit_{\mathcal{L}}$. We denote $SimOrdCl_{\mathcal{L}} = \{C \mid C \in OrdCl_{\mathcal{L}}, C \subseteq SimOrdLit_{\mathcal{L}}\} \subseteq OrdCl_{\mathcal{L}}$. Let $\tilde{f_0} \notin Func_{\mathcal{L}}$; $\tilde{f_0}$ is a new function symbol. Let $\mathbb{I} = \mathbb{N} \times \mathbb{N}$; $\mathbb{I}$ is an infinite countable set of indices. Let $\tilde{\mathbb{P}} = \{\tilde{p_i} \mid i \in \mathbb{I}\}$ such that $\tilde{\mathbb{P}} \cap Pred_{\mathcal{L}} = \emptyset$; $\tilde{\mathbb{P}}$ is an infinite countable set of new predicate symbols. From a computational point of view, the worst case time and space complexity will be estimated using the logarithmic cost measurement. Let $\mathcal{A}$ be an algorithm. $\#\mathcal{O}_{\mathcal{A}}(.) \geq 1$ denotes the number of all elementary operations executed by $\mathcal{A}$.

2.3 Substitutions

We assume the reader to be familiar with the standard notions and notation of substitutions. We introduce a few definitions and denotations; some of them are slightly different from the standard ones, but found to be more convenient.[4] Let $X = \{x_i \mid 1 \leq i \leq n\} \subseteq Var_{\mathcal{L}}$. A substitution ϑ of $\mathcal{L}$ is a mapping $\vartheta : X \longrightarrow Term_{\mathcal{L}}$. ϑ may be written in the form $x_1/\vartheta(x_1), \ldots, x_n/\vartheta(x_n)$. We denote $dom(\vartheta) = X \subseteq Var_{\mathcal{L}}$ and $range(\vartheta) = \bigcup_{x \in X} vars(\vartheta(x)) \subseteq_{\mathcal{F}} Var_{\mathcal{L}}$. The set of all substitutions of $\mathcal{L}$ is designated as $Subst_{\mathcal{L}}$. Let $Qx\,a \in QAtom_{\mathcal{L}}$. ϑ is applicable to $Qx\,a$ iff $dom(\vartheta) \supseteq freevars(Qx\,a)$ and $x \notin range(\vartheta|_{freevars(Qx\,a)})$. We define the application of ϑ to $Qx\,a$ as $(Qx\,a)\vartheta = Qx\,a(\vartheta|_{freevars(Qx\,a)} \cup x/x) \in QAtom_{\mathcal{L}}$. Let ε and ε' be expressions. ε' is an instance of ε of $\mathcal{L}$ iff there exists $\vartheta^* \in Subst_{\mathcal{L}}$ such that $\varepsilon' = \varepsilon\vartheta^*$. ε' is a variant of ε of $\mathcal{L}$ iff there exists a variable renaming $\rho^* \in Subst_{\mathcal{L}}$ such that $\varepsilon' = \varepsilon\rho^*$. Let $C \in OrdCl_{\mathcal{L}}$ and $S \subseteq OrdCl_{\mathcal{L}}$. C is an instance | a variant of S of $\mathcal{L}$ iff there exists $C^* \in S$ such that C is an instance | a variant of C^* of $\mathcal{L}$. We denote $Inst_{\mathcal{L}}(S) = \{C \mid C$ is an instance of S of $\mathcal{L}\} \subseteq OrdCl_{\mathcal{L}}$ and $Vrnt_{\mathcal{L}}(S) = \{C \mid C$ is a variant of S of $\mathcal{L}\} \subseteq OrdCl_{\mathcal{L}}$.

ϑ is a unifier of $\mathcal{L}$ for E iff $E\vartheta$ is a singleton set. Let $\theta \in Subst_{\mathcal{L}}$. θ is a most general unifier of $\mathcal{L}$ for E iff θ is a unifier of $\mathcal{L}$ for E, and for every unifier ϑ of $\mathcal{L}$ for E, there exists $\gamma^* \in Subst_{\mathcal{L}}$ such that $\vartheta|_{freevars(E)} = \theta|_{freevars(E)} \circ \gamma^*$. By $mgu_{\mathcal{L}}(E) \subseteq Subst_{\mathcal{L}}$ we denote the set of all most general unifiers of $\mathcal{L}$ for E. Let $\overline{E} = E_0, \ldots, E_n$, $E_i \subseteq A_i$, either $A_i = Term_{\mathcal{L}}$ or $A_i = Atom_{\mathcal{L}}$ or $A_i = QAtom_{\mathcal{L}}$ or $A_i = OrdLit_{\mathcal{L}}$. ϑ is a unifier of $\mathcal{L}$ for $\overline{E}$ iff, for all $i \leq n$, ϑ is a unifier of $\mathcal{L}$ for E_i. θ is a most general unifier of $\mathcal{L}$ for $\overline{E}$ iff θ is a unifier of $\mathcal{L}$ for $\overline{E}$, and for every unifier ϑ of $\mathcal{L}$ for $\overline{E}$, there exists $\gamma^* \in Subst_{\mathcal{L}}$ such that $\vartheta|_{freevars(\overline{E})} = \theta|_{freevars(\overline{E})} \circ \gamma^*$. By $mgu_{\mathcal{L}}(\overline{E}) \subseteq Subst_{\mathcal{L}}$ we denote the set of all most general unifiers of $\mathcal{L}$ for $\overline{E}$.

---

[4]cf. http://ii.fmph.uniba.sk/~guller/sci14.pdf, Sect. 2.3.

Theorem 1 (Unification Theorem) *Let* $\overline{E} = E_0, \ldots, E_n$, *either* $E_i \subseteq_{\mathcal{F}} Term_{\mathcal{L}}$ *or* $E_i \subseteq_{\mathcal{F}} Atom_{\mathcal{L}}$. *If there exists a unifier of* $\mathcal{L}$ *for* $\overline{E}$, *then there exists* $\theta^* \in mgu_{\mathcal{L}}(\overline{E})$ *such that* $range(\theta^*|_{vars(\overline{E})}) \subseteq vars(\overline{E})$.

Proof By induction on $\|vars(\overline{E})\|$. ☐

Theorem 2 (Extended Unification Theorem) *Let* $\overline{E} = E_0, \ldots, E_n$, *either* $E_i \subseteq_{\mathcal{F}}$ *Term*$_{\mathcal{L}}$ *or* $E_i \subseteq_{\mathcal{F}}$ *Atom*$_{\mathcal{L}}$ *or* $E_i \subseteq_{\mathcal{F}}$ *QAtom*$_{\mathcal{L}}$ *or* $E_i \subseteq_{\mathcal{F}}$ *OrdLit*$_{\mathcal{L}}$, *and* $boundvars(\overline{E}) \subseteq V \subseteq_{\mathcal{F}} Var_{\mathcal{L}}$. *If there exists a unifier of* $\mathcal{L}$ *for* $\overline{E}$, *then there exists* $\theta^* \in mgu_{\mathcal{L}}(\overline{E})$ *such that* $range(\theta^*|_{freevars(\overline{E})}) \cap V = \emptyset$.

Proof A straightforward consequence of Theorem 1. ☐

2.4 A Formal Treatment

Translation of a formula or a theory to *CNF* and clausal form, is based on the following lemma:

Lemma 1 *Let* $n_\phi, n_0 \in \mathbb{N}$, $\phi \in Form_{\mathcal{L}}$, $T \subseteq Form_{\mathcal{L}}$.

(1) *There exist either* $J_\phi = \emptyset$ *or* $J_\phi = \{(n_\phi, j) \mid j \leq n_{J_\phi}\}$, $J_\phi \subseteq \{(n_\phi, j) \mid j \in \mathbb{N}\}$; *a CNF* $\psi \in Form_{\mathcal{L} \cup \{\tilde{p}_j \mid j \in J_\phi\}}$, $S_\phi \subseteq_{\mathcal{F}} SimOrdCl_{\mathcal{L} \cup \{\tilde{p}_j \mid j \in J_\phi\}}$ *such that*

   (a) $\|J_\phi\| \leq 2 \cdot |\phi|$;
   (b) *there exists an interpretation* $\mathfrak{A}$ *for* $\mathcal{L}$ *and* $\mathfrak{A} \models \phi$ *if and only if there exists an interpretation* $\mathfrak{A}'$ *for* $\mathcal{L} \cup \{\tilde{p}_j \mid j \in J_\phi\}$ *and* $\mathfrak{A}' \models \psi$, *satisfying* $\mathfrak{A} = \mathfrak{A}'|_{\mathcal{L}}$;
   (c) *there exists an interpretation* $\mathfrak{A}$ *for* $\mathcal{L}$ *and* $\mathfrak{A} \models \phi$ *if and only if there exists an interpretation* $\mathfrak{A}'$ *for* $\mathcal{L} \cup \{\tilde{p}_j \mid j \in J_\phi\}$ *and* $\mathfrak{A}' \models S_\phi$, *satisfying* $\mathfrak{A} = \mathfrak{A}'|_{\mathcal{L}}$;
   (d) $|\psi| \in O(|\phi|^2)$; *the number of all elementary operations of the translation of* ϕ *to* ψ, *is in* $O(|\phi|^2)$; *the time and space complexity of the translation of* ϕ *to* ψ, *is in* $O(|\phi|^2 \cdot (\log(1 + n_\phi) + \log |\phi|))$;
   (e) $|S_\phi| \in O(|\phi|^2)$; *the number of all elementary operations of the translation of* ϕ *to* S_ϕ, *is in* $O(|\phi|^2)$; *the time and space complexity of the translation of* ϕ *to* S_ϕ, *is in* $O(|\phi|^2 \cdot (\log(1 + n_\phi) + \log |\phi|))$;
   (f) *for all* $a \in qatoms(\psi)$, *there exists* $j^* \in J_\phi$ *and* $preds(a) = \{\tilde{p}_{j^*}\}$;
   (g) *for all* $j \in J_\phi$, *there exist a sequence* $\bar{x}$ *of variables of* $\mathcal{L}$ *and* $\tilde{p}_j(\bar{x}) \in atoms(\psi)$ *satisfying, for all* $a \in atoms(\psi)$ *and* $preds(a) = \{\tilde{p}_j\}$, $a = \tilde{p}_j(\bar{x})$; *if there exists* $a^* \in qatoms(\psi)$ *and* $preds(a^*) = \{\tilde{p}_j\}$, *then there exists* $Qx\, \tilde{p}_j(\bar{x}) \in qatoms(\psi)$ *satisfying, for all* $a \in qatoms(\psi)$ *and* $preds(a) = \{\tilde{p}_j\}$, $a = Qx\, \tilde{p}_j(\bar{x})$;
   (h) *for all* $a \in qatoms(S_\phi)$, *there exists* $j^* \in J_\phi$ *and* $preds(a) = \{\tilde{p}_{j^*}\}$;
   (i) *for all* $j \in J_\phi$, *there exist a sequence* $\bar{x}$ *of variables of* $\mathcal{L}$ *and* $\tilde{p}_j(\bar{x}) \in atoms(S_\phi)$ *satisfying, for all* $a \in atoms(S_\phi)$ *and* $preds(a) = \{\tilde{p}_j\}$, $a = \tilde{p}_j(\bar{x})$; *if there exists* $a^* \in qatoms(S_\phi)$ *and* $preds(a^*) = \{\tilde{p}_j\}$, *then there exists* $Qx\, \tilde{p}_j(\bar{x}) \in qatoms(S_\phi)$ *satisfying, for all* $a \in qatoms(S_\phi)$ *and* $preds(a) = \{\tilde{p}_j\}$, $a = Qx\, \tilde{p}_j(\bar{x})$.

(II) There exist $J_T \subseteq \{(i, j) \mid i \geq n_0\}$ and $S_T \subseteq SimOrdCl_{\mathcal{L} \cup \{\tilde{p}_j \mid j \in J_T\}}$ such that

(a) there exists an interpretation $\mathfrak{A}$ for $\mathcal{L}$ and $\mathfrak{A} \models T$ if and only if there exists an interpretation $\mathfrak{A}'$ for $\mathcal{L} \cup \{\tilde{p}_j \mid j \in J_T\}$ and $\mathfrak{A}' \models S_T$, satisfying $\mathfrak{A} = \mathfrak{A}'|_{\mathcal{L}}$;

(b) if $T \subseteq_{\mathcal{F}} Form_{\mathcal{L}}$, then $J_T \subseteq_{\mathcal{F}} \{(i, j) \mid i \geq n_0\}$, $\|J_T\| \leq 2 \cdot |T|$; $S_T \subseteq_{\mathcal{F}} SimOrdCl_{\mathcal{L} \cup \{\tilde{p}_j \mid j \in J_T\}}$, $|S_T| \in O(|T|^2)$; the number of all elementary operations of the translation of T to S_T, is in $O(|T|^2)$; the time and space complexity of the translation of T to S_T, is in $O(|T|^2 \cdot \log(1 + n_0 + |T|))$;

(c) for all $a \in qatoms(S_T)$, there exists $j^* \in J_T$ and $preds(a) = \{\tilde{p}_{j^*}\}$;

(d) for all $j \in J_T$, there exist a sequence $\bar{x}$ of variables of $\mathcal{L}$ and $\tilde{p}_j(\bar{x}) \in atoms(S_T)$ satisfying, for all $a \in atoms(S_T)$ and $preds(a) = \{\tilde{p}_j\}$, $a = \tilde{p}_j(\bar{x})$; if there exists $a^* \in qatoms(S_T)$ and $preds(a^*) = \{\tilde{p}_j\}$, then there exists $Qx\, \tilde{p}_j(\bar{x}) \in qatoms(S_T)$ satisfying, for all $a \in qatoms(S_T)$ and $preds(a) = \{\tilde{p}_j\}$, $a = Qx\, \tilde{p}_j(\bar{x})$.

Proof Technical, using the interpolation rules in Tables 2, 3, 4 and 5.

$$\text{Let } n_\theta \in \mathbb{N} \text{ and } \theta \in Form_{\mathcal{L}}. \text{ There exists } \theta' \in Form_{\mathcal{L}} \text{ such that} \quad (13)$$

(a) $\theta' \equiv \theta$;

(b) $|\theta'| \leq 2 \cdot |\theta|$; θ' can be built up via a postorder traversal of θ with $\#\mathcal{O}(\theta) \in O(|\theta|)$ and the time, space complexity in $O(|\theta| \cdot (\log(1 + n_\theta) + \log|\theta|))$;

(c) θ' does not contain $\neg$;

(d) either $\theta' = 0$, or 0 is a subformula of θ' if and only if 0 is a subformula of a subformula of θ' of the form $\vartheta \to 0$, $\vartheta \neq 0$;

(e) either $\theta' = 1$ or 1 is not a subformula of θ'.

The proof is by induction on the structure of θ.

$$\text{Let } l \in Lit_{\mathcal{L}}. \text{ There exists } C \in SimOrdCl_{\mathcal{L}} \text{ such that} \quad (14)$$

(a) for every interpretation $\mathfrak{A}$ for $\mathcal{L}$, for all $e \in S_{\mathfrak{A}}$, $\mathfrak{A} \models_e l$ if and only if $\mathfrak{A} \models_e C$;

(b) $|C| \leq 3 \cdot |l|$, C can be built up from l with $\#\mathcal{O}(l) \in O(|l|)$.

In Table 1, for every form of l, C is assigned so that for every interpretation $\mathfrak{A}$ for $\mathcal{L}$, for all $e \in S_{\mathfrak{A}}$, $\mathfrak{A} \models_e l$ if and only if $\mathfrak{A} \models_e C$.

cf. http://ii.fmph.uniba.sk/~guller/sci14.pdf, Sect. 2.4. □

The described translation produces order clausal theories in some restrictive form, which will be utilised in inference using our order hyperresolution calculus. Let $P \subseteq \tilde{\mathbb{P}}$ and $S \subseteq OrdCl_{\mathcal{L} \cup P}$. S is admissible iff

(a) for all $a \in qatoms(S)$, $preds(a) \subseteq P$;

(b) for all $\tilde{p} \in P$, there exist a sequence $\bar{x}$ of variables of $\mathcal{L}$ and $\tilde{p}(\bar{x}) \in atoms(S)$ satisfying, for all $a \in atoms(S)$ and $preds(a) = \{\tilde{p}\}$, a is an instance of $\tilde{p}(\bar{x})$ of $\mathcal{L} \cup P$; if there exists $a^* \in qatoms(S)$ and $preds(a^*) = \{\tilde{p}\}$, then there exists $Qx\, \tilde{p}(\bar{x}) \in qatoms(S)$ satisfying, for all $a \in qatoms(S)$ and $preds(a) = \{\tilde{p}\}$, a is an instance of $Qx\, \tilde{p}(\bar{x})$ of $\mathcal{L} \cup P$.

Table 1 Translation of l to C, $a, b \in Atom_{\mathcal{L}} - \{0, 1\}$, $c \in QAtom_{\mathcal{L}}$

Case	l	C	$|l|$	$|C|$
1	a	$a = 1$	$|a|$	$|a| + 2 \leq 3 \cdot |l|$
2	$a \to 0$	$a = 0$	$|a| + 2$	$|a| + 2 \leq 3 \cdot |l|$
3	$a \to b$	$a \prec b \vee a = b$	$|a| + |b| + 1$	$2 \cdot |a| + 2 \cdot |b| + 2 \leq 3 \cdot |l|$
4	$(a \to 0) \to 0$	$0 \prec a$	$|a| + 4$	$|a| + 2 \leq 3 \cdot |l|$
5	$(a \to b) \to b$	$b \prec a \vee b = 1$	$|a| + 2 \cdot |b| + 2$	$|a| + 2 \cdot |b| + 3 \leq 3 \cdot |l|$
6	$c \to a$	$c \prec a \vee c = a$	$|a| + |c| + 1$	$2 \cdot |a| + 2 \cdot |c| + 2 \leq 3 \cdot |l|$
7	$a \to c$	$a \prec c \vee a = c$	$|a| + |c| + 1$	$2 \cdot |a| + 2 \cdot |c| + 2 \leq 3 \cdot |l|$

Table 2 Binary interpolation rules for $\wedge$ and $\vee$

Case		Laws
$\theta = \theta_1 \wedge \theta_2$		
Positive interpolation $\dfrac{\tilde{p}_i(\bar{x}) \to \theta_1 \wedge \theta_2}{(\tilde{p}_i(\bar{x}) \to \tilde{p}_{i_1}(\bar{x})) \wedge (\tilde{p}_i(\bar{x}) \to \tilde{p}_{i_2}(\bar{x})) \wedge (\tilde{p}_{i_1}(\bar{x}) \to \theta_1) \wedge (\tilde{p}_{i_2}(\bar{x}) \to \theta_2)}$ $|\text{Consequent}| = 9 + 4 \cdot |\bar{x}| + |\tilde{p}_{i_1}(\bar{x}) \to \theta_1| + |\tilde{p}_{i_2}(\bar{x}) \to \theta_2| \leq 13 \cdot (1 + |\bar{x}|) + |\tilde{p}_{i_1}(\bar{x}) \to \theta_1| + |\tilde{p}_{i_2}(\bar{x}) \to \theta_2|$	(6)	(15)
Positive interpolation $\dfrac{\tilde{p}_i(\bar{x}) \to \theta_1 \wedge \theta_2}{\left\{ \begin{array}{l} \tilde{p}_i(\bar{x}) \prec \tilde{p}_{i_1}(\bar{x}) \vee \tilde{p}_i(\bar{x}) = \tilde{p}_{i_1}(\bar{x}),\ \tilde{p}_i(\bar{x}) \prec \tilde{p}_{i_2}(\bar{x}) \vee \tilde{p}_i(\bar{x}) = \tilde{p}_{i_2}(\bar{x}), \\ \tilde{p}_{i_1}(\bar{x}) \to \theta_1, \tilde{p}_{i_2}(\bar{x}) \to \theta_2 \end{array} \right\}}$ $|\text{Consequent}| = 12 + 8 \cdot |\bar{x}| + |\tilde{p}_{i_1}(\bar{x}) \to \theta_1| + |\tilde{p}_{i_2}(\bar{x}) \to \theta_2| \leq 15 \cdot (1 + |\bar{x}|) + |\tilde{p}_{i_1}(\bar{x}) \to \theta_1| + |\tilde{p}_{i_2}(\bar{x}) \to \theta_2|$		(16)
Negative interpolation $\dfrac{\theta_1 \wedge \theta_2 \to \tilde{p}_i(\bar{x})}{(\tilde{p}_{i_1}(\bar{x}) \to \tilde{p}_i(\bar{x}) \vee \tilde{p}_{i_2}(\bar{x}) \to \tilde{p}_i(\bar{x})) \wedge (\theta_1 \to \tilde{p}_{i_1}(\bar{x})) \wedge (\theta_2 \to \tilde{p}_{i_2}(\bar{x}))}$ $|\text{Consequent}| = 9 + 4 \cdot |\bar{x}| + |\theta_1 \to \tilde{p}_{i_1}(\bar{x})| + |\theta_2 \to \tilde{p}_{i_2}(\bar{x})| \leq 13 \cdot (1 + |\bar{x}|) + |\theta_1 \to \tilde{p}_{i_1}(\bar{x})| + |\theta_2 \to \tilde{p}_{i_2}(\bar{x})|$	(8)	(17)
Negative interpolation $\dfrac{\theta_1 \wedge \theta_2 \to \tilde{p}_i(\bar{x})}{\left\{ \begin{array}{l} \tilde{p}_{i_1}(\bar{x}) \prec \tilde{p}_i(\bar{x}) \vee \tilde{p}_{i_1}(\bar{x}) = \tilde{p}_i(\bar{x}) \vee \tilde{p}_{i_2}(\bar{x}) \prec \tilde{p}_i(\bar{x}) \vee \tilde{p}_{i_2}(\bar{x}) = \tilde{p}_i(\bar{x}), \\ \theta_1 \to \tilde{p}_{i_1}(\bar{x}), \theta_2 \to \tilde{p}_{i_2}(\bar{x}) \end{array} \right\}}$ $|\text{Consequent}| = 12 + 8 \cdot |\bar{x}| + |\theta_1 \to \tilde{p}_{i_1}(\bar{x})| + |\theta_2 \to \tilde{p}_{i_2}(\bar{x})| \leq 15 \cdot (1 + |\bar{x}|) + |\theta_1 \to \tilde{p}_{i_1}(\bar{x})| + |\theta_2 \to \tilde{p}_{i_2}(\bar{x})|$		(18)
$\theta = \theta_1 \vee \theta_2$		
Positive interpolation $\dfrac{\tilde{p}_i(\bar{x}) \to (\theta_1 \vee \theta_2)}{(\tilde{p}_i(\bar{x}) \to \tilde{p}_{i_1}(\bar{x}) \vee \tilde{p}_i(\bar{x}) \to \tilde{p}_{i_2}(\bar{x})) \wedge (\tilde{p}_{i_1}(\bar{x}) \to \theta_1) \wedge (\tilde{p}_{i_2}(\bar{x}) \to \theta_2)}$ $|\text{Consequent}| = 9 + 4 \cdot |\bar{x}| + |\tilde{p}_{i_1}(\bar{x}) \to \theta_1| + |\tilde{p}_{i_2}(\bar{x}) \to \theta_2| \leq 13 \cdot (1 + |\bar{x}|) + |\tilde{p}_{i_1}(\bar{x}) \to \theta_1| + |\tilde{p}_{i_2}(\bar{x}) \to \theta_2|$	(5)	(19)
Positive interpolation $\dfrac{\tilde{p}_i(\bar{x}) \to (\theta_1 \vee \theta_2)}{\left\{ \begin{array}{l} \tilde{p}_i(\bar{x}) \prec \tilde{p}_{i_1}(\bar{x}) \vee \tilde{p}_i(\bar{x}) = \tilde{p}_{i_1}(\bar{x}) \vee \tilde{p}_i(\bar{x}) \prec \tilde{p}_{i_2}(\bar{x}) \vee \tilde{p}_i(\bar{x}) = \tilde{p}_{i_2}(\bar{x}), \\ \tilde{p}_{i_1}(\bar{x}) \to \theta_1, \tilde{p}_{i_2}(\bar{x}) \to \theta_2 \end{array} \right\}}$ $|\text{Consequent}| = 12 + 8 \cdot |\bar{x}| + |\tilde{p}_{i_1}(\bar{x}) \to \theta_1| + |\tilde{p}_{i_2}(\bar{x}) \to \theta_2| \leq 15 \cdot (1 + |\bar{x}|) + |\tilde{p}_{i_1}(\bar{x}) \to \theta_1| + |\tilde{p}_{i_2}(\bar{x}) \to \theta_2|$		(20)
Negative interpolation $\dfrac{(\theta_1 \vee \theta_2) \to \tilde{p}_i(\bar{x})}{(\tilde{p}_{i_1}(\bar{x}) \to \tilde{p}_i(\bar{x})) \wedge (\tilde{p}_{i_2}(\bar{x}) \to \tilde{p}_i(\bar{x})) \wedge (\theta_1 \to \tilde{p}_{i_1}(\bar{x})) \wedge (\theta_2 \to \tilde{p}_{i_2}(\bar{x}))}$ $|\text{Consequent}| = 9 + 4 \cdot |\bar{x}| + |\theta_1 \to \tilde{p}_{i_1}(\bar{x})| + |\theta_2 \to \tilde{p}_{i_2}(\bar{x})| \leq 13 \cdot (1 + |\bar{x}|) + |\theta_1 \to \tilde{p}_{i_1}(\bar{x})| + |\theta_2 \to \tilde{p}_{i_2}(\bar{x})|$	(7)	(21)
Negative interpolation $\dfrac{(\theta_1 \vee \theta_2) \to \tilde{p}_i(\bar{x})}{\left\{ \begin{array}{l} \tilde{p}_{i_1}(\bar{x}) \prec \tilde{p}_i(\bar{x}) \vee \tilde{p}_{i_1}(\bar{x}) = \tilde{p}_i(\bar{x}) \vee \tilde{p}_{i_2}(\bar{x}) \prec \tilde{p}_i(\bar{x}) \vee \tilde{p}_{i_2}(\bar{x}) = \tilde{p}_i(\bar{x}), \\ \theta_1 \to \tilde{p}_{i_1}(\bar{x}), \theta_2 \to \tilde{p}_{i_2}(\bar{x}) \end{array} \right\}}$ $|\text{Consequent}| = 12 + 8 \cdot |\bar{x}| + |\theta_1 \to \tilde{p}_{i_1}(\bar{x})| + |\theta_2 \to \tilde{p}_{i_2}(\bar{x})| \leq 15 \cdot (1 + |\bar{x}|) + |\theta_1 \to \tilde{p}_{i_1}(\bar{x})| + |\theta_2 \to \tilde{p}_{i_2}(\bar{x})|$		(22)

Table 3 Binary interpolation rules for $\to$

Case	Laws	
$\theta = \theta_1 \to \theta_2, \theta_2 \neq 0$		
Positive interpolation $\dfrac{\tilde{p}_{\mathtt{i}}(\bar{x}) \to (\theta_1 \to \theta_2)}{\begin{array}{c}(\tilde{p}_{\mathtt{i}}(\bar{x}) \to \tilde{p}_{\mathtt{i}_2}(\bar{x}) \vee \tilde{p}_{\mathtt{i}_1}(\bar{x}) \to \tilde{p}_{\mathtt{i}_2}(\bar{x})) \wedge \\ (\theta_1 \to \tilde{p}_{\mathtt{i}_1}(\bar{x})) \wedge (\tilde{p}_{\mathtt{i}_2}(\bar{x}) \to \theta_2)\end{array}}$ $|\text{Consequent}| = 9 + 4 \cdot |\bar{x}| + |\theta_1 \to \tilde{p}_{\mathtt{i}_1}(\bar{x})| + |\tilde{p}_{\mathtt{i}_2}(\bar{x}) \to \theta_2| \leq 13 \cdot (1 + |\bar{x}|) + |\theta_1 \to \tilde{p}_{\mathtt{i}_1}(\bar{x})| + |\tilde{p}_{\mathtt{i}_2}(\bar{x}) \to \theta_2|$	(8), (9)	(23)
Positive interpolation $\dfrac{\tilde{p}_{\mathtt{i}}(\bar{x}) \to (\theta_1 \to \theta_2)}{\left[\begin{array}{c}\tilde{p}_{\mathtt{i}}(\bar{x}) \prec \tilde{p}_{\mathtt{i}_2}(\bar{x}) \vee \tilde{p}_{\mathtt{i}_2}(\bar{x}) \vee \tilde{p}_{\mathtt{i}_1}(\bar{x}) \prec \tilde{p}_{\mathtt{i}_2}(\bar{x}) \vee \tilde{p}_{\mathtt{i}_1}(\bar{x}) = \tilde{p}_{\mathtt{i}_2}(\bar{x}), \\ \theta_1 \to \tilde{p}_{\mathtt{i}_1}(\bar{x}), \tilde{p}_{\mathtt{i}_2}(\bar{x}) \to \theta_2\end{array}\right]}$ $|\text{Consequent}| = 12 + 8 \cdot |\bar{x}| + |\theta_1 \to \tilde{p}_{\mathtt{i}_1}(\bar{x})| + |\tilde{p}_{\mathtt{i}_2}(\bar{x}) \to \theta_2| \leq 15 \cdot (1 + |\bar{x}|) + |\theta_1 \to \tilde{p}_{\mathtt{i}_1}(\bar{x})| + |\tilde{p}_{\mathtt{i}_2}(\bar{x}) \to \theta_2|$		(24)
Negative interpolation $\dfrac{(\theta_1 \to \theta_2) \to \tilde{p}_{\mathtt{i}}(\bar{x})}{\begin{array}{c}((\tilde{p}_{\mathtt{i}_1}(\bar{x}) \to \tilde{p}_{\mathtt{i}_2}(\bar{x})) \to \tilde{p}_{\mathtt{i}_2}(\bar{x}) \vee \tilde{p}_{\mathtt{i}}(\bar{x})) \wedge (\tilde{p}_{\mathtt{i}_2}(\bar{x}) \to \tilde{p}_{\mathtt{i}}(\bar{x})) \wedge \\ (\tilde{p}_{\mathtt{i}_1}(\bar{x}) \to \theta_1) \wedge (\theta_2 \to \tilde{p}_{\mathtt{i}_2}(\bar{x}))\end{array}}$ $|\text{Consequent}| = 13 + 6 \cdot |\bar{x}| + |\tilde{p}_{\mathtt{i}_1}(\bar{x}) \to \theta_1| + |\theta_2 \to \tilde{p}_{\mathtt{i}_2}(\bar{x})| \leq 13 \cdot (1 + |\bar{x}|) + |\tilde{p}_{\mathtt{i}_1}(\bar{x}) \to \theta_1| + |\theta_2 \to \tilde{p}_{\mathtt{i}_2}(\bar{x})|$	(1), (3), (11)	(25)
Negative interpolation $\dfrac{(\theta_1 \to \theta_2) \to \tilde{p}_{\mathtt{i}}(\bar{x})}{\left[\begin{array}{c}\tilde{p}_{\mathtt{i}_2}(\bar{x}) \prec \tilde{p}_{\mathtt{i}_1}(\bar{x}) \vee \tilde{p}_{\mathtt{i}_2}(\bar{x}) = 1 \vee \tilde{p}_{\mathtt{i}}(\bar{x}) = 1, \\ \tilde{p}_{\mathtt{i}_2}(\bar{x}) \prec \tilde{p}_{\mathtt{i}}(\bar{x}) \vee \tilde{p}_{\mathtt{i}_2}(\bar{x}) = \tilde{p}_{\mathtt{i}}(\bar{x}), \tilde{p}_{\mathtt{i}_1}(\bar{x}) \to \theta_1, \theta_2 \to \tilde{p}_{\mathtt{i}_2}(\bar{x})\end{array}\right]}$ $|\text{Consequent}| = 15 + 8 \cdot |\bar{x}| + |\tilde{p}_{\mathtt{i}_1}(\bar{x}) \to \theta_1| + |\theta_2 \to \tilde{p}_{\mathtt{i}_2}(\bar{x})| \leq 15 \cdot (1 + |\bar{x}|) + |\tilde{p}_{\mathtt{i}_1}(\bar{x}) \to \theta_1| + |\theta_2 \to \tilde{p}_{\mathtt{i}_2}(\bar{x})|$		(26)

Table 4 Unary interpolation rules for $\to$

Case	Laws	
$\theta = \theta_1 \to 0$		
Positive interpolation $\dfrac{\tilde{p}_{\mathtt{i}}(\bar{x}) \to (\theta_1 \to 0)}{(\tilde{p}_{\mathtt{i}}(\bar{x}) \to 0 \vee \tilde{p}_{\mathtt{i}_1}(\bar{x}) \to 0) \wedge (\theta_1 \to \tilde{p}_{\mathtt{i}_1}(\bar{x}))}$ $|\text{Consequent}| = 8 + 2 \cdot |\bar{x}| + |\theta_1 \to \tilde{p}_{\mathtt{i}_1}(\bar{x})| \leq 13 \cdot (1 + |\bar{x}|) + |\theta_1 \to \tilde{p}_{\mathtt{i}_1}(\bar{x})|$	(8), (9)	(27)
Positive interpolation $\dfrac{\tilde{p}_{\mathtt{i}}(\bar{x}) \to (\theta_1 \to 0)}{\{\tilde{p}_{\mathtt{i}}(\bar{x}) = 0 \vee \tilde{p}_{\mathtt{i}_1}(\bar{x}) = 0, \theta_1 \to \tilde{p}_{\mathtt{i}_1}(\bar{x})\}}$ $|\text{Consequent}| = 6 + 2 \cdot |\bar{x}| + |\theta_1 \to \tilde{p}_{\mathtt{i}_1}(\bar{x})| \leq 15 \cdot (1 + |\bar{x}|) + |\theta_1 \to \tilde{p}_{\mathtt{i}_1}(\bar{x})|$		(28)
Negative interpolation $\dfrac{(\theta_1 \to 0) \to \tilde{p}_{\mathtt{i}}(\bar{x})}{((\tilde{p}_{\mathtt{i}_1}(\bar{x}) \to 0) \to 0 \vee \tilde{p}_{\mathtt{i}}(\bar{x})) \wedge (\tilde{p}_{\mathtt{i}_1}(\bar{x}) \to \theta_1)}$ $|\text{Consequent}| = 8 + 2 \cdot |\bar{x}| + |\tilde{p}_{\mathtt{i}_1}(\bar{x}) \to \theta_1| \leq 13 \cdot (1 + |\bar{x}|) + |\tilde{p}_{\mathtt{i}_1}(\bar{x}) \to \theta_1|$	(11)	(29)
Negative interpolation $\dfrac{(\theta_1 \to 0) \to \tilde{p}_{\mathtt{i}}(\bar{x})}{\{0 \prec \tilde{p}_{\mathtt{i}_1}(\bar{x}) \vee \tilde{p}_{\mathtt{i}}(\bar{x}) = 1, \tilde{p}_{\mathtt{i}_1}(\bar{x}) \to \theta_1\}}$ $|\text{Consequent}| = 6 + 2 \cdot |\bar{x}| + |\tilde{p}_{\mathtt{i}_1}(\bar{x}) \to \theta_1| \leq 15 \cdot (1 + |\bar{x}|) + |\tilde{p}_{\mathtt{i}_1}(\bar{x}) \to \theta_1|$		(30)

Theorem 3 *Let $n_0 \in \mathbb{N}$, $\phi \in Form_{\mathcal{L}}$, $T \subseteq Form_{\mathcal{L}}$. There exist $J_T^\phi \subseteq \{(i, j) \mid i \geq n_0\}$ and $S_T^\phi \subseteq SimOrdCl_{\mathcal{L} \cup \{\tilde{p}_{\mathtt{j}} \mid \mathtt{j} \in J_T^\phi\}}$ such that*

(i) there exists an interpretation $\mathfrak{A}$ for $\mathcal{L}$ and $\mathfrak{A} \models T$, $\mathfrak{A} \not\models \phi$ if and only if there exists an interpretation $\mathfrak{A}'$ for $\mathcal{L} \cup \{\tilde{p}_{\mathtt{j}} \mid \mathtt{j} \in J_T^\phi\}$ and $\mathfrak{A}' \models S_T^\phi$, satisfying $\mathfrak{A} = \mathfrak{A}'|_{\mathcal{L}}$;

(ii) if $T \subseteq_{\mathcal{F}} Form_{\mathcal{L}}$, then $J_T^\phi \subseteq_{\mathcal{F}} \{(i, j) \mid i \geq n_0\}$, $\|J_T^\phi\| \in O(|T| + |\phi|)$; $S_T^\phi \subseteq_{\mathcal{F}} SimOrdCl_{\mathcal{L} \cup \{\tilde{p}_{\mathtt{j}} \mid \mathtt{j} \in J_T^\phi\}}$, $|S_T^\phi| \in O(|T|^2 + |\phi|^2)$; the number of all elementary operations of the translation of T and ϕ to S_T^ϕ, is in $O(|T|^2 + |\phi|^2)$; the time

Table 5 Unary interpolation rules for $\forall$ and $\exists$

Case	
$\forall x\, \theta_1$	
Positive interpolation $\dfrac{\tilde{p}_{\mathrm{i}}(\bar{x}) \rightarrow \forall x\, \theta_1}{(\tilde{p}_{\mathrm{i}}(\bar{x}) \rightarrow \forall x\, \tilde{p}_{\mathrm{i}_1}(\bar{x})) \wedge (\tilde{p}_{\mathrm{i}_1}(\bar{x}) \rightarrow \theta_1)}$ $\|\text{Consequent}\| = 6 + 2 \cdot |\bar{x}| + |\tilde{p}_{\mathrm{i}_1}(\bar{x}) \rightarrow \theta_1| \leq 13 \cdot (1 + |\bar{x}|) + |\tilde{p}_{\mathrm{i}_1}(\bar{x}) \rightarrow \theta_1|$	(31)
Positive interpolation $\dfrac{\tilde{p}_{\mathrm{i}}(\bar{x}) \rightarrow \forall x\, \theta_1}{\{\tilde{p}_{\mathrm{i}}(\bar{x}) \prec \forall x\, \tilde{p}_{\mathrm{i}_1}(\bar{x}) \vee \tilde{p}_{\mathrm{i}}(\bar{x}) \approx \forall x\, \tilde{p}_{\mathrm{i}_1}(\bar{x}),\, \tilde{p}_{\mathrm{i}_1}(\bar{x}) \rightarrow \theta_1\}}$ $\|\text{Consequent}\| = 10 + 4 \cdot |\bar{x}| + |\tilde{p}_{\mathrm{i}_1}(\bar{x}) \rightarrow \theta_1| \leq 15 \cdot (1 + |\bar{x}|) + |\tilde{p}_{\mathrm{i}_1}(\bar{x}) \rightarrow \theta_1|$	(32)
Negative interpolation $\dfrac{\forall x\, \theta_1 \rightarrow \tilde{p}_{\mathrm{i}}(\bar{x})}{(\forall x\, \tilde{p}_{\mathrm{i}_1}(\bar{x}) \rightarrow \tilde{p}_{\mathrm{i}}(\bar{x})) \wedge (\theta_1 \rightarrow \tilde{p}_{\mathrm{i}_1}(\bar{x}))}$ $\|\text{Consequent}\| = 6 + 2 \cdot |\bar{x}| + |\theta_1 \rightarrow \tilde{p}_{\mathrm{i}_1}(\bar{x})| \leq 13 \cdot (1 + |\bar{x}|) + |\theta_1 \rightarrow \tilde{p}_{\mathrm{i}_1}(\bar{x})|$	(33)
Negative interpolation $\dfrac{\forall x\, \theta_1 \rightarrow \tilde{p}_{\mathrm{i}}(\bar{x})}{\{\forall x\, \tilde{p}_{\mathrm{i}_1}(\bar{x}) \prec \tilde{p}_{\mathrm{i}}(\bar{x}) \vee \forall x\, \tilde{p}_{\mathrm{i}_1}(\bar{x}) \approx \tilde{p}_{\mathrm{i}}(\bar{x}),\, \theta_1 \rightarrow \tilde{p}_{\mathrm{i}_1}(\bar{x})\}}$ $\|\text{Consequent}\| = 10 + 4 \cdot |\bar{x}| + |\theta_1 \rightarrow \tilde{p}_{\mathrm{i}_1}(\bar{x})| \leq 15 \cdot (1 + |\bar{x}|) + |\theta_1 \rightarrow \tilde{p}_{\mathrm{i}_1}(\bar{x})|$	(34)
$\exists x\, \theta_1$	
Positive interpolation $\dfrac{\tilde{p}_{\mathrm{i}}(\bar{x}) \rightarrow \exists x\, \theta_1}{(\tilde{p}_{\mathrm{i}}(\bar{x}) \rightarrow \exists x\, \tilde{p}_{\mathrm{i}_1}(\bar{x})) \wedge (\tilde{p}_{\mathrm{i}_1}(\bar{x}) \rightarrow \theta_1)}$ $\|\text{Consequent}\| = 6 + 2 \cdot |\bar{x}| + |\tilde{p}_{\mathrm{i}_1}(\bar{x}) \rightarrow \theta_1| \leq 13 \cdot (1 + |\bar{x}|) + |\tilde{p}_{\mathrm{i}_1}(\bar{x}) \rightarrow \theta_1|$	(35)
Positive interpolation $\dfrac{\tilde{p}_{\mathrm{i}}(\bar{x}) \rightarrow \exists x\, \theta_1}{\{\tilde{p}_{\mathrm{i}}(\bar{x}) \prec \exists x\, \tilde{p}_{\mathrm{i}_1}(\bar{x}) \vee \tilde{p}_{\mathrm{i}}(\bar{x}) \approx \exists x\, \tilde{p}_{\mathrm{i}_1}(\bar{x}),\, \tilde{p}_{\mathrm{i}_1}(\bar{x}) \rightarrow \theta_1\}}$ $\|\text{Consequent}\| = 10 + 4 \cdot |\bar{x}| + |\tilde{p}_{\mathrm{i}_1}(\bar{x}) \rightarrow \theta_1| \leq 15 \cdot (1 + |\bar{x}|) + |\tilde{p}_{\mathrm{i}_1}(\bar{x}) \rightarrow \theta_1|$	(36)
Negative interpolation $\dfrac{\exists x\, \theta_1 \rightarrow \tilde{p}_{\mathrm{i}}(\bar{x})}{(\exists x\, \tilde{p}_{\mathrm{i}_1}(\bar{x}) \rightarrow \tilde{p}_{\mathrm{i}}(\bar{x})) \wedge (\theta_1 \rightarrow \tilde{p}_{\mathrm{i}_1}(\bar{x}))}$ $\|\text{Consequent}\| = 6 + 2 \cdot |\bar{x}| + |\theta_1 \rightarrow \tilde{p}_{\mathrm{i}_1}(\bar{x})| \leq 13 \cdot (1 + |\bar{x}|) + |\theta_1 \rightarrow \tilde{p}_{\mathrm{i}_1}(\bar{x})|$	(37)
Negative interpolation $\dfrac{\exists x\, \theta_1 \rightarrow \tilde{p}_{\mathrm{i}}(\bar{x})}{\{\exists x\, \tilde{p}_{\mathrm{i}_1}(\bar{x}) \prec \tilde{p}_{\mathrm{i}}(\bar{x}) \vee \exists x\, \tilde{p}_{\mathrm{i}_1}(\bar{x}) \approx \tilde{p}_{\mathrm{i}}(\bar{x}),\, \theta_1 \rightarrow \tilde{p}_{\mathrm{i}_1}(\bar{x})\}}$ $\|\text{Consequent}\| = 10 + 4 \cdot |\bar{x}| + |\theta_1 \rightarrow \tilde{p}_{\mathrm{i}_1}(\bar{x})| \leq 15 \cdot (1 + |\bar{x}|) + |\theta_1 \rightarrow \tilde{p}_{\mathrm{i}_1}(\bar{x})|$	(38)

and space complexity of the translation of T and ϕ to S_T^ϕ, is in $O(|T|^2 \cdot \log(1 + n_0 + |T|) + |\phi|^2 \cdot (\log(1 + n_0) + \log|\phi|))$;

(iii) S_T^ϕ is admissible.

Proof cf. http://ii.fmph.uniba.sk/~guller/sci14.pdf, Sect. 2.4. $\qquad\square$

Corollary 1 *Let* $n_0 \in \mathbb{N}$, $\phi \in Form_{\mathcal{L}}$, $T \subseteq Form_{\mathcal{L}}$. *There exist* $J_T^\phi \subseteq \{(i, j) \mid i \geq n_0\}$ *and* $S_T^\phi \subseteq SimOrdCl_{\mathcal{L} \cup \{\tilde{p}_j \mid j \in J_T^\phi\}}$ *such that*

(i) $T \models \phi$ *if and only if* S_T^ϕ *is unsatisfiable;*

(ii) *if* $T \subseteq_{\mathcal{F}} Form_{\mathcal{L}}$, *then* $J_T^\phi \subseteq_{\mathcal{F}} \{(i, j) \mid i \geq n_0\}$, $\|J_T^\phi\| \in O(|T| + |\phi|)$; $S_T^\phi \subseteq_{\mathcal{F}} SimOrdCl_{\mathcal{L} \cup \{\tilde{p}_j \mid j \in J_T^\phi\}}$, $|S_T^\phi| \in O(|T|^2 + |\phi|^2)$; *the number of all elementary operations of the translation of* T *and* ϕ *to* S_T^ϕ, *is in* $O(|T|^2 + |\phi|^2)$; *the time* and space complexity of the translation of T and ϕ to S_T^ϕ, is in $O(|T|^2 \cdot \log(1 + n_0 + |T|) + |\phi|^2 \cdot (\log(1 + n_0) + \log|\phi|))$;

(iii) S_T^ϕ is admissible.

Proof An immediate consequence of Theorem 3. $\qquad\square$

2.5 Order Hyperresolution Rules

At first, we introduce some basic notions and notation concerning chains of order literals. A chain Ξ of $\mathcal{L}$ is a sequence $\Xi = \varepsilon_0 \diamond_0 \upsilon_0, \ldots, \varepsilon_n \diamond_n \upsilon_n, \varepsilon_i \diamond_i \upsilon_i \in OrdLit_{\mathcal{L}}$, such that for all $i < n$, $\upsilon_i = \varepsilon_{i+1}$. ε_0 is the beginning element of Ξ and υ_n the ending element of Ξ. $\varepsilon_0 \, \Xi \, \upsilon_n$ denotes Ξ together with its respective beginning and ending element. Let $\Xi = \varepsilon_0 \diamond_0 \upsilon_0, \ldots, \varepsilon_n \diamond_n \upsilon_n$ be a chain of $\mathcal{L}$. Ξ is an equality chain of $\mathcal{L}$ iff, for all $i \leq n$, $\diamond_i = \text{=}$. Ξ is an increasing chain of $\mathcal{L}$ iff there exists $i^* \leq n$ such that $\diamond_{i^*} = \prec$. Ξ is a contradiction of $\mathcal{L}$ iff Ξ is an increasing chain of $\mathcal{L}$ of the form $\varepsilon_0 \, \Xi \, 0$ or $1 \, \Xi \, \upsilon_n$ or $\varepsilon_0 \, \Xi \, \varepsilon_0$. Let $S \subseteq OrdCl_{\mathcal{L}}$ be unit and $\Xi = \varepsilon_0 \diamond_0 \upsilon_0, \ldots, \varepsilon_n \diamond_n \upsilon_n$ be a chain | an equality chain | an increasing chain | a contradiction of $\mathcal{L}$. Ξ is a chain | an equality chain | an increasing chain | a contradiction of S iff, for all $i \leq n$, $\varepsilon_i \diamond_i \upsilon_i \in S$.

Let $\widetilde{W} = \{\tilde{w}_i \mid i \in \mathbb{I}\}$ such that $\widetilde{W} \cap (Func_{\mathcal{L}} \cup \{\tilde{f}_0\}) = \emptyset$; $\widetilde{W}$ is an infinite countable set of new function symbols. Let $\mathcal{L}$ contain a constant (nullary function) symbol. Let $P \subseteq \widetilde{\mathbb{P}}$ and $S \subseteq OrdCl_{\mathcal{L} \cup P}$. We denote $GOrdCl_{\mathcal{L}} = \{C \mid C \in OrdCl_{\mathcal{L}} \text{ is closed}\} \subseteq OrdCl_{\mathcal{L}}$ and $GInst_{\mathcal{L}}(S) = \{C \mid C \in GOrdCl_{\mathcal{L}} \text{ is an instance of } S \text{ of } \mathcal{L}\} \subseteq GOrdCl_{\mathcal{L}}$. A basic order hyperresolution calculus is defined as follows:

$$\text{(Basic order hyperresolution rule)} \qquad (39)$$

$$\frac{l_0 \vee C_0, \ldots, l_n \vee C_n \in S_{\kappa-1}}{\bigvee_{i=0}^{n} C_i \in S_{\kappa}};$$

$l_0, \ldots, l_n$ is a contradiction of $\mathcal{L}_{\kappa-1}$.

$$\text{(Basic order trichotomy rule)} \qquad (40)$$

$$\frac{a, b \in atoms(S_{\kappa-1}) - \{0, 1\}, qatoms(S_{\kappa-1}) \neq \emptyset}{a \prec b \vee a = b \vee b \prec a \in S_{\kappa}}.$$

$$\text{(Basic order } \forall - \text{quantification rule)} \qquad (41)$$

$$\frac{\forall x \, a \in qatoms^{\forall}(S_{\kappa-1})}{\forall x \, a \prec a\gamma \vee \forall x \, a = a\gamma \in S_{\kappa}};$$

$t \in GTerm_{\mathcal{L}_{\kappa-1}}, \gamma = x/t, dom(\gamma) = \{x\} = vars(a)$.

$$\text{(\textit{Basic order} } \exists - \textit{quantification rule})\qquad(42)$$

$$\frac{\exists x\, a \in qatoms^{\exists}(S_{\kappa-1})}{a\gamma \prec \exists x\, a \vee a\gamma \doteq \exists x\, a \in S_{\kappa}};$$

$$t \in GTerm_{\mathcal{L}_{\kappa-1}}, \gamma = x/t, dom(\gamma) = \{x\} = vars(a).$$

$$\text{(\textit{Basic order} } \forall - \textit{witnessing rule})\qquad(43)$$

$$\frac{\forall x\, a \in qatoms^{\forall}(S_{\kappa-1}), b \in atoms(S_{\kappa-1}) \cup qatoms(S_{\kappa-1})}{a\gamma \prec b \vee b \doteq \forall x\, a \vee b \prec \forall x\, a \in S_{\kappa}};$$

$$\tilde{w} \in \tilde{\mathbb{W}}, \tilde{w} \notin Func_{\mathcal{L}_{\kappa-1}}, ar(\tilde{w}) = |freetermseq(\forall x\, a), freetermseq(b)|,$$

$$\gamma = x/\tilde{w}(freetermseq(\forall x\, a), freetermseq(b)), dom(\gamma) = \{x\} = freevars(a).$$

$$\text{(\textit{Basic order} } \exists - \textit{witnessing rule})\qquad(44)$$

$$\frac{\exists x\, a \in qatoms^{\exists}(S_{\kappa-1}), b \in atoms(S_{\kappa-1}) \cup qatoms(S_{\kappa-1})}{b \prec a\gamma \vee \exists x\, a \doteq b \vee \exists x\, a \prec b \in S_{\kappa}};$$

$$\tilde{w} \in \tilde{\mathbb{W}}, \tilde{w} \notin Func_{\mathcal{L}_{\kappa-1}}, ar(\tilde{w}) = |freetermseq(\exists x\, a), freetermseq(b)|,$$

$$\gamma = x/\tilde{w}(freetermseq(\exists x\, a), freetermseq(b)), dom(\gamma) = \{x\} = freevars(a).$$

The basic order hyperresolution calculus can be generalised to an order hyperresolution one.

$$\text{(\textit{Order hyperresolution rule})}\qquad(45)$$

$$\frac{\bigvee_{j=0}^{k_0} \varepsilon_j^0 \diamond_j^0 v_j^0 \vee \bigvee_{j=1}^{m_0} l_j^0, \ldots, \bigvee_{j=0}^{k_n} \varepsilon_j^n \diamond_j^n v_j^n \vee \bigvee_{j=1}^{m_n} l_j^n \in S_{\kappa-1}^{Vr}}{\left(\bigvee_{i=0}^{n} \bigvee_{j=1}^{m_i} l_j^i \right)\theta \in S_{\kappa}};$$

for all $i < i' \leq n$,

$$freevars\left(\bigvee_{j=0}^{k_i} \varepsilon_j^i \diamond_j^i v_j^i \vee \bigvee_{j=1}^{m_i} l_j^i \right) \cap freevars\left(\bigvee_{j=0}^{k_{i'}} \varepsilon_j^{i'} \diamond_j^{i'} v_j^{i'} \vee \bigvee_{j=1}^{m_{i'}} l_j^{i'} \right) = \emptyset,$$

$$\theta \in mgu_{\mathcal{L}_{\kappa-1}}\left(\bigvee_{j=0}^{k_0} \varepsilon_j^0 \diamond_j^0 v_j^0, l_1^0, \ldots, l_{m_0}^0, \ldots, \bigvee_{j=0}^{k_n} \varepsilon_j^n \diamond_j^n v_j^n, l_1^n, \ldots, l_{m_n}^n, \right.$$

$$\left. \{v_0^0, \varepsilon_0^1\}, \ldots, \{v_0^{n-1}, \varepsilon_0^n\}, \{a, b\} \right),$$

$$dom(\theta) = freevars(\{\varepsilon_j^i \diamond_j^i v_j^i \mid j \leq k_i, i \leq n\}, \{l_j^i \mid 1 \leq j \leq m_i, i \leq n\}),$$

$$a = \varepsilon_0^0, b = 1 \text{ or } a = v_0^n, b = 0 \text{ or } a = v_0^n, b = \varepsilon_0^0,$$

there exists $i^* \leq n$ such that $\diamond_0^{i^*} = \prec$.

$$(Order\ trichotomy\ rule) \qquad (46)$$

$$\frac{a, b \in atoms(S_{\kappa-1})^{Vr} - \{0, 1\}, qatoms(S_{\kappa-1}) \neq \emptyset}{a \prec b \vee a \doteq b \vee b \prec a \in S_\kappa};$$

$vars(a) \cap vars(b) = \emptyset.$

$$(Order\ \forall - quantification\ rule) \qquad (47)$$

$$\frac{\forall x\, a \in qatoms^\forall(S_{\kappa-1})}{\forall x\, a \prec a \vee \forall x\, a \doteq a \in S_\kappa}.$$

$$(Order\ \exists - quantification\ rule) \qquad (48)$$

$$\frac{\exists x\, a \in qatoms^\exists(S_{\kappa-1})}{a \prec \exists x\, a \vee a \doteq \exists x\, a \in S_\kappa}.$$

$$(Order\ \forall - witnessing\ rule) \qquad (49)$$

$$\frac{\forall x\, a \in qatoms^\forall(S_{\kappa-1}^{Vr}), b \in atoms(S_{\kappa-1}^{Vr}) \cup qatoms(S_{\kappa-1}^{Vr})}{a\gamma \prec b \vee b \doteq \forall x\, a \vee b \prec \forall x\, a \in S_\kappa};$$

$freevars(\forall x\, a) \cap freevars(b) = \emptyset,$

$\tilde{w} \in \tilde{\mathbb{W}}, \tilde{w} \notin Func_{\mathcal{L}_{\kappa-1}}, ar(\tilde{w}) = |freetermseq(\forall x\, a), freetermseq(b)|,$

$\gamma = x/\tilde{w}(freetermseq(\forall x\, u), freetermseq(b)) \cup id|_{vars(a)-\{x\}},$

$dom(\gamma) = \{x\} \cup (vars(a) - \{x\}) = vars(a).$

$$(Order\ \exists - witnessing\ rule) \qquad (50)$$

$$\frac{\exists x\, a \in qatoms^\exists(S_{\kappa-1}^{Vr}), b \in atoms(S_{\kappa-1}^{Vr}) \cup qatoms(S_{\kappa-1}^{Vr})}{b \prec a\gamma \vee \exists x\, a \doteq b \vee \exists x\, a \prec b \in S_\kappa};$$

$freevars(\exists x\, a) \cap freevars(b) = \emptyset,$

$\tilde{w} \in \tilde{\mathbb{W}}, \tilde{w} \notin Func_{\mathcal{L}_{\kappa-1}}, ar(\tilde{w}) = |freetermseq(\exists x\, a), freetermseq(b)|,$

$\gamma = x/\tilde{w}(freetermseq(\exists x\, a), freetermseq(b)) \cup id|_{vars(a)-\{x\}},$

$dom(\gamma) = \{x\} \cup (vars(a) - \{x\}) = vars(a).$

Let $\mathcal{L}_0 = \mathcal{L} \cup P$ and $S_0 = \emptyset \subseteq GOrdCl_{\mathcal{L}_0} \mid OrdCl_{\mathcal{L}_0}$. Let $\mathcal{D} = C_1, \ldots, C_n$, $C_\kappa \in GOrdCl_{\mathcal{L} \cup \tilde{\mathbb{W}} \cup P} \mid OrdCl_{\mathcal{L} \cup \tilde{\mathbb{W}} \cup P}, n \geq 1$. $\mathcal{D}$ is a deduction of C_n from S by basic order hyperresolution iff, for all $1 \leq \kappa \leq n$, $C_\kappa \in \{0 \prec 1\} \cup GInst_{\mathcal{L}_{\kappa-1}}(S)$, or there exist $1 \leq j_k^* \leq \kappa - 1$, $k = 1, \ldots, m$, such that C_κ is a basic order resolvent of $C_{j_1^*}, \ldots, C_{j_m^*}$ using Rules (39)–(44); $\mathcal{D}$ is a deduction of C_n from S by order hyperresolution iff, for all $1 \leq \kappa \leq n$, $C_\kappa \in \{0 \prec 1\} \cup S$, or there exist $1 \leq j_k^* \leq \kappa - 1$, $k = 1, \ldots, m$, such that C_κ is an order resolvent of $C'_{j_1^*}, \ldots, C'_{j_m^*}$ using Rules (45)–(50) where $C'_{j_k^*}$ is a variant of $C_{j_k^*}$ of $\mathcal{L}_{\kappa-1}$; $\mathcal{L}_\kappa$ and S_κ are defined by recursion on $1 \leq \kappa \leq n$ as follows:

$$
\mathcal{L}_\kappa = \begin{cases} \mathcal{L}_{\kappa-1} \cup \{\tilde{w}\} & \text{in case of Rule (43), (44) | (49), (50),} \\ \mathcal{L}_{\kappa-1} & \text{else;} \end{cases}
$$

$$
S_\kappa = S_{\kappa-1} \cup \{C_\kappa\} \subseteq GOrdCl_{\mathcal{L}_\kappa} \mid OrdCl_{\mathcal{L}_\kappa},
$$

$$
S_\kappa^{Vr} = Vrnt_{\mathcal{L}_\kappa}(S_\kappa) \subseteq OrdCl_{\mathcal{L}_\kappa}.
$$

$\mathcal{D}$ is a refutation of S iff $C_n = \square$. We denote

$$
clo^{\mathcal{BH}}(S) = \{C \mid \text{there exists a deduction of } C \text{ from } S
$$
$$
\text{by basic order hyperresolution}\} \subseteq GOrdCl_{\mathcal{L} \cup \tilde{\mathbb{W}} \cup P},
$$
$$
clo^{\mathcal{H}}(S) = \{C \mid \text{there exists a deduction of } C \text{ from } S
$$
$$
\text{by order hyperresolution}\} \subseteq OrdCl_{\mathcal{L} \cup \tilde{\mathbb{W}} \cup P}.
$$

Lemma 2 (Lifting Lemma) *Let $\mathcal{L}$ contain a constant symbol. Let $P \subseteq \tilde{\mathbb{P}}$ and $S \subseteq OrdCl_{\mathcal{L} \cup P}$. Let $C \in clo^{\mathcal{BH}}(S)$. There exists $C^* \in clo^{\mathcal{H}}(S)$ such that C is an instance of C^* of $\mathcal{L} \cup \tilde{\mathbb{W}} \cup P$.*

Proof Straightforward. □

Lemma 3 (Reduction Lemma) *Let $\mathcal{L}$ contain a constant symbol. Let $P \subseteq \tilde{\mathbb{P}}$ and $S \subseteq OrdCl_{\mathcal{L} \cup P}$. Let $\{\bigvee_{j=0}^{k_i} \varepsilon_j^i \diamond_j^i v_j^i \vee C_i \mid i \leq n\} \subseteq clo^{\mathcal{BH}}(S)$ such that for all $\mathcal{S} \in Sel(\{\{j \mid j \leq k_i\}_i \mid i \leq n\})$, there exists a contradiction of $\{\varepsilon_{\mathcal{S}(i)}^i \diamond_{\mathcal{S}(i)}^i v_{\mathcal{S}(i)}^i \mid i \leq n\}$. There exists $\emptyset \neq I^* \subseteq \{i \mid i \leq n\}$ such that $\bigvee_{i \in I^*} C_i \in clo^{\mathcal{BH}}(S)$.*

Proof Straightforward. □

Lemma 4 (Unit Lemma) *Let $\mathcal{L}$ contain a constant symbol. Let $P \subseteq \tilde{\mathbb{P}}$ and $S \subseteq OrdCl_{\mathcal{L} \cup P}$. Let $\square \notin clo^{\mathcal{BH}}(S) = \{\bigvee_{j=0}^{k_\iota} \varepsilon_j^\iota \diamond_j^\iota v_j^\iota \mid \iota < \gamma\}$, $\gamma \leq \omega$. There exists $\mathcal{S}^* \in Sel(\{\{j \mid j \leq k_\iota\}_\iota \mid \iota < \gamma\})$ such that there does not exist a contradiction of $\{\varepsilon_{\mathcal{S}^*(\iota)}^\iota \diamond_{\mathcal{S}^*(\iota)}^\iota v_{\mathcal{S}^*(\iota)}^\iota \mid \iota < \gamma\}$.*

Proof An immediate consequence of König's Lemma and Lemma 3. □

We are in position to prove the refutational soundness and completeness of the order hyperresolution calculus.

Theorem 4 (Refutational Soundness and Completeness) *Let $\mathcal{L}$ contain a constant symbol. Let $P \subseteq \tilde{\mathbb{P}}$ and $S \subseteq OrdCl_{\mathcal{L} \cup P}$. $\square \in clo^{\mathcal{H}}(S)$ if and only if S is unsatisfiable.*

Proof cf. http://ii.fmph.uniba.sk/~guller/sci14.pdf, Sect. 2.5. □

3 A DPLL Procedure

3.1 Propositional Product Logic

We shall use the standard notions and notation of the propositional Product logic.[5] The set of propositional atoms of Product logic will be denoted as *PropAtom*. By *PropForm* we designate the set of all propositional formulae of Product logic built up from *PropAtom* using the propositional constants *0*, false, *1*, true, and the connectives: ¬, negation, ∧, conjunction, ∨, disjunction, &, strong conjunction, →, implication. In addition, we introduce new binary connectives =, equality, and ≺, strict order. By *OrdPropForm* we designate the set of all so-called order propositional formulae of Product logic built up from *PropAtom* using the propositional constants *0*, *1*, and the connectives: ¬, ∧, ∨, &, →, =, ≺.[6] We shall assume that *PropAtom* is countable; hence, *PropForm* and *OrdPropForm* are countable.

Product logic is interpreted by the standard Π-algebra augmented by binary operators $\rightleftharpoons$ and $\prec$ for = and ≺, respectively.

$$\Pi = ([0, 1], \leq, \vee, \wedge, \cdot, \Rightarrow, {}^{-}, \rightleftharpoons, \prec, 0, 1)$$

where $\vee \mid \wedge$ denotes the supremum | infimum operator on $[0, 1]$;

$$a \Rightarrow b = \begin{cases} 1 & \text{if } a \leq b, \\ \frac{b}{a} & \text{else}; \end{cases} \qquad \overline{a} = \begin{cases} 1 & \text{if } a = 0, \\ 0 & \text{else}; \end{cases}$$

$$a \rightleftharpoons b = \begin{cases} 1 & \text{if } a = b, \\ 0 & \text{else}; \end{cases} \qquad a \prec b = \begin{cases} 1 & \text{if } a < b, \\ 0 & \text{else}. \end{cases}$$

We recall that Π is a complete linearly ordered lattice algebra; $\vee \mid \wedge$ is commutative, associative, idempotent, monotone; $0 \mid 1$ is its neutral element; $\cdot$ is commutative, associative, monotone; 1 is its neutral element; the residuum operator $\Rightarrow$ of $\cdot$ satisfies the condition of residuation:

$$\text{for all } a, b, c \in \Pi, a \cdot b \leq c \Longleftrightarrow a \leq b \Rightarrow c; \tag{51}$$

Product (Gödel) negation $^{-}$ satisfies the condition:

$$\text{for all } a \in \Pi, \overline{a} = a \Rightarrow 0; \tag{52}$$

the following properties, which will be exploited later, hold[7]:
for all a, b, c ∈ Π,

---

[5] cf. http://ii.fmph.uniba.sk/~guller/sci14.pdf, Sect. 3.1.
[6] We assume a decreasing connective precedence: ¬, &, ∧, →, =, ≺, ∨.
[7] We assume a decreasing operator precedence: $^{-}$, $\cdot$, ∧, $\Rightarrow$, $\rightleftharpoons$, $\prec$, ∨.

$$a \vee b \wedge c = (a \vee b) \wedge (a \vee c), \quad \text{(distributivity of } \vee \text{ over } \wedge) \tag{53}$$

$$a \wedge (b \vee c) = a \wedge b \vee a \wedge c, \quad \text{(distributivity of } \wedge \text{ over } \vee) \tag{54}$$

$$a \cdot (b \vee c) = a \cdot b \vee a \cdot c, \quad \text{(distributivity of } \cdot \text{ over } \vee) \tag{55}$$

$$a \Rightarrow (b \vee c) = a \Rightarrow b \vee a \Rightarrow c, \tag{56}$$

$$a \Rightarrow b \wedge c = (a \Rightarrow b) \wedge (a \Rightarrow c), \tag{57}$$

$$(a \vee b) \Rightarrow c = (a \Rightarrow c) \wedge (b \Rightarrow c), \tag{58}$$

$$a \wedge b \Rightarrow c = a \Rightarrow c \vee b \Rightarrow c, \tag{59}$$

$$a \Rightarrow (b \Rightarrow c) = a \cdot b \Rightarrow c, \tag{60}$$

$$((a \Rightarrow b) \Rightarrow b) \Rightarrow b = a \Rightarrow b. \tag{61}$$

3.2 Translation to Order Clausal Form

We now describe some translation of a formula to a finite order clausal theory. To have the output theory of polynomial size, our translation exploits interpolation using new atoms. The output theory will be of linearithmic size at the cost of being only equivalent satisfiable to the input formula. A similar approach exploiting the renaming subformulae technique can be found in [3–7]. At first, we introduce notions of a to the power of n and of conjunction of propositional atoms. Let $a \in PropAtom$ and $n > 0$. a to the power of n is the pair (a, n), written as a^n. The power a^1 is denoted as a; if it does not cause the ambiguity with the denotation of the single propositional atom a in given context. We define the size of a^n as $|a^n| = n > 0$. A conjunction Cn of propositional atoms is a non-empty finite set of powers such that for all $a^m \neq b^n \in Cn$, $a \neq b$. A conjunction $\{a_0^{m_0}, \ldots, a_n^{m_n}\}$ of propositional atoms is written in the form $a_0^{m_0} \& \cdots \& a_n^{m_n}$. A conjunction $\{p\}$ of propositional atoms is called a unit conjunction of propositional atoms and denoted as p. The set of all conjunctions of propositional atoms is designated as $PropConj$. Let $\mathcal{V}$ be a (partial) valuation; p be a power, $Cn \in PropConj$, $Cn_1, Cn_2 \in PropConj \cup \{\emptyset\}$. Let $atoms(Cn) \subseteq dom(\mathcal{V})$ in case of $\mathcal{V}$ being a partial valuation. The truth value of $Cn = a_0^{m_0} \& \cdots \& a_n^{m_n}$ in $\mathcal{V}$ is defined by

$$\|Cn\|^{\mathcal{V}} = \underbrace{\|a_0\|^{\mathcal{V}} \cdots \cdot \|a_0\|^{\mathcal{V}}}_{m_0} \cdots \cdot \underbrace{\|a_n\|^{\mathcal{V}} \cdots \cdot \|a_n\|^{\mathcal{V}}}_{m_n}.$$

We define the size of Cn as $|Cn| = \sum_{p \in Cn} |p| > 0$. By $p \& Cn$ we denote $\{p\} \cup Cn$ where $p \notin Cn$. Cn_1 is a subconjunction of Cn_2, in symbols $Cn_1 \sqsubseteq Cn_2$, iff, for all $a^m \in Cn_1$, there exists $a^n \in Cn_2$ such that $m \leq n$. We define $Cn_1 \sqcap Cn_2 = \{a^{min(m,n)} \mid a^m \in Cn_1, a^n \in Cn_2\} \in PropConj \cup \{\emptyset\}$. Cn_1 and Cn_2 are disjoint iff

$Cn_1 \sqcap Cn_2 = \emptyset$. We finally introduce order clauses in Product logic. l is an order literal of Product logic iff $l = \varepsilon_1 \diamond \varepsilon_2$ where $\varepsilon_1 \in PropAtom \cup \{0, 1\}$, $\varepsilon_2 \in \{0, 1\}$, or $\varepsilon_1 \in \{0, 1\}$, $\varepsilon_2 \in PropAtom \cup \{0, 1\}$, or $\varepsilon_i \in PropConj$, $\varepsilon_1 \sqcap \varepsilon_2 = \emptyset$, $\diamond \in \{=, \prec\}$. The set of all order literals of Product logic is designated as $OrdLit$. Let $l = \varepsilon_1 \diamond \varepsilon_2 \in OrdLit$. We define the size of l as $|l| = 1 + |\varepsilon_1| + |\varepsilon_2| > 0$. An order clause of Product logic is a finite set of order literals of Product logic. An order clause $\{l_1, \ldots, l_n\}$ is written in the form $l_1 \vee \cdots \vee l_n$. The order clause $\emptyset$ is called the empty order clause and denoted as $\square$. An order clause $\{l\}$ is called a unit order clause and denoted as l. We designate the set of all order clauses of Product logic as $OrdCl$. Let $l, l_0, \ldots, l_n \in OrdLit$ and $C, C' \in OrdCl_{\mathcal{L}}$. We define the size of C as $|C| = \sum_{l \in C} |l|$. By $l \vee C$ we denote $\{l\} \cup C$ where $l \notin C$. Analogously, by $l_0 \vee \cdots \vee l_n \vee C$ we denote $\{l_0\} \cup \cdots \cup \{l_n\} \cup C$ where, for all $i, i' \leq n$, $i \neq i'$, $l_i \notin C$ and $l_i \neq l_{i'}$. By $C \vee C'$ we denote $C \cup C'$. C is a subclause of C', in symbols $C \sqsubseteq C'$, iff $C \subseteq C'$. An order clausal theory is a set of order clauses. A unit order clausal theory is a set of unit order clauses.

Let $\phi, \phi' \in PropOrdForm, T, T' \subseteq PropOrdForm, S, S' \subseteq OrdCl, \mathcal{V}$ be a (partial) valuation. Let $atoms(C), atoms(S) \subseteq dom(\mathcal{V})$ in case of $\mathcal{V}$ being a partial valuation. $\mathcal{V}$ is a (partial) propositional model of C, in symbols $\mathcal{V} \models C$, iff there exists $l^* \in C$ such that $\mathcal{V} \models l^*$. $\mathcal{V}$ is a (partial) propositional model of S, in symbols $\mathcal{V} \models S$, iff, for all $C \in S, \mathcal{V} \models C$. $\phi' \mid T' \mid C' \mid S'$ is a propositional consequence of $\phi \mid T \mid C \mid S$, in symbols $\phi \mid T \mid C \mid S \models_P \phi' \mid T' \mid C' \mid S'$, iff, for every propositional model $\mathcal{V}$ of $\phi \mid T \mid C \mid S, \mathcal{V} \models \phi' \mid T' \mid C' \mid S'$. $\phi \mid T \mid C \mid S$ is satisfiable iff there exists a propositional model of $\phi \mid T \mid C \mid S$. $\phi \mid T \mid C \mid S$ is equisatisfiable to $\phi' \mid T' \mid C' \mid S'$ iff $\phi \mid T \mid C \mid S$ is satisfiable if and only if $\phi' \mid T' \mid C' \mid S'$ is satisfiable. Let $S \subseteq_{\mathcal{F}} OrdCl$. We define the size of S as $|S| = \sum_{C \in S} |C|$. Let $l \in OrdLit$. l is a simplified order literal of Product logic iff $l = \varepsilon_1 \diamond \varepsilon_2$ where $\varepsilon_1 \in PropAtom, \varepsilon_2 \in \{0, 1\}$, or $\varepsilon_1 \in \{0, 1\}$, $\varepsilon_2 \in PropAtom$, or $\varepsilon_i \in PropAtom$, or $\varepsilon_1 \in PropAtom, \varepsilon_2 = a\& b \in PropConj$, or $\varepsilon_1 = a\& b \in PropConj, \varepsilon_2 \in PropAtom, a, b \in PropAtom$. The set of all simplified order literals of Product logic is designated as $SimOrdLit$. We denote $SimOrdCl = \{C \mid C \in OrdCl, C \subseteq SimOrdLit\}$. Let $\tilde{\mathbb{A}} = \{\tilde{a}_i \mid i \in \mathbb{I}\} \subseteq PropAtom$ such that $PropAtom - \tilde{\mathbb{A}}$ is infinite; $\tilde{\mathbb{A}}$ is an infinite countable set of new propositional atoms. Let E be a set of expressions and $A \subseteq \tilde{\mathbb{A}}$. We denote $\mathsf{E}_A = \{\varepsilon \mid \varepsilon \in \mathsf{E}, atoms(\varepsilon) \cap \tilde{\mathbb{A}} \subseteq A\}$. The translation to order clausal form is based on the following lemma.

Lemma 5 *Let $n_\phi, n_0 \in \mathbb{N}, \phi \in PropForm_\emptyset, T \subseteq PropForm_\emptyset$.*

(I) There exist either $J_\phi = \emptyset$ or $J_\phi = \{(n_\phi, j) \mid j \leq n_{J_\phi}\}, J_\phi \subseteq \{(n_\phi, j) \mid j \in \mathbb{N}\}$, and $S_\phi \subseteq_{\mathcal{F}} SimOrdCl_{\{\tilde{a}_j \mid j \in J_\phi\}}$ such that

(a) $\|J_\phi\| \leq 2 \cdot |\phi|$;

(b) there exists a partial valuation $\mathcal{V}, dom(\mathcal{V}) = atoms(\phi)$, and $\mathcal{V} \models \phi$ if and only if there exists a partial valuation $\mathcal{V}', dom(\mathcal{V}') = atoms(\phi) \cup \{\tilde{a}_j \mid j \in J_\phi\}$, and $\mathcal{V}' \models S_\phi$, satisfying $\mathcal{V} = \mathcal{V}'|_{atoms(\phi)}$;

(c) $|S_\phi| \in O(|\phi|)$; *the number of all elementary operations of the translation of* ϕ *to* S_ϕ, *is in* $O(|\phi|)$; *the time and space complexity of the translation of* ϕ *to* S_ϕ, *is in* $O(|\phi| \cdot (\log(1 + n_\phi) + \log |\phi|))$.

(II) *There exist* $J_T \subseteq \{(i, j) \,|\, i \geq n_0\}$ *and* $S_T \subseteq SimOrdCl_{\{\tilde{a}_j \,|\, j \in J_T\}}$ *such that*

(a) *there exists a partial valuation* $\mathcal{V}$, $dom(\mathcal{V}) = atoms(T)$, *and* $\mathcal{V} \models T$ *if and only if there exists a partial valuation* $\mathcal{V}'$, $dom(\mathcal{V}') = atoms(T) \cup \{\tilde{a}_j \,|\, j \in J_T\}$, *and* $\mathcal{V}' \models S_T$, *satisfying* $\mathcal{V} = \mathcal{V}'|_{atoms(T)}$;

(b) *if* $T \subseteq_\mathcal{F} Form_\mathcal{L}$, *then* $J_T \subseteq_\mathcal{F} \{(i, j) \,|\, i \geq n_0\}$, $\|J_T\| \leq 2 \cdot |T|$; $S_T \subseteq_\mathcal{F}$ $SimOrdCl_{\{\tilde{a}_j \,|\, j \in J_T\}}$, $|S_T| \in O(|T|)$; *the number of all elementary operations of the translation of* T *to* S_T, *is in* $O(|T|)$; *the time and space complexity of the translation of* T *to* S_T, *is in* $O(|T| \cdot \log(1 + n_0 + |T|))$.

Proof Technical, using the interpolation rules in Table 6.
cf. http://ii.fmph.uniba.sk/~guller/sci14.pdf, Sect. 3.2. □
We conclude this section by the following theorem.

Theorem 5 *Let* $n_0 \in \mathbb{N}$, $\phi \in PropForm_\emptyset$, $T \subseteq PropForm_\emptyset$. *There exist* $J_T^\phi \subseteq \{(i, j) \,|\, i \geq n_0\}$ *and* $S_T^\phi \subseteq SimOrdCl_{\{\tilde{a}_j \,|\, j \in J_T^\phi\}}$ *such that*

(i) $T \models_P \phi$ *if and only if* S_T^ϕ *is unsatisfiable;*

(ii) *if* $T \subseteq_\mathcal{F} PropForm_\emptyset$, *then* $J_T^\phi \subseteq_\mathcal{F} \{(i, j) \,|\, i \geq n_0\}$, $\|J_T^\phi\| \in O(|T| + |\phi|)$; $S_T^\phi \subseteq_\mathcal{F} SimOrdCl_{\{\tilde{a}_j \,|\, j \in J_T^\phi\}}$, $|S_T^\phi| \in O(|T| + |\phi|)$; *the number of all elementary operations of the translation of* T *and* ϕ *to* S_T^ϕ, *is in* $O(|T| + |\phi|)$; *the time and space complexity of the translation of* T *and* ϕ *to* S_T^ϕ, *is in* $O(|T| \cdot \log(1 + n_0 + |T|) + |\phi| \cdot (\log(1 + n_0) + \log |\phi|))$.

Proof cf. http://ii.fmph.uniba.sk/~guller/sci14.pdf, Sect. 3.2. □

3.3 DPLL Procedure Rules

We devise a basic variant of the *DPLL* procedure over order clausal theories. Let $a, \ldots, f \in PropAtom$, $Cn, Cn_1, \ldots, Cn_4 \in PropConj$, $\diamond_1, \diamond_2 \in \{\eqcirc, \prec\}$, $l, l_1, l_2, l_3 \in OrdLit$, $C \in OrdCl$, $S \subseteq OrdCl$. l is a contradiction iff either $l = 0 \eqcirc 1$ or $l = 0 \prec 0$ or $l = 1 \prec 0$ or $l = 1 \prec 1$ or $l = a \prec 0$ or $l = 1 \prec a$ or $l = Cn \prec Cn$. l is a tautology iff either $l = 0 \eqcirc 0$ or $l = 1 \eqcirc 1$ or $l = 0 \prec 1$ or $l = Cn \eqcirc Cn$. $0 \eqcirc a \vee 0 \prec a$ is a 0-dichotomy. $a \prec 1 \vee a \eqcirc 1$ is a 1-dichotomy. $Cn_1 \prec Cn_2 \vee Cn_1 \eqcirc Cn_2 \vee Cn_2 \prec Cn_1$ is a trichotomy. Some auxiliary operations are defined in Tables 7, 8 and 9.

Table 6 Interpolation rules for $\wedge$, $\vee$, &, $\rightarrow$

Case	Laws	
$\theta = \theta_1 \wedge \theta_2$		
Positive interpolation $\dfrac{\tilde{a}_i \rightarrow \theta_1 \wedge \theta_2}{\left\{ \tilde{a}_i \prec \tilde{a}_{i_1} \vee \tilde{a}_i \equiv \tilde{a}_{i_1}, \tilde{a}_i \prec \tilde{a}_{i_2} \vee \tilde{a}_i \equiv \tilde{a}_{i_2}, \tilde{a}_{i_1} \rightarrow \theta_1, \tilde{a}_{i_2} \rightarrow \theta_2 \right\}}$ $|\text{Consequent}| = 12 + |\tilde{a}_{i_1} \rightarrow \theta_1| + |\tilde{a}_{i_2} \rightarrow \theta_2| \leq 20 + |\tilde{a}_{i_1} \rightarrow \theta_1| + |\tilde{a}_{i_2} \rightarrow \theta_2|$	(57)	(62)
Negative interpolation $\dfrac{\theta_1 \wedge \theta_2 \rightarrow \tilde{a}_i}{\left\{ \tilde{a}_{i_1} \prec \tilde{a}_i \vee \tilde{a}_{i_1} \equiv \tilde{a}_i \vee \tilde{a}_{i_2} \prec \tilde{a}_i \vee \tilde{a}_{i_2} \equiv \tilde{a}_i, \theta_1 \rightarrow \tilde{a}_{i_1}, \theta_2 \rightarrow \tilde{a}_{i_2} \right\}}$ $|\text{Consequent}| = 12 + |\theta_1 \rightarrow \tilde{a}_{i_1}| + |\theta_2 \rightarrow \tilde{a}_{i_2}| \leq 20 + |\theta_1 \rightarrow \tilde{a}_{i_1}| + |\theta_2 \rightarrow \tilde{a}_{i_2}|$	(59)	(63)
$\theta = \theta_1 \vee \theta_2$		
Positive interpolation $\dfrac{\tilde{a}_i \rightarrow (\theta_1 \vee \theta_2)}{\left\{ \tilde{a}_i \prec \tilde{a}_{i_1} \vee \tilde{a}_i \equiv \tilde{a}_{i_1} \vee \tilde{a}_i \prec \tilde{a}_{i_2} \vee \tilde{a}_i \equiv \tilde{a}_{i_2}, \tilde{a}_{i_1} \rightarrow \theta_1, \tilde{a}_{i_2} \rightarrow \theta_2 \right\}}$ $|\text{Consequent}| = 12 + |\tilde{a}_{i_1} \rightarrow \theta_1| + |\tilde{p}_{i_2} \rightarrow \theta_2| \leq 20 + |\tilde{a}_{i_1} \rightarrow \theta_1| + |\tilde{p}_{i_2} \rightarrow \theta_2|$	(56)	(64)
Negative interpolation $\dfrac{(\theta_1 \vee \theta_2) \rightarrow \tilde{a}_i}{\left\{ \tilde{a}_{i_1} \prec \tilde{a}_i \vee \tilde{a}_{i_1} \equiv \tilde{a}_i, \tilde{a}_{i_2} \prec \tilde{a}_i \vee \tilde{a}_{i_2} \equiv \tilde{a}_i, \theta_1 \rightarrow \tilde{a}_{i_1}, \theta_2 \rightarrow \tilde{a}_{i_2} \right\}}$ $|\text{Consequent}| = 12 + |\theta_1 \rightarrow \tilde{a}_{i_1}| + |\theta_2 \rightarrow \tilde{a}_{i_2}| \leq 20 + |\theta_1 \rightarrow \tilde{a}_{i_1}| + |\theta_2 \rightarrow \tilde{a}_{i_2}|$	(58)	(65)
$\theta = \theta_1 \,\&\, \theta_2$		
Positive interpolation $\dfrac{\tilde{a}_i \rightarrow \theta_1 \,\&\, \theta_2}{\left\{ \tilde{a}_i \prec \tilde{a}_{i_1} \,\&\, \tilde{a}_{i_2} \vee \tilde{a}_i \equiv \tilde{a}_{i_1} \,\&\, \tilde{a}_{i_2}, \tilde{a}_{i_1} \rightarrow \theta_1, \tilde{a}_{i_2} \rightarrow \theta_2 \right\}}$ $|\text{Consequent}| = 8 + |\tilde{a}_{i_1} \rightarrow \theta_1| + |\tilde{a}_{i_2} \rightarrow \theta_2| \leq 20 + |\tilde{a}_{i_1} \rightarrow \theta_1| + |\tilde{a}_{i_2} \rightarrow \theta_2|$		(66)
Negative interpolation $\dfrac{\theta_1 \,\&\, \theta_2 \rightarrow \tilde{a}_i}{\left\{ \tilde{a}_{i_1} \,\&\, \tilde{a}_{i_2} \prec \tilde{a}_i \vee \tilde{a}_{i_1} \,\&\, \tilde{a}_{i_2} \equiv \tilde{a}_i, \theta_1 \rightarrow \tilde{a}_{i_1}, \theta_2 \rightarrow \tilde{a}_{i_2} \right\}}$ $|\text{Consequent}| = 8 + |\theta_1 \rightarrow \tilde{a}_{i_1}| + |\theta_2 \rightarrow \tilde{a}_{i_2}| \leq 20 + |\theta_1 \rightarrow \tilde{a}_{i_1}| + |\theta_2 \rightarrow \tilde{a}_{i_2}|$		(67)
$\theta = \theta_1 \rightarrow 0$		
Positive interpolation $\dfrac{\tilde{a}_i \rightarrow (\theta_1 \rightarrow 0)}{\left\{ \tilde{a}_i \equiv 0 \vee \tilde{a}_{i_1} \equiv 0, \theta_1 \rightarrow \tilde{a}_{i_1} \right\}}$ $|\text{Consequent}| = 6 + |\theta_1 \rightarrow \tilde{a}_{i_1}| \leq 20 + |\theta_1 \rightarrow \tilde{a}_{i_1}|$	(60)	(68)
Negative interpolation $\dfrac{(\theta_1 \rightarrow 0) \rightarrow \tilde{a}_i}{\left\{ 0 \prec \tilde{a}_{i_1} \vee \tilde{a}_i = 1, \tilde{a}_{i_1} \rightarrow \theta_1 \right\}}$ $|\text{Consequent}| = 6 + |\tilde{a}_{i_1} \rightarrow \theta_1| \leq 20 + |\tilde{a}_{i_1} \rightarrow \theta_1|$		(69)
$\theta = \theta_1 \rightarrow \theta_2, \theta_2 \neq 0$		
Positive interpolation $\dfrac{\tilde{a}_i \rightarrow (\theta_1 \rightarrow \theta_2)}{\left\{ \tilde{a}_i \,\&\, \tilde{a}_{i_1} \prec \tilde{a}_{i_2} \vee \tilde{a}_i \,\&\, \tilde{a}_{i_1} \equiv \tilde{a}_{i_2}, \theta_1 \rightarrow \tilde{a}_{i_1}, \tilde{a}_{i_2} \rightarrow \theta_2 \right\}}$ $|\text{Consequent}| = 8 + |\theta_1 \rightarrow \tilde{a}_{i_1}| + |\tilde{a}_{i_2} \rightarrow \theta_2| \leq 20 + |\theta_1 \rightarrow \tilde{a}_{i_1}| + |\tilde{a}_{i_2} \rightarrow \theta_2|$	(60)	(70)
Negative interpolation $\dfrac{(\theta_1 \rightarrow \theta_2) \rightarrow \tilde{a}_i}{\left\{ \begin{array}{l} \tilde{a}_{i_1} \prec \tilde{a}_{i_2} \vee \tilde{a}_{i_1} \equiv \tilde{a}_{i_2} \vee \tilde{a}_{i_2} \prec \tilde{a}_{i_1} \,\&\, \tilde{a}_i \vee \tilde{a}_{i_2} \equiv \tilde{a}_{i_1} \,\&\, \tilde{a}_i, \\ \tilde{a}_{i_2} \prec \tilde{a}_{i_1} \vee \tilde{a}_i = 1, \tilde{a}_{i_1} \rightarrow \theta_1, \theta_2 \rightarrow \tilde{a}_{i_2} \end{array} \right\}}$ $|\text{Consequent}| = 20 + |\tilde{a}_{i_1} \rightarrow \theta_1| + |\theta_2 \rightarrow \tilde{a}_{i_2}| \leq 20 + |\tilde{a}_{i_1} \rightarrow \theta_1| + |\theta_2 \rightarrow \tilde{a}_{i_2}|$		(71)

Table 7 Auxiliary operations, $Cn_1, Cn_2 \in PropConj \cup \{\emptyset\}$

$Cn_1 \odot Cn_2 = \{a^{m+n} \mid a^m \in Cn_1, a^n \in Cn_2\} \cup \{a^m \mid a^m \in Cn_1, a \notin atoms(Cn_2)\} \cup$
$\qquad\qquad \{a^n \mid a^n \in Cn_2, a \notin atoms(Cn_1)\} \in PropConj \cup \{\emptyset\},$

$Cn_1 \Downarrow Cn_2 = \{a^{m-n} \mid a^m \in Cn_1, a^n \in Cn_2, m > n\} \cup \{a^m \mid a^m \in Cn_1, a \notin atoms(Cn_2)\} \in$
$PropConj \cup \{\emptyset\}$
$\qquad\qquad$ if $Cn_2 \sqsubseteq Cn_1,$

$Cn_1 \rhd Cn_2 = \{a^{n-m} \mid a^m \in Cn_1, a^n \in Cn_2, n > m\} \cup \{a^n \mid a^n \in Cn_2, a \notin atoms(Cn_1)\} \in$
$PropConj \cup \{\emptyset\}$

Table 8 Transitivity operation, $Cn_1, \ldots, Cn_4 \in PropConj$, $\diamond_1, \diamond_2 \in \{\doteq, \prec\}$

$$(Cn_1 \diamond_1 Cn_2) \blacktriangleright (Cn_3 \diamond_2 Cn_4) = \begin{cases} 1 \diamond 1 & \text{if } Cn_7 = Cn_8 = \emptyset, \\ \square & \text{if } Cn_7 = \emptyset, Cn_8 \neq \emptyset, \\ \square & \text{if } Cn_7 \neq \emptyset, Cn_8 = \emptyset, \diamond = \doteq, \\ 1 \doteq 1 & \text{if } Cn_7 \neq \emptyset, Cn_8 = \emptyset, \diamond = \prec, \\ Cn_7 \diamond Cn_8 & \text{if } Cn_7 \neq \emptyset, Cn_8 \neq \emptyset, \end{cases}$$

$$Cn_5 = (Cn_1 \odot (Cn_2 \rhd Cn_3)),$$
$$Cn_6 = ((((Cn_2 \odot (Cn_2 \rhd Cn_3)) \Downarrow Cn_3) \odot Cn_4),$$
$$Cn_7 = (Cn_5 \Downarrow (Cn_5 \sqcap Cn_6)),$$
$$Cn_8 = (Cn_6 \Downarrow (Cn_5 \sqcap Cn_6)),$$
$$\diamond = \begin{cases} \doteq & \text{if } \diamond_1 = \diamond_2 = \doteq, \\ \prec & \text{else}, \end{cases}$$
$(Cn_1 \diamond_1 Cn_2) \blacktriangleright (Cn_3 \diamond_2 Cn_4) \in OrdCl$

Basic rules are defined as follows:

$$\text{(Contradiction simplification rule)} \qquad (72)$$

$$\frac{S}{S - \{l \vee C\} \cup \{C\}}$$

$$l \vee C \in S, l \text{ is a contradiction.}$$

$$\text{(One literal 0-simplification rule)} \qquad (73)$$

$$\frac{S}{S - \{l \vee C\} \cup simpl(a \doteq 0, l \vee C)}$$

$$a \doteq 0, l \vee C \in S, a \in atoms(l).$$

$$\text{(One literal 1-simplification rule)} \qquad (74)$$

$$\frac{S}{S - \{l \vee C\} \cup simpl(a \doteq 1, l \vee C)}$$

$$a \doteq 1, l \vee C \in S, a \in atoms(l).$$

Table 9 Auxiliary simplification function, $a \in PropAtom$, $\varepsilon \in \{0, 1\}$, $Cn_1, Cn_2 \in PropConj$, $l \in \{a = 0, a = 1\}$, $C \in OrdCl$

$simpl(a = 0, a \diamond \varepsilon \vee C) = \{0 \diamond \varepsilon \vee C\}$ if $a = 0 \neq a \diamond \varepsilon \vee C$,
$simpl(a = 0, \varepsilon \diamond a \vee C) = \{\varepsilon \diamond 0 \vee C\}$ if $a = 0 \neq \varepsilon \diamond a \vee C$,
$simpl(a = 0, Cn_1 = Cn_2 \vee C) = \left\{ \bigvee_{b \in atoms(Cn_2)} b = 0 \vee C \right\}$ if $a \in atoms(Cn_1)$,
$simpl(a = 0, Cn_1 \prec Cn_2 \vee C) = \{0 \prec b \vee C \mid b \in atoms(Cn_2)\}$ if $a \in atoms(Cn_1)$,
$simpl(a = 0, Cn_1 \prec Cn_2 \vee C) = \{C\}$ if $a \in atoms(Cn_2)$;
$simpl(a = 1, a \diamond \varepsilon \vee C) = \{1 \diamond \varepsilon \vee C\}$ if $a = 1 \neq a \diamond \varepsilon \vee C$,
$simpl(a = 1, \varepsilon \diamond a \vee C) = \{\varepsilon \diamond 1 \vee C\}$ if $a = 1 \neq \varepsilon \diamond a \vee C$,
$simpl(a = 1, Cn_1 = Cn_2 \vee C) = \{(Cn_1 - \{a^n\}) = Cn_2 \vee C\}$ if $\{a\} \subset atoms(Cn_1), a^n \in Cn_1$,
$simpl(a = 1, Cn_1 = Cn_2 \vee C) = \{b = 1 \vee C \mid b \in atoms(Cn_2)\}$ if $\{a\} = atoms(Cn_1)$,
$simpl(a = 1, Cn_1 \prec Cn_2 \vee C) = \{(Cn_1 - \{a^n\}) \prec Cn_2 \vee C\}$ if $\{a\} \subset atoms(Cn_1), a^n \in Cn_1$,
$simpl(a = 1, Cn_1 \prec Cn_2 \vee C) = \{C\}$ if $\{a\} = atoms(Cn_1)$,
$simpl(a = 1, Cn_1 \prec Cn_2 \vee C) = \{Cn_1 \prec (Cn_2 - \{a^n\}) \vee C\}$ if $\{a\} \subset atoms(Cn_2), a^n \in Cn_2$,
$simpl(a = 1, Cn_1 \prec Cn_2 \vee C) = \left\{ \bigvee_{b \in atoms(Cn_1)} b \prec 1 \vee C \right\}$ if $\{a\} = atoms(Cn_2)$;
$simpl(l, C) \subseteq_{\mathcal{F}} OrdCl$

$$(\textit{0-dichotomy branching rule}) \qquad (75)$$

$$\frac{S}{S \cup \{l_1\} \mid S \cup \{l_2\}}$$

$l_1 \vee l_2$ is a 0-dichotomy, $atoms(l_1 \vee l_2) \subseteq atoms(S)$.

$$(\textit{1 − dichotomy branching rule}) \qquad (76)$$

$$\frac{S}{S \cup \{l_1\} \mid S \cup \{l_2\}}$$

$l_1 \vee l_2$ is a $1 -$ dichotomy, $atoms(l_1 \vee l_2) \subseteq atoms(S)$.

$$(\textit{One literal transitivity rule}) \qquad (77)$$

$$\frac{S}{S \cup \{(Cn_1 \diamond_1 Cn_2) \blacktriangleright (Cn_3 \diamond_2 Cn_4)\}}$$

S is a unit order clausal theory, $Cn_1 \diamond_1 Cn_2, Cn_3 \diamond_2 Cn_4 \in S$,
for all $a \in atoms(Cn_1, \ldots, Cn_4), 0 \prec a, a \prec 1 \in S$.

$$(Trichotomy\ branching\ rule) \qquad (78)$$

$$\frac{S}{S - \{l_1 \vee C\} \cup \{l_1\} \mid S - \{l_1 \vee C\} \cup \{C\} \cup \{l_2\} \mid S - \{l_1 \vee C\} \cup \{C\} \cup \{l_3\}}$$

$$l_1 \vee C \in S, C \neq \square, l_1 \vee l_2 \vee l_3\ is\ a\ trichotomy,$$
$$for\ all\ a \in atoms(l_1, l_2, l_3), 0 \prec a, a \prec 1 \in S.$$

Rules (72)–(78) are sound in view of satisfiability. Using the basic rules, one can construct a finitely generated tree with the input theory as the root in the usual manner, so as the classical *DPLL* procedure does.[8] The *DPLL* procedure is refutation sound, and complete in the case of finite order clausal theory.

Theorem 6 (Refutational Soundness and Completeness of the *DPLL* Procedure) *Let $S \subseteq OrdCl$.*

(i) *If there exists a closed tree Tree with the root S constructed using Rules (72)–(78), then S is unsatisfiable.*
(ii) *If $S \subseteq_{\mathcal{F}} OrdCl$, then there exists a finite tree Tree with the root S constructed using Rules (72)–(78) with the following properties:*

$$if\ S\ is\ unsatisfiable,\ then\ Tree\ is\ closed; \qquad (79)$$
$$if\ S\ is\ satisfiable,\ then\ Tree\ is\ open\ and\ there\ exists\ a\ partial\ model\ \mathfrak{A}$$
$$(80)$$

$$of\ S, dom(\mathfrak{A}) = atoms(S), related\ to\ Tree.$$

Proof cf. http://ii.fmph.uniba.sk/~guller/sci14.pdf, Sect. 3.3 ☐

The refutational completeness of the *DPLL* procedure can be generalised to the countable case. Let $S \subseteq OrdCl$ and $A \subseteq PropAtom$. We denote $S|_A = \{C \mid C \in S, atoms(C) \subseteq A\} \subseteq S$, $atoms(S|_A) \subseteq atoms(S) \cap A$.

Theorem 7 (Compactness Theorem) *Let $S \subseteq OrdCl$, $\gamma \leq \omega$, $\delta : \gamma \longrightarrow atoms(S)$ be a sequence of atoms(S). If, for all $\alpha < \gamma$, there exists a partial model $\mathfrak{A}_\alpha$ of $S|_{\delta[\alpha]} \subseteq_{\mathcal{F}} S$, $dom(\mathfrak{A}_\alpha) = \delta[\alpha] \subseteq_{\mathcal{F}} atoms(S)$, then there exists a partial model $\mathfrak{A}$ of $S, dom(\mathfrak{A}) = atoms(S)$.*

Proof cf. http://ii.fmph.uniba.sk/~guller/sci14.pdf, Sect. 3.3. ☐

Corollary 2 (Refutational Completeness of the *DPLL* Procedure (The Countable Case)) *Let $S \subseteq OrdCl$. If S is unsatisfiable, then there exists a closed tree Tree with the root S constructed using Rules (72)–(78).*

Proof An immediate consequence of Theorems 6 and 7. ☐

We conclude the treatment with the following corollary.

---

[8]cf. http://ii.fmph.uniba.sk/~guller/sci14.pdf, Sect. 3.3.

Corollary 3 *Let* $\phi \in PropForm_\emptyset$ *and* $T \subseteq PropForm_\emptyset$. *There exist* $A_T^\phi \subseteq \tilde{\mathbb{A}}$, $S_T^\phi \subseteq$ *SimOrdCl* $_{A_T^\psi}$, *a finite tree Tree with the root* S_T^ϕ *constructed using Rules* (72)–(78) *with the following properties:*

$$if \ T \models_P \phi, \ then \ Tree \ is \ closed; \tag{81}$$

$$if \ T \not\models \phi, then \ Tree \ is \ open \ and \ there \ exists \ a \ partial \ model \ \mathfrak{A} \ of \ T, \tag{82}$$
$$dom(\mathfrak{A}) = atoms(T, \phi), related \ to \ Tree \ such \ that \ \mathfrak{A} \not\models \phi.$$

Proof An immediate consequence of Theorems 5, 6 and Corollary 2. □

4 Conclusions

We have investigated the deduction problem in Gödel and Product logics, both equipped with Gödel negation, in the countable case. Our approach is based on translation of a formula to an equivalent satisfiable finite order clausal theory, consisting of order clauses. An order clause is a finite set of order literals of the form $\varepsilon_1 \diamond \varepsilon_2$ where $\diamond$ is a connective either $=$ or $\prec$. $=$ and $\prec$ are interpreted by the equality and standard strict linear order on [0, 1], respectively. We have proposed a hyperresolution calculus in the first-order Gödel logic and a *DPLL* procedure in the propositional Product logic, operating over order clausal theories. Both the calculi have been proved to be refutation sound and complete for the countable case.

References

1. Metcalfe, G., Olivetti, N., Gabbay, D.M.: Analytic calculi for product logics. Arch. Math. Log. **43**, 859–890 (2004)
2. Savický, P., Cignoli, R., Esteva, F., Godo, L., Noguera, C.: On product logic with truth-constants. J. Log. Comput. **16**, 205–225 (2006)
3. Plaisted, D.A., Greenbaum, S.: A structure-preserving clause form translation. J. Symb. Comput. **2**, 293–304 (1986)
4. de la Tour, T.B.: An optimality result for clause form translation. J. Symb. Comput. **14**, 283–302 (1992)
5. Hähnle, R.: Short conjunctive normal forms in finitely valued logics. J. Log. Comput. **4**, 905–927 (1994)
6. Nonnengart, A., Rock, G., Weidenbach, C.: On generating small clause normal forms. In: Kirchner, C., Kirchner, H. (eds.) Automated Deduction—CADE-15, Proceedings of 15th International Conference on Automated Deduction., Lindau, Germany. Lecture Notes in Computer Science, vol. 1421, pp. 397–411. Springer, 5–10 July 1998
7. Guller, D.: A DPLL procedure for the propositional Gödel logic. In: Filipe, J., Kacprzyk, J. (eds.) ICFC-ICNC 2010—Proceedings of the International Conference on Fuzzy Computation and International Conference on Neural Computation (Parts of the International Joint Conference on Computational Intelligence IJCCI 2010), Valencia, Spain, pp. 31–422. SciTePress, 4–26 Oct 2010

Fuzzy Optimization Models for Seaside Port Logistics: Berthing and Quay Crane Scheduling

Christopher Expósito-Izquiero, Eduardo Lalla-Ruiz, Teresa Lamata,
Belén Melián-Batista and J. Marcos Moreno-Vega

Abstract The service time of container vessels is the main indicator of the competitiveness of a maritime container terminal. Vessels have to be berthed along the quay, a subset of quay cranes must be assigned to them and work schedules have to be planned for unloading the import containers and loading the export containers. This work addresses the Tactical Berth Allocation Problem, in which the vessels are assigned to a given berth, and the Quay Crane Scheduling Problem, for which the work schedules of the quay cranes are determined. The nature of this environment gives rise to inaccurate knowledge about the information related to the incoming vessels. Therefore, the aforementioned optimization problems can be tackled by considering fuzzy arrival times for the vessels and fuzzy processing times for the loading/unloading operations. Two fuzzy mathematical models are provided to solve the problems at hand. The computational experiments carried out in this work corroborate the effectiveness of the proposed methodologies.

This work has been partially funded by the Spanish Ministry of Economy and Competitiveness (TIN2012-32608). It has also been partially supported by FEDER founds, the DGICYT and Junta de Andalucía under the projects (TIN2011-27696-C02-01) and (P11-TIC-8001), respectively. Christopher Expósito and Eduardo Lalla thank the Canary Government for the financial support they receive through their post-graduate grants.

C. Expósito-Izquiero (✉) · E. Lalla-Ruiz · B. Melián-Batista · J.M. Moreno-Vega
Department of Computer Engineering and Systems,
University of La Laguna, Santa Cruz de Tenerife, Spain
e-mail: cexposit@ull.es

E. Lalla-Ruiz
e-mail: elalla@ull.es

B. Melián-Batista
e-mail: mbmelian@ull.es

J.M. Moreno-Vega
e-mail: jmmoreno@ull.es

T. Lamata
Department of Computer Science and Artificial Intelligence,
University of Granada, Granada, Spain
e-mail: mtl@decsai.ugr.es

© Springer International Publishing Switzerland 2016
K. Madani et al. (eds.), *Computational Intelligence*,
Studies in Computational Intelligence 613,
DOI 10.1007/978-3-319-23392-5_18

323

Keywords Fuzzy optimization model · Seaside operations · Maritime container terminal · Variable neighbourhood search.

1 Introduction

The global container trade has grown over the decades.[1] The maritime container terminals are highlighted infrastructures built with the goal of facing the technical requirements arising from the increasing volume of containers in the international sea freight trade. They are aimed at transferring and storing containers within multimodal transportation networks. The main transport modes found at a maritime container terminal are container vessels, trucks and trains. In this regard, a maritime container terminal can be considered as an open system that brings together different container flows, those stemming from freight sources to destinations [1].

The layout of a maritime container terminal is usually split into three different functional areas: seaside, yard, and landside [2]. The seaside is the area of the terminal where the container vessels arriving at port are berthed in order to be loaded or unloaded. An exhaustive analysis concerning the seaside operations planning problems is provided in the book [3]. The yard is the part of the terminal in which the containers are temporarily stored until their later retrieval [4]. Finally, the landside is the area which connects the container terminal with the land transport modes [5].

The main goal of a maritime container terminal is to serve appropriately those container vessels that arrive at port. In this regard, the service of a container vessel can be modeled through a well-defined sequence of steps. Firstly, it is required to provide a specific berthing position along the quay and berthing time for each container vessel according to its particular characteristics (dimensions, expected service time, draft, arrival time, etc.) and contractual agreements [6]. Afterwards, a subset of the available quay cranes at the terminal is allocated to each container vessel to perform the loading and unloading tasks established by its stowage plan [7]. Finally, the work schedules associated with the allocated quay cranes are determined [8].

The aforementioned planning decisions can be modeled by means of several optimization problems in maritime container terminals. The Tactical Berth Allocation Problem (TBAP) seeks to define the berthing position, berthing time and quay cranes allocated to each container vessel over a given planning horizon. On the other hand, the Quay Crane Scheduling Problem (QCSP) is aimed at determining the work schedules for the quay cranes allocated to a container vessel. It is worth mentioning that solving the TBAP and the QCSP for each container vessel provides an overall service planning for a container terminal.

Algorithms and model formulations to solve the TBAP and QCSP have usually assumed that the data is known accurately. However, this is not true in real-world applications and it is particularly problematic for data representing the arrival times of vessels and processing times that cannot be precisely estimated, but that need to be taken into account in order to provide the decision makers with real solutions.

---

[1] United Nations Conference on Trade and Development, http://unctad.org

The inherent imprecision that appears in the data involved in real-world problems can have different natures: randomness, subjectivity, vagueness, etc. If the goal is to tackle the problems at hand without altering their nature, what seems more desirable is to consider the approach that fits the best to the origin of the imprecision, either stochastic, interval-based, or fuzzy. Among these possibilities, the imprecision considered in this work is not random but linguistic and therefore vague (for instance, "the processing time will be large"). Although vague information appears in these real-world applications, to the best of our knowledge, this is the first work that considers this nature of imprecision in maritime container terminals when solving the TBAP and QCSP. Other imprecision natures are out of the scope of this paper.

In those scenarios in which subjectivity in the interpretation of the data is related to randomness, the fuzzy sets provide us a theoretical framework to solve a wide range of problems in different research areas [9, 10] with a high degree of efficacy and efficiency. In this paper, we are particularly interested in solving the TBAP with fuzzy arrival times of the vessels and the QCSP with fuzzy processing times for the loading/unloading operations. With the purpose of solving these problems, we propose mathematical models in which some coefficients in the constraints are not known accurately. In both cases, in order to solve the optimization problems derived from these situations, we make use of models which are well known in the area of Fuzzy Mathematical Programming [11–13].

The remainder of this paper is structured as follows. Section 2 describes the main logistic problems arising in the seaside. Section 3 presents the fuzzy coefficients used in the mathematical formulation of the TBAP. Section 4 describes the fuzzy constraints that appear in the mathematical model of the QCSP. Section 5 proposes two Variable Neighbourhood Search algorithms for the TBAP and QCSP. Section 6 shows the computational experiments performed in this work. Finally, Sect. 7 draws forth the main conclusions extracted from the work and indicates several directions for further research.

2 Seaside Operations

The seaside operations are those related to the service of container vessels that arrive at port. The turnaround time of container vessels constitutes the main indicator of the competitiveness of maritime container terminals. With this fact in mind, terminal managers are particularly interested in reducing the service times and maximizing the usage of the available resources: berths and quay cranes.

2.1 Tactical Berth Allocation Problem

The Tactical Berth Allocation Problem (TBAP) seeks to determine the berthing position, berthing time, and allocation of quay cranes for the container vessels arriving at port over a well-defined time horizon.

In the TBAP, we are given a set of incoming vessels N, berths M, and quay crane profiles $p \in P_i$ per each vessel $i \in N$. Each container vessel $i \in N$ must be assigned to an available berth $k \in M$ within the vessel and berth time window $[a_i, b_i]$ and $[a^k, b^k]$, respectively. The berthing position of a vessel should be close to the departure position of its containers. In this regard, the housekeeping cost represents the cost derived from moving a given container among different berthing positions of the quay. Moreover, for each vessel $i \in N$, a quay crane profile, $p \in P_i$, determines the distribution of quay cranes used for serving it. Q denotes the number of quay cranes at the terminal. The service time of a vessel depends on the quay crane profile associated with it. Each profile, $p \in P_i$, has an associated value v_i^p, which reflects the usage of quay cranes.

The goal of the TBAP is to maximize the value of the quay crane profiles used to serve the vessels and minimize the housekeeping costs derived from the transshipment of containers among vessels.

An example of the TBAP is depicted in Fig. 1. The example shows 3 container vessels, 3 berths and a maximum number of 8 quay cranes. For each container vessel $i \in N$, the number of QC hours required to perform its loading and unloading operations is termed as m_i. The profiles assigned to the vessels determine the distribution of quay cranes and their service time. For instance, the vessel 3 is served by 2, 3, 3, and 2 quay cranes and its service time is 4 h. It is worth pointing out that, despite vessels 1 and 3 require the same number of quay crane hours, their service times are different because they have assigned different profiles. In this regard, the quay crane profile assigned to vessel 1 is more expensive than that assigned to vessel 3.

In order to make this work self-contained, this subsection describes the mixed integer linear program formulation for the TBAP proposed in [14].

The following notations are used in the model:

- N, Set of container vessels
- M, Set of berths
- H, Set of time steps
- P_i, Set of feasible quay crane profiles for the container vessel $i \in N$
- t_i^p, Service time of container vessel $i \in N$ under QC profile $p \in P_i$
- v_i^p, The value of serving the container vessel $i \in N$ with the QC profile $p \in P_i$

Fig. 1 Example of TBAP with 3 vessels and 3 berths

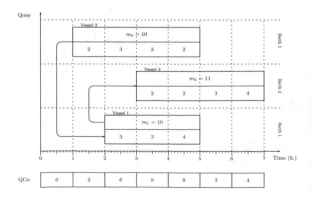

- Q^h, Maximum number of quay cranes available at the time step $h \in H$
- q_i^{pu}, Number of quay cranes assigned to the container vessel $i \in N$ in profile $p \in P_i$ at the time step $u \in H$
- f_{ij}, Flow of containers exchanged between container vessels $i, j \in N$
- d_{kw}, Housekeeping cost per unit of container between yard slots in berths $k, w \in M$
- $[a_i, b_i]$, [earliest, latest] arrival time of container vessel $i \in N$
- $[a^k, b^k]$, [start, end] of the availability time of the berth $k \in M$
- $[a^h, b^h]$, [start, end] of the time step $h \in H$

A graph $G^k = (V^k, A^k) \; \forall \, k \in M$, where $V^k = N \cup \{o(k), d(k)\}$, with $o(k)$ and $d(k)$ additional vertices representing berth k and $A^k \subseteq V^k \times V^k$ is generated. The decision variables are shown below.

- $x_{ij}^k \in \{0, 1\}$, $\forall \, k \in M$, $\forall \, (i, j) \in A^k$, set to 1 if container vessel j is scheduled after container vessel i in berth k, and 0 otherwise.
- $y_i^k \in \{0, 1\}$, $\forall \, k \in M$, $\forall \, i \in N$, set to 1 if container vessel i is assigned to berth k, and 0 otherwise.
- $z_{ij}^{kw} \in \{0, 1\}$, $\forall \, k, w \in M$, $\forall \, i, j \in N$, set to 1 if $y_i^k = y_j^w$, and 0 otherwise.
- $\gamma_i^h \in \{0, 1\}$, $\forall \, h \in H$, $\forall \, i \in N$, set to 1 if the container vessel i arrives in time step h, and 0 otherwise.
- $\lambda_p^i \in \{0, 1\}$, $\forall \, p \in P_i$, $\forall \, i \in N$, set to 1 if container vessel i is served under profile p, and 0 otherwise.
- $\rho_i^{ph} \in \{0, 1\}$, $\forall \, p \in P_i$, $\forall \, h \in H$, $\forall \, i \in N$, set to 1 if container vessel i is served under profile p and arrives at time step h, and 0 otherwise.
- $T_i^k \geq 0$, $\forall \, k \in M$, $\forall \, i \in N$, berthing time of container vessel i at berth k.
- $T_{o(k)}^k \geq 0$, $\forall \, k \in M$, $\forall \, i \in N$, starting operation time of berth k.
- $T_{d(k)}^k \geq 0$, $\forall \, k \in M$, $\forall \, i \in N$, ending operation time of berth k.

The corresponding MILP formulation for the TBAP is stated as follows:

$$\max \sum_{i \in N} \sum_{p \in P_i} \lambda_i^p \nu_i^p - \frac{1}{2} \sum_{i \in N} \sum_{j \in N} \sum_{k \in M} \sum_{w \in M} f_{ij} d_{kw} z_{ij}^{kw} \tag{1}$$

$$\sum_{k \in M} y_i^k = 1, \quad \forall i \in N \tag{2}$$

$$\sum_{j \in N \cup \{d(k)\}} x_{o(k)j}^k = 1, \quad \forall k \in M \tag{3}$$

$$\sum_{i \in N \cup \{o(k)\}} x_{id(k)}^k = 1, \quad \forall k \in M \tag{4}$$

$$\sum_{j \in N \cup \{d(k)\}} x_{ij}^k - \sum_{j \in N \cup \{o(k)\}} x_{ji}^k = 0, \quad \forall k \in M, \; \forall i \in N \tag{5}$$

$$\sum_{j \in N \cup \{d(k)\}} x_{ij}^k = y_i^k, \quad \forall k \in M, \ \forall i \in N \tag{6}$$

$$T_i^k + \sum_{p \in P_i} t_i^p \lambda_i^p - T_j^k \leq (1 - x_{ij}^k)M1, \forall k \in M \forall i \in N, \forall j \in N \cup \{d(k)\} \tag{7}$$

$$T_{o(k)}^k - T_j^k \leq (1 - x_{o(k)j}^k)M2, \quad \forall k \in M, \ \forall j \in N \tag{8}$$

$$a_i y_i^k \leq T_i^k, \quad \forall k \in M, \forall i \in N \tag{9}$$

$$T_i^k \leq b_i y_i^k, \quad \forall k \in M, \forall i \in N \tag{10}$$

$$a^k \leq T_{o(k)}^k, \quad \forall k \in M \tag{11}$$

$$T_{d(k)}^k \leq b^k, \quad \forall k \in M \tag{12}$$

$$\sum_{p \in P_i} \lambda_i^p = 1, \quad \forall i \in N \tag{13}$$

$$\sum_{h \in H^s} \gamma_h^i = \sum_{p \in P_i^s} \lambda_i^p, \quad \forall i \in N, \ \forall s \in S \tag{14}$$

$$\sum_{k \in M} T_i^k - b^h \leq (1 - \gamma_i^h)M3, \quad \forall h \in H, \ \forall i \in N \tag{15}$$

$$a^h - \sum_{k \in M} T_i^k \leq (1 - \gamma_i^h)M4, \quad \forall h \in H, \ \forall i \in N \tag{16}$$

$$\rho_i^{ph} \geq \lambda_i^p + \gamma_i^h - 1, \quad \forall h \in H, \ \forall i \in N, \ \forall p \in P_i \tag{17}$$

$$\sum_{i \in N} \sum_{p \in P_i} \sum_{u=max(h-t_i^p+1;1)}^{h} \rho_i^{pu} q_i^{p(h-u+1)} \leq Q^h, \quad \forall h \in H^{\bar{s}} \tag{18}$$

$$\sum_{k \in M} \sum_{w \in M} z_{ij}^{kw} = g_{ij}, \quad \forall i, j \in N \tag{19}$$

$$z_{ij}^{kw} \leq y_i^k \quad \forall i, j \in N, \quad \forall k, w \in M \tag{20}$$

$$z_{ij}^{kw} \leq y_j^w \quad \forall i, j \in N, \quad \forall k, w \in M \tag{21}$$

In this model, $M1$, $M2$, $M3$, and $M4$ represent sufficiently large constants. The objective function (1) maximizes the sum of the values of the chosen quay crane assignment profiles over all the container vessels and, at the same time, minimizes the housekeeping costs generated by the flows of containers exchanged between container vessels. Constraints (2) establish that every container vessel must be assigned to one and only one berth. Constraints (3) and (4) define the outcoming and incoming flows to the berths, whereas flow conservation for the remaining vertices is ensured by constraints (5). Constraints (6) establish the link between variables x_{ij}^k, whereas precedences in every sequence are ensured by constraints (7) and (8). The time windows of the container vessels are defined by the constraints (9) and (10), whereas berths time windows are defined by constraints (11) and (12). Constraints (13) ensure that one and only one QC profile is assigned to every container vessel. Constraints (14) define the link between variables γ_i^h and λ_i^p, whereas constraints (15) and (16) link binary variables γ_i^h and T_i^k. Variables ρ_i^{ph} are linked to variables λ_i^p and γ_i^h by constraints (17). Finally, constraints (18) ensure that, at every time step, the total number of assigned QCs does not exceed the number of maximum QCs available in the terminal. Constraints (19), (20), and (21) are included to linearize the quadratic objective function.

2.2 Quay Crane Scheduling Problem

The QCSP seeks to define the sequences of transshipment operations performed by a set of quay cranes in order to load and unload the containers associated with a given vessel berthed at the container terminal.

The input data for the QCSP are composed of the set of tasks $\Omega = \{1, \ldots, n\}$ and the set of quay cranes $Q = \{1, \ldots, m\}$ allocated to the vessel. Each task $t \in \Omega$ represents a set of containers with similar characteristics (weight, dimensions, destination port, etc.) located adjacent to each other in the same bay, l_t. The processing time of the task $t \in \Omega$ is denoted by p_t. Two dummy tasks 0 and T with $p_0 = p_T = 0$ are considered with the goal of representing the beginning and the ending of the vessel service, respectively. In addition, we define the set $\bar{\Omega} = \Omega \cup \{0, T\}$. The structure of the vessel imposes limitations in the transshipment operations order [15]. For instance, unloading operations have to be performed before loading operations. The precedence relationships among tasks located in the same bay are defined by the set Φ, in such a way that, $(i, j) \in \Phi$ if and only if task i has to be finished before the starting of task j. On the other hand, each quay crane $q \in Q$ is located in the bay l_0^q and it is available after time r^q. The time required by the quay crane $q \in Q$ to move between the bays in which the tasks $i, j \in \Omega$ are currently located is denoted by t_{ij}^q. For safety reasons, the quay cranes must keep a minimum distance between them, δ, and measured in bay units. The safety distance gives rise to that same pairs of tasks cannot be performed simultaneously due to the fact that they are close. These pairs of tasks are gathered into the set Ψ.

The optimization criterion of the QCSP is to minimize the makespan of the schedule, that is, the finishing time of the last task performed by the quay cranes, c_T. The QCSP is already known to be an NP-hard problem [16].

An example of the QCSP is depicted in Fig. 2. The example represents a vessel berthed at the quay with 10 bays, ranged from the bay 0 up to the bay 9, for which the bays 1, 3, 4, 6 and 8 have at least one task to perform by a quay crane. For each bay, the tasks are sorted according to their precedence relationships. For instance, the task 1 has to be performed before the starting of task 2 in the bay 1. The location and processing time of each task are reported in the associated table. A schedule with 2 quay cranes for this example is shown in Fig. 3. The quay crane 1 performs the tasks 1, 2, 3, 5, 6, and 7, whereas the quay crane 2 performs the tasks 4 and 8. As can be seen, the quay cranes keep a safety distance of at least 2 bays and move with a speed of one bay per time unit. In this case, the makespan is 52 time units.

In the following, we present the mathematical formulation proposed in [8] for the QCSP. The following notation is used by the formulation:

- Δ_{ij}^{vw}, Minimum temporal span to elapse between the processing of the tasks i and j if they are processed by the quay cranes v and w, respectively.
- Θ, Set of all combinations of tasks and quay cranes that potentially lead to quay crane interference.

Fig. 2 Example of a QCSP instance composed of 8 tasks

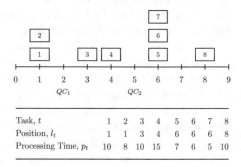

Task, t	1	2	3	4	5	6	7	8
Position, l_t	1	1	3	4	6	6	6	8
Processing Time, p_t	10	8	10	15	7	6	5	10

Fig. 3 Schedule with 2 quay cranes for the example depicted in Fig. 2

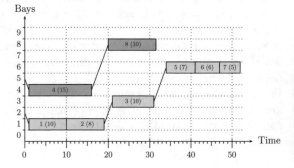

The decision variables of the model are:

- $x_{ij}^k \in \{0, 1\}$, set to 1 if tasks i and j are processed consecutively by quay crane k, 0 otherwise.
- $z_{ij} \in \{0, 1\}$, set to 1 if task j starts after the completion of task i, 0 otherwise.

The mathematical formulation is as follows:

$$\min c_T \tag{22}$$

$$\sum_{j \in \Omega^T} x_{0j}^k = 1, \forall k \in Q \tag{23}$$

$$\sum_{j \in \Omega^0} x_{jT}^k = 1, \forall k \in Q \tag{24}$$

$$\sum_{k \in Q} \sum_{j \in \Omega^T} x_{ij}^k = 1, \forall i \in \Omega \tag{25}$$

$$\sum_{j \in \Omega^0} x_{ji}^k - \sum_{i \in \Omega^T} x_{ij}^k = 0, \forall i \in \Omega, \forall \in Q \tag{26}$$

$$c_i + t_{ij} + p_j - c_j \leq M(1 - x_{ij}^k), \forall i, j \in \bar{\Omega}, \forall k \subset Q \tag{27}$$

$$c_i + p_j - c_j \leq 0, \forall (i, j) \in \Phi \tag{28}$$

$$c_i + p_j - c_j \leq M(1 - z_{ij}), \forall i, j \in \Omega \tag{29}$$

$$c_j - p_j - c_i \leq M z_{ij}, \forall i, j \in \Omega \tag{30}$$

$$z_{ij} + z_{ji} = 1, \forall (i, j) \in \Psi \tag{31}$$

$$\sum_{u \in \Omega^0} x_{ui}^v + \sum_{u \in \Omega^0} x_{uj}^w \leq 1 + z_{ij} + z_{ji}, \forall (i, j, v, w) \in \Theta \tag{32}$$

$$c_i + \Delta_{ij}^v w + p_j - c_j \leq M(3 - z_{ij} - \sum_{u \in \Omega^0} x_{ui}^v - \sum_{u \in \Omega^0} x_{uj}^w), \forall (i, j, v, w) \in \Theta \tag{33}$$

$$c_j + \Delta_{ij}^v w + p_i - c_i \leq M(3 - z_{ji} - \sum_{u \in \Omega^0} x_{ui}^v - \sum_{u \in \Omega^0} x_{uj}^w), \forall (i, j, v, w) \in \Theta \tag{34}$$

$$r + t_{0j}^k + p_j - c_j \leq M(1 - x_{0j}^k), \forall j \in \Omega, \forall k \in Q \tag{35}$$

$$c_i \geq 0, \forall i \in \bar{\Omega} \tag{36}$$

The minimization of the makespan is modeled by (22). Each quay crane starts performing the task 0 and finishes performing the task T. This is modeled by constraints (23) and (24). All the non-dummy tasks have to be performed exactly once as set by constraints (25) and have a previous task and a succeeding task, constraint (26). Constraints (27) set the finishing time of each task. Constraints (28) ensure the precedence relationships among tasks. The values of variables z_{ij} are established by constraints (29) and (30). Constraints (31) ensure the non-simultaneity of tasks. The interference between quay cranes are avoided by constraints (32)–(34). The ready times of the quay cranes are handled by constraints (35). Finally, the domain of the finishing times of each task is defined by constraints (36).

3 Fuzzy Coefficients in the Constraints for the TBAP

In the TBAP, the arrival times of the vessels are considered to be uncertain values. Modeling the uncertainty in the values of the parameters can be tackled by considering that they are fuzzy numbers. In this sense, independently of the wide range of different models that may be used, the problems above can be addressed by the following model.

$$\max\{cx/A^f x \leq_I b^f, x \geq 0\},$$

where A^f and b^f refer to the fact that we are considering fuzzy numbers in the coefficients that define the restrictions (thereby allowing, as a trivial case, them to also be real numbers when there are no ambiguities), and the symbol $\leq_I$ means that the way of comparing both members in the inequality, due to formal coherence, must be done by using a relationship for ordering the fuzzy numbers. This comparison relation $\leq_I$ may be any one from the extensive list available [17], which in turn would also allow the decision-maker to have a greater degree of freedom when it comes to establish preferences. In more specific terms, in order to provide that theoretical model with a way for operating, let us briefly refer back to the different indices for comparing fuzzy numbers that have been described in the literature [17]. Amongst the different approaches described for comparing them, for the sake of simplicity, in this paper we shall only deal with the one that is derived from the use of indices for comparison. Hence, by denoting as $F(R)$ the set of fuzzy numbers, if

$$I : F(R) \rightarrow [0, 1]$$

is a comparison index for this kind of numbers, then

$$\forall P^f, Q^f \in F(R), P^f \leq_I Q^f \Leftrightarrow I(P^f) \leq I(Q^f)$$

whereby, according to the index I that is used, different auxiliary models may be obtained for effectively solving the problems described above from the practical point of view. Therefore, in general, the auxiliary models used to solve the problems described above from the practical viewpoint, would be approached as follows.

$$\max\{cx/I(A^f x) \leq I(b^f), x \geq 0\}$$

Using an index I or another depends on the decision-maker, and hence what index I to choose to be used is not the matter here. In order to illustrate the approach, and as a trivial example, let us consider two triangular fuzzy numbers $P^f, Q^f \in F(R)$, usually denoted as $P^f = (P, P_i, P_d)$ and $Q^f = (Q, Q_i, Q_d)$, and as form of comparison, the one given by Yager's First Index [17],

$$P^f \leq_I Q^f \Leftrightarrow (1/3)(P + P_i + P_d) \leq (1/3)(Q + Q_i + Q_d)$$

The membership function corresponding to a triangular fuzzy number $A^f = (A, A_i, A_d)$ is stated as follows:

$$\mu_{A^f}(x) = \begin{cases} 0, & x \leq A_i \\ \frac{x-A_i}{A-A_i}, & A_i < x \leq A \\ \frac{A_d-x}{A_d-A}, & A < x \leq A_d \\ 0, & x \geq A_d \end{cases}$$

Then, the previous model takes the following operating form,

$$\max\{cx/(A + A_i + A_d)x \leq (b + b_i + b_d), x \geq 0\}$$

from which we can obtain a solution for the previous models in a straightforward way.

Since in the TBAP model described above, the arrival times of the vessels, $a_i, i \in N$, which cannot be estimated accurately, appear as coefficients in the constraints, the methodology explained in this section can be used to provide the decision makers with adequate solutions. When applying this methodology with the Yager's First Index as a simple example, the original fuzzy model can be converted into a mixed integer linear model that can be solved using any effective optimization technique from the literature.

4 Fuzzy Constraints for the QCSP

In the QCSP, the processing times of the loading/unloading operations, p_t, of the different tasks $t \in \Omega$ are also considered to be uncertain values. Even though the number of quay cranes assigned to carry out the loading and unloading operations of a container vessel is fixed, the real service time, will depend on several factors, such as interferences or breaks of the quay cranes assigned to that container vessel. In this case, we consider that the constraints are fuzzy, so that it is possible to state the problem as follows:

$$
\begin{aligned}
&\min z = cx \\
&s.t. : \\
&\quad (Ax)_i \leq^f b_i, i \in I \\
&\quad x_j \geq 0, x_j \in \mathbf{N}, j \in J
\end{aligned}
\tag{37}
$$

where the symbol "$\leq^f$" means that the decision maker allows violations in compliance with the constraints and considers fuzzy constraints defined by the following membership functions:

$$
\mu_i : \mathbf{R}^n \to (0, 1], i \in I.
$$

Each membership function provides the satisfaction degree with which any $x \in \mathbf{R}^n$ satisfies the corresponding fuzzy constraint on which it is defined. This degree is equal to 1 when the constraint is satisfied without any violation, and decreases to zero as violations are larger. In the linear case, the membership functions can be formulated as follows:

$$
\mu_i(x, b_i) = \begin{cases} 1 & (Ax)_i \leq b_i \\ 1 - ((Ax)_i - b_i)/d_i) & b_i \leq (Ax)_i \leq b_i + d_i \\ 0 & (Ax)_i > b_i + d_i \end{cases}
$$

If we apply the methodology proposed by Herrera and Verdegay [12], the original model with fuzzy constraints can be solved by means of the following auxiliar parametric mixed integer linear model:

$$
\begin{aligned}
&\min z = cx \\
&s.t. : \\
&\quad (Ax)_i \leq b_i + d_i(1 - \alpha), i \in I \\
&\quad x_j \geq 0, x_j \in \mathbf{N}, j \in J, \alpha \in (0, 1]
\end{aligned}
\tag{38}
$$

In the particular application to the QCSP, b_i corresponds to the processing times of the cranes and we consider $d_i = 25\% \times b_i$ to be the greatest deviation in the processing times.

5 Optimization Techniques

In order to generate suitable solutions for the fuzzy considerations of the TBAP and QCSP, the solution approaches proposed by [18] are considered. These approaches are based on the Variable Neighbourhood Search (VNS) which has demonstrated to be a high competitive metaheuristic when solving combinatorial and global optimization problems [19]. In general terms, the foundation of a VNS is to perform a systematic change of neighbourhood structures within a local search algorithm.

With the goal of providing a self-contained paper, in the following subsections we present the VNS approaches used for solving the TBAP and QCSP, respectively. For further details the interested reader is referred to [18].

5.1 VNS for Solving the TBAP

Algorithm 1 depicts the pseudocode of the VNS used for solving the TBAP. Given a solution ω, it considers two neighbourhood structures based upon the reinsertion movement, $N_a(\omega, \lambda)$, in which λ vessels and their assigned profiles are removed from the berth $b \in B$ and reinserted into another berth b', where $b \neq b'$, and the interchange movement, $N_b(\omega)$, which consists of exchanging a vessel $v \in V$ assigned to berth $b \in B$ with another vessel v' assigned to berth b', where $b \neq b'$.

The starting solution of the VNS, ω, is generated by assigning the profile $p \in P$ with the highest usage cost to each container vessel. The berthing position of each vessel is selected at random, whereas the starting of its service time is selected as the earliest possible within its time window (line 1). The value of the parameter k is set to 1 (line 2). The shaking process (line 4) allows to escape from those local optima found along the search by using the neighbourhood structure N_a. The solution exploitation phase of the VNS is based on a Variable Neighbourhood Descent Search (VND) (lines 6–14). Given a solution ω', it explores one neighbourhood at a time until a local optimum with respect to the neighbourhood structures N_a and N_b is found. The application of the neighbourhoods structures in the VND is carried out according to the value of the parameter k_1, initially set to 1 (line 5). The first neighbourhood structure explored is N_a and later N_b. The best solution found by means of the VND is denoted by ω'. The objective function value of ω' allows to update the best solution found along the search (denoted by ω) and restart the value of k (lines $15 - 17$). Otherwise, the value of k is increased (line 19). These steps are carried out until $k = k_{max}$ (line 21).

Algorithm 1. VNS for the TBAP.

1: $\omega \leftarrow$ Generate initial solution
2: $k \leftarrow 1$
3: **repeat**
4:    $\omega' \leftarrow$ Shake(ω, k)
5:    $k_1 \leftarrow 1$
6:    **repeat**
7:       $\omega'' \leftarrow$ Local Search(ω', k_1)
8:       **if** $f(\omega'') > f(\omega')$ **then**
9:          $\omega' \leftarrow \omega''$
10:          $k_1 \leftarrow 1$
11:       **else**
12:          $k_1 \leftarrow k_1 + 1$
13:       **end if**
14:    **until** $k_1 = k_{1max}$
15:    **if** $f(\omega') > f(\omega)$ **then**
16:       $\omega \leftarrow \omega'$
17:       $k \leftarrow 1$
18:    **else**
19:       $k \leftarrow k + 1$
20:    **end if**
21: **until** $k = k_{max}$

5.2 VNS for Solving the QCSP

The pseudocode of the proposed VNS for solving the QCSP is depicted in Algorithm 2. It is based upon two neighbourhood structures, the reassignment (N_1) and interchange of tasks (N_2). The search starts generating an initial schedule, σ, by assigning each task $t \in \Omega$ to its nearest quay crane (line 1). The value of the parameter k is also set to 1 (line 4). A shaking procedure allows to reach unexplored regions of the search space by means of the reassignment of k tasks to another quay crane. The reassigned tasks are selected on the basis of a frequency memory. In this way, at each step, a neighbour schedule, σ', is generated at random from σ within the neighbourhood structure N_k (line 6). A local optimum, σ'', is reached through a local search based on the proposed neighbourhood structures (line 7). An improvement in the value of σ'' allows to update σ and restart k (lines 9, 10 and 11). Otherwise, the value of k is increased (line 13). These steps are carried out until $k = k_{max}$ (line 15).

An elite set, ES, is included into the VNS with the goal of collecting the promising schedules found during the search process. It is composed of those schedules with the lowest objective function value and those local optima with the highest diversity in the ES. The diversity of two schedules is measured as the number of tasks performed by different quay cranes. At each step, ES provides a pair of schedules σ and σ' selected at random (line 16) in order to be combined (line 17) and restart the search. The combination process keeps those tasks performed by the same quay crane, whereas the remaining ones are randomly assigned to one quay crane on the basis of the objective function values of σ and σ'.

Algorithm 2. VNS for the QCSP.

1: $\sigma \leftarrow$ Generate initial solution
2: $ES \leftarrow \emptyset$
3: **repeat**
4:　　$k \leftarrow 1$
5:　　**repeat**
6:　　　$\sigma' \leftarrow$ Shake(σ, k)
7:　　　$\sigma'' \leftarrow$ Local Search(σ')
8:　　　Update ES
9:　　　**if** $f(\sigma'') < f(\sigma)$ **then**
10:　　　　$\sigma \leftarrow \sigma''$
11:　　　　$k \leftarrow 1$
12:　　　**else**
13:　　　　$k \leftarrow k + 1$
14:　　　**end if**
15:　　**until** $k = k_{max}$
16:　　$\sigma', \sigma'' \leftarrow$ Select schedules from ES
17:　　$\sigma \leftarrow$ Combine(σ', σ'')
18: **until** Stopping Criteria

6　Computational Experiments

This section is devoted to assess and analyze the performance of the VNSs described in Sect. 5 for solving the TBAP and the QCSP under imprecise scenarios. All the computational experiments have been carried out on a computer equipped with a CPU Intel 3.16 GHz and 4 GB of RAM.

6.1　Computational Experiments for the TBAP

The computational tests aimed at evaluating the behaviour of the VNS introduced in Sect. 5.1 for the TBAP were conducted by using the problem instances proposed in the work by [14]. These instances are based upon real data provided by the Medcenter Container Terminal of Gioia Tauro (Italy). In this case, only a subset of 9 of these instances are used during this computational experiment. The size of the problem instances ranges from 20 up to 40 container vessels that must be located in 5 berths over a time horizon of one week.

The Table 1 reports the computational results obtained by means of the VNS over the group of instances taken up from the benchmark suite described above. The first column (*Instance*) shows the instances to solve. For each instance, the name (*Name*),

Table 1 Computational results obtained by means of the Variable Neighbourhood Search considering different fuzzy numbers.

Instance				CPLEX	VNS		Gap (%)	VNS_{S-I}		VNS_{S-II}	
Name	N	M	P	UB	f_{VNS}	t. (s.)		f_{VNS}	t. (s.)	f_{VNS}	t. (s.)
A_{p10}	20	5	10	1,383,614	1,344,035	8.70	2.86	1,348,244	16.70	1,345,272	71.87
A_{p20}	20	5	20	1,384,765	1,349,455	18.67	2.55	1,344,308	11.62	1,346,265	24.81
A_{p30}	20	5	30	1,385,119	1,343,119	13.09	3.03	1,344,374	10.56	1,342,836	13.14
B_{p10}	30	5	10	1,613,252	1,546,878	64.78	4.11	1,557,261	80.69	1,542,448	21.39
B_{p20}	30	5	20	1,613,769	1,550,839	26.26	3.90	1,551,015	27.90	1,546,422	28.03
B_{p30}	30	5	30	1,613,805	1,553,673	36.73	3.73	1,550,520	13.12	1,548,215	27.67
C_{p10}	40	5	10	2,289,660	2,229,807	195.76	2.61	2,231,800	170.91	2,235,877	176.04
C_{p20}	40	5	20	2,290,662	2,234,516	330.21	2.45	2,231,730	263.82	2,229,076	310.01
C_{p30}	40	5	30	2,291,301	2,232,260	228.64	2.58	2,227,989	221.56	2,226,721	134.12

the number of container vessels (N), the number of berths (M) and the maximum number of quay crane profiles per each vessel (P) are presented. The second column $(CPLEX)$ shows the upper bound (UB) obtained by the CPLEX Optimizer with a maximum computational time of 2 h. The column VNS shows the results obtained when the arrival time of the vessels is deterministic. Under this heading, it is reported the objective value of the best solution found by the VNS (f_{VNS}), the execution time $(t.\ (s.))$ and the relative error $(Gap(\%))$ regarding the upper bound. Lastly, the next columns $(VNS_{S-I}$ and $VNS_{S-II})$ show the results by considering scenarios with fuzzy numbers to model the arrival times of the container vessels. In this context, we have evaluated the performance of the VNS concerning the following fuzzy numbers:

- S-I $= (a_i, a_i - 1, a_i + 4)$
- S-II $= (a_i, a_i - 2, a_i + 8)$

These fuzzy scenarios (S-I and S-II) model, on one hand, the anticipation or delay of the arrival time of the container vessel arrived at port regarding their expected times. These scenarios represent a common problem in maritime container terminals, since the vessels are subject to tidal, traffic or contractual changes. In this regard, these issues are frequently translated into a delay of the expected arrival time of the container vessel, due to that, the fuzzy numbers of both scenarios S-I and S-II consider a higher delay than anticipation of their arrival time. For each scenario, the best solution value found by the VNS (f_{VNS}) and its required computational time $(t.\ (s.))$, measured in seconds, are reported.

In spite of the change of scenario there is not a clear trend in the target values, as reported in Table 1. It is expected that the uncertainty in the arrival times of the container vessels has a direct impact on the feasibility of the solutions due to the reduction in the time window constraints and the availability of quay cranes. Moreover, it may implicitly affect to the objective function value if the late arrival of the container vessel forces to allocate it in another berth than the expected one. The reason is found in that this fact would increase the housekeeping cost derived from the transshipment operations. In this regard, the anticipation or delay of the arrival time of the vessels may also impact on the assignment of quay crane profiles, namely, early arrival times would allow to assign longer quay cranes profiles, whereas late arrival times would require shorter quay crane profiles if one is willing to keep the vessels assigned to their initial assigned berth. These facts and further analysis of the structure of the final solutions would be a topic of future work. It is worth mentioning that for the instances from the literature are obtained feasible solutions for fuzzy arrival times considered in this computational experiment.

6.2 Computational Experiments for the QCSP

In order to check the suitability of the VNS for solving the QCSP, we have considered a representative subset of the problem instances proposed by [8]. The original set of instances is composed of 90 instances grouped into 9 groups with 10 instances each one. Each group has a different number of tasks (from 10 up to 50) and quay cranes (from 2 up to 6) which allows to cover real-world scenarios. In this case, we have selected one instance from each original group in such a way that our benchmark suite is composed of 9 instances. It is worth mentioning that, as done in previous works, we have established in this experiment that the quay cranes are available from the starting of the service time (r^q, $\forall q \in Q$) and they have to keep a safety distance of one bay ($\delta = 1$) among them.

Table 2 shows the computational results we have obtained by means of the proposed VNS when solving the aforementioned instances. The first column (*Instance*) reports the characteristics of the instances used during the experiment: name (*Name*), number of tasks (n) and number of quay cranes (m). The second column (*Optimal*) shows the objective function value of the optima schedules for the instances at hand reported in [8]. The third column (*VNS*) shows the computational results obtained by means of the VNS under deterministic scenarios. In this case the objective function value of the best schedule found during the search (f under the heading $VNS_{\alpha=1}$) and the computational time ($t. (m.)$), measured in minutes, are reported. Finally, as described in Sect. 4, the original fuzzy model for this problem can be converted into a parametric model that depends on α. In this context, we have evaluated the performance of the VNS using the values $\alpha \in 1, 0.8, 0.6, 0.4, 0.2$. This parametric model let us take into account the fact that delays in the processing time of the tasks are the most common in maritime container terminals. The case in which $\alpha = 1$ corresponds to the original processing times for which there are not delays. The case in which $\alpha = 0.2$ corresponds to the largest delays allowed for the processing times.

In spite of the fact that analyzing the performance of the VNS under deterministic scenarios is not a major goal of this work, with the aim of providing an overall study, we firstly focus on this issue. In this regard, the computational results reported in Table 2 indicate that the proposed VNS is highly effective at finding optimal or near-optimal schedules for the QCSP under deterministic scenarios (column $VNS_{\alpha=1}$). As can be seen, it provides the optimal schedules for 7 instances from the benchmark suite at hand, whereas the gap is below 1.2 % in the worst case (*k94*). Moreover, a time below 4.5 min has been required for finishing the search process in all the cases. An exhaustive analysis of the performance of the VNS under deterministic scenarios is described in [18]. Additionally, the efficiency of our optimization technique has been successfully applied to integrated approaches as described in the same work.

On the other hand, as might be expected, the uncertainty on the processing times of the tasks defined by the stowage plan of a given container vessel has a direct impact on its service time (makespan). This is evidenced by the increment in the objective function value of the schedules obtained by the VNS for the different values of the parameter α. The makespan of the schedules reported by the VNS_α are larger than those of the schedules found under the deterministic scenario.

Table 2 Computational results obtained by means of the Variable Neighbourhood Search for different values of α.

Instance			Optimal	$VNS_{\alpha=1}$		Gap (%)	$VNS_{\alpha=0.8}$		$VNS_{\alpha=0.6}$		$VNS_{\alpha=0.4}$		$VNS_{\alpha=0.2}$	
Name	n	m	Obj.	f	t. (m.)		f	t. (m.)	f	t. (m.)	f	t. (m.)	f	t. (m.)
k13	10	2	453	453	<0.01	0.00	477	<0.01	498	<0.01	519	<0.01	540	<0.01
k27	15	2	657	657	0.02	0.00	687	0.01	720	0.01	753	0.01	783	0.01
k42	20	3	573	573	0.11	0.00	603	0.05	636	0.05	654	0.05	684	0.05
k48	25	3	639	639	0.23	0.00	669	0.12	699	0.12	729	0.129	759	0.12
k53	30	4	717	717	0.64	0.00	753	0.34	789	0.35	822	0.38	861	0.35
k69	35	4	807	807	0.99	0.00	846	0.56	885	0.62	927	0.58	963	0.61
k82	40	5	717	723	2.22	0.84	780	1.67	807	1.54	843	1.73	864	1.60
k89	45	5	843	843	2.83	0.00	885	1.83	930	1.69	966	1.97	1005	1.70
k94	50	6	786	795	4.42	1.15	840	2.68	876	2.82	924	2.82	966	3.16

7 Conclusions

The maritime container terminals are large infrastructures aimed at serving the container vessels arrived at port. In this regard, a berthing time and berthing position along the quay is assigned to each vessel. For loading and unloading its containers, a subset of quay cranes is allocated to it. These quay cranes perform the loading and unloading operations associated with the containers of the vessel. In this context, two relevant logistical problems have to be highlighted: the Tactical Berth Allocation Problem (TBAP) and the Quay Crane Scheduling Problem (QCSP). Unfortunately, due to the inherent imprecision that appears in the data involved in this environment, terminal managers are particularly interested in solving these problems by considering the uncertainty arising in the terminals.

In this paper, the TBAP and the QCSP are tackled under imprecise scenarios. For this purpose, two fuzzy models that consider the uncertainty of the arrival time of the vessels and the processing time of the quay cranes are proposed. Moreover, in order to effectively solve these models, two solutions approaches based on the Variable Neighbourhood Search metaheuristic are introduced. Both methods are able to provide high-quality solutions by means of reasonable computational times.

References

1. Stahlbock, R.: Voβ, S.: Operations research at container terwminals: a literature update. OR Spectr **30**, 1–52 (2008)
2. Petering, M.: Decision support for yard capacity, fleet composition, truck substitutability, and scalability issues at seaport container terminals. Transp. Res. Part E: Logist. Transp. Rev. **47**, 85–103 (2011)
3. Meisel, F.: Seaside Operations Planning in Container Terminals. Physica-Verlag HD, Contributions to Management Science (2010)
4. Kim, K., Park, Y.M., Jin, M.J.: An optimal layout of container yards. OR Spectr. **30**, 675–695 (2008)
5. Froyland, G., Koch, T., Megow, N., Duan, E., Wren, H.: Optimizing the landside operation of a container terminal. OR Spectr. **30**, 53–75 (2008)
6. Lalla-Ruiz, E., Melián-Batista, B., Moreno-Vega, J.M.: Artificial intelligence hybrid heuristic based on tabu search for the dynamic berth allocation problem. Eng. Appl. Artif. Intell. **25**, 1132–1141 (2012)
7. Bierwirth, C., Meisel, F.: A survey of berth allocation and quay crane scheduling problems in container terminals. Eur. J. Oper. Res. **202**, 615–627 (2010)
8. Bierwirth, C., Meisel, F.: A fast heuristic for quay crane scheduling with interference constraints. J. Sched. **12**, 345–360 (2009)
9. Tiwari, A., Knowles, J., Avineri, E., Dahal, K., Roy, R.: Applications of Soft Computing : Recent Trends (Advances in Soft Computing Series). Springer, Heidelberg (2006)
10. Verdegay, J.: Fuzzy Sets based Heuristics for Optimization. Springer, Studies in Fuzziness and Soft Computing (2003)
11. Cadenas, J., Verdegay, J.: A primer on fuzzy optimization models and methods. Iran. J. Fuzzy Syst. **3**, 1–21 (2006)
12. Herrera, F., Verdegay, J.: Three models of fuzzy integer linear programming. Eur. J. Oper. Res. **83**, 581–593 (1995)

13. Sancho-Royo, A., Pelta, D., Verdegay, J.: A proposal of metaheuristics based in the cooperation between operators in combinatorial optimization problems. In: Proceedings of the IEEE 20-th International Parallel and Distributed Processing Symposium, Rhodes, Greece (2006)
14. Giallombardo, G., Moccia, L., Salani, M., Vacca, I.: Modeling and solving the tactical berth allocation problem. Transp. Res. Part B: Methodol. **44**, 232–245 (2010)
15. Kim, K., Park, Y.M.: A crane scheduling method for port container terminals. Eur. J. Oper. Res. **156**, 752–768 (2004)
16. Sammarra, M., Cordeau, J., Laporte, G., Monaco, M.: A tabu search heuristic for the quay crane scheduling problem. J. Sched. **10**, 327–336 (2007)
17. Wang, X., Kerre, E.: On the classification and the dependencies of the ordering methods. In Ruan, D., ed.: Fuzzy Logic Foundation and Industrial Applications. International Series in Intelligent Technologies, pp. 73–90. Kluwer, Dordrecht (1996)
18. Lalla-Ruiz, E., Expósito-Izquierdo, C., Melián-Batista, B., Moreno-Vega, J.M.: A metaheuristic approach for the seaside operations in maritime container terminals. In: IWANN (Part II). Volume 7903 of Lecture Notes in Computer Science, Springer, Berlin (2013)
19. Hansen, P., Mladenović, N., Moreno-Pérez, J.A.: Variable neighbourhood search: methods and applications. Ann. Op. Res. **175**, 367–407 (2010)

Obtaining the Decision Criteria and Evaluation of Optimal Sites for Renewable Energy Facilities Through a Decision Support System

Juan M. Sánchez-Lozano, Jose Angel Jiménez-Pérez, M. Socorro García-Cascales and M. Teresa Lamata

Abstract In projects regarding renewable energy facilities, decision making is an essential activity that provides greater consistency and viability to the project. The first step that any promoter of such facilities should face is to select an optimal location. To do so, it is necessary to consider all the criteria that influence the decision. However, not all the criteria are equally important, which means that determining their weights is extremely important. The objective of this chapter is to obtain the weights of the decision criteria that influence the location problems of wind farms and solar photovoltaic and thermoelectric plants. For this, a Decision Support System (DSS) has been designed that allows to carry out the extraction of knowledge from an expert group by Fuzzy AHP methodology. Finally, DSS will sort the viable locations based on the importance of the criteria that influence the decision.

Keywords Decision support systems (DSS) · Optimal location · Renewable energy facilities · Fuzzy AHP

J.M. Sánchez-Lozano (✉)
Centro Universitario de la Defensa. Academia General Del Aire de San Javier,
Universidad Politécnica de Cartagena, Murcia, Spain
e-mail: juanmi.sanchez@cud.upct.es

J.A. Jiménez-Pérez · M.S. García-Cascales
Depto de Electrónica Tecnología de Computadoras y Proyectos,
Universidad Politécnica de Cartagena, Murcia, Spain
e-mail: socorro.garcia@upct.es

M.T. Lamata
Depto de Ciencias de la Computación e Inteligencia Artificial,
Universidad de Granada, Granada, Spain
e-mail: mtl@decsai.ugr.es

© Springer International Publishing Switzerland 2016
K. Madani et al. (eds.), *Computational Intelligence*,
Studies in Computational Intelligence 613,
DOI 10.1007/978-3-319-23392-5_19

345

1 Introduction

Renewable energy is the energy obtained from virtually inexhaustible natural sources, either due to the vast amount of energy they contain, or because they are able to regenerate by natural media. One of the great problems of humanity's dependence on fossil fuels is their depletion and the environmental impact they cause [1, 2].

When implementing renewable energy facilities, the promoter must find and select the best location in order to obtain a better use of energy and reduce the risks that, in facilities of this size, can cause serious economic and environmental damage [3]. It is, however, not unusual that in choosing the right site among various sites, there is a degree of uncertainty. If the knowledge and experience of the decision group are combined with methodologies and tools to assist in decision making [4], this uncertainty could be avoided.

Decision Support Systems DSS [5] appeared in the 1970s as solutions which could be used to help with complex decision-making and problem solving in a structured manner. The DSS are particularly suitable for solving the same complex problem several times. In location problems in industrial plants and specifically in the problems of locating renewable energy facilities a set of decision criteria exist which affect the decision on the location of these facilities. These criteria will depend on the type of technology (solar, wind ...) to be installed on the facilities. Therefore it is of great interest to have a DSS to help obtain the weights of criteria when deciding on the optimal locations for renewable energy installations [6].

Thus, this chapter focuses on the design of a DSS that facilitates the decision maker to obtain the weights of the criteria in a location problem of renewable energy facilities.

The chapter will be structured as follows: Sect. 2 will focus on the hierarchical structure of decision criteria for the case of wind facilities and solar photovoltaic and thermoelectric plants. Section 3 will focus on the design of the DSS algorithms to work with and the data entry into the system and the results of the DSS output for different renewable technologies. Section 4 presents an example of how it is possible to obtain a classification of suitable locations through the DSS and finally in Sect. 5 we present the main conclusions of the work.

2 Decision Criteria for the Optimal Location of Renewable Energy Facilities

It is necessary to know which criteria influence (and to what extent), the decision-making problem proposed. Although previous studies have been conducted indicating the features that these criteria should meet [7, 8], the fact of using one or another will depend mainly on the study area. However, it is possible to establish common generic criteria that subsequently may be decomposed into specific criteria, which will depend on the characteristics and nature of the area to be analyzed.

Therefore, following the guidelines established in [9], four groups of main criteria will be established (environment, location, orography and climatology criterion).

Through the environment criterion it is not intended to assess the impact that these renewable energy plants cause in certain sites, the description of this criterion is based on the suitability of installing renewable energy plants depending on the capacity that the land presents to host them. Location criteria will be composed on the one hand by those criteria that allow to evaluate the distances that the future renewable plants would have regarding infrastructures or areas in which they cannot be implemented (cities, airports, masts, etc.) and, on the other hand by those criteria that will not only allow to reduce the installation costs but will also favour its performance (distance to main roads, power lines, etc.). Orography criteria are based on both the extension and the orographic features that the land presents to implement this type of facilities in order to minimize the installation costs and increase efficiency, for example, to implement solar facilities it will not only be appropriate that the land has sufficient area but it must also have low slopes and a correct orientation. Finally climatology criteria will allow evaluating the production capacity of the renewable energy plants. Sites should be chosen where these criteria present appropriate values because these criteria are essential not only for the correct operation of the plant but also to optimize the production.

These criteria are common to the main renewable energy facilities, and especially to those which this paper is focused on: wind farms, solar photovoltaic plants and thermoelectric plants.

The difference between the different technologies exists in the definition of the criteria to be considered in the location, based on the type of technology used. So for wind farms the hierarchy of criteria is that shown in Fig. 1 [10]:

- C_1: Agrological capacity (Classes): Suitability of land for agricultural development, if the land presents excellent agrological capacity it will not be suitable to implement the renewable facility and vice versa.
- C_2: Slope (%): Inclination of the land, the higher the percentage of surface inclination, the worse fitness it will have to implement a wind farm.
- C_3: Area (m^2): Surface contained within a perimeter of land that can accommodate a renewable energy facility.
- C_4: Distance to main airports (m): Space of interval between the nearest airport and the different possible sites.
- C_5: Distance to main roads (m): Space of interval between the nearest main road and the different possible sites.
- C_6: Distance to power lines (m): Space of interval between the nearest power line and the different possible sites.
- C_7: Distance to cities (m): Space of interval between the population centers (cities and towns) and the different possible sites.
- C_8: Distance to electricity transformer substations (m): Space of interval between the nearest electricity transformer substation and the different possible sites.

348   J.M. Sánchez-Lozano et al.

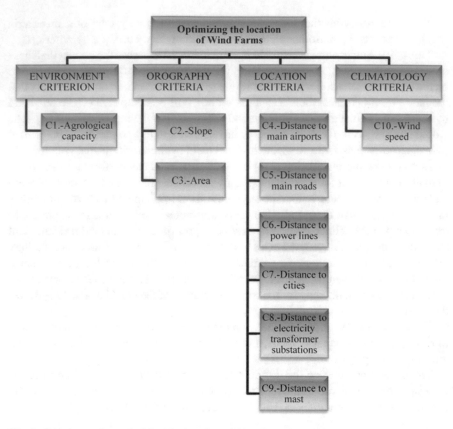

Fig. 1 Criteria tree for optimizing the location of wind farms

- C_9: Distance to mast (m): Space of interval between the nearest mast and the different possible sites.
- C_{10}: Wind speed (m/s): It corresponds to the wind speed at an elevation of 80 meters in the different possible sites.

In the case of solar photovoltaic and thermoelectric plants the criteria tree is as in Fig. 2 where we have some similar criteria (C_1, C_2, C_3, C_5, C_6, C_7, and C_8) but others which are different, due to the technology used [11]:

- C_4: Field Orientation (Cardinal points): Position or direction of the ground to a cardinal point.
- C_9: Potential solar radiation (kJ m^2/day): It corresponds to the amount of solar energy a ground surface receives over a period of time (day).
- C_{10}: Average temperature (C): Average temperatures measured on ground in the course of one year.

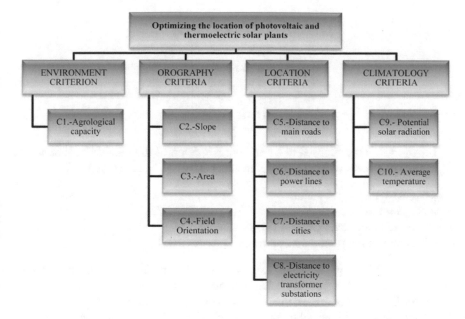

Fig. 2 Criteria tree for optimizing the location of solar photovoltaic and thermoelectric plants

3 Decision Support System for the Location of Renewable Energy Facilities

We have developed a Decision Support System DSS for the location of renewable energy facilities with the structure shown in Fig. 5 and called Optimal Location v1.0. Optimal Location v1.0 is formed by three sub-systems [5]:

- Data handling sub-system: Contains information about the problem. In this case, the Data Base is obtained by means of a Geographical Information System (GIS).
- Models' handling sub-system: Mathematical models that are used to solve the problem. Optimal Location v1.0 uses AHP and the TOPSIS method with or without fuzzy logic. By means of AHP we obtain the weights of the criteria.
- AHP estimates the impact of each one of the alternatives on the overall objective of the hierarchy. In this method the quantified judgments provided by experts in the field on pairs of criteria (C_i, C_j) are represented in an $n \times n$ matrix expressed by the following expression (1).

$$c_1 \quad c_2 \quad \dots \quad c_n$$

$$C = \begin{array}{c} C_1 \\ C_2 \\ . \\ . \\ . \\ C_n \end{array} \begin{bmatrix} c_{11} & c_{12} & \dots & c_{1n} \\ c_{21} & c_{22} & \dots & c_{2n} \\ . & . & \dots & . \\ . & . & \dots & . \\ c_{n1} & c_{n2} & \dots & c_{n3} \end{bmatrix} \qquad (1)$$

The c_{12} value is supposed to be an approximation of the relative importance of C_1 to C_2, i.e., $c_{12} \approx (w_1/w_2)$. The statements below can be concluded:

- $c_{ij} \approx (w_i/w_j)$ i, j = 1, 2, ..., n
- $c_{ii} = 1$, i = 1, 2, ..., n
- If $c_{ij} = \alpha$, $\alpha \neq 0$, then $c_{ji} = 1/\alpha$, i=1,2,...,n
- If C_i is more important than C_j then $c_{ij} \cong (w_i/w_j) > 1$

Matrix C should be a positive and reciprocal matrix with 1's in the main diagonal; so the expert needs only to provide value judgments in the upper triangle of the matrix. The TOPSIS method is applied to obtain the ranking of the alternatives. Nevertheless, this chapter has been focused on the aim of obtaining the weight of the criteria.

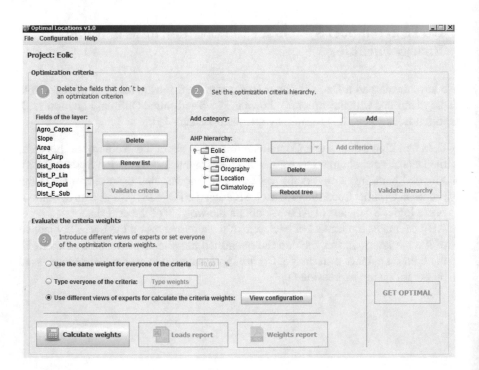

Fig. 3 Insertion of the criteria and categories in Optimal Location v1.0

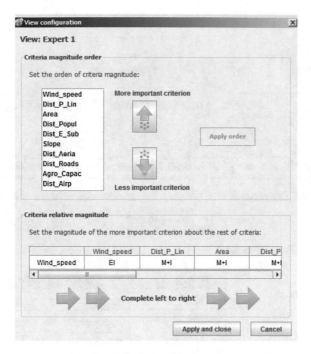

Fig. 4 Insertion of the order of importance for each criterion in Optimal Location v1.0

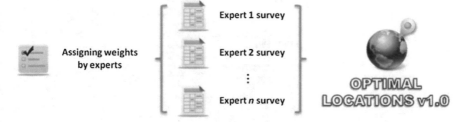

Fig. 5 Group decision making with Optimal Location v1.0

- *User Interface Sub-system*: It is the environment in which the user controls the DSS. By means of this interface, the input data can be introduced in order to apply the AHP method (see Figs. 3 and 4) and additionally the results (output of the DSS) can be shown, these results are shown in Figs. 7, 8 and 9.

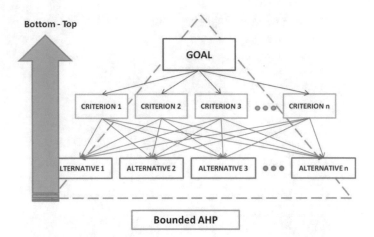

Fig. 6 *Bottom* to *Top* AHP method

Fig. 7 Weights of the
criteria for wind farms

Table of weights of the criteria		
Environment		
	Agro_Capac	[03,78, 04,36, 05,35] %
Orography		
	Slope	[05,08, 06,97, 09,36] %
	Area	[05,12, 07,17, 10,61] %
Location		
	Dist_Airp	[03,78, 04,36, 05,35] %
	Dist_Roads	[04,73, 06,35, 08,33] %
	Dist_P_Lin	[05,32, 07,66, 11,28] %
	Dist_Popul	[06,73, 10,15, 16,05] %
	Dist_E_Sub	[05,79, 08,40, 12,47] %
	Dist_Aeria	[04,26, 05,32, 06,91] %
Climatology		
	Wind_speed	[34,06, 39,26, 42,83] %

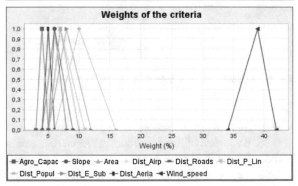

3.1 Data Input to the DSS

The DSS starts with a file format ESRI Shape file (.Shp.) to perform its functions.
This file must have been previously published and analyzed on professional GIS

Fig. 8 Weights of the criteria for photovoltaic plants

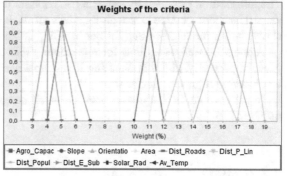

Table of weights of the criteria		
Environment		
	Agro_Capac	[03,84, 04,19, 04,93] %
Orography		
	Slope	[04,64, 05,86, 07,28] %
	Orientatio	[04,21, 05,13, 06,22] %
	Area	[11,45, 12,71, 14,14] %
Location		
	Dist_Roads	[04,27, 04,93, 05,99] %
	Dist_P_Lin	[12,42, 14,49, 17,78] %
	Dist_Popul	[17,25, 18,55, 19,31] %
	Dist_E_Sub	[14,58, 16,80, 18,71] %
Climatology		
	Solar_Rad	[10,97, 11,95, 12,93] %
	Av_Temp	[04,48, 05,38, 06,75] %

Fig. 9 Weights of the criteria for thermoelectric plants

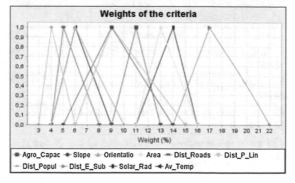

Table of weights of the criteria		
Environment		
	Agro_Capac	[09,93, 11,67, 13,24] %
Orography		
	Slope	[05,84, 09,16, 14,95] %
	Orientatio	[04,50, 06,74, 10,27] %
	Area	[11,01, 13,77, 16,57] %
Location		
	Dist_Roads	[04,23, 05,99, 08,81] %
	Dist_P_Lin	[06,07, 09,43, 15,12] %
	Dist_Popul	[03,62, 04,89, 06,55] %
	Dist_E_Sub	[14,07, 17,88, 22,14] %
Climatology		
	Solar_Rad	[11,10, 14,02, 16,56] %
	Av_Temp	[04,39, 06,46, 09,86] %

software. In this particular case, the gvSIG tool has been used because it is free software.

For optimization calculations it is necessary to establish the relative importance of each decision criterion. To do so, the DSS uses the AHP method [12, 13].

This seeks to establish the pairwise comparisons required by this method by conducting surveys to different experts in the field. It is a pseudo-Delphi technique, in which different independent experts without mutual interaction value judgments made for pairwise comparison. In this way, the aim is to obtain a vector of weights of the criteria from each expert and then to produce a single weight vector by performing an arithmetic mean between them, see Fig. 6.

The information provided by the experts is qualitative in character or is very vague since it has been obtained through linguistic terms; because of this the data obtained should be set modeled so that further handling is feasible and easy.

Among the various options for representing information and because the data is grouped perfectly, and that handling it is simple and effective, fuzzy numbers will be chosen to represent information [14, 15].

In the case studied, the data provided shall be represented by triangular fuzzy numbers [16–18].

3.2 Treatment of the Data

For that purpose, a questionnaire similar to that made by [19] was developed, which was given to experts with the aim of reducing uncertainty and imprecision of the proposed problem. The linguistic labels used in the Fuzzy AHP model are shown in Table 1.

In AHP problems, where the values are fuzzy, the geometric normalized average will be used, expressed by the following expression (2):

Table 1 Linguistic labels used in fuzzy AHP

Verbal judgments of preferences between criterion i and criterion j	Triangular fuzzy scale and reciprocals
Ci and Cj are equally important (II)	(1, 1, 1)/(1,1,1)
Ci is slightly more/less important than Cj (S+I/S-I)	(2, 3, 4)/(1/4,1/3,1/2)
Ci is strongly more/less important than Cj (+I/-I)	(4, 5, 6)/(1/6,1/5,1/4)
Ci is very strongly more/less important than Cj (VS+I/VS-I)	(6, 7, 8)/(1/8,1/7,1/6)
Ci is extremely more/less important than Cj (Ex+I/Ex-I)	(8, 9, 9)/(1/9,1/9,1/8)

$$W_i = \frac{\displaystyle\prod_{j=1}^{n} (a_{ij}, b_{ij}, c_{ij})}{\displaystyle\sum_{i=1}^{m} \prod_{j=1}^{n} (a_{ij}, b_{ij}, c_{ij})} \tag{2}$$

where (a_{ij}, b_{ij}, c_{ij}) is a fuzzy number

The group of experts involved in the decision process answer a survey based in the Fuzzy AHP model. In this case the way to obtain the weighted criteria is bottom to top type (see Fig. 6), this is to calculate all the weights of the criteria at the second level by comparing all the criteria with each other.

The survey is divided into two parts:

1. The decision problem is explained indicating what the goal to achieve is (optimal location of sites for renewable energy facilities), the methodology used, and the criteria that influence the decision making process. Thus, the basic elements of the decision problem are described through a hierarchical structure, as shown in the criteria trees (Figs. 1 and 2).
2. It is based on the hierarchical structure described and its purpose is to gather data to obtain the weight or coefficient of importance of criteria. The survey consists of a block of three questions:

   - Q_1: Do you believe that all the criteria have the same weight?
     - If the answer is yes, it will not be necessary to apply any MCDM to obtain the weights of the criteria, as these will have the same value. Otherwise, i.e., if experts consider that not all the criteria have equal importance, the second question in the survey will be posed:
   - Q_2: List the criteria in descending importance.
   - Q_3: Compare the approach to be considered first with respect to that considered secondly and successively, using the linguistic labels in Table 1.

In the particular case of wind farms, the answers for each of the criteria indicated in Fig. 2 were the following.

Answer Q1: NO

Answer Q2: The orders of importance for each of the experts are shown in Table 2.

Answer Q3: The pairwise comparisons among criteria by the experts are shown in Table 3

So, the weights of the criteria will be determined by pairwise comparison among criteria. As a result of the data collection used, a total of $(n-1)$ comparisons will be required against the complete AHP method $n(n-1)/2$ comparisons.

3.3 Weights of the Criteria in Wind, Solar Photovoltaic and Thermoelectric Plants

The results of the DSS output are discussed for the three types of technologies and with the hierarchical structure criteria according to Figs. 1 and 2 for the criteria.

DSS provides the results for the criteria as seen in Fig. 7, in the case of the decision criteria for the location of wind farms; Fig. 8 for the case of the decision criteria for the location of solar photovoltaic plants; and Fig. 9 in the case of decision criteria for locating thermoelectric plants.

In the case of wind farms the criterion (Fig. 7) which clearly stands out above the other criteria is the wind speed (C_{10}) with almost 40 % of the total weight. This result is logical since to implement a wind farm the wind speed plays a crucial role, and if this is not enough in a given area, that area is removed by any promoter of these facilities. The remainder of these criteria are further apart and grouped around weights between 5 and 10 % of the total.

In the case of solar technologies the situation is different since there is no single criterion whose weight or importance coefficient is so high that it allows to discard the rest. Analyzing Fig. 8, the criteria for photovoltaic plants, it is shown that the three best criteria for the location problem for solar plants are the distance to power lines (C_6); distance to electricity transformer substations (C_8); and distance to cities (C_7), with the latter being the highest rated. By contrast, the criteria that less influence the decision, that is to say, those with the lowest values, correspond to the criterion of agrological capacity (C_1) and to the criterion of distance to main roads (C_5).

The results are consistent since in the implementation of a photovoltaic solar plant, the fact of having a pour point to the nearest grid greatly reduces the initial investment costs, thus reducing the payback period of the facility. However, it should also be highlighted that the most important criterion presented corresponds to the distance to centers of population, the justification for this high weight can be found in both the potential environmental impact that this type of facility can generate and

Table 2 Order of importance of the criteria for each of the experts for the case of location of wind farms

Criteria	Expert 1	Expert 2	Expert 3
C_1	9	10	10
C_2	6	3	5
C_3	3	8	6
C_4	10	7	9
C_5	8	5	3
C_6	2	2	7
C_7	4	6	2
C_8	5	4	4
C_9	7	9	8
C_{10}	1	1	1

Table 3 Pairwise comparisons among criteria for the case of location of wind farms by linguistic labels

	Expert 1	Expert 2	Expert 3
$1° → 2°$	$S + I$	$VS+I$	$S + I$
$1° → 8°$	$VS+I$	$Ex+I$	$Ex+I$
$1° → 5°$	$S + I$	$VS+I$	$+I$
$1° → 3°$	$S + I$	$VS+I$	$+I$
$1° → 9°$	$Ex+I$	$Ex+I$	$Ex+I$
$1° → 7°$	$+I$	$Ex+I$	$VS+I$
$1° → 4°$	$S + I$	$VS+I$	$+I$
$1° → 6°$	$+I$	$VS+I$	$VS+I$
$1° → 10°$	$Ex+I$	$Ex+I$	$Ex+I$

in growth and expansion of cities because, given the useful life of photovoltaic solar plants, implementing these facilities in close proximity to centers of population can condition their expansion.

Analyzing Fig. 9, the criteria for thermoelectric plants, it is shown that the three best criteria for the location problem for solar thermoelectric plants are potential solar radiation (C_9); distance to electricity transformer substations (C_8); and area (C_3), with the latter being the highest rated. By contrast the criteria that have less influence in the decision in this case are distance to cities (C_7) and distance to roads (C_5).

The results are consistent as solar thermoelectric plants are facilities that not only require a territory covering a large area, but also, the installed capacity of them is usually very high (with the aim of reducing the payback period) therefore there is a need to have nearby transformer substations that allow to directly pour the electricity generated because, if not, the promoter himself should meet the additional

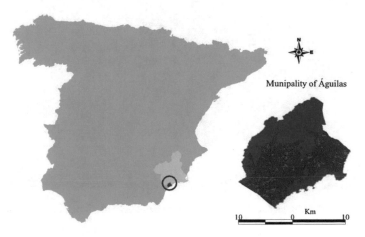

Fig. 10 Position and suitable locations in the municipality of Águilas

Fig. 11 Map of the capacity to accommodate solar photovoltaic farms in the municipality of Águilas

cost of building a transformer substation to discharge the energy generated in the thermoelectric plant.

Fig. 12 Map of the capacity to accommodate solar thermoelectric farms in the municipality of Águilas

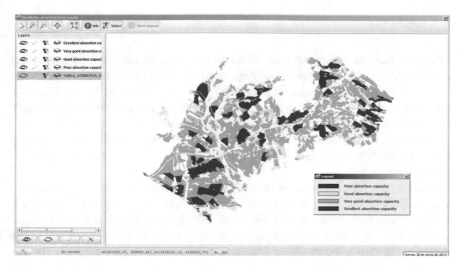

Fig. 13 Map of the capacity to accommodate wind farms in the municipality of Águilas

4 Obtaining Optimal Locations. Case Study: Municipality in Southeast of Spain

Once the weights of the criteria that influence in the decision have been obtained, the available locations to implement renewable energy facilities will be evaluated with DSS. To do so, the thematic layer obtained in [20] will be used, which will provide the suitable locations to implement wind farms, solar photovoltaic and thermoelectric plants in 13 municipalities in the Region of Murcia, in south-eastern Spain. As an example the thematic layers of one of the municipalities that compose the coast of this Region, specifically the town of Águilas, will be used (Fig. 10).

Introducing these thematic layers in the DSS, the software will be able to evaluate the plots of the municipality of Águilas according to the weights of the criteria. Once that evaluation has been made, the DSS will provide a map showing the ability of each plot to host renewable energy facilities, these capacities will be linked to a colour code (excellent: blue; very good: yellow; and regular: red). The evaluations obtained are shown in Figs. 11, 12 and 13.

Analyzing Figs. 11 and 12 it is observed that there is some similarity in the best rated locations for solar photovoltaic and thermoelectric facilities, since the categories of the available locations are very similar. Regarding wind farms it is observed that most of the locations have very good or excellent capacity for this type of facilities.

5 Conclusions

This study has shown that we must take into account a number of criteria to select which is the best location for renewable energy facilities (wind farms, solar photovoltaic plants and solar thermoelectric plants). Moreover, such criteria do not equally influence in decision making so it is very important to know beforehand the weights of these criteria for each technology when implementing such facilities.

Moreover, it is interesting to show that there are important differences among Wind and Solar technologies, while between the two solar technologies there is a greater similarity.

Carrying out the assessment of the facilities available in a case study, it is observed that the DSS is able to provide a classification of the locations according to their ability to host such facilities: it is observed that the optimal locations to host solar photovoltaic and thermoelectric farms coincide. It should also be noted that in the case of wind farms, the number of locations that have very good capacity for this type of facility increases.

It is of great interest for the promoters of renewable energy facilities to have a tool such as this, a DSS to model the importance of the decision criteria when locating renewable energy installations that aggregates all the information by different experts to be involved in decision making.

This DSS is simple and intuitive to manage for any expert in the field of renewable energy without any knowledge of soft computing, when experts only have to answer three simple questions to obtain the weights of the criteria involved in the decision making of the optimal location for renewable energy facilities.

Acknowledgments This work is partially supported by FEDER funds, the DGICYT and Junta de Andalucía under projects TIN2014-55024-P and P11-TIC-8001, respectively.

References

1. Intergovernmental Panel on Climate Change.: The IPCC 1990 and 1992 Assessments. World Meteorological Organization/United Nations Environment Program, Toronto (1992)
2. United Nations.: Framework Convention on Climatic Change: Report of the Conference of the Parties on Its Third Session. Adoption of the Kyoto Protocol, Kyoto (1997)
3. Kahraman, C., Kaya, I., Cebi, S.: A comparative analysis for multiattribute selection among renewable energy alternatives using fuzzy axiomatic design and fuzzy analytic hierarchy process. Energy **34**, 1603–1616 (2009)
4. Ramírez-Rosado, I.J., García-Garrido, E.G., Fernández-Jiménez, L.A., Zorzano-Santamaría, P.J., Monteiro, C., Miranda, V.: Promotion of new wind farms based on a decision support system. Renewable Energy **33**, 558–566 (2008)
5. Turban E., Aronson J.E., Liang, T.P., Sharda, R.: Decision Support and Business Intelligence Systems, 9th edn. Prentice Hall Press, Upper Saddle River (2006)
6. Sánchez-Lozano, J.M., Jiménez-Pérez, J.A., García-Cascales, M.S., Lamata, M.T.: Decision support systems to obtain decision criteria by fuzzy AHP for location of renewable energy facilities. In: 5th International Joint Conference on Computational Intelligence, pp. 300–308. IJCCI 2013, Vilamoura (2013)

7. Janke, J.R.: Multicriteria GIS modeling of wind and solar farms in Colorado. Renewable Energy **35**, 2228–2234 (2010)
8. Al-Yahyai, S., Charabi, Y., Gastli, A., Al-Badi, A.: Wind farm land suitability indexing using multi-criteria analysis. Renewable Energy **44**, 80–87 (2012)
9. Arán-Carrión, J., Espín-Estrella, A., Aznar-Dols, F., Zamorano-Toro, M., Rodríguez, M., Ramos-Ridao, A.: Environmental decision-support systems for evaluating the carrying capacity of land areas: optimal site selection for grid-connected photovoltaic power plants. Renew. Sustain. Energy Rev. **12**, 2358–2380 (2008)
10. Sánchez-Lozano, J.M., García-Cascales, M.S., Lamata, M.T., Sierra, C.: Decision criteria for optimal location of wind farms. In: IGI Global (ed.) *Exploring Innovative and Successful Applications of Soft Computing* , (pp. 199–215). Information Science Publishing , Hershey (2013)
11. Sánchez-Lozano, J.M., García-Cascales, M.S., Lamata, M.T.: Decision criteria for optimal location of solar plants: photovoltaic and thermoelectric. In: Cavallaro, F. (ed.) *Assessment and Simulation Tools for Sustainable Energy Systems, Green Energy and Technology*, vol. 129. Springer, London (2013)
12. Saaty, T.L.: The Analytic Hierarchy Process. McGraw-Hill, New York (1980)
13. Saaty, T.L.: Group Decision Making and the AHP. Springer, New York (1989)
14. Delgado, M., Verdegay, J.L., Vila, M.A.: Linguistic decision making models. Int. J. Intell. Syst. **7**, 479–492 (1992)
15. Herrera, F., Alonso, S., Chiclana, F., Herrera-Viedma, E.: Computing with Words in decision making: foundations, trends and prospects. Fuzzy Optim. Decis. Making **8**, 337–364 (2009)
16. Zadeh L.A.: Fuzzy sets. Information and control. In: Zeiler, M. (ed.) *Modeling Our World. The ESRI Guide to Geodatabase Design*, 2nd edn, vol. 8, pp. 338–353. Esri Press, Redlands (1965)
17. Klir, G.J., Yuan, B.: Fuzzy Sets and Fuzzy Logic: Theory and Applications. Prentice Hall, Upper Saddle River (1995)
18. Dubois, D., Prade, H.: Fuzzy Sets and Systems: Theory and Applications. Academic Press Inc, New York (1980)
19. García-Cascales, M.S., Lamata, M.T., Sánchez-Lozano, J.M.: Evaluation of photovoltaic cells in a multi-criteria decision making process. Ann. Oper. Res. **199**, 373–391 (2012)
20. Sánchez-Lozano, J.M.: Search and evaluation of optimal sites to implant renewable energies facilities, combination of geographic information systems (GIS) and soft computing: case study of the coast of the Region of Murcia. MS Thesis 2013, Technical University of Cartagena

Gene Priorization for Tumor Classification Using an Embedded Method

Jose M. Cadenas, M. Carmen Garrido, Raquel Martínez,
David Pelta and Piero P. Bonissone

Abstract The application of microarray technology to the diagnosis of cancer has been a challenge for computational techniques because the datasets obtained have high dimension and a few examples. In this paper two computational techniques are applied to tumor datasets in order to carry out the task of diagnosis of cancer (classification task) and identifying the most promising candidates among large list of genes (gene prioritization). Both techniques obtain good classification results but only one provides a ranking of genes as additional information and thus, more interpretable models, being more suitable for jointly addressing both tasks.

Keywords Fuzzy random forest · Gene priorization · Gene expression data · Tumor datasets

1 Tumor Classification from Gene Expression Data

The challenge of cancer treatment has been to target specific therapies to pathogenetically distinct tumor types, to maximize efficacy and minimize toxicity. Improvements in cancer classification have thus been central to advances in cancer treatment. Cancer classification is divided into two challenges: class discovery and class prediction. Class discovery refers to defining previously unrecognized tumor subtypes.

J.M. Cadenas (✉) · M.C. Garrido · R. Martínez
Department of Information Engineering and Communications,
University of Murcia, Murcia, Spain
e-mail: jcadenas@um.es; carmengarrido@um.es; raquel.m.e@um.es

D. Pelta
Department of Computer Science and Artificial Intelligence,
University of Granada, Granada, Spain
e-mail: dpelta@decsai.ugr.es

P.P. Bonissone
General Electric Global Research, One Research Circle,
Niskayuna, NY, U.S.A.
e-mail: bonissone@ge.com

© Springer International Publishing Switzerland 2016
K. Madani et al. (eds.), *Computational Intelligence*,
Studies in Computational Intelligence 613,
DOI 10.1007/978-3-319-23392-5_20

363

Class prediction refers to the assignment of particular tumor examples to already-defined classes. In the early days, cancer classification has been relying on subjective judgment from experienced pathologists. When microarray technology was discovered began to be applied to cancer diagnosis. The most important application of the microarray technique is to discriminate the normal and cancerous tissue samples according to their expression levels, identify a small subset of genes that are responsible for the disease and to discover potential drugs [15].

Experimental techniques based on oligonucleotide or cDNA arrays now allow the expression level of thousands of genes to be monitored in parallel [1]. To use the full potential of such experiments, it is important to develop the ability to process and extract useful information from large gene expression datasets.

Constantly improving gene expression profiling technologies are expected to provide understanding and insight into cancer related cellular processes. Gene expression data is also expected to significantly aid in the development of efficient cancer diagnosis and classification platforms. Gene expression data can help in better understanding of cancer. Normal cells can evolve into malignant cancer cells through a series of mutations in genes that control the cell cycle, apoptosis, and genome integrity, to name only a few. As determination of cancer type and stage is often crucial to the assignment of appropriate treatment [16], a central goal of the analysis of gene expression data is the identification of sets of genes that can serve, via expression profiling assays, as classification or diagnosis platforms.

Another important purpose of gene expression studies is to improve understanding of cellular responses to drug treatment. Expression profiling assays performed before, during and after treatment, are aimed at identifying drug responsive genes, indications of treatment outcomes, and at identifying potential drug targets [9]. More generally, complete profiles can be considered as a potential basis for classification of treatment progression or other trends in the evolution of the treated cells.

Data obtained from cancer related gene expression studies typically consists of expression level measurements of thousands of genes. This complexity calls for data analysis methodologies that will efficiently aid in extracting relevant biological information. Previous gene expression analysis work emphasizes clustering techniques (nonsupervised classification), which aim at partitioning the set of genes into subsets that are expressed similarly across different conditions. On the other hand, supervised classification techniques (also called class prediction or class discrimination) with the aim to assign examples to predefined categories [12, 16, 19].

The objectives of supervised classification techniques are: (1) to build accurate classifiers that enable the reliable discrimination between different cancer classes, (2) to identify biomarkers of diseases, i.e. a small set of genes that leads to the correct discrimination between different cancer states. This second purpose of supervised classification can be achieved by classifiers that provide understandable results and indicate which genes contribute to the discrimination.

Following this line, in this paper the goal is to apply two techniques with embedded capacity to discard input features and thus propose a subset of discriminative genes (embedded methods [20]). We apply them to classify and select features to tumor datasets in order to carry out an analysis of these datasets and to obtain the information that provide understandable results. These techniques are the Fuzzy Random Forest method (FRF) proposed in [3, 7] and the Feature Selection Fuzzy Random Forest method (FRF-fs) proposed in [6].

This paper is organized as follows. First, in Sect. 2 some techniques applied to gene expression data reported in literature are briefly described. Next, in Sect. 3, the applied methods are described. Then, in Sect. 4 we perform an analysis of two tumor datasets using these methods. Finally, in Sect. 5 remarks and conclusions are presented.

2 Machine Learning and Gene Expression Data

In this section, we describe some of the machine learning techniques used for the management of gene expression data.

2.1 Cluster Analysis Based Techniques

Clustering is one of the primary approaches to analyze such large amount of data to discover the groups of co-expressed genes. In [18] an attempt to improve a fuzzy clustering solution by using SVM classifier is presented. In this regard, two fuzzy clustering algorithm, VGA and IFCM have been used.

In [1] a clustering algorithm to organize the data in a binary tree is used. The algorithm was applied to both the genes and the tissues, revealing broad coherent patterns that suggest a high degree of organization underlying gene expression in these tissues. Coregulated families of genes clustered together. Clustering also separated cancerous from noncancerous tissue.

In [16] a SOM to divide the leukemia examples into cluster is used. First, they applied a two-cluster SOM to automatically discovering the two types of leukemia. Next, they applied a four-cluster SOM. They subsequently obtained immunophenotype data on the examples and found that the four classes largely corresponded to AML, T-lineage ALL, B-lineage ALL, and B-lineage ALL, respectively. The four-cluster SOM thus divided the examples along another key biological distinction.

In [2] a clustering based classifier is built. The clustering algorithm on which the classifier is constructed is the CAST algorithm that takes as input a threshold parameter t, which controls the granularity of the resulting cluster structure, and a similarity measure between the tissues. To classify a example they cluster the training data and example, maximizing compatibility to the labeling of the training data. Then they examine the labels of all elements of the cluster the example belongs to and use a simple majority rule to determine the unknown label.

2.2 Techniques for Feature Selection and Supervised Classification

Discovering novel disease genes is still challenging for constitutional genetic diseases (a disease involving the entire body or having a widespread array of symptoms) for which no prior knowledge is available. Performing genetic studies frequently result in large lists of candidate genes of which only few can be followed up for further investigation. Gene prioritization establishes the ranking of candidate genes based on their relevance with respect to a biological process of interest, from which the most promising genes can be selected for further analysis [19]. This is a special case of feature selection, a well-known problem in machine learning.

In [16] a procedure that uses a fixed subset of "informative genes" is developed. These "informative genes" are chosen based on their correlation with the class distinction.

In [12], a Random Forest ensemble is used to carry out the feature selection process for classification from gene expression data. The technique calculates a measure of importance for each feature based on how the permutation of the values of that feature in the dataset affects to the classification of the out-of-bag (OOB) dataset of each decision tree of ensemble [5]. Following this study, in [14], a Random forest ensemble which solves the problems existing in [12] is proposed.

In [13] a study of classification of gene expression data using metaheuristics is presented. The authors show that gene selection can be casted as a combinatorial search problem, and consequently be handled by these optimization techniques.

In [19], four different strategies to prioritize candidate genes are proposed. These strategies are based on network analysis of differential expression using distinct machine learning approaches to determine whether a gene is surrounded by highly differentially expressed genes in a functional association or protein-protein interaction network.

Another work to select genes is proposed in [10]. This paper shows that a systematic and efficient algorithm, mixed integer linear programming based hyper-box enclosure (HBE) approach, can be applied to classification of different cancer types efficiently.

3 Classification and Feature Selection by Fuzzy Random Forest

In this section, we describe the methods that we will use in this paper.

3.1 Fuzzy Random Forest for Classification

We briefly describe the Fuzzy Random Forest (FRF) ensemble proposed in [3, 7]. FRF ensemble was originally presented in [3], and then extended in [7], to handle imprecise and uncertain data. We describe the basic elements of the FRF ensemble and the types of data that are supported by this ensemble in both learning and classification phases.

Fuzzy Random Forest Learning

Let E be a dataset. FRF learning phase uses Algorithm 1 to generate the FRF ensemble whose base classifier is a Fuzzy Decision Tree (FDT). Algorithm 2 shows the FDT learning algorithm [8].

Algorithm 1: FRFlearning.

1: **Input:** E, *Fuzzy Partition*;   **Output:** FRF
2: **begin**
3:   **repeat**
4:     Take a random sample of $|E|$ examples with replacement from the dataset E
5:     Apply Algorithm 2 to the examples obtained in the previous step to construct a FDT
6:   **until** all FDTs are built to constitute the FRF ensemble
7: **end**

Algorithm 2 has been designed so that the FDTs can be constructed without considering all the features to split the nodes. Algorithm 2 is an algorithm to construct FDTs where the numerical features have been discretized by a fuzzy partition. The domain of each numerical feature is represented by trapezoidal fuzzy sets, $F_1, \ldots, F_f$ so each internal node of the FDTs, whose division is based on a numerical feature, generates a child node for each fuzzy set of the partition. Moreover, Algorithm 2 uses a function, denoted by $\chi_{t,N}(e)$, that indicates the degree with which the example e satisfies the conditions that lead to node N of FDT t. Each example e is composed of features which can be crisp, missing, interval, fuzzy values belonging (or not) to the fuzzy partition of the feature. Furthermore, we allow the class feature to be set-valued. These examples (according to the value of their features) have the following treatment:

Algorithm 2: FDecisionTree.

1: **Input:** E, *Fuzzy Partition*;   **Output:** FDT
2: **begin**
3: Start with the examples in E with values $\chi_{Fuzzy\_Tree,root}(e) = 1$ to all examples with a single class and replicate
the examples with set-valued class and initialize their weight according to the available knowledge about their class
4: Let A be the feature set (numerical features are partitioned according to Fuzzy Partition)
5: **repeat**
6:    Choose a feature to the split at the node N
7:    **loop**
8:       Make a random selection of features from the set A
9:       Compute the information gain for each selected feature using the values $\chi_{Fuzzy\_Tree,N}(e)$ of each e in node N
taking into account the function $\mu_{simil(e)}$ for the cases required
10:       Choose the feature such that information gain is maximal
11:    **end loop**
12:    Divide N in children nodes according to possible selected feature outputs in the previous step and remove it from
the set A. Let E_n be the dataset of each child node
13: **until** the stopping criteria is satisfied
14: **end**

- Each example e used in the training of the FDT t has assigned an initial value $\chi_{t,root}(e)$. If an example has a single class this value is 1. If an example has a set-valued class, it is replicated with a weight according to the available knowledge about the classes.
- According to the membership degree of the example e to different fuzzy sets of partition of a split based on a numerical feature:
  - If the value of e is crisp, the example e may belong to one or two children nodes, i.e., $\mu_{fuzzy\_set\_partition}(e) > 0$. In this case $\chi_{t,childnode}(e) = \chi_{t,node}(e) \cdot \mu_{fuzzy\_set\_partition}(e)$.
  - If the value of e is a fuzzy value matching with one of the sets of the fuzzy partition of the feature, e will descend to the child node associated. In this case, $\chi_{t,childnode}(e) = \chi_{t,node}(e)$.
  - If the value of e is a fuzzy value different from the sets of the fuzzy partition of the feature, or the value of e is an interval value, we use a similarity measure, $\mu_{simil}(\cdot)$, that, given the feature "$Attr$" to be used to split a node, measures the similarity between the values of the fuzzy partition of the feature and fuzzy values or intervals of the example in that feature. In this case, $\chi_{t,childnode}(e) = \chi_{t,node} \cdot \mu_{simil}(e)$.
  - When the example e has a missing value, the example descends to each child node $node_h, h = 1, \ldots, H_i$ with a modified value proportionately to the weight of each child node. The modified value for each $node_h$ is calculate as $\chi_{node_h}(e) = \chi_{node}(e) \cdot \frac{T\chi_{node_h}}{T\chi_{node}}$ where $T\chi_{node}$ is the sum of the weights of the examples with known value in the feature i at $node$ and $T\chi_{node_h}$ is the sum of the weights of the examples with known value in the feature i that descend to the node $node_h$.

Fuzzy Random Forest Classification

The fuzzy classifier module operates on FDTs of the FRF ensemble using one of these two possible strategies: Strategy 1—Combining the information from the different leaves reached in each FDT to obtain the decision of each individual FDT and then

Fig. 1 Framework of FRF-fs

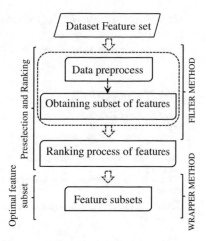

applying the same or another combination method to generate the global decision of the FRF ensemble; and Strategy 2—Combining the information from all leaves reached from all FDTs to generate the global decision of the FRF ensemble.

3.2 Fuzzy Random Forest for Feature Selection

The FRF-fs method [6] is classified as a hybrid method that combines the filter and wrapper methods. The framework (Fig. 1) consists of main steps: (1) Scaling and discretization process of the feature set; and feature pre-selection using the discretization process; (2) The feature pre-selection ranking process using information given by Fuzzy Random Forest ensemble; and (3) Wrapper feature selection using a classification technique. Starting from the ordered features, this wrapper method constructs an ascending sequence of sets of candidate features, by invoking and testing the features stepwise. The different feature subsets obtained by this process are evaluated by a machine learning method. In each step, the method obtains information useful to the user: pre-selected feature subset, feature subsets ranking and optimal feature subset.

In the filter method, we use the method proposed in [8]. From the feature subset and the dataset obtained with the filter method, we apply FRF method. Once FRF ensemble has been obtained, we have all the information about each FDT. Algorithm 3 describes how information provided for each FDT of the ensemble is compiled and used to measure the importance of each feature.

More specifically, the information we get from each FDT t for each feature a is the following:

- Information gain of node N for the feature a (IG_{Na}) where the feature a has been selected as the best candidate to split it.

- Depth level of node N (P_{Na}) where feature a has been selected as the best candidate to split it.
- Classification accuracy Acc_t of FDT t when classifying the dataset OOB_t.

Algorithm 3: INFFRF Information of the FRF.

1: **Input:** $E, Fuzzy\ Partition, TN$;　**Output:** INF
2: **begin**
3: Building a Fuzzy Random Forest (Algorithm 1 - 3.1)
4: **for** each FDT t=1 to TN of the FRF ensemble **do**
5:　Save the feature a chosen to split each node N, information gain of node, IG_{Na}, and the depth of that node P_{Na}, in INF_a.
6:　Obtain the classification accuracy Acc_t of the FDT t with its corresponding OOB_t dataset.
7: **end for**
8: **end**

Algorithm 4 details how the information INF obtained from the FRF ensemble is combined to obtain an importance measure of the features where p_i is the weight we assign to feature a depending on the place where it appears in the FDT t. After the information is combined, the output of this algorithm is a matrix (IMP) where for each FDT t and for each feature a, the importance value obtained in the FDT t for the feature a is stored.

Algorithm 4: IMPFRF Combining information INF.

1: **Input:** INF, TN;　**Output:** IMP
2: **begin**
3: **for** each FDT t=1 to TN **do**
4:　**for** each feature a=1 to $|Attr|$ **do**
5:　　**for** all nodes N where feature a appears **do**
6:　　　**if** $P_{Na} = i$ **then**
7:　　　　$IMP_{ta} = IMP_{ta} + p_i \cdot IG_{Na}$ with $i \geq 0$ and $P_{rootnode} = 0$
8:　　　**end if**
9:　　**end for**
10:　　**for** each feature a=1 to $|Attr|$ **do**
11:　　　$IMP_{ta} = \left(\frac{IMP_{ta} - min(IMP_t)}{max(IMP_t) - min(IMP_t)} \right) \cdot OOB_t$
12:　　**end for**
13:　　The vector IMP_t is ordered in descending order, $IMP_{t\sigma_t}$, where σ_t is the permutation obtained when ordering IMP_t
14:　**end for**
15: **end for**
16: **end**

The idea behind the measure of importance of each feature is that it uses the features of the FDTs obtained and the decision nodes built with them in the following way. The importance of a feature is determined by its depth in a FDT. Therefore a feature that appears on the top of a FDT is more important in that FDT than another feature that appears in the lower nodes. And, a FDT that has a classification accuracy greater than another to classify the corresponding OOB (dataset independent of the training dataset) is a better FDT. The final decision is agreed by the information obtained for all FDTs.

As a result of Algorithm 4, we obtain for each FDT of FRF ensemble an importance ranking of features. Specifically, we will have TN importance rankings for each feature a. Applying an operator OWA, we add them into one ranking. This final ranking indicates the definitive importance of the features.

OWA operators (Ordered Weighted Averaging) were introduced by Yager in 1988 [22]. OWA operators are known as compensation operators. They are aggregation operators of numerical information that consider the order of the assessments that will be added. In our case, we have TN ordered sets. Given a weight vector W, the vector $RANK$ represents the ranking of the pre-selected feature subset and is obtained as follows (the vector $RANK$ is ordered in descending order: $RANK_\sigma$):

$$OWAIMP_t = W \cdot IMP_{t_{\sigma_t}}, \text{ for } t = 1, \ldots, TN$$

$$RANK_a = \sum_{t=1}^{TN} OWAIMP_{t\sigma_t(a)}, \text{ for } a = 1, \ldots, |A|$$

3.3 Wrapper for Feature Final Selection

Once the ranking of the pre-selected feature subset, $RANK_\sigma$, is obtained, we have to find an optimal subset of features. One option to search the optimal subset is by adding a single feature at a time following a process that uses $RANK_\sigma$. The several feature subsets obtained by this process are evaluated by a machine learning method that supports low quality data (called $Classifier_{LQD}$) with a process of cross-validation. The detailed process of the proposed wrapper method is shown in Algorithm 5.

Starting from the ordered feature pre-selected, construct an ascending sequence of FRF models, by invoking and testing the features stepwise. We perform a sequential feature introduction in two phases:

- In the first phase two feature subsets are built: the feature subsets CF_{base} and CF_{comp}. A feature f_i is added to the CF_{base} subset only if the decrease of the error rate using the features of $CF_{base} \cup \{f_i\}$ subset exceeds a threshold δ_1. The idea is that the error decrease by adding f_i must be significant for that feature to belong to the CF_{base} subset. If when we classify using the subset $CF_{base} \cup \{f_i\}$, an error decrease smaller than a threshold δ_1 or an error increase smaller than a threshold δ_2 is obtained, f_i becomes part of the subset CF_{comp}.
- The second phase starts with both CF_{base} and CF_{comp} sets. We fix CF_{base} and add feature subgroups from CF_{comp} to build several FRF models. This phase determines the final feature set with minimum error according to the conditions reflected on line 22 of Algorithm 5. These conditions are interpreted as "select the subset that decrements the error in an amount over threshold δ_3 or decrements the error in an amount below δ_3 but using a smaller number of features."

4 FRF and Tumor Classification

In this section we examine the performance of the FRF ensemble for classification and feature selection from gene expression data.

4.1 Gene Expression Data

In this section, we describe the three datasets that we will analyze. The first dataset involves comparing tumor and normal examples of the same tissue, the second one involves examples from two variants of the same disease and the third one contains measurements of the gene expression of cancer patients and healty men.

Algorithm 5: Wrapper method.

Input: E, candidate feature set CF and information system $RANK_\sigma$;   **Output:** CF_{opt} selected feature set
begin
$CF_{comp} = \{\}$ and $CF_{base} = \{f_1\}$ where f_1 is the first feature of $RANK_\sigma$
$ERR_1 = Classifier(E, CF_{base})$ using cross-validation, $BE = ERR_1$
for each $f_i \in CF$, with $i = 2, \ldots, |CF|$ in the order determined by $RANK_\sigma$ **do**
  $ERR_B = Classifier_{LQD}(E, CF_{base} \cup \{f_i\})$ using cross-validation
  if $(BE - ERR_B) > \delta_1$ **then**
    $CF_{base} = CF_{base} \cup \{f_i\}$
  else
    if $(ERR_B - BE) < \delta_2$ **then**
      $CF_{comp} = CF_{comp} \cup \{f_i\}$
    end if
  end if
end for
$CF_{aux} = CF_{base}$
for each $f_i \in CF_{comp}$, with $i = 1, \ldots, |CF_{comp}|$ in the order determined by $RANK_\sigma$ **do**
  $B = CF_{base}$, $STOP = 0$, $j = i$
  while $(STOP < \delta_2)$ and $(j \leq |CF_{comp}|)$ **do**
    $B = B \cup \{f_j\}$
    $ERR_B = Classifier_{LQD}(D, B)$ using cross-validation
    if $((BE - ERR_B) \geq \delta_3)$ or $(0 \leq (BE - ERR_B) < \delta_3$ and $|CF_{aux}| > |B|)$ **then**
      $CF_{aux} = B$, $BE = ERR_B$
    else
      if $(ERR_B - BE) > \delta_2$ **then**
        $STOP = (ERR_B - BE)$
      end if
    end if
    $j = j + 1$
  end while
end for
Return: $CF_{opt} = CF_{aux}$
end

Colon Cancer, Leukemia and Prostate Datasets. Colon tumor is a disease in which cancerous growths are found in the tissues of the colon epithelial cells. The Colon dataset contains 62 examples. Among them, 40 tumor biopsies are from tumors (labeled as "negative") and 22 normal (labeled as "positive") biopsies are from healthy parts of the colons of the same patients. The final assignments of the status of biopsy examples were made by pathological examination. The total number of genes to be tested is 2000 [1].

In the 1960s was provided the first basis for classification of acute leukemias into those arising from lymphoid precursors (acute lymphoblastic leukemia, ALL) or from myeloid precursors (acute myeloid leukemia, AML). The Leukemia dataset is a collection of expression measurements reported by [16]. The dataset contains 72 examples. These examples are divided to two variants of leukemia: 25 examples of acute myeloid leukemia (AML) and 47 examples of acute lymphoblastic leukemia (ALL). The source of the gene expression measurements was taken from 63 bone marrow examples and 9 peripheral blood examples. Gene expression levels in these 72 examples were measured using high density oligonucleotide microarrays. The expression levels of 7129 genes are reported.

Prostate dataset contains gene expression data (6033 genes for 102 examples) from the microarray study reported by [21]. The obtained results support the notion that the clinical behavior of prostate cancer is linked to underlying gene expression differences that are detectable at the time of diagnosis. This dataset contains measurements of gene expression of 52 prostate patients and 50 healty men.

4.2 Estimating Prediction Errors

We apply the cross-validation method to evaluate the prediction accuracy of the classification method. To apply this method, we partition the dataset E into k sets of examples, $C_1, \ldots, C_k$. Then, we construct a data set $D_i = E - C_i$, and test the accuracy of a model obtained from D_i on the examples in C_i. We estimate the accuracy of the method by averaging the accuracy over the k cross-validation trials.

There are several possible choices of k. A common approach is to set k =number of examples. This method is known as leave one out cross validation (LOOCV). We will use the LOOCV method.

Although our purpose is not to compare the results with other methods, as a sample, in Table 1 we show the accuracy estimates for the different methods applied to the three datasets. The results obtained in [12, 14] are calculated with the 0.632+bootstrap method, and the Leukemia dataset has 38 examples and 3051 features.

Estimates of classification accuracy give only a partial insight on the performance of a method. Also, we treat all errors as having equal penalty. In the problems we handle, however, errors have asymmetric weights. We distinguish false positive error-normal tissues classified as tumor, and false negative errors - tumor tissues classified as normal. In diagnostic applications, false negative errors can be detrimental, while false positives may be tolerated.

ROC curves are used to evaluate the "power" of a classification method for different asymmetric weights [4, 17]. Since the area under the ROC curve (denoted by AUC) is a portion of the area of the unit square, its value will always be between 0.0 and 1.0. A realistic classifier should not have an AUC less than 0.5 (area under the diagonal line between (0,0) and (1,1)). The AUC has an important statistical prop-

Table 1 Accuracy of different methods on datasets

	Colon		Leukemia		Prostate	
	Correct	Unclassified	Correct	Unclassified	Correct	Unclassified
Clustering[A]	88.70	0.00	–	–	–	–
Nearest neighbor[A]	80.60	0.00	91.60	0.00	–	–
SVM, linear kernel[A]	77.40	9.70	93.00	5.60	–	–
SVM, quad. kernel[A]	74.20	11.30	94.40	4.20	–	–
Boosting, 100 iter.[A]	72.60	9.70	95.80	1.40	–	–
NN.vs[B]	84.20	0.00	94.40	0.00	91.9	0.00
RF.du (s.e.=0)[B]	84.10	0.00	91.30	0.00	93.9	0.00
RF.ge[C]	91.70	0.00	99.00	0.00	96.07	0.00
FRF	91.94	0.00	98.61	0.00	96.08	0.00

The results marked with A, B and C are obtained from [2, 12, 14], respectively

Table 2 Confusion matrixes obtained with FRF

		Colon			Leukemia			Prostate	
		Actual value			Actual value			Actual value	
		1	0		ALL	AML		1	0
Prediction	1	37	2	ALL	46	0	1	49	1
Outcome	0	3	20	AML	1	25	0	3	49

erty: the AUC of a classifier is equivalent to the probability that the classifier will rank a randomly chosen positive instance higher than a randomly chosen negative instance. This is equivalent to the Wilcoxon test of ranks [17].

The confusion matrixes obtained by applying FRF to the three datasets are shown in Table 2.

Confusion matrix of Colon dataset shows five errors, and a Specificity of 0.9091 and Sensibility of 0.9250. Confusion matrix of Leukemia dataset shows one error, and a Specificity of 1.0 and Sensibility of 0.9787. Confusion matrix of Prostate dataset shows four errors, and a Specificity of 0.98 and Sensibility of 0.9423.

ROC curves with all features are shown in Fig. 2 and AUC values for (a) Colon, (b) Leukemia and (c) Prostate datasets are 0.9761, 0.9991 and 0.9983 respectively.

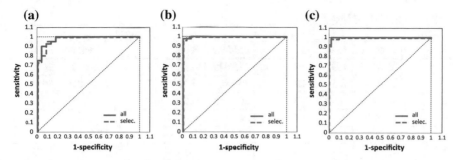

Fig. 2 ROC curves with all/selected features

Table 3 Accuracy with/without selected features with FRF method

Dataset	All features		Sel. features	
	Correct	Unclassified	Correct	Unclassified
Colon	91.40	0.00	93.55	0.00
Leukemia	98.61	0.00	98.61	0.00
Prostate	96.08	0.00	97.06	0.00

Table 4 Confusion matrixes obtained with FRF using selected features

	Colon			Leukemia			Prostate		
		Actual value			Actual value			Actual value	
		1	0		ALL	AML		1	0
Prediction	1	38	2	ALL	46	1	1	50	1
Outcome	0	2	20	AML	0	25	0	2	49

4.3 Gene Selection

It is clear that the expression levels of many of the genes in our datasets are irrelevant to the distinction between tumors. Taking such genes into account during classification increases the dimensionality of the classification problem, presents computational difficulties, and introduces noise to the process. Another issue with a large number of genes is the interpretability of the results. If our methods to distinguish tumor from normal tissues is encoded in the expression levels of few genes, then we might be able to understand the biological significance of these genes.

Thus, it is crucial to recognize whether a small number of genes can suffice for good classification. The gene expression datasets are problematic in that they contain a large number of genes (features) and thus methods that search over subsets of features can be expensive. Moreover, these datasets contain only a small number of examples, so the detection of irrelevant genes can suffer from statistical instabilities.

Table 5 Features ranking in datasets

Colon			Leukemia			Prostate		
	Ranking	Fe. n.		Ranking	Fe. n.		Ranking	Fe. n.
1	35.6266	419	1	31.2849	3252	1	72.6306	2619
2	17.0359	765	2	30.1804	1882	2	12.9096	5016
3	15.6419	1635	3	30.1763	1834	3	8.1136	1881
4	13.5216	824	4	26.5833	4847	4	7.9227	1359
5	13.4986	1168	5	23.9430	2288	5	7.7503	4335
6	13.4898	513	6	13.5707	2354	6	6.3471	4183
7	9.6363	1772	7	13.1465	6041	7	5.1158	4087
8	7.2361	571	8	9.8707	6376	8	5.0216	4287
9	7.0409	1546	9	4.8665	4644	9	4.2499	3616
10	6.8134	1423	10	1.4004	3623	10	4.2361	4136
11	6.7085	1761	..	.....	...	11	4.2223	3946
12	6.6085	1939	–	–	–	12	4.1510	3606
13	6.4989	1990	–	–	–	..	.....	...
14	5.9908	377	–	–	–	–	–	–
15	4.6654	1668	–	–	–	–	–	–
16	4.0917	1346	–	–	–	–	–	–
17	3.1929	1586	–	–	–	–	–	–
18	2.3743	548	–	–	–	–	–	–
19	2.0175	474	–	–	–	–	–	–
20	1.8373	802	–	–	–	–	–	–
21	1.7315	1867	–	–	–	–	–	–
..	.....	...	–	–	–	–	–	–

Significance of a Gene and Ranking. The FRF-fs method [6] to feature selection obtains a feature ranking based on an importance measurement of each feature, and from that ranking, an optimal feature subset. The vector $RANK$ (see Sect. 3.2) contains the importance measure of the features. In Table 5 a portion of that ranking of features and their importance values is shown.

Gene Prioritization in Cancer Data. In the final phase of the FRF-fs method [6] an optimal feature subset is obtained.

In the Colon dataset the optimal feature subset is {419, 765, 824, 1168, 513, 1772, 571, 1546, 1423, 1761, 1939, 1990, 377, 1668, 1346, 1586, 548, 474, 802, 1867}. In addition, to give more interpretability, FRF-fs method obtains a feature partition. In Table 6 the partition obtained for this optimal features subset is shown. The first column shows the gene number while the second one shows the different partitions for this gene.

Table 6 Features partition in colon dataset

Fe.n.	Partitions	–	–
377	(0,0,0.4046,0.5246)	(0.4046,0.5246,1,1)	–
419	(0,0,0.6981,0.7140)	(0.6981,0.7140,0.7241,0.7256)	(0.7241,0.7256,1,1)
474	(0,0,0.8360,0.9194)	(0.8360,0.9194,1,1)	–
513	(0,0,0.5625,0.5657)	(0.5625,0.5657,1,1)	–
548	(0,0,0.7852,0.9132)	(0.7852,0.9132,1,1)	–
571	(0,0,0.3579,0.4168)	(0.3579,0.4168,1,1) 7	–
765	(0,0,0.4869,0.5655)	(0.4869,0.5655,0.6270,0.6286)	(0.6270,0.6286,0.63,0.63)
–	(0.63,0.63,0.6543,0.6769)	(0.6543,0.6769,0.7320,0.7667)	(0.7320,0.7677,1,1)
802	(0,0,0.4227,0.7499)	(0.4227,0.7499,1,1)	–
824	(0,0,0.6009,0.6017)	(0.6009,0.6017,0.6026,0.6033)	(0.6026,0.6033,1,1)
1168	(0,0,0.5665,0.5793)	(0.5665,0.5793,1,1)	–
1346	(0,0,0.4839,0.5456)	(0.4839,0.5456,1,1)	–
1423	(0,0,0.8269,0.8730)	(0.8269,0.8730,1,1)	–
1546	(0,0,0.0792,0.3206)	(0.0792,0.3206,0.4904,0.5156)	(0.4904,0.5156,1,1)
1586	(0,0,0.9168,0.9753)	(0.9168,0.9753,1,1)	–
1668	(0,0,0.2804,0.6472)	(0.2804,0.6472,1,1)	–
1761	(0,0,0.5641,0.5764)	(0.5641,0.5764,0.5784,0.5902)	(0.5784,0.5902,1,1)
1772	(0,0,0.5156,0.5172)	(0.5156,0.5172,1,1	–
1867	(0,0,0.5292,0.6251)	(0.5292,0.6251,1,1)	–
1939	(0,0,0.8908,0.8934)	(0.8908,0.8934,1,1)	–
1990	(0,0,0.1022,0.3066)	(0.1022,0.3066,0.4484,0.5811)	(0.4484,0.5811,1,1)

In the Leukemia dataset the optimal feature subset is {3252, 4847, 2288, 2354, 6041, 6376, 4644}. In Table 7 the partition obtained for this optimal features subset is shown.

In the Prostate dataset the optimal feature subset is {2619, 5016, 1881, 1359, 4335, 4183, 4087, 4287, 3616, 4136, 3946, 3606}. In Table 8 the partition obtained for this optimal features subset is shown.

Classifying with Selected Subsets. Now, the classification procedure is applied using the training data restricted to the subset of selected genes.

In Table 3 we show the accuracy estimates for FRF method applied to the three datasets with/without the selected features.

The confusion matrixes obtained by applying FRF to the three datasets with the selected features are shown in Table 4.

Confusion matrix of Colon dataset shows four errors, and a Specificity of 0.9091 and Sensibility of 0.9500. Confusion matrix of Leukemia dataset shows one error, and a Specificity of 0.9600 and Sensibility of 1.0. Confusion matrix of Prostate dataset shows three errors, and a Specificity of 0.98 and Sensibility of 0.9615. ROC curves are shown in Fig. 2. AUC values for Colon, Leukemia and Prostate are 0.9710, 0.9987 and 0.9954 respectively.

Table 7 Features partition in Leukemia dataset

Fe.n.	Partitions	–	–
2288	(0,0,0.0733,0.0835)	(0.0733,0.0835,1,1)	–
2354	(0,0,0.1451,0.1931)	(0.1451,0.1931,1,1)	–
3252	(0,0,0.0681,0.0706)	(0.0681,0.0706,0.0738,0.0747)	(0.0738,0.0747,1,1)
4644	(0,0,0.2425,0.2427)	(0.2425,0.2427,1,1)	–
4847	(0,0,0.2116,0.2157)	(0.2116,0.2157,0.3479,0.3531)	(0.3479,0.3531,1,1)
6041	(0,0,0.1937,0.1963)	(0.1937,0.1963,0.2001,0.2037)	(0.2001,0.2037,1,1)
6376	(0,0,0.1408,0.1422)	(0.1408,0.1422,1,1)	–

Table 8 Features partition in prostate dataset

Fe.n.	Partitions	–	–
1359	(0,0,0.0662,0.0741)	(0.0662,0.0741,1,1)	–
1881	(0,0,0.4734,0.4959)	(0.4734,0.4959,1,1)	–
2619	(0,0,0.3212,0.3870)	(0.3212,0.3870,0.4740,0.4818)	(0.4740,0.4818,0.4873,0.4874)
	(0.4873,0.4874,0.5001,0.5062)	(0.5001,0.5062,0.5134,0.5139)	(0.5134,0.5139,0.5192,0.5199)
	(0.5192,0.5199,0.5801,0.5866)	(0.5801,0.5866,1,1)	–
3606	(0,0,0.1498,0.1540)	(0.1498,0.1540,0.1558,0.1614)	(0.1558,0.1614,1,1)
3616	(0,0,0.6545,0.6571)	(0.6545,0.6571,0.6810,0.6830)	(0.6810,0.6830,1,1)
3946	(0,0,0.9506,0.9573)	(0.9506,0.9573,1,1)	–
4087	(0,0,0.8361,0.8783)	(0.8361,0.8783,1,1)	–
4136	(0,0,0.4793,0.6177)	(0.4793,0.6177,1,1)	–
4183	(0,0,0.0173,0.0190)	(0.0173,0.0190,1,1)	–
4287	(0,0,0.0099,0.0100)	(0.0099,0.0100,0.0101,0.0103)	(0.0101,0.0103,1,1)
4335	(0,0,0.6304,0.6436)	(0.6304,0.6436,0.7509,0.7889)	(0.7509,0.7889,1,1)
5016	(0,0,0.3068,0.3075)	(0.3068,0.3075,0.3098,0.3098)	(0.3098,0.3098,0.3121,0.3134)
	(0.3121,0.3134,0.3249,0.3376)	(0.3249,0.3376,1,1)	–

Following the methods proposed in [11, 17], we conclude that there are no significant differences between the results obtained when using all features or the selected ones.

We can therefore conclude that the selection of features does not cause loss of accuracy but significantly decreases the number of features.

5 Conclusions

In this paper we have applied a fuzzy decision tree ensemble to tumor datasets with gene expression data.

On the one hand, we have applied the ensemble to the classification of examples described by the set of all features. On the other hand, we have applied the ensemble to select a gene subset and to classify examples only described with the selected genes. The classification accuracies, in both cases, are high. These results are validated statistically by the ROC curve and AUC area.

When we work with a fuzzy decision tree ensemble, in addition to achieve good results, these one are provided in a highly interpretable way.

As part of the solution, the method provides a partition of numerical features of the problem and a ranking of importance of these features which permits the identification of sets of genes that can serve as classification or diagnosis platforms.

Acknowledgments Supported by the projects TIN2011-27696-C02-01 and TIN2011-27696-C02-02 of the Ministry of Economy and Competitiveness of Spain. Thanks also to "Agencia de Ciencia y Tecnología de la Región de Murcia" (Spain) for the support given to Raquel Martínez by the scholarship program FPI.

References

1. Alon, U., Barkai, N., Notterman, D.A., Gish, K., Ybarra, S., Mack, D., Levine, A.J.: Broad patterns of gene expression revealed by clustering analysis of tumor and normal colon tissues probed by oligonucleotide arrays. Proc. Natl. Acad. Sci. U.S.A. **96**, 6745–6750 (1999)
2. Ben-Dor, A., Bruhn, L., Friedman, N., Nachman, I., Schummer, M., Yakhini, Z.: Tissue classification with gene expression profiles. J. Comput. Biol. **7**(3–4), 559–583 (2004)
3. Bonissone, P.P., Cadenas, J.M., Garrido, M.C., Díaz-Valladares, R.A.: A fuzzy random forest. Int. J. Approximate Reasoning **51**(7), 729–747 (2010)
4. Brandley, A.P.: The use of the area under the roc curve in the evaluation of machine learning algorithms. Pattern Recogn. **30**(7), 1145–1159 (1997)
5. Breiman, L.: Random forests. Mach. Learn. **45**, 5–32 (2001)
6. Cadenas, J.M., Garrido, M.C., Martínez, R.: Feature subset selection filter-wrapper based on low quality data. Expert Syst. Appl. **40**, 1–10 (2013)
7. Cadenas, J.M., Garrido, M.C., Martínez, R., Bonissone, P.P.: Extending information processing in a fuzzy random forest ensemble. Soft Comput. **16**(5), 845–861 (2012)
8. J.M. Cadenas, M.C. Garrido, R. Martínez, P.P. Bonissone, Ofp_class: a hybrid method to generate optimized fuzzy partitions for classification. Soft Comput. **16**(4), 667–682 (2012)
9. Clarke, P.A., George, M., Cunningham, D., Swift, I., Workman, P.: Analysis of tumor gene expression following chemotherapeutic treatment of patients with bowel cancer. Nat. Genet. **23**(3), 39–39 (1999)
10. Dagliyan, O., Uney-Yuksektepe, F., Kavakli, I.H., Turkay, M.: Optimization based tumor classification from microarray gene expression data. PLoS ONE **6**(2), e14579 (2011)
11. DeLong, E.R., DeLong, D.M., Clarke-Pearson, D.L.: Comparing the areas under two or more correlated receiver operating characteristic curves: a nonparametric approach. Biometrics **44**(3), 837–845 (1988)
12. Diaz-Uriarte, R., Alvarez de Andrés, S.: Gene selection and classification of microarray data using random forest. BMC Bioinform. **7**(3), (2006)
13. Duval, B., Hao, J.K.: Advances in metaheuristics for gene selection and classification of microarray data. Briefings Bioinform. **11**(1), 127–141 (2010)
14. Genuer, R., Poggi, J.M., Tuleau-Malot, C.: Variable selecting using random forest. Pattern Recogn. Lett. **31**(14), 2225–2236 (2010)
15. Ghoraia, S., Mukherjeeb, A., Duttab, P.K.: Gene expression data classification by VVRKFA. Procedia Technol. **4**, 330–335 (2012)

16. Golub, T.R., Slonim, D.K., Tamayo, P., Huard, C., Gaasenbeek, M., Mesirov, J.: P, Coller H., Loh M., Downing J. R., Caligiuri M. A., Bloomfield C. D., Lander E.S.: Molecular classification of cancer: class discovery and class prediction by gene expression monitoring. Science **286**(5439), 531–537 (1999)
17. Hanley, J.A., McNeil, B.J.: The meaning and use of the area under a receiver operating characteristic (roc) curve. Radiology **143**(1), 29–36 (1982)
18. Mukhopadhyaya, A., Maulikb, U.: Towards improving fuzzy clustering using support vector machine: Application to gene expression data. Pattern Recogn. **42**(11), 2744–2763 (2009)
19. Nitsch D., Gonzalves J. P., Ojeda F., De Moor B., Moreau Y.: Candidate gene prioritization by network analysis of differential expression using machine learning approaches. BMC Bioinform. **11**(460), (2010)
20. Saeys, Y., Inza, I., Larraaga, P.: A review of feature selection techniques in bioinformatics. Bioinformatics **23**(19), 2507–2517 (2007)
21. Singh D., Febbo P. G., Ross K., Jackson D. G. et all: Gene expression correlates of clinical prostate cancer behavior. Cancer Cell **1**(2), 203–209 (2002)
22. Yager, R.R.: On ordered weighted averaging aggregation operators in multicriteria decision making. IEEE Trans. Syst. Man Cybern. **18**(1), 183–190 (1988)

Generalizing and Formalizing Precisiation Language to Facilitate Human-Robot Interaction

Takehiko Nakama, Enrique Muñoz, Kevin LeBlanc
and Enrique Ruspini

Abstract We develop a formal logic as a generalized precisiation language. This formal logic can serve as a middle ground between the natural-language-based mode of human communication and the low-level mode of machine communication. Syntactic structures in natural language are incorporated in the syntax of the formal logic. As regards the semantics, we establish the formal logic as a many-valued logic. We present examples that illustrate how our formal logic can facilitate human-robot interaction.

Keywords Precisiated natural language · Precisiation language · Formal logic · Propositional logic · Predicate logic · Quantificational logic · Fuzzy logic · Fuzzy relation · Human-robot interaction

1 Introduction

Zadeh (e.g., [21–23]) introduced the concept of precisiated natural language (PNL), which is an integral part of his computational theory of perceptions. PNL refers to a set of natural-language propositions that can be linked to objects of computation and deduction. The propositions in PNL are assumed to describe human perceptions, and they allow artificial intelligence to operate on and reason with perception-based information, which is intrinsically imprecise, uncertain, or vague.

T. Nakama (✉) · E. Muñoz · K. LeBlanc · E. Ruspini
European Center for Soft Computing, C/Gonzalo Gutiérrez Quirós, s/n,
33600 Mieres, Spain
e-mail: nakama@jhu.edu

E. Muñoz
e-mail: enrique.munoz@softcomputing.es

K. LeBlanc
e-mail: kevin.leblanc@softcomputing.es

E. Ruspini
e-mail: enrique.ruspini@softcomputing.es

© Springer International Publishing Switzerland 2016   381
K. Madani et al. (eds.), *Computational Intelligence*,
Studies in Computational Intelligence 613,
DOI 10.1007/978-3-319-23392-5_21

Precisiation language plays an essential role in performing computations on PNL propositions (e.g., [21–23]). It is used to express each PNL proposition as a set or a sequence of computational objects that can be effectively processed by machines. Zadeh proposed a precisiation language in which each proposition is a generalized constraint on a variable. This precisation language is called a generalized-constraint language.

Since Zadeh considered the primary function of natural language as describing human perceptions, his PNL and precisiation language only deal with perceptual propositions ([21–23]). However, the importance of natural language is not limited to describing human perceptions. For instance, using a natural language, we describe not only perceptions but also actions. Therefore, it is important, both theoretically and practically, to extend PNL and precisiaton language to other types of proposition. Generalized constraints in Zadeh's precisiation language are suitable for precisiating perceptual propositions but not for precisiating action-related propositions (see Sect. 3).

Robotics is one of the major fields that require the precisiation of action-related propositions in natural language. Many studies (e.g., [2, 3, 9, 10, 13]) have been conducted to develop robotic systems in which humans and robots work as true team members, requiring peer-to-peer human-robot interaction. Such systems can be highly effective and efficient in performing a wide range of sophisticated and practical tasks—assistance to people with disabilities (e.g., [12]), search and rescue (e.g., [15]), and space exploration (e.g., [4]), for instance. One of the major challenges of developing these robotic systems is the increased complexity of the human-robot interactions (e.g., [7]). Although humans prefer natural language as a communication medium, it presents several major problems when used for human-robot communications; natural-language expressions tend to be notoriously underspecified, diverse, vague, or ambiguous, so they often lead to errors that are hard to overcome (e.g., [5, 6, 17, 18, 20]). Low-level sensory and motor signals and executable code are easy for machines to interpret, but they are cumbersome for humans and thus cannot, on their own, create an effective human-robot interface. Task descriptions or specifications for robotic systems typically involve action-related propositions, such as *bring the box to the room* and *keep the robot in the building if it rains*, so Zadeh's precisiation language (generalized-constraint language), which is designed to deal with perceptual propositions, is not suitable for processing them.

Recently, we [14] have taken a first step toward generalizing precisiation language by establishing a formal logic as a generalized precisiation language. The resulting precisiation language can serve as a middle ground between the natural-language-based mode of human communication and the low-level mode of machine communication, so it can effectively mediate human-robot interaction in robotic systems that employ a peer-to-peer communication mode. In this paper, we further develop and elaborate on the framework proposed in [14].

The remainder of this paper is organized as follows. In Sect. 2, we examine the properties of formal logic that are desirable for precisiating natural-language expressions. The syntax of our formal logic is explained in Sect. 3. In Sect. 4, we discuss the generality of our framework. In Sect. 5, we examine how to add a deductive apparatus to our formal logic so that we can infer and reason in it. In Sect. 6, we develop a hierarchy of propositions that enhances the expressive power and the interactivity of our formal logic. The semantics of the formal logic is explained in Sect. 7.

2 Suitability of Formal Logic as Precisiation Language

Formal logic has been applied to analyze the syntax and semantics of natural language; this paradigm is called logical analysis (e.g., [18]). Formal logic provides genuine insight into the syntactic structures of natural-language sentences and the consequential characters of their assertions. Thus the framework of logical analysis is quite useful for precisiating natural-language sentences.

Infinitely many sentences can be generated in natural language, and clearly this high expressive power is desirable for precisiation language. Using the recursive definition of the syntax of formal logic, we can ensure that our precisiation language can generate infinitely many precisiated propositions while ensuring that every proposition in it is precisiated.

As in other formal logics, we can reason logically in our formal logic by adding a deductive apparatus to it; the resulting analytical machinery allows us to determine when one sentence in the formal language follows logically from other sentences. Thus our formal logic is capable of precisiating the inference and the reasoning in which humans engage using a natural language. See Sect. 5.

Our scheme also reflects the theory of descriptions in formal logic, which was introduced by Russell [16]. He claimed that the reality consists of logical atoms, which can be considered indecomposable, self-contained building blocks of all propositions in formal logic, and that logical analysis ends when we arrive at logical atoms. In our precisiation language, precisiation ends when we arrive at logical atoms, which will be represented by atomic propositions at the lowest level of a hierarchy of propositions. See Sect. 6.

3 Syntax of the Formal Logic

First, we describe how to form propositions in our precisiation language. To generate examples of ordinary practice, we consider establishing task descriptions for human-robot interaction, but keep in mind that our scheme is not limited to precisiating task descriptions. We will discuss the generality of our formal logic in Sect. 4. Our precisiation language generalizes Zadeh's generalized-constraint language by incorporating multiple syntactic forms that can be observed in many natural languages so that it can deal with not only perceptual propositions but also action-related propositions.

In Sect. 3.1, we describe the components of such propositions. In Sect. 3.2, we describe how to form an atomic proposition. In Sect. 3.3, we describe how to form a compound proposition. In Sect. 3.4, we provide a recursive definition of well-

Table 1 Examples of component sets

Set	Elements
S	Agents that can perform tasks
	e.g., $S = \{robot1, robot2, user\}$
V	Verbs that characterize actions required by tasks
	e.g., $V = \{find, deliver, go, move, press\}$
O	Objects that may receive an action in V or compose an adverbial phrase
	e.g., $O = \{box, button, table, room1, room2, robot1, robot2, user, null\}$
A	Adverbial phrases that can be included in task descriptions
	e.g., $A = \{in\ \gamma, from\ \gamma, from\ \gamma_1\ to\ \gamma_2, to\ \gamma, null \mid \gamma, \gamma_1, \gamma_2 \in O\}$
C	Connectives that can be used to combine multiple propositions in forming compound propositions
	e.g., $C = \{and, if, or, then\}$

formed formulas that allows our formal logic to generate infinitely many well-formed formulas while ensuring that every formula in it is well-formed.

3.1 Component Sets

We generate propositions using elements in component sets. To provide concrete examples, we consider the sets S, V, O, A and C in Table 1 as component sets. The element labeled as "null," called the null element, is included in O and A. In Sect. 3.2, we will explain how the null element is used in forming atomic propositions. In Sect. 3.3, we will explain how to form compound propositions using the connectives in C.

3.2 Atomic Propositions

In our formal logic, an atomic proposition is defined to be a tuple in the Cartesian product of component sets, and the Cartesian product specifies each admissible tuple structure. To develop formal propositions that can be easily identified with natural-language sentences, we employ tuple structures that reflect syntactic structures observed in natural languages. For instance, using the component sets described in Sect. 3.1, we can define each atomic proposition in our formal logic to be an SVOA clause (The S, V, O, and A in SVOA stand for subject, verb, object, and adverbial phrase, respectively) by setting the admissible tuple structure to $S \times V \times O \times A$. Using the null element in O and A, we can also generate SVO, SVA, and SV clauses.

See the following examples of atomic propositions resulting from the component sets in Table 1:

- $\frac{robot1}{S} \frac{move}{V}$. (The actual form of this proposition is $\frac{robot1}{S} \frac{move}{V} \frac{null}{O} \frac{null}{A}$, but we will omit instances of the null element to simplify the resulting expressions.)
- $\frac{robot2}{S} \frac{find}{V} \frac{ball}{O}$.
- $\frac{robot1}{S} \frac{deliver}{V} \frac{box}{O} \frac{from\ room1\ to\ room2}{A}$.

The SVOA structure is used in many languages, including English, Russian, and Mandarin. For humans, these propositions (task descriptions) are easy to specify and understand. Meanwhile, the structural and lexical constraints noticeably limit the diversity and flexibility of everyday language to ensure that robots can unambiguously interpret the resulting propositions (i.e., the specified tasks can be precisely interpreted and executed by robots).

Atomic propositions can be considered building blocks of all propositions. As will be explained in Sect. 6, we establish a hierarchy of propositions. At the lowest level of the hierarchy, each atomic proposition is directly associated with an indecomposable, self-contained executable code, and atomic propositions at each level compose propositions at higher levels.

There are several ways to deal with the undesirable or nonsensical atomic propositions that can be formed in $S \times V \times O \times A$. (Note that in formal logics, there can be well-formed formulas that are self-contradictory.) We can remove all such propositions from the cartesian product to ensure that each resulting atomic proposition is a precisiated proposition. (In this case, we abuse the notation and let $S \times V \times O \times A$ denote the "cleaned" cartesian product.) Also, we can consider them as always false so that they will never be executed in practice (see Sect. 7).

In our formal logic, the atomic propositions need not be expressed as generalized constraints on variables. By incorporating the SV, SVO, SVA, and SVOA structures in the syntax, we can precisiate action-related propositions rather naturally and effectively. Clearly, other syntactic structures can be incorporated in our formal logic; see Sect. 4.

3.3 Compound Propositions

In our formal logic, we generate each compound proposition by combining multiple atomic propositions using one or more connectives in the component set C. For instance, using the component sets in Table 1, we can form the following compound proposition:

$$\bullet \quad \frac{\frac{robot1}{S} \ \frac{go}{V} \ \frac{to\ room1}{A}}{\text{atomic proposition}} \ \frac{if}{C} \ \left(\frac{\frac{user}{S} \ \frac{call}{V} \ \frac{robot1}{O}}{\text{atomic proposition}} \ \frac{or}{C} \ \frac{\frac{user}{S} \ \frac{press}{V} \ \frac{button}{O}}{\text{atomic proposition}} \right). \quad (1)$$

In formal logic, parentheses are used to indicate the scope of each connective. In our examples, parentheses disambiguate the manner in which atomic tasks are performed.

3.4 Recursive Definition of Well-Formed Formulas

As described in Sect. 2, we can attain high expressive power in our precisiation language by recursively defining its syntax; formally, our formal logic can generate infinitely many well-formed formulas while ensuring that every formula in it is well-formed.

The syntax of the formal logic described in Sects. 3.1–3.3 can be recursively defined as follows:

1. Any $x \in S \times V \times O \times A$ is an atomic well-formed formula.
2. If α and β are well-formed formulas, then $\alpha \, c \, \beta$, where $c \in C$, is also a well-formed formula.
3. Nothing else is a well-formed formula.

This recursive definition allows our precisiation language to generate infinitely many precisiated propositions while ensuring that every proposition in it is precisiated.

4 Generality of the Formal Logic

Our scheme is quite general. Each component set can be made as large as necessary, and a variety of component sets or clause structures can be incorporated in our formal logic. For instance, in addition to the SV, SVO, SVA, and SVOA structures described and used in Sect. 3 (and in Sect. 6), we can also incorporate other commonly observed clause structures (see, for instance, [1]), such as the SVC, SVOC, and SVOO structures, in the syntax of atomic propositions. Furthermore, we can extend the clause structures so that a phrase can be used as the subject or the object in an atomic proposition. Negation, a unary logical connective, can certainly be incorporated in the formal logic. We can also include Zadeh's generalized constraints, which are suitable for expressing perceptual propositions, in our formal logic; each generalized constraint can be considered an atomic proposition that has the SVC structure, and it can be combined with other propositions by connectives.

In our scheme, we can establish not only a propositional logic but also a quantificational logic, which fully incorporates quantifiers and predicates in well-formed formulas. Since propositions that describe perceptions often include quantifiers (see, for instance, [21, 22]), it is desirable to develop a quantificational logic as a precisiation language that covers both actions and perceptions.

5 Inference and Reasoning in the Formal Logic

We can infer and reason in our formal logic by adding a deductive apparatus to it. Typical introduction- and elimination-rules in formal logics, such as modus ponens and modus tollens, and axioms can be easily incorporated in our formal logic. (The hierarchy described in Sect. 6 represents non-logical, domain-dependent axioms.) Consequently, we can form a sequent, which consists of a finite set of well-formed formulas (the premises) and a single well-formed formula (the conclusion), and we can examine its provability (derivability) using proof theory; we can determine if a conclusion follows logically from a set of premises by examining whether there is a proof of that conclusion from just those premises in the formal logic.

As will be described in Sect. 7, we can employ fuzzy relations to establish the semantics of our formal logic. This semantics allows us to investigate the truth conditions and the semantic validity of each proposition or sequent. As in other formal logics, comparative truth tables can be used to determine semantic validity.

6 Hierarchy of Propositions

The importance of the hierarchy of propositions described in this section is three-fold. First, it enhances the expressive power of the formal logic by building up its vocabulary while ensuring the precisiability of each resulting proposition. Second, it augments the interactivity of our formal logic by allowing human-robot communications to take place at various levels of detail. Third, it strengthens the deductive apparatus of the formal logic by establishing domain-dependent axioms that can be used for inference and reasoning.

We will explain the hierarchy intuitively using the task description scheme described in Sect. 3. Consider the following task description:

$$\underset{S}{robot1} \ \underset{V}{examine} \ \underset{O}{patient1} \ \underset{A}{in\ room1}. \tag{2}$$

This atomic proposition can be reexpressed as a compound proposition that consists of three atomic propositions representing subtasks that must be performed to accomplish the task:

$$\underbrace{\underset{S}{robot1} \ \underset{V}{find} \ \underset{O}{patient1} \ \underset{A}{in\ room1}}_{\text{atomic proposition 1}} \ \underset{C}{then} \ \underbrace{\underset{S}{robot1} \ \underset{V}{check} \ \underset{O}{patient1}}_{\text{atomic proposition 2}}$$

$$\underset{C}{then} \ \underbrace{\underset{S}{robot1} \ \underset{V}{send} \ \underset{O}{data}}_{\text{atomic proposition 3}}. \tag{3}$$

Atomic propositions 1 and 2 in (3) can also be reexpressed as compound propositions that clarify how they are performed; atomic proposition 1 in (3) can be defined as

$$
\underset{\text{atomic proposition}}{\underbrace{\overset{robot1}{S}\ \overset{go}{V}\ \overset{to\ room1}{A}}}\ \underset{C}{then}\ \underset{\text{atomic proposition}}{\underbrace{\overset{robot1}{S}\ \overset{search}{V}\ \overset{patient1}{O}}},
\tag{4}
$$

and atomic proposition 2 in (3) can be defined as

$$
\underset{\text{atomic proposition}}{\underbrace{\overset{robot1}{S}\ \overset{go}{V}\ \overset{to\ patient1}{A}}}\ \underset{C}{then}\ \left(\underset{\text{atomic proposition}}{\underbrace{\overset{robot1}{S}\ \overset{measure}{V}\ \overset{heart\ rate}{O}}}\ \underset{C}{and}\right.
$$

$$
\left.\underset{\text{atomic proposition}}{\underbrace{\overset{robot1}{S}\ \overset{measure}{V}\ \overset{blood\ pressure}{O}}}\right).
\tag{5}
$$

Therefore, using (4)–(5) and atomic proposition 3 in (3), we can reexpress (2) as

$$
\underset{\text{atomic proposition}}{\underbrace{\overset{robot1}{S}\ \overset{go}{V}\ \overset{to\ room1}{A}}}\ \underset{C}{then}\ \underset{\text{atomic proposition}}{\underbrace{\overset{robot1}{S}\ \overset{search}{V}\ \overset{patient1}{O}}}
$$

$$
\underset{C}{then}\ \underset{\text{atomic proposition}}{\underbrace{\overset{robot1}{S}\ \overset{go}{V}\ \overset{to\ patient1}{A}}}\ \underset{C}{then}\ \left(\underset{\text{atomic proposition}}{\underbrace{\overset{robot1}{S}\ \overset{measure}{V}\ \overset{heart\ rate}{O}}}\ \underset{C}{and}\right.
$$

$$
\left.\underset{\text{atomic proposition}}{\underbrace{\overset{robot1}{S}\ \overset{measure}{V}\ \overset{blood\ pressure}{O}}}\right)
$$

$$
\underset{C}{then}\ \underset{\text{atomic proposition}}{\underbrace{\overset{robot1}{S}\ \overset{send}{V}\ \overset{data}{O}}}.
\tag{6}
$$

Figure 1 visualizes the underlying hierarchy, which consists of three levels (levels 0, 1, and 2). For simplicity, each atomic proposition is represented by its verb; for instance, the atomic proposition at the highest level (level 2), "Robot1 *examine* patient1 in room1," is represented by "examine." The task expressed by the atomic proposition at level 2 is described in more detail at the intermediate level (level 1), where the atomic propositions that involve the verbs "find," "check," and "send" describe the subtasks that constitute the task. These subtasks are described in more detail at the lowest level (level 0), where they are expressed by the atomic propositions that involve the verbs "go," "search," "measure," and "send."

The hierarchy clearly shows how atomic proposition (2) at level 2 is precisiated. At level 0, we have atomic propositions that are not decomposable; each of them is directly associated with a self-contained executable code that is run to perform the corresponding task. Thus, atomic propositions at level 0 can be considered logical atoms described in Sect. 2, and they precisiate each proposition at higher levels.

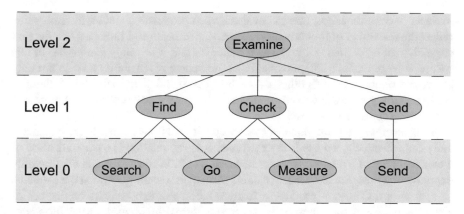

Fig. 1 Hierarchical task description. At the highest level (level 2), the task is expressed as atomic proposition (2), represented by "examine." At the intermediate level (level 1), the task is expressed as compound proposition (3), which consists of atomic propositions represented by "find," "check," and "send." At the lowest level (level 0), the task is expressed as compound proposition (6), which consists of atomic propositions represented by "go," "search," "measure," and "send"

The suitability of a given proposition depends on the level of granularity required for it. As regards the task description scheme, naive users will most likely prefer describing tasks at level 2, thus preferring (2). For expert users, there may be situations where they prefer specifying a given task step by step or reconfiguring its subtasks according to various circumstances; in such cases, interacting with robots at level 1 using (3) or at level 0 using (6) will be desirable. Thus, the hierarchy allows a variety of users to interact with robots at various levels of detail.

Formally, the hierarchy clearly shows the definition of each proposition by expressing it in terms of precisiated propositions at lower levels. Thus, non-logical, domain-dependent axioms result from the hierarchy, and they can be used for inference and reasoning in the formal logic (see Sect. 5).

Different levels of granularity may require different component sets, but the same syntactic structure is enforced at all levels. Using the hierarchy, we can ensure that all the resulting propositions remain precisiated at each level, and we can attain flexibility in the level of detail.

7 Semantics of the Formal Logic

The semantics of formal logic specifies how to determine the truth value of each proposition. In two-valued logics, for instance, the truth value is either 1 (true) or 0 (false). As described by Zadeh (e.g., [22, 23]), this bivalence is not suitable for PNL, so we develop a many-valued semantics for our formal logic. The meaning of the truth value depends on the context. For the task description scheme described in Sects. 3–6, for instance, one can evaluate each proposition and let its truth value reflect the feasibility of the corresponding task specification; 1 indicates that the task certainly can be carried out whereas 0 indicates that it certainly cannot be. In this case, it is more realistic and practical to let the degree of feasibility take on not only the values 0 and 1 but also other values between 0 and 1. In real-world problems, it can be highly practical to evaluate the feasibility of a task description before any serious attempt is made to execute it.

To determine the truth value of each proposition in our formal logic systematically and effectively, we use fuzzy relations. A fuzzy relation is a generalization of a classical ("crisp") relation (see, for instance, [11]). While a classical relation only expresses the presence or absence of some form of association between the elements of factors in a Cartesian product, a fuzzy relation can express various degrees or strengths of association between them. In our formal logic, each proposition consists of pre-specified components, so a function that assigns a truth value to each proposition can be represented by a fuzzy relation on the Cartesian product of the components. Using some of the operations defined on fuzzy relations, we can systematically and economically determine the truth value of each proposition.

We will explain the semantics of our formal logic using concrete examples of task descriptions so that the reader can understand it intuitively. To facilitate our exposition, we consider very simple task descriptions resulting from atomic propositions in $S \times V \times O$. Notice that even in this case, we need an efficient scheme for determining the truth value of each proposition. For instance, if each of the component sets S, V, and O contains ten elements, then there are 10^3 atomic propositions in $S \times V \times O$, and it may be impractical to determine the truth values of all the atomic propositions individually. Moreover, if the component set C consists of three connectives, then we can generate a total of $3 \cdot 10^6$ compound propositions that consist of two atomic propositions. In practice, it may be necessary to promptly evaluate and compare the truth values of a large number of task descriptions represented by such compound propositions in order to determine which option to execute, so even this simple case requires an efficient, systematic scheme for examining the truth conditions of propositions.

In Sect. 7.1, we explain how to determine the truth values of atomic propositions. In Sect. 7.2, we explain how to determine the truth values of compound propositions.

7.1 Truth Conditions of Atomic Propositions

We consider establishing a fuzzy relation on $S \times V \times O$, which is a mapping from the Cartesian product to a totally ordered set called a valuation set. In our formulation of many-valued logic, the valuation set is the unit interval [0, 1]. To facilitate the exposition of our scheme, we consider the following simple component sets: $S = \{robot1, robot2\}$, $V = \{recognize, hold\}$, $O = \{ball, pen\}$. Thus the Cartesian product $S \times V \times O$ consists of eight atomic propositions, which are shown in Table 2. We will use three operations on fuzzy relations: projection, cylindric extension,

Table 2 Atomic propositions resulting from $S = \{robot1, robot2\}$, $V = \{recognize, hold\}$, and $O = \{ball, pen\}$

$s \in S$	$v \in V$	$o \in O$
robot1	recognize	ball
robot1	recognize	pen
robot1	hold	ball
robot1	hold	pen
robot2	recognize	ball
robot2	recognize	pen
robot2	hold	ball
robot2	hold	pen

and cylindric closure (see, for instance, [11]). These operations are explained in Appendix.

Suppose that the truth conditions (for concreteness, we assume that they represent degrees of feasibility) of these atomic propositions are determined for a robotic system under the following conditions:

(a) Robot1 is equipped with a high-resolution camera that enables it to recognize various objects, including a ball and a pen. However, it does not have any arm, so it cannot hold any object.

(b) Robot2 has an arm that enables it to hold various objects, including a ball and a pen. However, it is not equipped with a high-resolution camera, so it is not fully capable of identifying objects.

(c) With the high-resolution camera, a ball is easier to recognize compared to a pen.

(d) With the arm, a pen is easier to hold compared to a ball.

Our strategy is to derive a fuzzy relation on $S \times V \times O$ from fuzzy relations on $S \times V$ and $V \times O$. Hence, we first establish fuzzy relations on $S \times V$ and $V \times O$. Let $R_{S \times V} : S \times V \to [0, 1]$ denote a fuzzy relation on $S \times V$. Based on conditions (a) and (b), we set the values of $R_{S \times V}$ as shown in Table 3. Recall that a fuzzy relation expresses various degrees or strengths of association between elements in component sets. The value assigned to $(robot1, recognize)$ is relatively large (0.9) because $robot1$ is equipped with a high-resolution camera and is thus suitable for

Table 3 Fuzzy relation $R_{S \times V}$ on $S \times V$ based on conditions (a) and (b)

$s \in S$	$v \in V$	$R_{S \times V}(s, v)$
robot1	recognize	0.9
robot1	hold	0
robot2	recognize	0.2
robot2	hold	0.8

Table 4 Fuzzy relation $R_{V \times O}$ on $V \times O$ based on conditions (c) and (d)

$v \in V$	$o \in O$	$R_{V \times O}(v, o)$
recognize	ball	0.9
recognize	pen	0.8
hold	ball	0.7
hold	pen	0.8

recognizing objects, whereas the value assigned to $(robot1, hold)$ is zero because $robot1$ is not equipped with an arm and is thus incapable of holding objects. Similarly, the value assigned to $(robot2, recognize)$ is relatively small (0.2) because $robot2$ is not equipped with a high-resolution camera and is thus unsuitable for recognizing objects, whereas the value assigned to $(robot2, hold)$ is relatively large (0.8) because $robot2$ is equipped with an arm and is thus suitable for holding objects. Technically, the fuzzy relation $R_{S \times V}$ is considered the underlying fuzzy relation on $S \times V \times O$ projected onto $S \times V$. (See Appendix for the operation of projection.)

Analogously, based on conditions (c) and (d), we set the values of a fuzzy relation $R_{V \times O} : V \times O \rightarrow [0, 1]$ as shown in Table 4. The value assigned to $(recognize, ball)$ is larger than that assigned to $(recognize, pen)$ because with a high-resolution camera, a ball is easier to recognize compared to a pen. Similarly, the value assigned to $(hold, pen)$ is larger than that assigned to $(hold, ball)$ because with an arm, a pen is easier to hold compared to a ball. Technically, the fuzzy relation $R_{V \times O}$ is considered the underlying fuzzy relation on $S \times V \times O$ projected onto $V \times O$.

We establish a fuzzy relation $R_{S \times V \times O} : S \times V \times O \rightarrow [0, 1]$ by combining the fuzzy relations $R_{S \times V}$ and $R_{V \times O}$. Formally, we obtain $R_{S \times V \times O}$ by first obtaining the cylindric extensions of $R_{S \times V}$ and $R_{V \times O}$ to $S \times V \times O$ and then computing their cylindric closure. (see Appendix for cylindric extension and cylindric closure). First, we obtain the cylindric extension of $R_{S \times V}$ to $S \times V \times O$, which we denote by $R_{S \times V \uparrow S \times V \times O}$, and the cylindric extension of $R_{V \times O}$ to $S \times V \times O$, which we denote by $R_{V \times O \uparrow S \times V \times O}$. See Table 5. The cylindric extensions can be characterized as maximizing nonspecificity in deriving a fuzzy relation on $S \times V \times O$ from fuzzy relations on $S \times V$ and $V \times O$.

Finally, we set $R_{S \times V \times O}$ to the cylindric closure of $R_{S \times V \uparrow S \times V \times O}$ and $R_{V \times O \uparrow S \times V \times O}$ on $S \times V \times O$, which is shown in Table 6. Notice that the resulting truth values (degrees of feasibility) assigned to the eight atomic propositions reflect the conditions (a)–(d). For example, the truth values clearly indicate that robot1 is highly capable of recognizing objects (because it is equipped with a high-resolution camera) but is incapable of holding objects (because it does not have any arm). Similarly, the truth values clearly indicate that robot2 is highly capable of holding objects (because it is equipped with an arm) but is rather incapable of recognizing objects (because it is not equipped with a high-resolution camera).

It is efficient to derive a fuzzy relation on $S \times V \times O$ from fuzzy relations on $S \times V$ and $V \times O$. Again, suppose that each of these component sets consists of

Table 5 Cylindric extensions $R_{S \times V \uparrow S \times V \times O}$ and $R_{V \times O \uparrow S \times V \times O}(s, v, o)$

$s \in S$	$v \in V$	$o \in O$	$R_{S \times V \uparrow S \times V \times O}(s, v, o)$	$R_{V \times O \uparrow S \times V \times O}(s, v, o)$
robot1	recognize	ball	0.9	0.9
robot1	recognize	pen	0.9	0.8
robot1	hold	ball	0	0.7
robot1	hold	pen	0	0.8
robot2	recognize	ball	0.2	0.9
robot2	recognize	pen	0.2	0.8
robot2	hold	ball	0.8	0.7
robot2	hold	pen	0.8	0.8

Table 6 Cylindric closure $R_{S \times V \times O}$ of $R_{S \times V \uparrow S \times V \times O}$ and $R_{V \times O \uparrow S \times V \times O}$ on $S \times V \times O$

$s \in S$	$v \in V$	$o \in O$	$R_{S \times V \times O}(s, v, o)$
robot1	recognize	ball	0.9
robot1	recognize	pen	0.8
robot1	hold	ball	0
robot1	hold	pen	0
robot2	recognize	ball	0.2
robot2	recognize	pen	0.2
robot2	hold	ball	0.7
robot2	hold	pen	0.8

ten elements. Then a total of 10^3 atomic propositions result from them, and it may be time-consuming to determine 10^3 truth values individually. With our scheme, we can derive the 10^3 truth values by determining $2 \cdot 10^2$ values of the fuzzy relations $R_{S \times V}$ and $R_{V \times O}$. This efficiency of the scheme becomes more notable as the size of each component set or the number of component sets increases.

Another important strength of our scheme lies in updating the truth values of the atomic propositions. In practice, the values shown in Tables 3 and 4 will be determined dynamically based on the conditions of the robots. For instance, if the high-resolution camera of robot1 becomes dysfunctional, then we will use the fuzzy relation $R'_{S \times V}$ shown in Table 7 instead of the fuzzy relation $R_{S \times V}$ in Table 3 in computing the truth values of the atomic propositions. Notice that the value of $R'(robot1, recognize)$ is 0.2, reflecting the fact that robot1 can no longer use its high-resolution camera to recognize objects (compare $R'_{S \times V}$ and $R_{S \times V}$ in Table 3). It is easy to verify that Table 8 shows the cylindric closure $R'_{S \times V \times O}$ of the cylindric extensions $R'_{S \times V \uparrow S \times V \times O}$ and $R_{V \times O \uparrow S \times V \times O}$. Comparing Tables 6 and 8, we can see that the updated truth values (shown in Table 8) reflect the condition that the high-resolution camera of robot1 has become dysfunctional. Notice that we have efficiently updated the fuzzy relation on $S \times V \times O$ by just updating the fuzzy relation on $S \times V$. With our scheme, it is

Table 7 Fuzzy relation $R'_{S\times V}$ on $S \times V$ (reflecting a damage to robot1's high-resolution camera; see $R_{S\times V}$ in Table 3)

$s \in S$	$v \in V$	$R'_{S\times V}(s, v)$
robot1	recognize	0.2
robot1	hold	0
robot2	recognize	0.2
robot2	hold	0.8

Table 8 Cylindric closure $R'_{S\times V\times O}$ of $R'_{S\times V\uparrow S\times V\times O}$ and $R_{V\times O\uparrow S\times V\times O}$ on $S \times V \times O$ (reflecting a damage to robot1's high-resolution camera)

$s \in S$	$v \in V$	$o \in O$	$R'_{S\times V\times O}(s, v, o)$
robot1	recognize	ball	0.2
robot1	recognize	pen	0.2
robot1	hold	ball	0
robot1	hold	pen	0
robot2	recognize	ball	0.2
robot2	recognize	pen	0.2
robot2	hold	ball	0.7
robot2	hold	pen	0.8

possible to keep the truth values of a large number of atomic propositions updated continually.

7.2 Truth Conditions of Compound Propositions

Our formal logic is many-valued, so we treat the connectives in C as logic primitives of many-valued logic or fuzzy logic. Here we examine thee typical logical primitives: conjunction (represented by "and" in C), disjunction (represented by "or" in C), and implication (also called conditional, represented by "if" in C).

In evaluating the truth value of a compound proposition, conjunction is often implemented as a t-norm, whereas disjunction is often implemented as a t-conorm (e.g., [8, 11]). Various forms of t-norm and t-conorm have been proposed. Some of the frequently used t-norms are the minimum t-norm, the product t-norm, and the Łukasiewicz t-norm, and some of the frequently used t-conorms are the maximum t-conorm, the probabilistic sum, and the Łukasiewicz t-conorm. In practice, the suitability of each of these t-norms or t-conorms depends on what the truth value represents. Also, there are various ways to implement implication in evaluating the truth value of a compound proposition (e.g., [19]). Some of the main forms of implication are the material implication, the conjunctive conditional, the residuated conditional, the Sasaki hook, the Dishkant hook, and the Mamdani-Larsen conditional. Again, in practice, the suitability of each conditional depends on what the truth value represents.

8 Conclusions

We have taken a first step toward establishing a formal logic as a generalized precisiation language, which is essential for generalizing PNL. Various syntactic structures in natural language can be incorporated in our formal logic so that it precisiates not only perceptual propositions but also action-related propositions. The syntax of the formal logic allows us to create infinitely many precisiated propositions while ensuring that every proposition in it is precisiated. As in other formal logics, we can infer and reason in our formal logic. The resulting generalized precisiation language serves as a middle ground between the natural-language-based mode of human communication and the low-level mode of machine communication and thus significantly facilitates human-machine interaction.

Acknowledgments This research is supported by the Spanish Ministry of Economy and Competitiveness through the project TIN2011-29824-C02-02 (ABSYNTHE).

Appendix

We describe three operations on fuzzy relations that are used in determining the truth conditions of atomic propositions in our formal logic: projection, cylindric extension, and cylindric closure. First, we establish notation. Let $X_1, X_2, \ldots, X_n$ be sets, and let $X_1 \times X_2 \times \cdots \times X_n$ denote their Cartesian product. We will also denote the Cartesian product by $\times_{i \in \mathbb{N}_n} X_i$, where $\mathbb{N}_n$ denotes the set of integers 1 through n. A fuzzy relation on $\times_{i \in \mathbb{N}_n} X_i$ is a function from the Cartesian product to a totally ordered set, which is called a valuation set. In our formulation, the unit interval [0, 1] is used as a valuation set. Each n-tuple $(x_1, x_2, \ldots, x_n)$ in $X_1 \times X_2 \times \cdots \times X_n$ (thus $x_i \in X_i$ for each $i \in \mathbb{N}_n$) will also be denoted by $(x_i \mid i \in \mathbb{N}_n)$. Let $I \subset \mathbb{N}_n$. A tuple $y := (y_i \mid i \in I)$ in $Y := \times_{i \in I} X_i$ is said to be a sub-tuple of $x := (x_i \mid i \in \mathbb{N}_n)$ in $\times_{i \in \mathbb{N}_n} X_i$ if $y_i = x_i$ for each $i \in I$, and we write $y \prec x$ to indicate that y is a sub-tuple of x.

Let $X := \times_{i \in \mathbb{N}_n} X_i$ and $Y := \times_{i \in I} X_i$ for some $I \subset \mathbb{N}_n$. Suppose that $R : X \to [0, 1]$ is a fuzzy relation on X. Then a fuzzy relation $R' : Y \to [0, 1]$ is called the projection of R on Y if for each $y \in Y$, we have $R'(y) = \max_{x \in X \,:\, y \prec x} R(x)$. We let $R_{\downarrow Y}$ denote the projection of R on Y.

We continue with $X := \times_{i \in \mathbb{N}_n} X_i$ and $Y := \times_{i \in I} X_i$ $(I \subset \mathbb{N}_n)$. Let $F : Y \to [0, 1]$ be a fuzzy relation on Y. A fuzzy relation $F' : X \to [0, 1]$ is said to be the cylindric extension of F to X if for all $x \in X$, we have $F'(x) = F(y)$, where y is the tuple in Y such that $y \prec x$. We let $F_{\uparrow X}$ denote the cylindric extension of F to X. The cylindric extension $F_{\uparrow X}$ of a fuzzy relation $F : Y \to [0, 1]$ is the "largest" fuzzy relation on X such that its projection on Y equals F; if we let $\mathscr{R}$ denote the set of all fuzzy relations $R' : X \to [0, 1]$ such that $R'_{\downarrow Y} = F$, then for all $x \in X$, we have $F_{\uparrow X}(x) = \max\{R'(x) \mid R' \in \mathscr{R}\}$.

For each j, let $Y_j := \times_{i \in I_j} X_i$, where $I_j \subset \mathbb{N}_n$. Let $R^{(j)} : Y_j \to [0, 1]$ denote a fuzzy relation on Y_j. Then a fuzzy relation $F : X \to [0, 1]$ is called the cylindric closure of $R^{(1)}, R^{(2)}, \ldots, R^{(m)}$ on X if for each $x \in X$, $F(x) = \min_{1 \le j \le m} R^{(j)}_{\uparrow X}(x)$.

References

1. Biber, D., Conrad, S., Leech, G.: A Student Grammar of Spoken and Written English. Pearson ESL, London (2002)
2. Dias, M.B., Kannan, B., Browning, B., Jones, E.G., Argall, B., Dias, M.F., Zinck, M., Veloso, M.M., Stentz, A.J.: Sliding autonomy for peer-to-peer human-robot teams. In: Proceedings of the 10th International Conference on Intelligent Autonomous Systems (2008)
3. Dias, M.B., Harris, T.K., Browning, B., Jones, E.G, Argall, B., Veloso, M.M., Stentz, A., Rudnicky, A.I.: Dynamically formed human-robot teams performing coordinated tasks. In: AAAI Spring Symposium: To Boldly Go Where No Human-Robot Team Has Gone Before, pp. 30–38 (2006)
4. Ferketic, J., Goldblatt, L., Hodgson, E., Murray, S., Wichowski, R., Bradley, A., Chun, W., Evans, J., Fong, T., Goodrich, M., Steinfeld, A., Stiles, R.: Toward human-robot interface standards: use of standardization and intelligent subsystems for advancing human-robotic competency in space exploration. In: Proceedings of the SAE 36th International Conference on Environmental Systems (2006)
5. Forsberg, M.: Why is Speech Recognition Difficult. Chalmers University of Technology, Gothenburg (2003)
6. Gieselmann, P., Stenneken, P.: How to talk to robots: evidence from user studies on human-robot communication. In: How People Talk to Computers, Robots, and Other Artificial Communication Partners, p. 68 (2006)
7. Goodrich, M.A., Schultz, A.C.: Human-robot interaction: a survey. Found. Trends Hum.-Comput. Interact. **1**, 203–275 (2007)
8. Hájek, P.: Metamathematics of Fuzzy Logic, vol. 4. Kluwer Academic, Dordrecht (1998)
9. Johnson, M., Feltovich, P.J., Bradshaw, J.M., Bunch, L.: Human-robot coordination through dynamic regulation. In: IEEE International Conference on Robotics and Automation, 2008. ICRA 2008. pp. 2159–2164 (2008)
10. Johnson, M., Intlekofer, K.: Coordinated operations in mixed teams of humans and robots. In: Proceedings of the IEEE International Conference on Distributed Human-Machine Systems (2008)
11. Klir, G.J., Folger, T.A.: Fuzzy Sets, Uncertainty, and Information. Prentice Hall, Englewood Cliffs (1988)
12. Kulyukin, V., Gharpure, C., Nicholson, J., Osborne, G.: Robot-assisted wayfinding for the visually impaired in structured indoor environments. Auton. Robot. **21**, 29–41 (2006)
13. Marble, J., Bruemmer, D., Few, D., Dudenhoeffer, D.: Evaluation of supervisory vs. peer-peer interaction with human-robot teams. In: Proceedings of the Hawaii International Conference on System Sciences (2004)
14. Nakama, T., Muñoz, E., Ruspini, E.: Generalizing precisiated natural language: a formal logic as a precisiation language. In: 8th Conference of the European Society for Fuzzy Logic and Technology (EUSFLAT-13). Atlantis Press (2013)

15. Norbakhsh, I.R., Sycara, K., Koes, M., Yong, M., Lewis, M., Burion, S.: Human-robot teaming for search and rescue. IEEE Pervasive Comput. **4**, 72–79 (2005)
16. Russell, B.: Lectures on the philosophy of logical atomism. In: Marsh, R.C. (ed.) Logic and Knowledge Essays 1901–1950. George Allen & Unwin, London (1984)
17. Shneiderman, B.: The limits of speech recognition. Commun. ACM **43**(9), 63–65 (2000)
18. Tomassi, I.: Logic. Routledge, London (1999)
19. Trillas, E., Alsina, C.: From Leibniz's shinning theorem to the synthesis of rules through Mamdani-Larsen conditionals. In: Combining Experimentation and Theory, pp. 247–258. Springer (2012)
20. Winograd, T., Flores, F.: Understanding Computers and Cognition: A New Foundation for Design. Ablex Pub, New Jersey (1986)
21. Zadeh, L.A.: Some reflections on information granulation and its centrality in granular computing, computing with words, the computational theory of perceptions and precisiated natural language. Stud. Fuzziness Soft Comput. **95**, 3–22 (2002)
22. Zadeh, L.A.: Precisiated natural language (PNL). AI Mag. **25**(3), 74–92 (2004)
23. Zadeh, L.A.: A new direction in ai: toward a computational theory of perceptions. AI Mag. **22**(1), 73 (2001)

Part III
Neural Computation Theory
and Applications

Growing Surface Structures: A Topology Focused Learning Scheme

Hendrik Annuth and Christian-A. Bohn

Abstract Iterative refinement approaches derived from unsupervised *artificial neural network* (ANN) methods, such as *Growing Cell Structures* (GCS), have proven very efficient for the application of surface reconstruction from scattered 3D points. The *Growing Surface Structures* (GSS) algorithm is a major conceptual change in the GCS approach. Instead of "adjusting" the learning behavior, the central learning scheme is shifted from optimizing the distribution of vertices to the creation of a valid surface model. Where in former GCS approaches the created topology is only implicitly represented in the process, it is explicitly integrated and represented in the refinement process of the GSS approach. Here the closest surface structure, such as a vertex, an edge or a triangle is found for a given sample and the actual *sample-to-surface* distance is measured. With this additional information the adaptation process can be focused on the created topology. We demonstrate the performance of the novel concept in the area of surface reconstruction.

Keywords Unsupervised learning · Competitive learning · Growing cell structures · Surface reconstruction · Surface fitting

1 Introduction

Due to the rapid development in 3D scanning technology, real world objects can be scanned faster, more accurately and at a higher resolution. This allows creating high quality virtual representations of these objects that can be utilized for many different purposes in digital data processing (see Fig. 1).

Laser scanning devices take samples of present surfaces as three dimensional data points. These points accumulate as an unorganized cloud of points. Such a point set typically includes noise, outliers, non-uniform sample densities, and holes. Since the

H. Annuth (✉) · C.-A. Bohn
Wedel University of Applied Sciences, Wedel, Fr. Germany
e-mail: annuth@fh-wedel.de

© Springer International Publishing Switzerland 2016   401
K. Madani et al. (eds.), *Computational Intelligence*,
Studies in Computational Intelligence 613,
DOI 10.1007/978-3-319-23392-5_22

Fig. 1 A photography of Michealangelo's David (*left*), a point cloud of the David statue (*middle*), and a surface fitted into that point cloud (*right*)

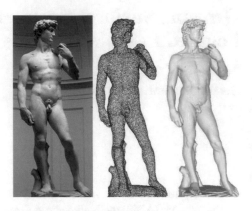

combination of these problems is often inherently ambiguous, a vast number of different reconstruction approaches, and pre- as well as post-processing methods exist.

Due to the remaining challenges in the field of reconstruction there is a strong tendency to focus on ANN based solutions since they are strong with incomplete and noisy data while being flexibly adaptable. Thus, there is hope that by a more intuitive, "ad-hoc" manner, ANN training can be modified to match the problems under consideration without the need of a deterministic mathematical model.

2 Previous Work

2.1 Classical Surface Reconstruction

Many surface reconstruction approaches have been suggested over the last decades. Range image methods can achieve very high resolutions and accuracy, while necessitating a very controlled scanning setup with a limited sensing area [8]. Region growing approaches like [12] or [5] extended an initial surface incrementally at its boundaries. Some methods reduce a 3D Delaunay triangulation of the samples to a final surface [9], some derive it from the Voronoi diagram of the points [1, 21]. Combined concepts use region-growing approaches for the triangulation and an additional global graph like a 3D Delaunay triangulation as a guidance [19] or a *medial scaffold* (MS) [7]. Balloon models construct a volumetric object surface by the "inflation" of a small surface, as if it would be a balloon, inside a point cloud [24]. Another huge class of reconstruction methods demands points that are augmented by its normals to define an incomplete distance function. This function is completed and the subspace in R^3 for which it returns zero—the zero-level-set—is the surface. The function can be composed of a multitude of linear functions [13], quadratic functions [17, 22] or radial base functions [6]. Model based reconstruction approaches compose a surface of a multitude of predefined models or components, which are recognized and fitted into the point cloud [11, 23]. Warping algorithms approximate the surface by deforming an initial surface to match the given points [4, 27].

2.2 Artificial Neural Network Based Reconstruction

Many neural computation techniques have been applied to the problem of surface reconstruction and are based on unsupervised learning concepts. Algorithms such as the k-means clustering approach [20] use reference vectors to accomplish classification and clustering tasks on huge and challenging data sets ("hard competitive learning"). Kohonen presented the *Self-Organizing-Map* (SOM) [18]—additionally reference vectors are connected adding a topology ("soft competitive learning") which enables the construction of a surface over the sample set. Kohonen's approach has the disadvantage of a fixed resolution, which strongly relates the results to the initial setting and size of the network. Fritzke presented the *Growing-Cells-Structure* (GCS) approach [10] where the network grows over time by dynamically adding reference vectors. The growing process can be determined by the approximation error toward a likelihood distribution or a quantization error, both of which are measured in relation to the reference vectors. Based on this, many convincing reconstruction methods have been presented [3, 14, 15, 26]. The main disadvantage of the methods above is the fact that 2D subspaces or surfaces are approximated by point distributions instead of surface models. This becomes most apparent when modeling flat surface areas where the granularity of the ANN surface depends on the distribution of samples and not on the complexity of the underlying surface.

In this work we present an approach which solves this problem. Basic ANN learning is changed to a surface oriented learning saving the advantages of neural networks but concurrently implementing a reasonable "surface learning". The difference to former approaches which modify ANNs by adding additional constraints to the learning rules our approach introduces an actual novel learning scheme.

3 Surface Reconstruction

Surface reconstruction creates a 2D subspace $\mathscr{S}$ in a 3D space $\mathbb{R}^3$ that represents a real world physical surface $\mathscr{S}_{phy}$. The information given about $\mathscr{S}_{phy}$ is a finite collection of surface samples $\mathscr{P} = \{p_1...p_n\} \subset \mathbb{R}^3$. If closest neighbors in $\mathscr{P}$ always indicate a connection on $\mathscr{S}_{phy}$, surface neighborhood relations can be investigated by accessing $\mathscr{P}$ in a 3D search pattern. Real world scenarios however, involving noise, non-uniform sample densities and incompletely sampled areas. This makes a 3D search unreliable (see Fig. 2), which is the basic problem to overcome in a reconstruction approach. Note that many of the following illustrations are in 2D and are therefore curve reconstructions, but to avoid confusions by swapping terminology, we proceed in using the terms of 3D surface reconstruction.

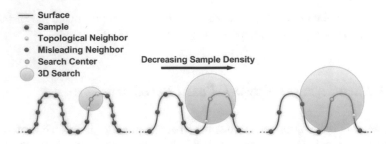

Fig. 2 The topology of a surface has to be derived from 3D samples. But searching in 3D for sample neighbors might produce misleading results. First, a highly sampled surface, where the two surface neighbors to a certain sample can be easily found (*left*). Then the same surface with a lower sampling, the search space for the two closest neighbors has grown (*middle*). And at last, a low sampling where the two closest neighbors are not the correct topological (on surface) neighbors (*right*)

4 Growing Cell Structures

A GCS network is composed of simplices of an initially chosen dimension. In case of surface reconstruction, a 2D surface $\mathscr{S}$ built of triangles as simplices. The initial surface is a very simple network of triangles such as a tetrahedron. It is positioned roughly at the center of $\mathscr{P}$. Since the GCS algorithm is inspired by growing organic tissue, the reference vectors are termed cells. GCS use an iterative refinement process to fit the current surface into the point data $\mathscr{P}$ (see Fig. 3). The refinement process randomly selects a sample of $\mathscr{P}$ and deforms the current surface in order to progressively minimize its distance. This basic step is repeated and in each iteration the local approximation error is measured. These errors are used to determine surface areas which need to be refined by local subdivision processes. Subdivision and further iterations lead to a better match of the surface to the sample distribution and the process is stopped when the chosen average approximation error reaches a threshold or a certain number of reference vectors is reached. In the following section we analyze the algorithm in detail for the application of surface reconstruction and show different kinds of handling the error such as likelihood distribution, error minimization and topology optimization which each lead to different results. Since the algorithm

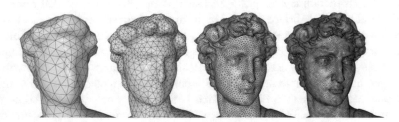

Fig. 3 Different successive surface stages in a GCS surface reconstruction

represents $\mathscr{S}$ as a triangular mesh, we will use the term vertex instead of reference vector, since it is more common in this context.

4.1 Likelihood Distribution

The approximation error (Algorithm 1 line 8) can be altered toward a likelihood distribution or a quantization error (see Sect. 4.2). The vertices create a likelihood distribution if for every given vertex $v \in \mathscr{S}$ the likelihood to be the closest neighbor to a randomly chosen sample $p \in \mathscr{P}$ is equal. If we see vertices being closest neighbors as the result of a probability experiment this approximation resembles entropy maximization. This means that the information carried by any given sample is of the same importance. These representations are especially important in pattern recognition and statistical analysis.

To achieve this every vertex carries a signal counter. To approximate the likelihood distribution (algorithm line 8) these counters are simply incremented when a vertex is closest to an input sample. If a new vertex is added (algorithm line 10) the highest error term refers to the space where most samples share the same vertex.

Since older signals tend to be less representative, all signal counters are decreased at every iteration cycle by a certain factor (algorithm line 8). By using a likelihood distribution signal counters can also be used to determine misplaced vertices in spaces that contain few or even no samples, since their signal counters are very low due to constant decreasing. This concept has therefore been used in most implementations for surface reconstruction.

These algorithms however determine the area for which the likelihood of the vertices is highest and not the surface. If a flat surface is approximated, the algorithm will create lots of vertices in relation to the amount of samples, although the area could be accurately approximated with a few triangles only.

Algorithm 1. An overview of the GCS algorithm. Conditions 1 and 2 can be defined as simple counters.

1: Given a point cloud $\mathscr{P} = \{\mathbf{p}_1...\mathbf{p}_n\}$ and an initial surface $\mathscr{S}$ in form of a tetrahedron represented as an interconnected network of vertices $\mathscr{S} = \{\mathbf{v}_1...\mathbf{v}_n\}$.
2: **repeat**
3:   **repeat**
4:     **repeat**
5:       Select random sample $\mathbf{p}_x$ of $\mathscr{P}$ and search the winning vertex $\mathbf{v}_x$ with smallest Euclidian distance to $\mathbf{p}_x$.
6:       Move $\mathbf{v}_x$ towards $\mathbf{p}_x$.
7:       Move all direct neighbors of $\mathbf{v}_x$ with a lesser factor towards $\mathbf{p}_x$.
8:       Adapt the approximation error of $\mathbf{v}_x$.
9:     **until** condition 1 holds.
10:    Add new vertex in the area of highest approximation error through a vertex split operation.
11: **until** condition 2 holds.
12: Search the least winning vertex in the network and delete it by an edge collapse operation.
13: **until** accuracy exceeds a certain threshold.

4.2 Distance Minimization

When the approximation error (algorithm line 8) is changed to account for a quantization error, vertices are placed exposing the smallest Euclidian distance to the samples in $\mathscr{P}$. If the samples $\mathscr{P}$ are equally distributed the goals of a likelihood distribution compared to an error minimization are nearly the same. If however some regions are represented by a denser sampling than others, these regions will be represented by less vertices in the error minimization scenario, since the error which is measured as the Euclidian distance can be lowered more significantly in regions where samples lay farther apart, hence vertices are more likely to be added there. This approximation is typically used for vector quantization in data compression. To implement this behavior every vertex carries an error value which is increased (algorithm line 8) by the distance or the squared distance between the winning vertex and the given sample. The highest approximation error refers to the space where the samples lay farthest away from a vertex, thus a new vertex will be added there (algorithm line 10). In contrast to a likelihood value, removing a vertex with low distance errors would make no sense, since these vertices indicate that they are well placed. However in case the created topology matters, as in surface reconstruction, it is reasonable to remove such vertices for memory efficiency reasons, since they might be redundant geometry wise.

The basic problem is the difference between the approximation of the right topology and achieving a lowest distance error. We will discuss this problem in more detail in section Sect. 4.3. When this approximation error is used, the deletion process of misplaced vertices need to be handled separately. Despite this disadvantage minimizing the distance error might be more convenient for surface reconstruction. But this is not the case if the approximation minimizes the distance to the vertices instead of the surface (see Fig. 4), since a surface approximation aims to fit $\mathscr{S}$ as close a possible to $\mathscr{P}$.

Many implementations have tackled this problem indirectly. Hoppe et al. [3] presents a roughness adaptation where the average surface curvature is compared to the one of a winning vertex and curved areas lead to higher signals leading to more subdivisions in such areas. In [16] vertices additionally have normals and the algorithm counts how much these normals are moved to increase subdivisions in such areas.

— Approximated Surface
◎ Vertex
→ Distance to closest Vertex
⇀ Distance to closest Surface

Fig. 4 Samples and an approximated surface (*top*), distance between vertices and samples (*middle*), distance between surface and samples (*bottom*). In case of surface approximation the distance to the surface is obviously more worthy

These changes lead to an implicit representation of the approximation error within the algorithm, since curved surface regions need more subdivisions to be correctly approximated. But the surface approximation error itself is not explicitly represented.

4.3 Topology Optimization

The SOM [18] introduced an unsupervised learning concept with an additional topology. In the given network $\mathscr{S}$ vertices are not allowed to move independently. When a vertex is moved toward a sample its neighbors are also moved to decrease the created surface tension (algorithm line 7). This principal adds elasticity to the network—the behavior of a continuous surface is manifested implicitly by creating dependencies between the vertices. A topology can increase the performance in placing reference vectors since the dependencies between them make their movements more stable and thereby make smoother distributions more likely. But the created topology itself can be used in many different ways as well. Data of a high dimensional input space can be mapped into a space of lower dimension and can then be visualized or analyzed with less computational effort (dimensionality reduction). The topology can also be used for regression analysis where $\mathscr{P}$ is known to originate from an unknown continuous function to be reconstructed from the data (function approximation). The SOM uses a static topology, which usually resembles a square shaped grid. The standard GCS algorithm also uses a static surface topology while the connectivity of the network can change (note that the network connectivity is often also referred to as topology, perceiving the network as a graph). This means a network area can be increased in resolution and thereby gather new vertices and connections, but the surface topology of a sphere, inherited from the initial tetrahedron shape, cannot be changed.

So functions that have a different topology can not be correctly resembled. GCS however has better surface approximation capabilities than the SOM, since it builds newly created surfaces by refining a former version of that surface. This adapts the vertex resolutions in different surface areas toward the target function and gives a surface a certain inertia when being modified, which avoids local failures if a surface is fitted into a challenging point constellation.

If the information of a sample placed in a 3D space is processed by the algorithm, this information is always set considering a pre-existing current surface $\mathscr{S}$. This is the strategy of the GCS algorithm to overcome the 3D search problem (see Sect. 3). But still the GCS can get stuck in local minima and the initial topology can mismatch the target surface.

In the standard algorithm a destructive method is presented which uses the average edge length as an indicator to cut out triangles [10]. In [14] triangles which are larger than the average size are cut out, boundaries that fall below a certain Hausdorff distance are joint, and in [3] high valences are used as an indicator to cut the surface and low distances in comparison to edge length to join boundaries. With these changes complex topologies can be created. The main problem of all presented GCS based algorithms concerning topology issues is the missing representation of the actual surface within the adaptation process, which makes many insufficient approximation states simply not measurable (see Fig. 5).

Although the given 3D information of samples is set in relation to the existing 2D surface, the surface is still represented as a collection of Voronoi regions of the vertices, since vertex-sample and not surface-sample distances are considered. This concept implicitly includes the assumption that the Voronoi regions of two connected vertices will not be interrupted within their attached surface. But for close or complex shaped surfaces, this is not the case (see Fig. 6). Here the actual representation of $\mathscr{S}$ becomes apparent being a permeable space of independent Voronoi volumes.

Topology optimization and error minimization are two different things. Error minimization tries to create the smallest possible distance to the samples in $\mathscr{P}$. Topology optimization however is concerned with reaching a topology with $\mathscr{S}$ that lies as close as possible to the topology of $\mathscr{S}_{phy}$ (see Fig. 7).

Fig. 5 Two surfaces with the same vertices, the same samples, and with the same approximation error, that expose an undesired (*left*) and a desired (*right*) solution. A triangle placed in an empty space (*top*) and an incorrect dent in the surface (*bottom*)

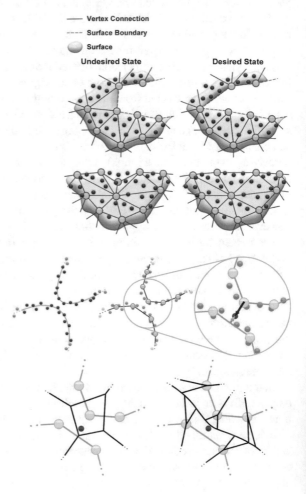

Fig. 6 Samples that originate from a curved surface (*top*, *left*), a fitting approximation of this surface and a magnification that shows the vertex-sample distance and the surface-sample distance (*top*, *right*), the vertex Voronoi regions which assign a sample at the wrong surface (*bottom*, *left*), the surface Voronoi regions which assign the same sample correctly (*bottom*, *right*)

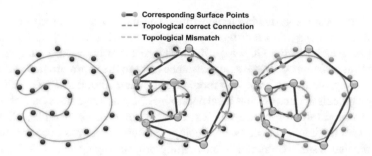

Fig. 7 Surface and samples (*left*), an approximation of that surface and the distance error between the surface and the samples (*middle*), another approximation of that surface with a topological error, but having a similar distance error (*right*)

In order to be topologically correct every point on one surface needs to have a unique equivalent on the other surface and vice versa, while neighbor relations are preserved, meaning the shortest on-surface path between any two given points projected onto $\mathscr{S}$ should always correspond to the one on $\mathscr{S}_{phy}$.

5 Approach

The GCS algorithm has proven to be a high quality surface reconstruction tool. However, in our analysis of the algorithm we saw that topology is only created implicitly and only accounted for through the additional adjustment of neighboring vertices. In the following section we will present our changes to the general approximation concept of the basic algorithm and then the improvements that can be made based on these changes.

5.1 Topology Focused Approximation

The basic algorithm concept focuses on placing vertices in positions likely to decrease the chosen approximation error. To put the actually created surface topology into focus, the approximation error needs to be set in relation to the surface-sample distance. The most important change is to search for the closest surface element (algorithm line 5) instead of the closest vertex. The adaptation process (algorithm line 6 and 7) can now also be set in relation to a sample being closest to a triangle or being closest to an edge, which gives rise to more different local surface modifications (see Sect. 5.2). In the basic GCS implementation the signal counter or error value is carried by the vertices. The surface structure element that most distance errors are measured towards and that is also the building block of the discretization of $\mathscr{S}_{phy}$ is the triangle and is therefore the structure that carries the local approximation error values in our implementation. The most sensible place for the error value of any topology focused function approximation is always the simplex of the highest dimension in the GCS algorithm. The distance between a sample and its closest structure represents the actual distance error to the approximated surface and gives this approximation error way more validity (algorithm line 8). This allows for better judgments about a current local approximation state and the choice of location for subdivision (algorithm line 10). The new approximation error also allows and demands for distinguishing between topology changing deletions to correct topologically misplaced surface structures and non-topology changing deletions that remove geometrically redundant structures from the surface (algorithm line 12). Topology changing deletions are realized by adding an "age" to every triangle and cutting them out when they reach a certain age.

5.2 Implementation of Growing Surface Structures

With the changes described in Sect. 5.1 additional and more accurate information about the current approximation state is available within the GCS process. This information can be used to create a better approximation result in accuracy and topology. In the following we will present our implementation details.

Search for Closest Element. Due to run time efficiency reasons we did not uses an actual triangle based spatial subdivision data structure, but still uses a vertex based octree. By searching for a number n_v of vertices and checking their surrounding triangles, we heuristically find the closest structure to a given sample. We used $n_v = 3$ which fails when the degree of curved and flat surface areas diverge too much and are close to each other, but we consider this case to be rare (see Fig. 8). The new search process has three possible outcomes: a vertex, an edge, or a triangle (algorithm line 5).

Surface Movement. Instead of having only a vertex as a closest element, we now can access additionally an edge and a triangle in our implementation. We modeled three different main movements (algorithm line 6). Since we now know the distance of a given sample p to the surface we can compare this distance d_p to the average sample-to-surface distance $\overline{d}_{\mathscr{P}}$. When d_p is only a fraction lim_{skip} of $\overline{d}_{\mathscr{P}}$ we entirely discard the adaptation since the sample already lies more or less close to the surface. When d_p is about the degree lim_{single} lower than $\overline{d}_{\mathscr{P}}$ we only move the closest vertex, since we consider the surface to be generally correct, but the "joints" of the divisions can be optimized.

Note that distance error distribution within the process is not Gaussian, thus we cannot describe the process in terms of standard deviations. For larger distances we move all vertices of the given structure towards the sample. Due to this, surface widening is less likely to cause spikes and the surface is moved more unified, which produces a smoother surface. The neighborhood movement (algorithm line 7) is kept

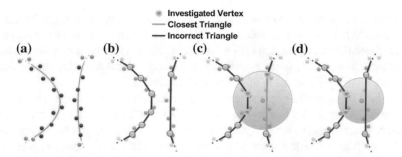

Fig. 8 The presented approach searches for the closest surface structure, such as a vertex, an edge or a triangle. However this process is emulated on a search investigating the surrounding elements of a number of vertices n_v, instead of having a search tree actually comprising edges and triangles. The figure shows a sampled curve and a flat surface (*top, left*), next to it its approximated surface (*top, right*). The search heuristic works correctly for $n_v = 3$, where three vertices are investigated (*bottom, left*). The search heuristic fails for $n_v = 2$ (*bottom, right*), where the vertex connected to the closest triangle is not investigated

unchanged and accomplished with the Laplace smoothing mechanism [25] for all first neighbors of the given structure.

Distance Error. A sample can be closest to a vertex, an edge, or a triangle. When it is closest to a triangle its distance error is changed. In case of an edge this is done for both triangles connected alongside that edge. In case of a vertex this is done for all triangles connected to this vertex.

If the error value was directly set to the given distance, all previous distance errors would be lost. If all distance errors are accumulated, old distance errors, exposing huge distances, would totally determine the subdivision process. If all error values are constantly decreased, as signal counters in a likelihood distribution, it would falsely imply a constant distance error improvement over the entire surface.

The formula proposed in this implementation creates a local *half-life* λ for a distance error update. Its influence on the error value halves after k additional updates (see Eq. 1).

$$\left(\frac{k-1}{k}\right)^\lambda = 0.5$$
$$k = \frac{-1}{\sqrt[\lambda]{0.5}-1} \tag{1}$$

$$err_{new} = \frac{d_x + err_{old}(k-1)}{k}$$

Refinement. Instead of the vertex we search the *triangle* with the highest approximation error. Subdivision is done by splitting the triangle's surface from one of its three vertices to the opposite edge and then also split the other triangle in the mesh with this edge. The edge with the additional triangle with the largest error term is taken. Four new triangles are added, four new edges, and one new vertex. The error value of each new triangle is set to the half of the error value of its predecessor.

Deletion. The deletion process is one of the most important changes in the new algorithm. When using a sample-to-surface distance error, the error values can be used to determine triangles that are geometrically redundant. In order to have a model representation that is as memory efficient as possible, these triangles can be deleted by an edge collapse operation of one of its three edges.

The best triangle edge for the collapse operation is the one which is surrounded by triangles with normals exposing the least differences to one another. A collapse of this edge changes the surface geometry the least. This edge can be determined as the one with the highest dot product of the normals of its two vertices, since these normals are calculated based on their surrounding triangles. It is reasonable to set a threshold to avoid decreasing the surface approximation quality when collapsing vertices actually exposing curvature.

In addition to the distance error value a triangle age is needed that indicates if a triangle reached a maximum age max_a. This should happen when the triangle has not been winning for a certain number of times γ. Those triangles are considered to be misplaced and topologically wrong. A misplaced triangle is detached from the rest of the mesh and then deleted. After the deletion process the mesh has to be cleaned with applied filters.

Since the likelihood for a triangle to win is proportional to its size, the aging process needs to be more differentiated. For every iteration the age of all triangles is increased by a tiny factor β, which has a relation to the overall number of triangles $|\mathcal{T}|$ (see Eq. 2). Instead of increasing the age of the triangles by a constant increment, it is done by multiplication, allowing the use of a tumble-tree [2]. The process reaches an upper bound, which resembles *aging*.

$$\beta = {}^{(\gamma \cdot |\mathcal{T}|)}\!\sqrt{max_a} - 1 \tag{2}$$

For all triangles in the basic step, whose error values are updated (see above), the age is renewed. Since small triangles are less likely to be winners, the initial age of a triangle is its $size(\mathbf{t})$ divided by the average triangle size $\overline{s_{\mathcal{T}}}$. Thus, the average triangle starts with an age of one, while small triangles start "younger" and big triangles "older".

The deletion process now explicitly distinguishes distance driven and topologically driven deletions.

Finalization. One of the assets of the GCS algorithm is the fact that $\mathcal{S}$ is an approximation of $\mathcal{S}_{phy}$ at any time during the running loop—the algorithm can be stopped and resumed at any given time. With the novel surface distance approximation error a potential stopping point for the algorithm can be chosen more sensibly.

6 Results

We accomplished different tests with the Stanford Dragon model as a good example for a point cloud that is relatively challenging by its shape and sample distribution, the hand model exposes sharp features, the Asian Dragon and the Thai Statue expose

Table 1 Parameter settings for GSS algorithm

Symbol	Setting	Meaning
λ	9	Half-life of a distance error
γ	7	Allowed misses before deletion theshold
max_a	10	Maximum age for triangle
$max_{\nabla n}$	0.9	Threshold before edge collapse
lim_{skip}	0.9	Distance threshold before mesh change
lim_{single}	1.2	Distance threshold before multiple vertex change
n_v	3	Number of investigated vertices in structure search

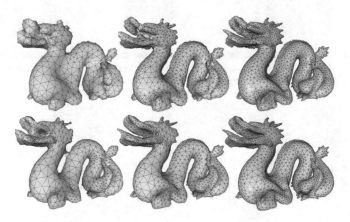

Fig. 9 A progression series of the dragon model from *left* to *right* with 2500, 5000 and 10000 triangles with the standard algorithm (*top*) and with the new algorithm (*bottom*). With the new algorithm the surface diverges faster toward the final topology

a lot of curved areas, the Heating Pipes model includes some extremely noisy areas, non-uniform sample densities and open surface areas, the Happy Buddha has regions of surfaces lying close together. From the basic algorithm we used [3] but deactivated *roughness adaptation* and *sharp feature detection* (for the parameter settings see [3]). For the GSS algorithm the parameter settings are listed in Table 1.

A reasonable aspect to judge efficiency is the time to reach a certain accuracy. In our experiments with the new approach, for instance the topology of the dragon model was approximated much faster, shown in Fig. 9.

Close surfaces can be handled correctly with our presented method and a vast number of additional iterations to avoid permeating Voronoi regions is not required any longer as we show in Fig. 10.

Although the standard algorithm is already quite robust when dealing with noise, we could show that spikes and rough surface gradients could be greatly reduced with our presented method. The moving of entire substructures seems to have a smoothing effect on the surface (see Fig. 11).

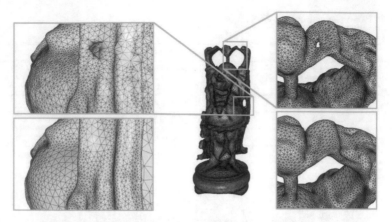

Fig. 10 Some thin areas of the Happy Buddha model, reconstructed with 200 K triangles with the old (*top*) and the new (*bottom*) algorithm. The new algorithm is able to build a correct topology in thin areas in an earlier algorithm stage

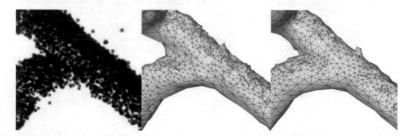

Fig. 11 Very noisy section of the Heating Pipes model (*left*); pointy vertices or spikes on the surface of the standard algorithm (*middle*) and a smoother surface with our approach (*right*)

In Table 2 the new and the old algorithm are compared. Generally the old algorithm creates a lower average point sample-to-surface distance, since it evenly distributes its subdivisions over $\mathscr{S}$, whereas the new algorithm focuses its subdivisions on areas of high approximation error rates. This is visible through a 25 % decrease of the mean squared error. Especially for curved models such as the Asian Dragon and the Thai Statue this effect is very salient.

Although the search process is more complex, the extra time costs are nearly leveled by the discarded operations which are for the Dragon model 43.3 % of discarded adaptations and a rate of 26.3 % inside the surface movements of vertices only with our setting for lim_{skip} and lim_{single}.

When the square distance was used as approximation error, the results for both the average distance error as well as the square distance error were worse than the results of the GCS algorithm. The reason is that most triangles are used up to model tiny but steep curvature. In addition, triangles tended to clump even for low error half-lifes λ. For the Happy Buddha model this led to many clump-like artifacts. Due to this results using the square distance error can be generally considered impractical.

Table 2 Our results for different models

Model (# triangles)	GCS			GCS error = d_p			GCS error = d_p^2		
	Time(s)	Dist	Dist^2	Time(s)	Dist	Dist^2	Time	Dist	Dist^2
Hand (20 K)	6	3.19	4.38	7	4.29	4.11	6	4.70	5.00
Dragon (100 K)	61	2.29	3.42	65	2.55	2.97	66	3.05	4.34
Asian Dragon (100 K)	55	2.36	3.62	61	2.71	2.53	61	2.83	2.78
Thai Statue (200 K)	146	3.21	15.4	147	3.00	4.02	148	4.05	7.17
Happy Buddha (200 K)	150	1.48	14.1	158	1.89	11.7	160	4.02	63.0
	$n_v = 1$			$n_v = 5$			$n_v = 10$		
Model (# triangles)	Time(s)	Dist	Dist^2	Time(s)	Dist	Dist^2	Time	Dist	Dist^2
Dragon (100 K)	62	2.35	3.01	71	2.55	3.05	121	2.57	2.95

We expose the time for the reconstruction process (time), the average distance to $\mathscr{P}$ (dist) times 10^4 and the square distance to $\mathscr{P}$ (dist^2) times 10^7. We also tested different values for n_v. Note that models are normalized by setting the cubic diagonal of their bounding box to one

Test Hardware. A *Dell*®Precision M6400 with *Intel*®QX9300 (2.53 GHz) processor with 8 GB 1066 MHz DDR3 Dual Channel RAM.

7 Conclusions and Future Work

In this paper we focused on the behavior of the growing cell structures approach as a function approximation algorithm. We analyzed the GCS with the classical adaption algorithm for matching requirements of surface reconstruction. Derived from these observations we presented our new GCS learning model and proved theoretically and by examples that it outperforms the classical GCS approach. The basic idea of the presented approach is to incorporate the constructed topology into the GCS learning scheme. GCS creates ideal distribution matching, clustering, or classical dimensionality reduction by an *implicit* representation of a topology. In our work, we introduced an *explicit* topology to model the approximation behavior according to it, while saving the valuable ANN capabilities mentioned above. As result, we got a new ANN learning strategy, which showed several advantages compared to classical models. We see this paper as proof of our conceptual change of the GCS algorithm, giving rise to many improvements for the algorithm in future work.

The differentiation of approximation errors in the form of sample-to-surface distances and topological misconstructions is the single most important ability and at the same time the most important task, when implementing a GSS based reconstruction algorithm. The separate deletion processes for geometric redundancy and for topological misconstructions are examples of considering this dichotomy.

Although the presented GSS implementation works well, the concept by itself does not solve this general differentiation problem, but rather offers the necessary information to enable a differentiated handling.

Acknowledgments The authors would like to thank the Stanford University Computer Graphics Laboratory and the Hamburg Hafen City University for making their laser scanned data available to us. We also would like to thank Kai Burjack for his assistance in rendering.

References

1. Amenta, N., Bern, M., Kamvysselis, M.: A new voronoi-based surface reconstruction algorithm. In: Proceedings of the 25th Annual Conference on Computer Graphics and Interactive Techniques, SIGGRAPH '98, pp. 415–421, ACM, New York, NY, USA (1998)
2. Annuth, H., Bohn, C. A.: Tumble tree: reducing complexity of the growing cells approach. In Proceedings of the 20th International Conference on Artificial Neural Networks: Part III, ICANN'10, pp. 228–236, Springer, Berlin, Heidelberg (2010)
3. Annuth, H., Bohn, C.-A.: Smart growing cells: supervising unsupervised learning. In: Computational Intelligence. Studies in Computational Intelligence, Vol. 399, pp. 405–420. Springer, Berlin/Heidelberg (2012)
4. Baader, A., Hirzinger, G.: Three-dimensional surface reconstruction based on a self-organizing feature map. In: Proceedings of 6th International Conference on Advantage Robotics, pp. 273–278 (1993)
5. Bernardini, F., Mittleman, J., Rushmeier, H., Silva, C., Taubin, G.: The ball-pivoting algorithm for surface reconstruction. IEEE Trans. Vis. Comput. Graph. **5**(4), 349–359 (1999)
6. Carr, J. C., Beatson, R. K., Cherrie, J. B., Mitchell, T. J., Fright, W. R., McCallum, B. C., Evans, T. R. Reconstruction and representation of 3d objects with radial basis functions. In: Proceedings of the 28th Annual Conference on Computer Graphics and Interactive Techniques, SIGGRAPH '01, pp. 67–76, ACM, New York, NY, USA (2001)
7. Chang, M.-C., Leymarie, F.F., Kimia, B.B.: Surface reconstruction from point clouds by transforming the medial scaffold. Comput. Vis. Image Underst. **113**(11), 1130–1146 (2009)
8. Curless, B., Levoy, M.: A volumetric method for building complex models from range images. In: Proceedings of the 23rd Annual Conference on Computer Graphics and Interactive Techniques, SIGGRAPH '96, pp. 303–312, ACM, New York, NY, USA (1996)
9. Edelsbrunner, H., Mücke, E. P. Three-dimensional alpha shapes. In: Volume Visualization, pp. 75–82 (1992)
10. Fritzke, B.: Growing cell structures—a self-organizing network for unsupervised and supervised learning. Neural Netw. **7**, 1441–1460 (1993)
11. Gal, R., Shamir, A., Hassner, T., Pauly, M., Cohen-Or, D. Surface reconstruction using local shape priors. In: Proceedings of the Fifth Eurographics Symposium on Geometry Processing, SGP '07, pp. 253–262, Eurographics Association, Aire-la-Ville, Switzerland, Switzerland (2007)
12. Gopi, M., Krishnan, S.: A fast and efficient projection-based approach for surface reconstruction. In: Proceedings of the 15th Brazilian Symposium on Computer Graphics and Image Processing, SIBGRAPI '02, pp. 179–186, IEEE Computer Society, Washington, DC, USA (2002)
13. Hoppe, H., DeRose, T., Duchamp, T., McDonald, J. A., Stuetzle, W.: Surface reconstruction from unorganized points. In: Thomas, J. J. (ed.) SIGGRAPH, pp. 71–78. ACM (1992)

14. Ivrissimtzis, I., Jeong, W.-K., Seidel, H.-P.: Neural meshes: statistical learning methods in surface reconstruction. In: Technical Report MPI-I-2003-4-007, Max-Planck-Institut f"ur Informatik, Saarbrücken (2003a)
15. Ivrissimtzis, I. P., Jeong, W.-K., and Seidel, H.-P. (2003b). Using growing cell structures for surface reconstruction. In SMI '03: Proceedings of the Shape Modeling International 2003, page 78, Washington, DC, USA. IEEE Computer Society
16. Jeong, W.-K., Ivrissimtzis, I., Seidel, H.-P.: Neural meshes: statistical learning based on normals. In: Computer Graphics and Applications, Pacific Conference on 404 (2003)
17. Kazhdan, M., Bolitho, M., Hoppe, H.: Poisson surface reconstruction. In: Proceedings of the fourth Eurographics symposium on Geometry processing, SGP '06, pp. 61–70, Eurographics Association, Aire-la-Ville, Switzerland, Switzerland (2006)
18. Kohonen, T.: Self-organized formation of topologically correct feature maps. Biol. Cybern. **43**, 59–69 (1982)
19. Kuo, C.-C., Yau, H.-T.: A delaunay-based region-growing approach to surface reconstruction from unorganized points. Comput. Aided Des. **37**(8), 825–835 (2005)
20. MacQueen, J. B.: Some methods for classification and analysis of multivariate observations. In: Proceedings of 5th Berkeley Symposium on Mathematical Statistics and Probability, pp. 281–297 (1967)
21. Mederos, B., Amenta, N., Velho, L., de Figueiredo, L. H.: Surface reconstruction from noisy point clouds. In: Proceedings of the Third Eurographics Symposium on Geometry Processing, SGP '05, Eurographics Association, Aire-la-Ville, Switzerland, Switzerland (2005)
22. Ohtake, Y., Belyaev, A., Alexa, M., Turk, G., Seidel, H.-P.: Multi-level partition of unity implicits. In: ACM SIGGRAPH 2003 Papers, SIGGRAPH '03, pp. 463–470, ACM, New York, NY, USA (2003)
23. Schnabel, R., Degener, P., Klein, R.: Completion and reconstruction with primitive shapes. Comput. Graph. Forum (Proc. of Eurographics) **28**(2), 503–512 (2009)
24. Sharf, A., Lewiner, T., Shamir, A., Kobbelt, L., Cohen-Or, D.: Competing fronts for coarse-to-fine surface reconstruction. In: Eurographics 2006 (Computer Graphics Forum), Vol. 25, pp. 389–398. Eurographics, Vienna (2006)
25. Taubin, G.: A signal processing approach to fair surface design. In: Proceedings of the 22nd Annual Conference on Computer Graphics and Interactive Techniques, SIGGRAPH '95, pp. 351–358, ACM, New York, NY, USA (1995)
26. Várady, L., Hoffmann, M., Kovács, E.: Improved free-form modelling of scattered data by dynamic neural networks. J. Geom. Graph. **3**, 177–183 (1999)
27. Yu, Y.: Surface reconstruction from unorganized points using self-organizing neural networks. In: IEEE Visualization 99, Conference Proceedings, pp. 61–64 (1999)

Selective Image Compression Using MSIC Algorithm

Enrique Pelayo, David Buldain and Carlos Orrite

Abstract This paper presents a new algorithm, Magnitude Sensitive Image Compression (MSIC), as a reliable and efficient approach for selective image compression. The algorithm uses MSCL neural networks (in direct and masked versions). These kind of neural networks tend to focus the learning process in data space zones with high values of a user-defined magnitude function. This property can be used for image compression to divide the image in irregular blocks, with higher resolution in areas of interest. These blocks are compressed by Vector Quantization in a later step, giving as a result that different areas of the image receive distinct compression ratios. Results in several examples demonstrate the better performance of MSIC compared to JPEG or other SOM based image compression algorithms.

Keywords Image compression · Competitive learning · Neural networks · Saliency · Self organizing maps · JPEG · DCT · MSCL

1 Introduction

In the human vision system the attention is attracted to visually salient stimuli, and therefore only scene locations sufficiently different from their surroundings are processed in detail. This provides the necessary motivation to devise a novel image compression method capable of applying distinct compression ratios to different zones of the image according to their saliency.

In this paper we make use of the Magnitude Sensitive Competitive Learning Algorithm (MSCL) [1]. MSCL is a Vector Quantization method based on competitive learning, where units compete not only by distance but also by a user defined magnitude. Using saliency as the magnitude, units tends to model more accurately

This work is partially supported by Spanish Grant TIN2010-20177 (MICINN) and FEDER and by the regional government DGA-FSE.

E. Pelayo (✉) · D. Buldain · C. Orrite
Aragon Institute for Engineering Research, University of Zaragoza, Zaragoza, Spain
e-mail: epelayoc@gmail.com
url: http://i3a.unizar.es/en

© Springer International Publishing Switzerland 2016
K. Madani et al. (eds.), *Computational Intelligence*,
Studies in Computational Intelligence 613,
DOI 10.1007/978-3-319-23392-5_23

the salient areas of the images, and therefore the neural network behaviour imitates the human vision system.

Vector quantization (VQ) is a classical quantization method. In the context of image processing, basic vector quantization consists in dividing the input image into regular blocks of pixels of a pre-defined size, where each block is considered as a *D-dimensional* vector. Each of these input vectors from the original image is replaced by the index of its nearest codeword, so only this index is transmitted through the media. The whole codebook serve as a database known on the reconstruction site. This scheme reduces the transmission rate while maintaining a good visual quality. Figure 1a shows this scheme.

In VQ, compression level depends on two factors, the number of blocks and the level of compression of each block. Both factors are related in an inverse way. Lower number of blocks means that they are higher in size, and therefore higher is the bit depth necessary to codify each block for a similar quality.

Some authors [2–5] have already used some VQ variants, such as Kohonen neural network [6] for image compression. These algorithms use a fixed block size and concentrate in several ways to get a smaller codification of each block or to improve the quality of the codification. Laha [2] uses surface fitting of data assigned to each codeword instead of the codeword itself, which improves the visual quality of the results. [3–5] apply DCT filtering [7] to each block previous to the quantization step to lower the dimension of the input data. On the other hand, [3] takes advantage of the topological ordering property of the SOM neural network to codify indexes with a few bytes.

In this paper blocks may have different size, chosen according to its relevance (which is selected following the image saliency). Blocks located in areas of high image saliency are smaller than those assigned with low saliency. As bit depth used in the quantization step is the same for all blocks, quantization error increases directly with the block size in areas of low image saliency. Therefore, a lower number of blocks are used to represent the whole image increasing the overall image compression and preserving at the same time the quality of most relevant areas.

Another important difference with the above mentioned methods is that, in our approach, block shapes are, in general, irregular, i.e., neither rectangular nor squared. Therefore, quantization has to take into account samples that may have invalid components. Figure 1b shows the basic idea of the proposed algorithm for grayscale images. It requires to transmit the block centers and index. At the receiver, it is possible to regenerate the shape and mask of each block and locate it with it center and magnitude. Then, with its index, the block image is regenerated and summed up to form the whole image. In Sect. 4 we present the complete algorithm, more complex to reduce the amount of data to be transmitted.

The remainder of this paper is organized as follows. Section 2 describes the MSCL algorithm. Section 4 shows its use to achieve selective image compression focused on the most salient regions of an image with the method that we call Magnitude Sensitive Image Compression (MSIC). A comparative between MSCL and classical JPEG and SOM based VQ algorithms for a high compression ratio task is carried out in Sect. 5. Finally, Sect. 6 concludes with a discussion and ideas for future work.

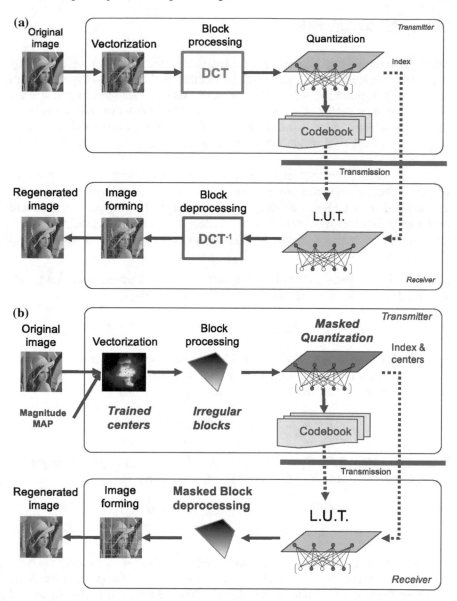

Fig. 1 Basic idea of Competitive Learning algorithms in the task of image compression for grayscale images. *Top* Common CL algorithm. *Bottom* MSIC algorithm. Differences with other CL algortihms are the use of a MSCL to get block centers (centers are trained weights of MSCL units), the use of irregular blocks and the masked quantization/deprocessing

2 The MSCL Algorithm

MSCL is a type of artificial neural network that is trained using unsupervised learning to produce a representation of the input space of the training samples depending on a magnitude. Codebook $\mathcal{M}$ is formed by M weight vectors. Prototype of unit $m \in \mathcal{M}$ is described by a vector of weights $\mathbf{w}_m(t) \in \mathbb{R}^D$ and the magnitude value mu_m. This value is calculated with the function $MF(i, t)$, that is a measure of any feature or property of the data inside the Voronoi region of unit m, or a function of its parameters.

The idea behind the use of this magnitude term is that, during competition between two units situated at similar distance from the input sample, the winner will be the unit with the lowest magnitude value. As a result of the training process units will be forced to move from the data regions with low $MF(i, t)$ values to regions where this magnitude function is higher. MSCL follows next steps, that are repeated until a termination condition is achieved:

Global Unit Competition. At this point, we form the local winner set $\mathcal{S}$, $(\mathcal{S} \subset \mathcal{M})$ with the M_{local} units closest to the input sample as: $\mathcal{S} = \{s_1, s_2, ..., s_{M_{local}}\}$.

$$\|\mathbf{x}(t) - \mathbf{w}_s(t)\| \le \|\mathbf{x}(t) - \mathbf{w}_m(t)\| \ \forall m \notin \mathcal{S} \wedge s \in \mathcal{S} \tag{1}$$

Local Unit Competition. Winner unit j is selected from units belonging to $\mathcal{S}$, as the one that minimizes the product of its magnitude value with the distance of its weights to the input data vector, following this equation:

$$j = \operatorname*{argmin}_{s \in \mathcal{S}}(mu_s(t)^\gamma \cdot \|\mathbf{x}(t) - \mathbf{w}_s(t)\|) \tag{2}$$

Winner and Magnitude Updating. For all units in the map, weights and magnitude are adjusted iteratively for each training sample, following:

$$\mathbf{w}_j(t + 1) = \mathbf{w}_j(t) + \alpha(t) \left(\mathbf{x}(t) - \mathbf{w}_j(t)\right) \tag{3}$$

$$mu_j(t + 1) = MF(j, t) \tag{4}$$

In the above equations, γ defines the strength of the magnitude during the competition. $\alpha(t)$ is the learning factor, calculated as $\alpha(t) = (1/h_j(t))^\beta$, where $h_j(t)$ stands for the number of input signals for which unit j has been winner so far, and β is a scalar value between 0 and 1. The winner j is also called the best matching unit (BMU).

3 The Masked MSCL Algorithm

The new proposed image compression algorithm will require the capability of dealing with incomplete data, as blocks to be compressed are irregular (in shape and size). Here we present a masked version of MSCL that is able to deal with data samples of different size (we speak of 'masked' data). To use this algorithm we will consider that each data sample consists in two vectors, $\mathbf{x} = (x_1, \ldots, x_D) \in \mathbb{R}^D$ the data vector itself (with the maximal possible dimension of a data sample D), and its corresponding mask $\mathbf{msk} = (msk_1, \ldots, msk_D) \in \mathbb{R}^D$. The mask is a vector with ones in the the valid components of $\mathbf{x}$ and zeros in the other components.

The algorithm follows the same steps than MSCL, but both competition and updates are slightly more complex as it has to be considered the mask. Changes are the following:

1. Only valid components (corresponding mask is one) are considered for global and local competition:

$$\|\mathbf{msk}(t) \circ (\mathbf{x}(t) - \mathbf{w}_s(t))\| \leq \|\mathbf{msk}(t) \circ (\mathbf{x}(t) - \mathbf{w}_m(t))\| \; \forall m \notin \mathcal{S}, s \in \mathcal{S} \quad (5)$$

$$j = \underset{s \in \mathcal{S}}{\operatorname{argmin}}(mu_s(t)^\gamma \cdot \|\mathbf{msk}(t) \circ (\mathbf{x}(t) - \mathbf{w}_s(t))\|) \quad (6)$$

2. Instead of scalar α, algorithm uses vector $\mathbf{alpha} = (\alpha_1, \ldots, \alpha_D) \in \mathbb{R}^D$ as the learning factor, where α_d is:

$$\alpha_d = \begin{cases} 1/h_{jd} & \text{if } msk_d = 1 \\ 0 & \text{otherwise.} \end{cases} \quad (7)$$

Here h_{jd} is the number of times up to the moment that component d has taken a valid value when unit j was a winner.

3. Only valid components (those with $msk_d = 1$) of winner weights are updated:

$$w_{jd}(t+1) = w_{jd}(t) + \alpha_d \left(x_d(t) - w_{jd}(t) \right) \quad (8)$$

4 Magnitude Sensitive Image Compression

Figure 2 shows the whole MSIC algorithm applied to grayscale images, where image compression, in the transmitter, is represented on the top and the image restoration process at the receiver is depicted on the bottom. Image is compressed with different quality according to a selected user magnitude. Section 4.6 explain how to extent this methodology to color images.

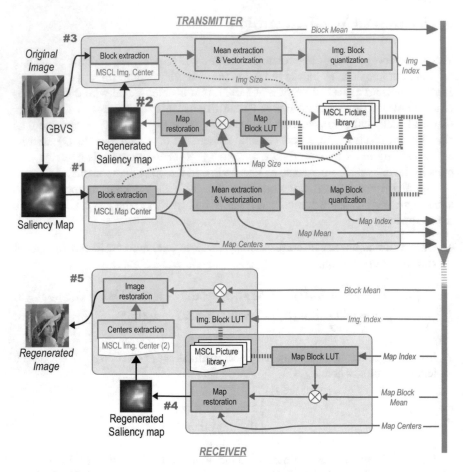

Fig. 2 Global algorithm for *grayscale* images. *Marked* with **#n** the corresponding subsection with the detailed explanation and, also showing the order of processing steps in the transmitter and receiver

In this work we use as magnitude the saliency map, with the same size as the processed image, provided by a user function. Section 5 explain these functions.

The results of the compression are a group of image blocks encoded by indexes. Unlike other image compression methods, our algorithm uses blocks of different sizes, which are located at any position of the image. Therefore, this implies that block centers and sizes has to be sent to the receiver, apart from the corresponding index. As this approach would mean the transmission of huge quantity of information, we have adopted an alternative solution.

We use the saliency map to train a MSCL network, using as inputs the coordinates (x_1, x_2) of each pixel and the saliency as magnitude. After training, the weights of its units (codewords) are the block centers $(bc(k), k = 1...N_{bc})$. The surrounding assigned to the Voronoi region of each block-center configure the corresponding

blocks. The image is so fragmented in so many blocks as units in this network (N_{bc}).
In Sect. 4.1 we will show how to determine the block sizes (and block limits) for
each codeword or unit. This process encodes the saliency map with low quality, and
both the encoded image and the encoded map are transmitted.

At the receiver first the saliency map is regenerated, and with it, the image block
limits and centers can be calculated. They are used with the image indexes to restore
the image.

It is worth noting that it is necessary an additional step at the transmitter. Instead
of using directly the saliency map to extract the image blocks, we first decode a
saliency map from the encoded map that has to be transmitted. Then we calculate
the image centers and limits of image blocks using this Regenerated Saliency Map
that will be also regenerated by the receiver.

Summarizing the MSIC algorithm steps are:

1. Map quantization (at transmitter).
2. Map restoration (at transmitter).
3. Image quantization (at transmitter).
4. Map restoration (at receiver).
5. Image restoration (at receiver).

MSIC algorithm uses several MSCL networks: $MSCL_{MC}$ (map center) to extract
map blocks, $MSCL_{IC}$ (image center) to extract image blocks, and a pool of MSCLs
that he call MSCL picture library ($MSCL_{PIC}$) to generate indexes that encode each
block pixels, and act as Look-Up-Table to decode the block shapes with these indexes.
This library is calculated using the masked version of MSCL (see Sect. 3) as blocks
may have irregular shapes. The first and second neural networks are trained online
during map and image quantization. Their codewords are the block centers. However
$MSCL_{PICT}$ form a codebook database that is trained offline. It is known by the
transmitter and the receiver as a library of the method. Finally receiver uses another
MSCL ($MSCL_{IC2}$), that becomes identical to $MSCL_{IC}$ when trained at receiver.
Following sections explain the process in detail.

4.1 Saliency Map Quantization

The idea is to consider the saliency map as an image and apply the same compression
steps that will be applied to the image.

First step corresponds to the block extraction from the saliency map according to
the saliency values. We train a MSCL network ($MSCL_{MC}$) using the 2D coordinates
of each pixel(x) as inputs and the following magnitude function:

$$MF(i, t) = \frac{\sum_{Vi} saliency(\mathbf{x}_{Vi})}{V_i(t)} \qquad (9)$$

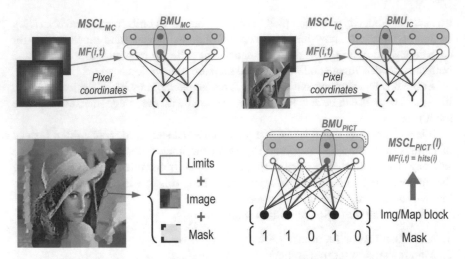

Fig. 3 Neural networks used in the MSIC algorithm: *Top* BMU_{MC} and BMU_{IC}. It is important to mention that this last MSCL is used also in receiver (BMU_{IC2}). *Bottom* Block extraction phase. Each block delivers the block limits, the image and a binary mask. $MSCL_{PICT}(l)$ neural network, where a input sample (vectorized block from the extraction phase) has several masked components

where $\mathbf{x}_{Vi}$ are the data samples belonging to the Voronoi region of unit i at time t, $V_i(t)$ is the number of samples in the Voronoi region, and $saliency(\mathbf{x})$ is the pixel saliency of the corresponding sample. Trained unit weights correspond to the coordinates of the unit in the image, and the magnitude value is the mean of the saliency inside its Voronoi region. Once trained, it is possible to find the best matching unit (BMU_{MC}) assigned to every pixel (using magnitude during competition). The block assigned to each unit is the rectangle wrapping its Voronoi region. A block mask of equal size than the block is also provided in order to mark the pixels belonging to that irregular Voronoi region, see Fig. 3. We used 40 units for $MSCL_{MC}$ in our experiment. With this small number of units a coarse saliency map is obtained, but it is enough to define areas with high saliency.

To codify each of the blocks by VQ, we first resize the block to a squared shape with side value as the maximum between its horizontal and vertical block sizes. The block and the mask are inserted in the squared image filling with zeros the void rows or columns. After that, both are resized to a vector form. We use mean-removed vectors to have a better quantification. Mean value of saliency in each block of pixels, that we call mean block-saliency ($m_b(x, y)$), is sent encoded by 7 bits.

The resulting vector is separated according to its size and dispatched for training or testing to the MSCL picture library ($MSCL_{PICT}(l)$). This pool of codebooks are trained separately only once and become a lookup table in the algorithm. In order to avoid the transmission of the whole codebook pool it is known by both the transmitter and the receiver.

Each codebook of the pool, with 256 codewords, is dedicated to a precise input-vector length. This election of the same number of codewords for different block sizes forces that larger blocks present less detail in pictorial content than smaller blocks. We have chosen a limited group of sizes that model several size possibilities (the value of l is the length of the square edge to which we have resized the block): $l = [4, 6, 7, 8, 10, 15, 29]$.

This pool of codebooks can be specialized in the type of images considered in the transmission task, or can be generated using an universal library of training images. The images for training are processed following previous described steps, but the magnitude function chosen for these $MSCL_{PICT}(l)$ networks is the hit frequency of each unit, that is:

$$MF(i, t) = hits(i, t) \tag{10}$$

During competition the BMU_{PICT} is calculated using the masked version of MSCL in order to avoid the zero-padding mentioned before. Each time a sample is presented to each neural network of the pool, the corresponding mask is also presented, and only masked weight components are used to compete (see Fig. 3, Right). Each sample might have different masked components. In this way, only pixels corresponding to the Voronoi region of a block are used to find its BMU_{PICT}.

At the end of this step, the magnitude map has been divided in 40 blocks. We have to send to the receiver the following information of each block: Map indexes (BMU_{PICT}) (1 byte), Map mean (7 bits) and Map Centers (2 bytes). Size of each block is not necessary because it is calculated with the block centers.

4.2 Map Restoration at Transmitter

Map representing the saliency of the image is also restored at transmitter with the information generated at the previous step. This is because the restored map will be used at both transmitter and receiver to define the block centers of the image, so results are the same in both sides. Map restoration is accomplished following the previous step in inverse order. First we calculate Voronoi regions assigned to each of the Map Centers by searching the BMU_{MC} of each pixel in $MSCL_{MC}$. The codewords of this neural network are the Map Centers. Additionally, we calculate block limits and mask wrapping by a rectangle the area corresponding to the Voronoi region of each center.

With the i index of the new block, it is converted again into an image block by the look-up table created with $MSCL_{PICT}(l)$. The codeword of the BMU_{PICT} consists of the pictorical content of the block image, but needs to be displaced with the mean block-saliency value of the corresponding block. After summing the mean, it is masked by the binary mask and added to the regenerated saliency map. Repeating the process for all the blocks we obtain the regenerated saliency map, that will represent the saliency values of pixels for the reconstructed image.

4.3 Image Quantization

A similar strategy to the previously described step is followed for image quantization. Blocks are extracted training the $MSCL_{IC}$ (with the coordinates at each pixel) to get the image block centers according to the Regenerated Saliency Map at the transmitter. Then the Voronoi regions of each of these centers are calculated. Blocks are extracted and vectorized. After removing the mean, each image block is processed using the masked version of MSCL with the $MSCL_{PICT}(l)$ (once again using the masked version of MSCL) that corresponds its size, in order to use the most similar pictorial content of the library that will be included in the reconstructed image. It is only necessary to send the corresponding block mean and index from the $MSCL_{PICT}(l)$ for each block.

4.4 Map Restoration at Receiver

Map restoration at receiver is accomplished following exactly the same process than map restoration at transmitter. To do it, the receiver uses for each block its Map index, mean block-saliency, block-center and the same offline $MSCL_{PICT}(l)$ picture library. As operations are the same and they are applied to the same data, the Regenerated Saliency Map at receiver is exactly the same than the one at the transmitter.

4.5 Image Restoration

Last step in the whole process is image restoration, using the received means of block-saliency, the pixel indexes and the regenerated saliency map. This step is similar to the previous described Map restoration with small changes.

The main difference is that the image block centers are not available (they have not been transmitted). They are calculated training a new MSCL ($MSCL_{IC2}$), with the coordinates of each pixel, and the magnitude values in the Regenerated Saliency Map (magnitude that was calculated with (9) at the emitter). This neural network becomes identical to $MSCL_{IC}$. The weights of $MSCL_{IC2}$ are the centers of the image blocks, and their Voronoi regions define the masks and limits.

Once again, image indexes are presented to the look-up table created with $MSCL_{PICT}(l)$ (according to the block size) that returns the block shape. Final image is regenerated by adding means of block-saliency, masking each block and positioning it in the image (adding it to the regenerated image as we had done before with the saliency map).

4.6 Extension to Color Images

Figure 4 defines the flowchart to use MSCL in the case of color images. The process is similar to the used in the case of grayscale images, but applied to each of the color components of the image.

First, we calculate the saliency map from the color image. With this saliency map we extract and quantify blocks as described in Sect. 4.1, blocks which are restored at transmitter as mentioned in Sect. 4.2. As a result of this step we get the map block-centers, block-means and indexes. Encoding is made with the previously trained $MSCL_{PICT}(l)$ picture library.

Then, original RGB image is transformed to the L-a-b color space. The reason of selecting this color codification is that it has been demonstrated its suitability for interpreting the real world [8].

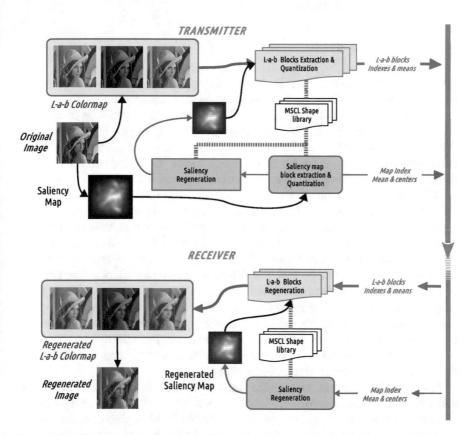

Fig. 4 Global algorithm for color images. Each color component is processed separately as in the *grayscale* method. However this process is exemplified with a different magnitude definition for the saliency map, oriented to preserve the detail of the image for certain colors selected by the user

Now with these L-a-b color components of the image, we follow the process indicated in Sect. 4.3. Each of them will be trained with a MSCL neural network $(MSCL_{IC-L}, MSCL_{IC-a}, MSCL_{IC-b},)$ and it will return the block sizes and indexes for each component. The indexes of the blocks are also encoded with $MSCL_{PICT}(l)$.

Once at receiver saliency map is restored (see Sect. 4.4). Then, we follow the image restoration step, applied to each L-a-b component. Its centers are calculated training three MSCL networks $(MSCL_{IC2-L}, MSCL_{IC2-a}, MSCL_{IC2-b},)$, with the coordinates of each pixel, and the regenerated saliency map. These neural networks becomes identical to those at the transmitter.

To get the final image, we transform the restored L-a-b image to RGB.

5 Experimental Results

5.1 Grayscale Images

Simulations were conducted on four 256×256 gray scaled images (65536 bytes), all of them are typical in image compression benchmarking tasks.

We applied the MSIC algorithm, with the following MSCL training parameters: 15 cycles and learning factor varying along the training process from 0.9 to 0.05. We used Graph-Based Visual Saliency $GBVS(\mathbf{x})$ ([9]) as the pixel saliency of the corresponding sample. However, it is possible to use other kind of magnitudes to define which areas of the image are compressed more or less deeply.

JPEG was applied with the standard Matlab implementation and a compression Quality of $Q = 3$ or $Q = 5$ (i.e., with a high compression ratio).

We also compare with the algorithm described in [5], whose main steps are followed for all the mentioned SOM based algorithms: The original image is divided into small blocks (we select a size of 8×8 to achieve a similar compression ratio to JPEG or MSCL). Then, 2-D DCT is first performed on each block. The DC term is directly send for reconstruction, and the AC terms after low-pass filtering (we only consider 8 AC coefficients) is fed to a SOM network for training or testing. All experiments were carried out with the following parameters: 256 units, 5 training cycles and β calculated so the learning factor decreases from 0.9 to 0.05.

The number of bytes used to compress each image was the same for MSCL and JPEG (see Table 1) and fixed to 2048 for SOM.

For evaluation purpose, we use the mean squared error (MSE) as an objective measurement for the performance. Table 1 shows the resulting mean of the MSE in 10 tests using our algorithm compared to JPEG and SOM applied to 4 test images. We present a second column showing the value of MSE but only calculated in those pixels which saliency is over 50 %. Standard deviation is also shown (in brackets).

To obtain the generic pictorial library $MSCL_{PICT}(l)$ we used three additional images different to the images used in testing from [10] with the same training

Table 1 Mean MSE for the whole image as well as for areas with saliency over 50% (grayscale example)

Image	Q/Bytes	JPEG(Tot/50%)	SOM(Tot/50%)	MSIC(Tot/50%)
Lena	Q5/2010	212.3/340.4	**205.4/374.0**	501.1(18.2)/**211.0(6.1)**
Street	Q5/2127	**302.3/369.0**	322.1/465.3	466.2(7.8)/**210.6(4.2)**
Boat	Q5/1988	**263.9/383.7**	280.4/486.6	436.4(12.3)/**282.0(5.6)**
Fish	Q3/2090	485.7/597.7	**466.3/904.3**	895.8(15.8)/**254.2(9.6)**

Standard deviation is also shown (in brackets)

parameters. This number is quite low, but enough to show the good performance of our proposal. However, in a real scenario it would be necessary to use a higher number of images to get a suitable pictorial library. Moreover, we have not used any entropic coding applied to indexes which would have result in a further compression.

As expected, the MSE value calculated for the whole image area given by JPEG is lower than the one provided by MSIC, because prototypes tend to focus on zones with high saliency while other areas in the image are under-represented.

However, when MSE was calculated taking into account only those pixels with high saliency, MSIC obtained better results than JPEG or SOM. This effect can be clearly appreciated by visual inspection of the images represented in Fig. 5. They show how MSIC achieves a higher detail level at image areas of high saliency. In the case of JPEG, it tends to fill up big portions of the image with plain blocks, being unable to obtain a good detail at any part of the image. On the other hand, SOM produces slightly blurred images due to the low frequency filtering.

The new algorithm could also be used in compression applications with other magnitude functions instead of saliency. Figure 6 shows the compressed results of applying MSIC using different Magnitude Functions to the street image. From left to right, first image is the original one, second image is MSIC using the same Magnitude Functions than the one used in (9). The Magnitude function in third image is the same of equation (9), but using $1 - GBVS(x)$ instead of the pixel saliency of the corresponding sample. The fourth image uses the value of the vertical coordinate (normalized to one) and finally the fifth one uses the value of the vertical coordinate (normalized to one) minus one. It can be clearly seen that depending on the defined Magnitude Function, certain areas are compressed in with quality (foreground, background, top or bottom of the image).

This toy example was only presented to show the possibilities of achieving selective compression in different areas of the image just by varying the Magnitude Function.

MSIC algorithm is much more slower than JPEG. In a serial execution on single core computer, JPEG processing takes only 0.11% of the total processing time of MSIC (that in our tests it take 6.8 s for compressing each of the grayscale test images). Most of the time (91.6%) is spent on block extraction (34% of which is used in extracting blocks from the saliency map and 66% in extracting blocks from the

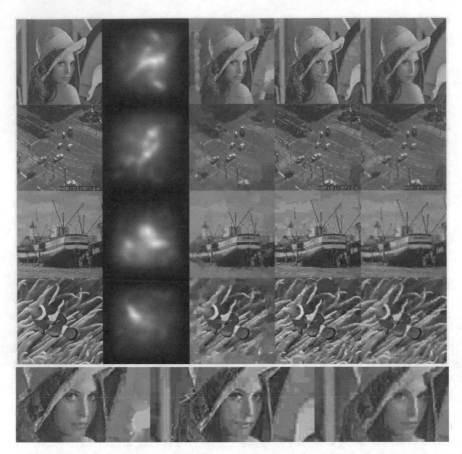

Fig. 5 *Top in columns* Original image, saliency map, MSIC, JPEG and SOM compression for the test images. *Bottom* Lena detail in the three methods. It can be clearly seen that the Lena face, compressed with MSIC shows a more natural view (almost like painted with Pointillism technique) than the other methods that have *square block borders*

Fig. 6 Original 'Street' image and the compressed images using MSIC with four different magnitude functions

image). Block encoding and decoding takes 6.7 % of the time, and 1.7 % the rest of the algorithm.

However processing time can be reduced using parallel processing and compiled libraries (now simulated in Matlab). The slowest task is finding the best matching unit for both, defining the Voronoi region to extract a block, and for encoding-decoding. This task represents the 68 % of the block extraction time, and the 51 % of the encoding-decoding time. It is a slow process because in our sequential implementation we must, for each sample, calculate the distances from sample to each of the units. In a parallel implementation, this processing could be applied simultaneously for all units. Then using for instance 1000 units, block extraction time could be only 29.3 % of initial total time. Using similar approach for encoding-decoding the final processing time can be reduced to be 2.3 s (34.3 % of the original processing time).

5.2 Color Images

In the color experiments, it is applied the same method explained in Sect. 4.6, with the same parameters used in the grayscale case.

We use a different saliency definition focused in those image zones with colors selected by the user. This type of compression, preserving with more detail image zones with certain color selection, may have different applications. For instance, in medical images, the specialist may define the colors of those areas that has to be well preserved. Other application is in video transmission limited by narrow bandwidths, as in underwater image transmission. In that case it is possible to work with a highly compressed global image, and if the user wants a higher definition in areas of a specific color, MSIC could get to a better definition of those areas, obviously degrading others to keep the limited bandwidth.

To calculate the saliency map with the magnitude values for the pixels, we first calculate the saliency map for each color in the set of colors. The saliency map of a selected color is obtained by binarizing the image, based on thresholding the distance of the pixel color and the selected color. Then we apply a border detection algorithm to get the edges of the image zones painted in that color.

The saliency map of the image is obtained as the maximum of the filtered edge images for all the set of colors. Using this value of magnitude, we get more units in the interesting regions whose colors are similar to the defined set. JPEG was implemented using Matlab and different compression qualities.

The experiments use the four test images depicted in the first column of Fig. 7. The second column shows the resulting saliency maps for the images. To maintain the details of the fish in the first image, it is used as color set: orange and white. The flower image uses dark and clear pink, the boat image uses only brown and the parachute image uses pink and black from the parachutist.

Table 2 shows in the first column the resulting mean of the MSE in 10 tests using MSCL compared to JPEG. Second column shows the value MSE calculated in

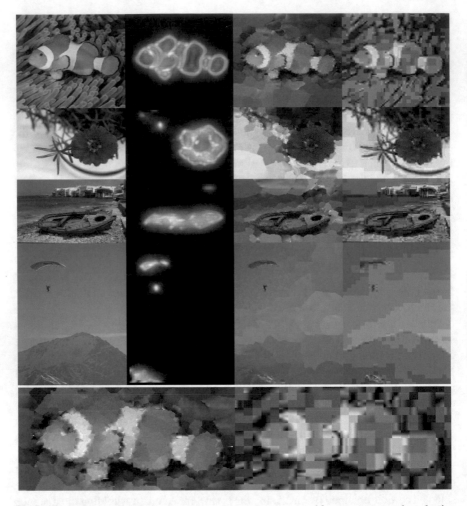

Fig. 7 *Top in columns* Original color image, saliency map generated for a one or two-color selection (fish with *orange* and *white*; flower with *dark* and *clear pink*; boat with *brown*; parachute with *pink* and *black*), MSIC and JPEG compression for the test images. *Bottom* Fish image detail in both compression methods

those pixels with saliency over 50 %. Standard deviation is also shown (in brackets). Number of bytes and quality are also shown.

As expected, the MSE value calculated for the whole image area is lower using JPEG than the one provided by MSIC. However, when MSE was calculated taking into account only those pixels exhibiting a high saliency, MSIC obtained the best results.

Table 2 Mean MSE for the whole image as well as for areas with saliency over 50% (color example)

Image	Q/Bytes	JPEG(Tot/50%)	MSIC(Tot/50%)
Fish	Q3/1702	1328/2695	2193(20.7)/**1789(40.3)**
Flower	Q5/1722	862/1299	3540(227.1) /**1167(49.4)**
Boat	Q6/1720	1303/1570	2366(87.4)/**1190(25.3)**
Sky	Q5/1706	967/2312	**240(58.2)** /**468(19.7)**

Standard deviation is also shown (in brackets)

6 Conclusions

In this paper we have shown how grayscale and color images compressed with MSIC exhibit a higher quality in relevant areas of the image when compared to other compression methods such as JPEG or SOM based algorithms.

MSIC has been proved to be a reliable and efficient approach to achieve selective Vector Quantization. This selectivity can be used in image compression to set the block centers focused on certain areas of the image to be compressed in a further step by Vector Quantization. The novelty of the algorithm is that areas of interest, which can be defined by a magnitude function, would receive lower compression than the rest of the image. Another novelty of the algorithm is that the image composition uses irregular blocks of pixels that tend to be smaller in zones of high interest and broader in zones of low interest.

These properties of the algorithm may be modulated for different applications by choosing the adequate magnitude function according to the desired task. For instance, it could be a good choice to use the Viola-Jones algorithm instead of GBSV to highlight some particular areas when dealing with facial areas in images with people. Another potential application is the compression of satellite and aerial imagery of the Earth. In that case, Automatic Building Extraction from Satellite Imagery algorithms may be used to define the areas of interest. Then, MSIC may compress the images keeping higher detail in the built areas. In a similar way, medical image storage tools might use MSIC to save images compressed with higher detail in certain biological tissues or anatomical structures.

Several applications that require image transmission with low bandwidth may use the algorithm, as in underwater image transmission, where there are low data rates compared to terrestrial communication. Another example of magnitude would be simply the predicted position of the user's fovea on the image in the next frame. This magnitude is useful for application in virtual reality glasses, where the image zone, that is predicted the user is going to focus his fovea, will present the highest detail, while surrounding zones can be more compressed.

Future work comprises several research lines such as the use of entropy coding for the information of each compressed image block, filtering each image with DCTs, and comparison against other compression algorithms. Another point to be analysed is the kind of images used to generate the generic pictorial codebooks used for compression and restoration, as the library of training images can be selected for the chosen task. The test of the algorithm in different tasks as mentioned in the previous paragraph is another research line left for future.

References

1. Pelayo, E., Buldain, D., Orrite, C.: Magnitude sensitive competitive learning. Neurocomputing **112**, 4–18 (2013)
2. Laha, A., Pal, N., Chanda, B.: Design of vector quantizer for image compression using self-organizing feature map and surface fitting. IEEE Trans. Image Process. **13**(10), 1291–1303 (2004)
3. Amerijckx, C., Legat, J.D., Verleysen, M.: Image compression using self-organizing maps. Syst. Anal. Modell. Simul. **43**(11), 1529–1543 (2003)
4. Harandi, M., Gharavi-Alkhansari, M.: Low bitrate image compression using self-organized kohonen maps. In: Proceedings 2003 International Conference on Image Processing, ICIP'03, vol. 3, pp. 267–270 (2003)
5. Liou, R.J., Wu, J.: Image compression using sub-band DCT features for self-organizing map system. J. Comput. Sci. Appl. **3**(2) (2007)
6. Kohonen, T.: The self-organizing map. Neurocomputing **21**(1), 1–6 (1998)
7. Ahmed, N., Natarajan, T., Rao, K.R.: Discrete cosine transform. IEEE Trans. Comput. **100**(1), 90–93 (1974)
8. Cheung, Y.: On rival penalization controlled competitive learning for clustering with automatic cluster number selection. IEEE Trans. Knowl. Data Eng. **17**, 1583–1588 (2005)
9. Harel, J., Koch, C., Perona, P.: Graph-based visual saliency. In: NIPS'06, pp. 545–552 (2006)
10. Computer Vision Group, U.o.G.: Dataset of standard 512×512 grayscale test images. http://decsai.ugr.es/cvg/CG/base.htm (2002)

Unsupervised Analysis of Morphological ECG Features for Attention Detection

Carlos Carreiras, André Lourenço, Helena Aidos,
Hugo Plácido da Silva and Ana L.N. Fred

Abstract Physiological Computing augments the information bandwidth between a computer and its user by continuous, real-time monitoring of the user's physiological traits and responses. This is especially interesting in a context of emotional assessment during human-computer interaction. The electroencephalogram (EEG) signal, acquired on the scalp, has been extensively used to understand cognitive function, and in particular emotion. However, this type of signal has several drawbacks, being susceptible to noise and requiring the use of impractical head-mounted apparatuses. For these reasons, the electrocardiogram (ECG) has been proposed as an alternative source to assess emotion, which is continuously available, and related with the psychophysiological state of the subject. In this paper we analyze morphological features of the ECG signal acquired from subjects performing an attention-demanding task. The analysis is based on various unsupervised learning techniques, which are validated against evidence found in a previous study by our team, where EEG signals collected for the same task exhibit distinct patterns as the subjects progress in the task.

Keywords Physiological computing · Attention · ECG · EEG · Unsupervised learning · Cluster validation

C. Carreiras (✉) · A. Lourenço · H. Aidos · H.P. da Silva · A.L.N. Fred
Instituto de Telecomunicações, Instituto Superior Técnico, Av. Rovisco Pais 1,
1049-001 Lisbon, Portugal
e-mail: carlos.carreiras@lx.it.pt

H. Aidos
e-mail: haidos@lx.it.pt

H.P. da Silva
e-mail: hugo.silva@lx.it.pt

A.L.N. Fred
e-mail: afred@lx.it.pt

A. Lourenço
Instituto Superior de Engenharia de Lisboa, Rua Conselheiro Emídio Navarro 1,
1959-007 Lisbon, Portugal
e-mail: arlourenco@lx.it.pt

437

1 Introduction

Physiological Computing, as a research area, integrates psychophysiological information into computer systems by continuous, real-time monitoring of the user [1]. These systems augment the information bandwidth between the user and the computer, enabling a better interpretations of the user's psychophysiological state. Indeed, in natural human communication, the speaker's attitude, posture, tone, and facial expressions, among others, strongly influence the semantic interpretation done by the receiver [2].

Straightforward approaches to physiological computing, requiring no extra hardware, are, for example, keystroke dynamics [3], speech analysis [4], and automatic facial expression recognition [5]. However, all these examples exhibit serious problems to their usefulness. Keystroke dynamics and speech analysis both require continuous voluntary activity, while the usefulness of facial expression analysis for behavioral science has been recently questioned in [6]. One possible alternative to these modalities, although requiring extra hardware, is the use of the subject's biosignals (e.g. electrodermal activity, peripheral temperature, blood volume pulse, electrocardiogram, electroencephalogram signals), acquiring them during normal human-computer interaction tasks [7, 8]. These signals have the twofold advantage of being always available, and measuring the natural physiological responses of the body to a given affective state, which cannot be voluntarily masked.

The electroencephalogram (EEG) signal, acquired on the scalp, has been extensively used to understand cognitive function, and in particular emotion [9, 10], being a noninvasive, cost-effective technique, with good temporal resolution [11]. However, it has various drawbacks, such as susceptibility to noise (in particular motion artifacts and eye blinks) and, most importantly, requires the use of some kind of head-mounted equipment to support the (typically wet) electrodes, which becomes impractical for continued use. In this context, the electrocardiogram (ECG) signal has been suggested as a possible option [12, 13]. Nevertheless, the usefulness of the EEG as source of ground-truth information has not been discarded [14].

In this paper, we make a morphological analysis, using unsupervised learning techniques, of the ECG acquired from subjects performing a task that demands high levels of attention over a long period of time. This experiment simulates what may happen, for instance, during an interactive educational game, extended work hours, repetitive daily tasks, or sleep deprivation, where attention levels fluctuate throughout the execution of the task. This is particularly important in various professions, such as doctors, pilots, drivers or industrial equipment operators, for which momentary or prolonged lapses of attention may be catastrophic [14]. In addition, we compare the results obtained with the ECG signal to our previous work using the EEG, which provided evidence that the subjects indeed exhibit distinct affective states throughout the completion of the task [15].

The remainder of the paper is organized as follows: Sect. 2 describes the experimental setup. Section 3 details the proposed methodology, including the description of the clustering methods used, as well as several clustering validation metrics. Section 4 presents the obtained results, which are discussed in Sect. 5. Finally, Sect. 6 concludes the paper.

2 Affective Elicitation and Data Acquisition

The ECG signal presents several attributes that make it especially interesting in a physiological computing framework. Specifically, it is continuously available, providing a rich wellbeing indicator, is related with the psychophysiological state of the subject, and is easy to acquire unobtrusively with wearable devices. This is further enhanced by following an *off-the-person* approach, where the sensors are seamlessly integrated into objects with which subjects regularly interact, such as a keyboard, a video game controller, or a mobile device, without the need to change normal interaction patterns [16].

It is widely known that the basic function of the heart is to pump blood throughout the body, demanding a highly synchronized sequence of muscular contractions. These contractions are initiated by small electrical currents that propagate through the heart's muscle cells, generating an electrical signal that can be recorded at the body surface (the ECG). In healthy individuals, the electrical activity of the heart is guided by the self-excitatory nature of the sinus node on the left atrium (see Fig. 1), which naturally produces electrical depolarizations at a rate of about 100 beats per minute. However, the sinus node is under systemic control by the endocrine system and the Autonomic Nervous System (ANS). The ANS is composed by two complementing, self-balancing subsystems, the Sympathetic and Parasympathetic Nervous Systems (SNS and PSNS, respectively). While the SNS is typically responsible for the promotion of *fight-or-flight* responses in the organism (e.g. by increasing the heart rate), the PSNS is responsible for the promotion of *rest-and-digest* responses, which induce relaxation and a return to normal function. As a whole, the ANS

Fig. 1 Schematic representation of the heart compartments and its electric system, showing the contribution of each component to the prototypical heartbeat signal recorded at the body surface (used with permission from [18])

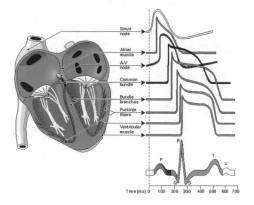

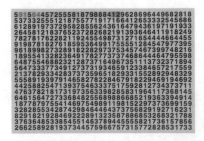

Fig. 2 Example matrix of the concentration test; the user selects, line by line, the pairs of consecutive numbers that add to 10

provides an access route to the affective state of a person [17], by analyzing the patterns of physiological activity initiated by both the SNS and PSNS. In particular for the ECG, the amplitude and latency of the P-QRS-T complexes is influenced by multiple psychophysiological factors, and some changes in the user's behavior result in slight variations in the heart rate and waveform morphology.

The ECG and EEG signals analyzed here were acquired in the context of the HiMotion project [19], an experiment to acquire information related to human-computer interaction and physiological signals on different cognitive activities. During the experimental session, the subjects were asked to execute various interactive cognitive tasks. Particularly, a concentration task was performed, adapted from a similar test from the MENSA set [20]. In this test, the subject is presented with a matrix of 800 integers (20 lines by 40 columns), as shown in Fig. 2. The goal of the game is to identify, line by line, all the pairs of consecutive numbers that add to 10. This task requires high levels of attention, as the pairs may overlap (i.e. the same number may belong to two pairs), measuring the capacity of the subject to maintain an attentive state over a long period of time.

Biosignal data was obtained from 24 subjects (17 males and 7 females) with ages in the range 23.3 ± 2.4 years, using a *Thought Technology ProComp2* acquisition system, with a sampling rate of 256 Hz. The ECG was acquired with Ag/AgCl electrodes placed on the chest (4th intercostal space in the mid clavicular line), while the EEG was acquired at four scalp locations according to the 10–20 system (F_{p1}, F_z, F_{p2}, and O_z), as shown in Fig. 3.

Fig. 3 Locations of the acquired EEG electrodes on the scalp (*red*)

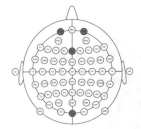

3 Proposed Methodology

It should be noted that each person has different characteristics in terms of attention span and, as such, aside from the temporal information regarding the start and the end of each line of the attention game, no more information is available for this data set. Particularly, there is no ground-truth information regarding the time instants in which the affective state of each test subject has in fact changed. For this reason, we propose the use of unsupervised learning techniques to analyze the ECG data.

The proposed methodology is presented in Fig. 4 and it is divided in three main stages: feature extraction, clustering, and validation of the clustering results. We start by filtering and segmenting the raw ECG, and then we apply clustering techniques to analyze the data. Subsequently, the results of those clustering algorithms are validated using several metrics, exploiting our previous analysis of the same data set with the EEG signal [15]. This somewhat follows the methodology proposed in [14], where the EEG signal is used as a benchmark against which the performance of attention recognition via the ECG is compared. All these stages are explained in the following subsections.

3.1 ECG Feature Extraction

Raw ECG signals are typically affected by various noise sources such as motion arti-facts, power line interference, and electromyographic noise. To enhance the signal-to-noise ratio (SNR), and to reduce the influence of the cited noise sources, we used a

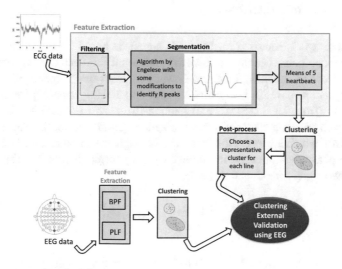

Fig. 4 The proposed methodology

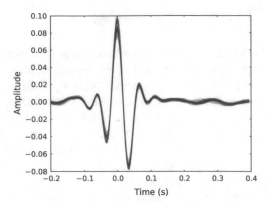

Fig. 5 ECG templates obtained for subject 11

band-pass Finite Impulse Response (FIR) filter with a Hamming window of 300 ms, and cutoff frequencies of 5–20 Hz. The filtered signal was then fed to a segmentation algorithm, with the purpose of identifying the locations of the R peaks. For that we used the algorithm by Engelse and Zeelenberg [21], with the modifications proposed in [22]. Individual heartbeat segments of 600 ms were extracted from the filtered signal, between 200 ms before and 400 ms after the R peak. Finally, in order to further improve the SNR, heartbeat templates were formed using sequences of 5 consecutive heartbeats, computing their element-wise mean (an example of these templates can be seen in Fig. 5). These templates form the feature space used by the clustering algorithms, described in Sect. 3.3.

3.2 EEG Feature Extraction

Our previous work, focusing on the EEG signal, is based on two distinct feature extraction techniques. The first follows the traditional approach of analyzing the various EEG frequency bands, the Band Power Features (BPF). Specifically, we used the theta (from 4 to 8 Hz), lower alpha (from 8 to 10 Hz), upper alpha (from 10 to 13 Hz), beta (from 13 to 25 Hz), and gamma (from 25 to 40 Hz) bands. The second approach uses a method of synchronization quantification, the Phase-Locking Factor (PLF), which leverages the fact that EEG signals exhibit an oscillatory behavior whose phase dynamics are modulated by the neurological tasks [23]. The PLF between two signals is defined as [24]:

$$\varrho_{ik} = \left| \frac{1}{T} \sum_{n=1}^{T} e^{j(\phi_i[n] - \phi_k[n])} \right|, \tag{1}$$

where $\phi_i[n]$ and $\phi_k[n]$, $n = 1, ..., T$ are the phases of the signals, T is the number of discrete time samples, and $j = \sqrt{-1}$ is the imaginary unit. This measure ranges from

0 to 1, with a value of $\varrho_{ik} = 1$ corresponding to perfect synchronization between the two signals (constant phase lag), while the value $\varrho_{ik} = 0$ corresponds to no synchronization. These two feature extraction methods form distinct feature spaces, upon which clustering methods were applied.

3.3 Unsupervised Learning

Clustering is one of the central problems in Pattern Recognition and Machine Learning. Hundreds of clustering algorithms exist, differently handling issues such as cluster shape, density, and noise, among other aspects. These techniques require the definition of a similarity measure between patterns, be it geometrical or probabilistic, which is not easy to specify in the absence of any prior knowledge about cluster shapes and structure.

One of the classical approaches for clustering is the use of hierarchical agglomerative algorithms [25], which produce a tree of nested objects (the dendrogram) that establishes the hierarchy between the clusters. These methods only require a measure of (dis)similarity and a linkage criterion between instances, while partitional methods (e.g. k-means or k-medoids) also require a priori the number of clusters, and an initial assignment of data to clusters. The linkage criterion specifies how intergroup similarity is defined. In particular, we apply the Average Link (AL) and Ward's Linkage (WL) criteria [26]. Furthermore, to obtain a partition of the data from a dendrogram, we use the largest lifetime criterion [27].

Moreover, we use a new high order dissimilarity measure, called *dissimilarity increments*, proposed in [28]. This measure is computed over triplets of nearest neighbor patterns and is defined as:

$$d_{inc}(x_i, x_j, x_k) = |D_*(x_i, x_j) - D_*(x_j, x_k)|, \tag{2}$$

where x_j is the nearest neighbor of x_i, and x_k is the nearest neighbor of x_j, different from x_i. In equation (2), $D_*(\cdot, \cdot)$ can be any dissimilarity measure, such as the Euclidean distance. The dissimilarity increments measure can give more information about patterns belonging to the same cluster, since it changes smoothly if the patterns are in the same cluster. In [29], an agglomerative hierarchical algorithm, called SLDID, was proposed. This algorithm is a variant of the Single Link (SL) criterion using the dissimilarity increments distribution (DID), which was derived under mild approximations in [30], to modify the way that clusters are merged. In this paper we used a family of DID algorithms: ALDID and WLDID. They are variants of the traditional hierarchical clustering algorithms AL and WL, respectively. The main difference between AL and ALDID is that in AL, in each iteration the pair of clusters with the highest cohesion is always merged; in ALDID some tests are made using the minimum description length (MDL) criterion between two possibilities. These two possibilities consist in the DID of the two clusters combined, and the DID of the two clusters separated. One advantage in using an algorithm from this family is that it stops merging clusters before all the data is merged into one cluster, revealing intrinsic cluster structure in the data when the true number of clusters is unknown.

Consensus Clustering Consensus clustering, also known as Clustering combination, is a powerful technique that combines the information of multiple clustering partitions, forming a *clustering ensemble* ($\mathbb{P}$), and creating a consensus partition that leverages the results of individual clustering algorithms. Recent surveys present an overview on this research topic [31, 32]. One of the significant approaches is the Evidence Accumulation Clustering (EAC) [33]. This framework is based on the aggregation of object *co-occurences*, and the consensus partition is obtained through a voting process among the objects. Specifically, the consensus clustering problem is addressed by summarizing the information of the ensemble into a pairwise *co-association matrix*, where each entry holds the fraction of clusterings in the ensemble in which a given pair of objects is placed in the same cluster:

$$C(i, j) = \frac{n_{ij}}{N}, i, j \in 1, \ldots, N. \tag{3}$$

For the construction of the ensemble, we use the k-means algorithm [25] with different parameters and initializations. We created a set of $N = 100$ partitions[1] by randomly choosing the number of clusters, following the work in [34] where the minimum and maximum number of clusters per partition depends on the number of objects n, and is bound to the interval $[\frac{\sqrt{n}}{2}, \sqrt{n}]$.

The extraction of the consensus partition can be performed using several approaches based on the induced co-association matrix: (i) as a new (dis)similarity-based representation of objects, where the intrinsic structure of the data is enhanced through the evidence accumulation process, enabling the determination of the consensus partition using algorithms that explicitly use similarities as input, such as hierarchical linkage methods (as classically performed in [33]); (ii) as a new vector-based object description, considering each line of the matrix a new feature vector representation, and using it as input to a clustering algorithm such as the k-means [35]; (iii) as a new probabilistic distribution characterized by the probability of pairs of objects being in the same cluster [36].

Application to EEG and ECG The focus of this work was the unsupervised analysis of the ECG signals, and for that we applied all the described techniques: (i) hierarchical agglomerative algorithms; (ii) hierarchical agglomerative algorithms with dissimilarity increments; (iii) consensus clustering based on evidence accumulation clustering, using as extraction criterion the average linkage method with the number of clusters automatically determined by the life-time criterion.

The clustering of the ECG heartbeats was performed over the means of 5 consecutive heartbeats. Since we are willing to compare these partitions with the ones obtained on the context of EEG, where for each line of the test there is only one cluster,

---

[1]This is the number typically proposed in the reference literature.

it was necessary to post-process the obtained partitions, choosing as representative cluster for each line the one with highest cardinality (largest time span).

In the context of EEG clustering, we applied the hierarchical agglomerative methods with and without dissimilarity increments.

3.4 Cluster Validation

Cluster validation techniques have been developed to guide the design of clustering experiments and to assess the quality of the outcome. There are three types of cluster validity measures [37–41]: (i) *External*: used to measure the goodness of a clustering structure with respect to external information; (ii) *Internal*: used to measure the goodness of a clustering structure without supplying any class labels; and (iii) *Relative*: used to compare different clusterings.

We adopt an external clustering validation perspective, using as external source of information the clusterings obtained with the EEG. There is a long list of external validation indices proposed in the literature [39, 40, 42, 43], which can be categorized as follows: (i) *Counting Pairs Methods*: a class of criteria based on counting the pairs of points on which two clusterings agree/disagree, Wallace [44], Fowlkes and Mallows [42], and Rand's [45] are the most representatives of this class; (ii) *Set Matching*: based on set matching cardinality, $\mathcal{H}$ criterion [39], and consistency index (Ci) [46, 47] are representative of this class; (iii) *Information Theoretic*: based on information theoretic concepts (entropy and mutual information); representatives of this class of criteria are the Variation of Information (VI) index [39] and Dom's index [43].

In this work, we compare the partitions obtained with the ECG with the ones obtained with the EEG (taken as ground-truth), and following the idea proposed in [14]. We use indices of the three categories, to verify the consistency of the results in several perspectives, namely: Rand [45], a modified version of the Consistency Index entitled Average Cluster Consistency (ACC) [47], and VI [39]. All the three indices take values between 0 and 1. Rand's index and the ACC take the value 1 for a perfect match between partitions, and for the VI index, 0 corresponds to a perfect match.

4 Experimental Results

Figure 6 exemplifies the clustering of the ECG templates obtained for one of the subjects, using the clustering combination (CC) method. It shows, for each line of the concentration task, the clusters to which the templates in that line belong to. The first observation to note is that the lines are not characterized by a single cluster, but rather by two or three clusters that alternate between them. However, it is possible to perceive the existence of different groups of lines. In this particular case, lines 0–2 are

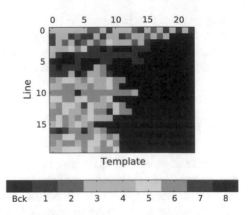

Fig. 6 Clustering obtained for subject 11, using the clustering combination method, where each color represents a cluster, with a total of 8 clusters; *Bck* denotes the background color of the matrix

mainly composed by clusters 1, 2, and 3, lines 4–7 are composed by clusters 7 and 8, and the remaining lines are composed by clusters 4, 5, and 6. Another interesting note is the fact that the number of templates per line decreases throughout the completion of the task, implying that the first few lines of the task take longer to complete than the last lines. These observations are valid for the majority of the subjects, although the number of clusters and their distribution differs from subject to subject, forming different groups of lines.

Inter-subject variability is evidenced in Fig. 7, where the clustering obtained, across all subjects, with the EEG (using PLF features and ALDID clustering—Fig. 7a) is compared to the clustering obtained with the ECG (using CC clustering—Fig. 7b). Remember that, in the case of the ECG, each line is represented by the most frequent cluster in that line. It is possible to observe that the ECG produces a higher number of clusters than the EEG, where each cluster tends to form groups

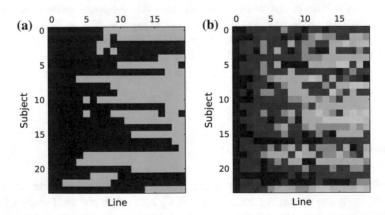

Fig. 7 Comparison of the clustering obtained with the EEG to the one obtained with the ECG, across all subjects; each color represents one cluster

Table 1 Cluster validation results ($\mu \pm \sigma$) using the Average Cluster Consistency (ACC) metric; higher values suggest a stronger agreement

		ECG clustering				
EEG clustering		AL	WL	ALDID	WLDID	CC
AL	PLF	0.70 ± 0.16	0.72 ± 0.17	0.71 ± 0.14	0.73 ± 0.15	**0.78** ± 0.16
	BPF	0.70 ± 0.15	0.74 ± 0.15	0.70 ± 0.18	0.72 ± 0.17	**0.79** ± 0.15
WL	PLF	0.61 ± 0.12	0.65 ± 0.16	0.62 ± 0.11	0.67 ± 0.13	**0.76** ± 0.17
	BPF	0.68 ± 0.11	0.72 ± 0.14	0.68 ± 0.13	0.70 ± 0.15	**0.78** ± 0.15
ALDID	PLF	0.71 ± 0.16	0.73 ± 0.15	0.73 ± 0.14	0.74 ± 0.15	**0.79** ± 0.16
	BPF	0.55 ± 0.13	0.60 ± 0.16	0.55 ± 0.14	0.60 ± 0.14	**0.69** ± 0.14
WLDID	PLF	0.62 ± 0.16	0.66 ± 0.16	0.64 ± 0.14	0.69 ± 0.16	**0.77** ± 0.18
	BPF	0.51 ± 0.14	0.56 ± 0.14	0.52 ± 0.14	0.57 ± 0.16	**0.65** ± 0.15

Table 2 Cluster validation results ($\mu \pm \sigma$) using the Variation of Information (VI) metric; lower values suggest a stronger agreement

ECG clustering			AL	WL	ALDID	WLDID	CC
EEG clustering	AL	PLF	**0.21** ± 0.07	0.33 ± 0.13	0.27 ± 0.11	0.34 ± 0.12	0.55 ± 0.11
		BPF	**0.22** ± 0.09	0.32 ± 0.11	0.28 ± 0.09	0.35 ± 0.14	0.55 ± 0.11
	WL	PLF	**0.26** ± 0.06	0.36 ± 0.11	0.32 ± 0.08	0.37 ± 0.11	0.54 ± 0.13
		BPF	**0.21** ± 0.06	0.32 ± 0.11	0.27 ± 0.06	0.35 ± 0.12	0.55 ± 0.11
	ALDID	PLF	**0.20** ± 0.07	0.32 ± 0.13	0.26 ± 0.11	0.34 ± 0.11	0.55 ± 0.11
		BPF	**0.34** ± 0.10	0.40 ± 0.13	0.38 ± 0.09	0.43 ± 0.12	0.55 ± 0.10
	WLDID	PLF	**0.25** ± 0.08	0.36 ± 0.12	0.31 ± 0.10	0.36 ± 0.12	0.55 ± 0.11
		BPF	**0.35** ± 0.10	0.42 ± 0.12	0.40 ± 0.09	0.45 ± 0.11	0.56 ± 0.09

Table 3 Cluster validation results ($\mu \pm \sigma$) using Rand's metric; higher values suggest a stronger agreement

			ECG clustering				
			AL	WL	ALDID	WLDID	CC
EEG clustering	AL	PLF	**0.59 ± 0.15**	**0.59 ± 0.14**	**0.59 ± 0.12**	0.54 ± 0.10	0.49 ± 0.11
		BPF	**0.58 ± 0.16**	0.56 ± 0.12	0.54 ± 0.13	0.57 ± 0.11	0.50 ± 0.14
	WL	PLF	0.49 ± 0.10	0.54 ± 0.11	0.50 ± 0.08	0.54 ± 0.09	**0.57 ± 0.08**
		BPF	0.56 ± 0.11	**0.57 ± 0.13**	0.52 ± 0.06	0.55 ± 0.08	0.51 ± 0.11
	ALDID	PLF	**0.61 ± 0.16**	0.59 ± 0.14	0.60 ± 0.13	0.53 ± 0.11	0.48 ± 0.12
		BPF	0.42 ± 0.13	0.55 ± 0.13	0.44 ± 0.10	0.53 ± 0.11	**0.61 ± 0.10**
	WLDID	PLF	0.51 ± 0.14	**0.55 ± 0.11**	0.52 ± 0.12	0.54 ± 0.13	0.54 ± 0.12
		BPF	0.39 ± 0.13	0.53 ± 0.13	0.42 ± 0.11	0.51 ± 0.12	**0.63 ± 0.10**

of contiguous lines. Contrastingly, in the ECG it is frequent to have transitions to clusters seen in previous lines.

The results of the cluster validation are shown in Tables 1, 2, and 3 for the Average Cluster Consistency (ACC), Variation of Information (VI), and Rand's index, respectively. For the ACC criterion, the highest agreement is obtained between the ECG clustering with CC and both the EEG clustering using AL (BPF) and ALDID (PLF), with a value of 0.79. Regarding the VI measure, the strongest agreements are seen for the ECG clustering using the AL algorithm, in particular with the ALDID method applied to the PLF features from the EEG, with a value of 0.20. Concerning Rand's index, the highest value, 0.63, is obtained between the ECG clustering through CC with the EEG clustering using WLDID (BPF).

5 Discussion

Our work addresses the following questions: (i) "Is ECG morphological analysis capable of identifying affective states throughout the realization of a task that demands a high attention span?"; (ii) "Are the obtained states related to the ones found while analyzing EEG data?"; and (iii) "What techniques can be considered to be more suitable for the analysis of the ECG?"

The validation of the partitions found using ECG, when considering the EEG partitions as ground-truth, shows that there is evidence of correlation between them, revealing that ECG can be used to infer affective states. The ECG partitions have a much higher number of partitions than the EEG ones, leading to distinct results over the various validation criteria (considering the different perspectives), associated with moderate to high matching. This was mainly due to small variations over time of the ECG heartbeats, that lead to slow time transitions between the different clusters.

The clustering technique that presents the best results varies depending on the validation index. When considering the average cluster consistency (ACC), the consensus clustering (CC) obtains partitions that lead to a best match; when using variation of information (VI) criterion, the Average Link (AL) method is the one that leads to best match; and when using the Rand's index there is not a method which can be considered a clear winner. The situations with best results are partitions with high number of clusters, which correspond to Average linkage and Consensus Clustering.

6 Conclusions

In this work we present a methodology for attention detection based on the morphological analysis of ECG signals, using data collected during the course of a task requiring a high level of attention span. We compare the ECG morphology results with the analysis performed using the EEG. This comparison was accomplished using clustering validation indices.

The ECG analysis was divided into several steps. For the feature extraction step, the signal was first digitally filtered, segmented based on the peaks found by a modification of the Engelse and Zeelenberg algorithm, and templates were formed using means of 5 consecutive heart beats. For the clustering step, several state of the art techniques were used, since the ECG heartbeats have very small variations over time, leading to touching clusters.

Several clustering validation indices were used, trying to compare the partitions using different perspectives. Each of the validation indices showed that there is a high evidence of correlation between the partitions obtained by the ECG and the EEG. There is not a clear winner method, but Average Linkage and Consensus Clustering can be considered suitable methods for this kind of analysis.

Acknowledgments This work was partially funded by the Portuguese Science Foundation (FCT) under grants PTDC/EEI-SII/2312/2012, SFRH/BD/65248/2009 and SFRH/PROTEC/49512/2009, and by Área Departamental de Engenharia Electrónica e Telecomunicações e de Computadores (ISEL), whose support the authors gratefully acknowledge.

References

1. Fairclough, S.H.: Fundamentals of physiological computing. Interact. Comput. **21**(1), 133–145 (2009)
2. Pell, M.D., Jaywant, A., Monetta, L., Kotz, S.A.: Emotional speech processing: disentangling the effects of prosody and semantic cues. Cogn. Emot. **25**(5), 834–853 (2011)
3. Epp, C., Lippold, M., Mandryk, R.L.: Identifying emotional states using keystroke dynamics. In: Proceedings of the 2011 Annual Conference on Human Factors in Computing Systems, pp. 715–724. ACM (2011)
4. Murray, I.R., Arnott, J.L.: Toward the simulation of emotion in synthetic speech: a review of the literature on human vocal emotion. J. Acoust. Soc. Am. **93**, 1097 (1993)
5. Zheng, W., Zhou, X., Zou, C., Zhao, L.: Facial expression recognition using kernel canonical correlation analysis (kcca). IEEE Trans. Neural Networks **17**(1), 233–238 (2006)
6. Aviezer, H., Trope, Y., Todorov, A.: Body cues, not facial expressions, discriminate between intense positive and negative emotions. Science **338**(6111), 1225–1229 (2012)
7. Canento, F., Fred, A., Silva, H., Gamboa, H., Lourenço, A.: Multimodal biosignal sensor data handling for emotion recognition. In: Proceedings of the IEEE Sensors Conference, pp. 647–650. IEEE Press (2011)
8. Silva, H., Fred, A., Eusebio, S., Torrado, M., Ouakinin, S.: Feature extraction for psychophysiological load assessment in unconstrained scenarios. In: Annual International Conference of the IEEE Engineering in Medicine and Biology Society (EMBC), pp. 4784–4787. IEEE Press (2012)
9. Ahern, G.L., Schwartz, G.E.: Differential lateralization for positive and negative emotion in the human brain: EEG spectral analysis. Neuropsychologia **23**(6), 745–755 (1985)
10. Coan, J.A., Allen, J.J.: Handbook of emotion elicitation and assessment. Oxford University Press, Oxford (2007)
11. Mak, J.N., Wolpaw, J.R.: Clinical applications of brain-computer interfaces: current state and future prospects. IEEE Rev. Biomed. Eng. **2**, 187–199 (2009)
12. Medina, L.: Identification of stress states from ECG signals using unsupervised learning methods. Master's thesis, Universidade Técnica de Lisboa, Instituto Superior Técnico (2009)

13. Belle, A., Ji, S.Y., Ansari, S., Hakimzadeh, R., Ward, K., Najarian, K.: Frustration detection with electrocardiograph signal using wavelet transform. In: IEEE International Conference on Biosciences (BIOSCIENCESWORLD), pp. 91–94. IEEE Press (2010)
14. Belle, A., Hargraves, R.H., Najarian, K.: An automated optimal engagement and attention detection system using electrocardiogram. Comput. Math. Meth. Med. **2012** (2012)
15. Carreiras, C., Aidos, H., Silva, H., Fred, A.: Exploratory EEG analysis using clustering and phase-locking factor. In: Proceedings of the 6th International Conference on Bio-Inspired Systems and Signal Processing (BIOSIGNALS 2013), pp. 79–88. SCITEPRESS (2013)
16. Silva, H., Lourenço, A., Lourenço, R., Leite, P., Coutinho, D., Fred, A.: Study and evaluation of a single differential sensor design based on electro-textile electrodes for ECG biometrics applications. In: Proceedings of the IEEE Sensors Conference, pp. 1764–1767. IEEE Press (2011)
17. Levenson, R.W.: Autonomic nervous system differences among emotions. Psychol. Sci. **3**(1), 23–27 (1992)
18. Malmivuo, J., Plonsey, R.: Bioelectromagnetism: Principles and Applications of Bioelectric and Biomagnetic Fields. Oxford University Press, Oxford (1995)
19. Gamboa, H., Silva, H., Fred, A.: HiMotion Project. Technical report, Instituto Superior Técnico, Lisbon, Portugal (2007)
20. Fulton, J.: Mensa Book of Total Genius. Barnes & Noble Books, Totowa (1999)
21. Engelse, W.A.H., Zeelenberg, C.: A single scan algorithm for QRS-detection and feature extraction. Comput. Cardiol. **6**, 37–42 (1979)
22. Canento, F., Lourenço, A., Silva, H., Fred, A., Raposo, N.: On real time ECG algorithms for biometric applications. In: Proceedings of the 6th International Conference on Bio-Inspired Systems and Signal Processing (BIOSIGNALS 2013), pp. 228–235. SCITEPRESS (2013)
23. Pfurtscheller, G., Lopes da Silva, F.H.: Event-related EEG/MEG synchronization and desynchronization: basic principles. Clin. Neurophysiol. **110**, 1842–1857 (1999)
24. Almeida, M., Bioucas-Dias, J., Vigário, R.: Source separation of phase-locked subspaces. In: Proceedings of the International Conference on Independent Component Analysis and Signal Separation, vol. 5441, pp. 203–210 (2009)
25. Jain, A.K., Dubes, R.C.: Algorithms for Clustering Data. Prentice-Hall Inc, Upper Saddle River (1988)
26. Theodoridis, S., Koutroumbas, K.: Patern Recognition. Academic Press (1999)
27. Fred, A., Jain, A.: Evidence accumulation clustering based on the k-means algorithm. In: Structural, Syntactic, and Statistical Pattern Recognition, pp. 303–333 (2002)
28. Fred, A., Leitão, J.: A new cluster isolation criterion based on dissimilarity increments. IEEE Trans. Pattern Anal. Mach. Intell. **25**(8), 944–958 (2003)
29. Aidos, H., Fred, A.: Hierarchical clustering with high order dissimilarities. In: Proceedings of the 7th International Conference on Machine Learning and Data Mining (MLDM 2011). LNCS, vol. 6871, pp. 280–293. New York, USA (2011)
30. Aidos, H., Fred, A.: Statistical modeling of dissimilarity increments for d-dimensional data: application in partitional clustering. Pattern Recogn. **45**(9), 3061–3071 (2012)
31. Ghosh, J., Acharya, A.: Cluster ensembles. WIREs Data Mining Knowled. Discovery **1**(4), 305–315 (2011)
32. Vega-Pons, S., Ruiz-Shulcloper, J.: A survey of clustering ensemble algorithms. Int. J. Pattern Recogn. Artifical Intell. (IJPRAI) **25**(3), 337–372 (2011)
33. Fred, A., Jain, A.K.: Combining multiple clustering using evidence accumulation. IEEE Trans. Pattern Anal. Mach. Intell. **27**(6), 835–850 (2005)
34. Lourenço, A., Fred, A., Jain, A.K.: On the scalability of evidence accumulation clustering. In: Proceedings of the 20th International Conference on Pattern Recognition (ICPR), pp. 782–785. IEEE Press (2010)
35. Kuncheva, L.I., Vetrov, D.P.: Evaluation of stability of k-means cluster ensembles with respect to random initialization. IEEE Trans. Pattern Anal. Mach. Intell. **28**(11), 1798–1808 (2006)

36. Lourenço, A., Rota Bulò, S., Rebagliati, N., Figueiredo, M., Fred, A., Pelillo, M.: Probabilistic evidence accumulation for clustering ensembles. In: Proceedings of the International Conference on Pattern Recognition Applications and Methods (ICPRAM), pp. 58–67. SCITEPRESS (2013)
37. Dubes, R., Jain, A.: Validity studies in clustering methodologies. Pattern Recogn. **11**, 235–254 (1979)
38. Halkidi, M., Batistakis, Y., Vazirgiannis, M.: On clustering validation techniques. J. Intell. Inform. Syst. **17**, 107–145 (2001)
39. Meilă, M.: Comparing clusterings–an information based distance. J. Multivar. Anal. **98**(5), 873–895 (2007)
40. Ben-Hur, A., Elisseeff, A., Guyon, I.: A stability based method for discovering structure in clustered data. In: Pacific Symposium on Biocomputing, pp. 6–17 (2002)
41. Luo, P., Xiong, H., Zhan, G., Wu, J., Shi, Z.: Information-theoretic distance measures for clustering validation: generalization and normalization. IEEE Trans. Knowl. Data Eng. **21**(9), 1249–1262 (2009)
42. Fowlkes, E.B., Mallows, C.L.: A method for comparing two hierarchical clusterings. J. Am. Statis. Assoc. **78**(383), 553–569 (1983)
43. Dom, B.E.: An information-theoretic external cluster-validity measure. Technical Report IBM Research Report RJ 10219 (2001)
44. Wallace, D.L.: A method for comparing two hierarchical clusterings: comment. J. Am. Statis. Assoc. **78**(383), 569–576 (1983)
45. Rand, W.M.: Objective criteria for the evaluation of clustering methods. J. Am. Statis. Assoc. **66**(336), 846–850 (1971)
46. Fred, A.: Finding consistent clusters in data partitions. In: Proceedings of the Second International Workshop on Multiple Classifier Systems, pp. 309–318. Springer-Verlag (2001)
47. Duarte, F., Duarte, J., Fred, A., Rodrigues, M.: Average cluster consistency for cluster ensemble selection. In: Fred, A., Dietz, J., Liu, K., Filipe, J. (eds.) Knowledge Discovery, Knowlege Engineering and Knowledge Management. Communications in Computer and Information Science, vol. 128, pp. 133–148. Springer, Berlin (2011)

Autonomous Learning Needs a Second Environmental Feedback Loop

Hazem Toutounji and Frank Pasemann

Abstract Deriving a successful neural control of behavior of autonomous and embodied systems poses a great challenge. The difficulty lies in finding suitable learning mechanisms, and in specifying under what conditions learning becomes necessary. Here, we provide a solution to the second issue in the form of an additional feedback loop that augments the sensorimotor loop in which autonomous systems live. The second feedback loop provides proprioceptive signals, allowing the assessment of behavior through self-monitoring, and accordingly, the control of learning. We show how the behaviors can be defined with the aid of this framework, and we show that, in combination with simple stochastic plasticity mechanisms, behaviors are successfully learned.

Keywords Neuromodulation · Learning · Plasticity · Sensorimotor loop · Auton-omous systems

1 Introduction

Only autonomous systems can learn autonomously. We use *animats* [1–3] as paradigmatic examples of autonomous systems. They are represented by simulated or physical robots. The animat approach is focusing on emergent behaviors and self-organizing processes which generate the life-sustaining interactions of an animat with its dynamically changing environment. It places emphasis on key features of autonomy to which learning is one of the basic properties. In addition, it takes into account the embodied and situated nature of relevant cognitive processes [4].

H. Toutounji (✉) · F. Pasemann
AG Neurocybernetics, Institute of Cognitive Science, University of Osnabrück,
Albrechtstr 28, 49069 Osnabrück, Germany
e-mail: htoutounji@uos.de
url: http://ikw.uni-osnabrueck.de/~neurokybernetik/

F. Pasemann
e-mail: frank.pasemann@uos.de

© Springer International Publishing Switzerland 2016
K. Madani et al. (eds.), *Computational Intelligence*,
Studies in Computational Intelligence 613,
DOI 10.1007/978-3-319-23392-5_25

An animat is equipped with sensors to perceive the properties of its environment, with proprioceptors to perceive its body's internal (metabolic, physiological) states, with actuators to act in its environment, as well as with a behavioral control that relates its sensory signals and internal states to its actions such that it is able to satisfy its needs for survival.

Survival of a system depends upon some essential internal variables that are monitored and maintained within a given viability zone, i.e. on homeostatic properties [5]. The assumption here is that the primary role of autonomous learning is to establish and to enhance the homeostatic properties of the body. In other words, there will be a close interplay between learning mechanisms and proprioception. In the context of embodied cognition and neuronal plasticity, homeostasis has been examined by e.g., [6–8]. Regulating homeostatic properties is often applied for exploring the system's behavior space, and usually results in a behavior that is not goal-directed [9]. Here, however, goal-directed behavior is considered to be the essential starting point for any kind of learning.

With respect to autonomous learning one is then left with three basic questions: What to learn? When to learn? How to learn? The last question refers to internal mechanisms, such as synaptic plasticity rules [10, 11] and regulatory mechanisms of neuronal excitability [12], which will change dynamical properties of the neural control. But by now, there is no definite general answer or optimal method to generate such a faculty in the neural control of animats. Known learning rules like backprop-agation [13] and variants of Hebbian rules [10] refer to specific network structures like feedforward networks or Hopfield networks [14], and to specific problems like pattern recognition or reconstruction. Thus, and since these methods are inadequate for learning a life-sustaining behavior in animats, in this paper, two simple stochastic plasticity mechanisms are deployed for testing the proposed framework.

On the other hand, the first of our questions seems easy to answer: A life-sustaining behavior has to be learned. But again, since environmental conditions and situations are changing frequently, the second of our questions can be rephrased as follows: What signals drive internal mechanisms and corresponding interactions towards a life-sustaining behavior?

A possible answer is to suggest a second environmental feedback loop. This idea can be traced back as far as the work of H. S. Jennings and his studies of lower-order animals [15], and was reformulated by W. R. Ashby in the early days of cybernetics [5]. The second environmental feedback loop is associated with our second question, namely, when an autonomous system has to learn a new behavior. This is assumed to be the case, for instance, when, during the interaction with the environment, there is a situation where "it hurts", or a situation which produces pleasant or unpleasant "feelings". These metabolic or physiological states stimulate the signals from the proprioceptors. For instance, those signals may be generated if joint angles of a legged animat exceed their limits, a motor gets hot, the system bumps into an obstacle, or the "low" state of an energy neuron signals "hunger". In all such cases, proprioceptors mediate corresponding internal, non-neural processes.

To systematically examine these problems, we implemented similar scenar-ios where proprioceptors are combined with artificial neuromodulators to form modulator subnetworks. These networks monitor the behavior of the animat, and stimulate the artificial *neuromodulator cells* in response to undesired or beneficial behavior. Stimulated neuromodulator cells then produce neuromodulators to trigger or inhibit plastic changes in the control subnetworks of the animat.

The paper is organized as follows. Section 2 describes the modulator network model with a simple random plasticity method, and an alternative Gaussian walk plasticity method. Section 3 introduces the two simulated robots that are used to test the method, followed by a description of the experiments by which we test the neuromodulation framework in Sect. 4. Finally, the results are presented, and the finding are discussed in Sect. 5, followed by final conclusions on the advantages and limitations of this learning approach.

2 Methods

2.1 Modulated Neural Networks (MNN)

A MNN can be any kind of standard artificial neural networks extended by a *neuromodulator layer*. Some related approaches, though more specialized, are e.g., Gas-Nets [16], Artificial Endocrine Systems [17], and Artificial Hormone Systems [18].

Our variant of a neuromodulator layer provides *neuromodulator cells* (NMCs) that maintain spatial distributions of neuromodulator (NM) concentrations as part of the network. NM produced by a NMC usually diffuses into the surrounding tissue and influences nearby network structures. Due to this spatial nature of NMs, a MNN must provide a spatial representation, i.e. neurons and other network elements (e.g. NMCs) must have a location in space. Each NMC represents a single source for a specific NM type and maintains its own concentration level and distribution within the network. The NM concentration $c(t, x, y)$ at each point in the network at time t is the sum of all locally maintained concentration levels $c_i(t, x, y)$ at that position.

$$c(t, x, y) = \sum_{i=1}^{n} c_i(t, x, y), \qquad x, y \in \mathbb{R} \tag{1}$$

NMCs are always in one of two modes: In *production mode* the cell may increase its modulator concentration, in *reduction mode* it may decrease it. To enter the *production mode*, a NMC must be stimulated for some time, whereas it falls back into *reduction mode* when it *is not* stimulated for a while. The actual model that determines when and how the stimulation happens can be chosen freely for each NMC. The same holds for the production, distribution, diffusion and decay of NMs. Usually, the concentration of the NM and its area of influence increase and decrease depending on the current stimulation and mode. But the characteristics of the diffusion area and gradient are specifics of the chosen models and depend on the MNN variant that is used for an experiment.

The effect of NM exposure on network elements can be various, such as affecting the synaptic plasticity or the function of neurons. Therefore, the actual choice of these effects strongly depends on the experiments and the planned interaction between NMs and network components.

2.2 Linearly-Modulated Neural Networks (LMNN)

The specific variant of the MNN used for the first presented experiments is based on the standard discrete-time neuron model given by

$$o_i(t + 1) = \tau_i(\theta_i + \sum_{j=1}^{n} w_{ij} o_j(t)) \quad \text{with} \quad i, j = 1, \dots, n, \tag{2}$$

where $o_i(t)$ is the output of the neuron i at a discrete time step t, w_{ij} is the weight of the synapse from neuron j to neuron i, θ_i is a bias term of neuron i and τ_i a transfer function, for instance $tanh$.

In LMNNs, the stimulation of NMCs follows a simple linear model. The mechanism by which the presented framework guides plasticity is demonstrated schematically in Fig. 1. Each neuromodulator cell (NMC) is attached to a carrier neuron within a modulatory subnetwork (MSN), and is stimulated when the output of this neuron is within a specified range $[S^{min}, S^{max}]$. At each time step t in which the NMC is

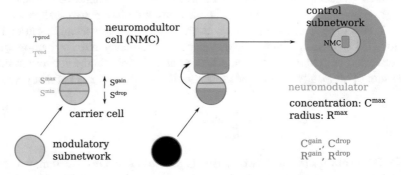

Fig. 1 Schematic representation of Linearly-Modulated Neural Networks. Each neuromodulator cell (NMC) is attached to a carrier neuron within a modulatory subnetwork, and is stimulated when the output of this neuron is within a specified range. At each time step in which the NMC is stimulated, its stimulation level increases, and it decreases If not stimulated. If the stimulation level exceeds a given threshold, the NMC enters the *production mode*. If the level decreases below a second threshold, the NMC re-enters the *reduction mode*. When in production mode, the neuromodulator defuses in time to the surrounding area of a control subnetwork, and initiates plasticity in rates that depend on its concentration at the locale of the synapse

stimulated, its stimulation level s_i increases by a small amount given by parameter S^{gain}. If not stimulated, it decreases by S^{drop}:

$$s_i(t+1) = \begin{cases} min(1, s_i(t) + S_i^{gain}) & \text{if } S_i^{min} \le o_i(t) \le S_i^{max} \\ max(0, s_i(t) - S_i^{drop}) & \text{otherwise.} \end{cases} \tag{3}$$

If the stimulation level exceeds a given threshold T^{prod}, the NMC enters the *production mode*. If the level decreases below a second threshold T^{red}, the NMC re-enters the *reduction mode*.

In *production mode*, the modulator concentration c and the radius r of a *circular* diffusion area are increased from 0 to C^{max} and R^{max} respectively. During *reduction mode* both decrease again. The rate of change of the concentration is given by parameters C^{gain} and C^{drop}, that of the radius similarly by R^{gain} and R^{drop}. The following formula shows this for the concentration level c_i; the area radius r_i is defined analogously.

$$c_i(t+1) = \begin{cases} min(C_i^{max}, c_i(t) + C_i^{gain}) & \text{if in } production \ mode \\ & \text{and } still \text{ stimulated} \\ max(0, c_i(t) - C_i^{drop}) & \text{if in } reduction \ mode \\ & \text{and } not \text{ stimulated} \\ c_i(t) & \text{otherwise.} \end{cases} \tag{4}$$

Due to NM diffusion, learning is triggered in control subnetworks (CSN), according to a particular learning rule whose dynamics depends on the NM concentration. The diffusion mode of each NMC can be chosen, so that the NM concentration is either constant across the diffusion area, or decays according to a linear or nonlinear function of the distance to the NMC. The inhomogeneous distributions are interesting for scenarios with local learning. However, in the shown examples, we will restrict the experiments to a homogeneous, global modulation to demonstrate that successful controllers can develop even in this simple case.

2.3 Plasticity via Modulated Random Search

The synapses of the network react to NM exposure with plastic changes. To demonstrate the viability of using neuromodulation to control the learning process, we choose one of the most simple plasticity methods available: *Random weight changes*. We chose this stochastic plasticity method because it is vastly unbiased and is capable of finding all kinds of network topologies and weight distributions within a given network substrate. Furthermore, the method does not require any heuristics for the choice of the network topology, except that solutions are possible with the given structure.

Table 1 Parameters of a
modulated random search
synapse

Parameter	Description
$Type$	The NM type the synapse is sensitive to
W	Weight change probability
D	Disable/enable probability
W^{min}, W^{max}	Min. and max. weight of the synapse
M	Max. NM sensitivity limit of the synapse

The parameters governing the modulated random search are summarized in Table 1. For a synapse i, the probability of a weight change p_i^w at time t is the product of an intrinsic weight change probability W_i and the current NM concentration $c(t, x, y)$ at the position (x_i, y_i) of the synapse. Hereby, each synapse may limit its sensitivity to NM to a maximal concentration level M_i to prevent too rapid changes when large amounts of overlapping NMs are present.

$$p_i^w(t) = min(M_i, \ c(t, x_i, y_i)) \ W_i, \qquad 0 < W_i \lll 1 \tag{5}$$

Stochastic weight changes may occur at any time step, therefore W_i must be very small. If a weight change is triggered, a new weight w_i is randomly chosen from the interval $[W_i^{min}, W_i^{max}]$, given as parameters of the synapse.

In addition to weight changes, synapses can also *disable* and *re-enable* themselves following a similar stochastic process. The probability p_i^d for a transition between the two states during each time step is the product of the modulator concentration $c(t, x, y)$ and the disable probability D_i.

$$p_i^d(t) = min(M_i, \ c(t, x_i, y_i)) \ D_i, \qquad 0 \le D_i < W_i \tag{6}$$

If a transition is triggered, an enabled synapse becomes disabled and vice versa. A disabled synapse is treated as a synapse with weight $w_i = 0$, but its actual weight is preserved until it is enabled again. This mechanism allows for a simple topology search within a given neural substrate.

2.4 Plasticity via Modulated Gaussian Walk

An alternative to using random search as a learning mechanism, we propose a learning mechanism that depends on small changes of synaptic efficacies when neuromodulation is released. We term this learning mechanism the *Modulated Gaussian Walk* (MGW), where, similarly to MRS, the probability of a weight change is the product of an intrinsic weight change probability and the neuromodulator concentration. However, unlike the MRS, no maximal concentration sensitivity is present.

Instead of randomly assigning a value to the synaptic weight in the interval $[W_i^{min}, W_i^{max}]$, the amount of weight change is drawn from a normal distribution with zero-mean and σ^2-variance. As such, the weight changes according to

$$w(t+1) = w(t) + \Delta w \text{ where } \Delta w \sim \mathcal{N}(0, \sigma^2). \tag{7}$$

To assure that the weight remains within its bound (since the term Δw can be infinitely large), sampling the normal distribution is repeated until the resulting weight is within the range.

A mechanism for disabling synapses in also implemented within the MGW learning rule. However, we do not elaborate on this feature here, since later experiments do not make use of it.

3 Robots

Later experiments on linearly-modulated neural networks use robot systems typical for classical neurorobotics problems: a *simple pendulum* (Fig. 2c) and a *differential drive robot* (Fig. 2f). In all cases, motor neurons with an activation range $(-1, +1)$

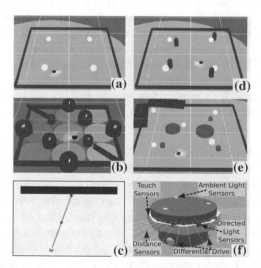

Fig. 2 a, b, d, e Environments for learning behavior of a differential drive robot. **a, b, d** The *white spheres* denote possible light source positions. Each light source is bright enough to cover the whole environment. **a** Light-tropism to one of four fixed light sources (E1). **b** Obstacle-avoidance with exploration (E2). **c** A simple pendulum simulator for learning oscillation to a target angle (E5). **d** Light-tropism to one of four fixed light sources, and avoiding nearby obstacles (E3). **e** Light-tropism to one of five randomly shifted light sources, and avoiding nearby obstacles, large obstacles, and a narrow corner (E4). **f** The differential drive robot with wheels and sensors shown

control the desired velocity of the motors. Negative activations are interpreted as backwards rotation.

The pendulum is equipped with an *angular sensor* for the current angle of the pendulum. The differential drive robot is equipped with *distance sensors* (DS) at the front, eight *touch sensors* (TS), three *ambient light sensors* (ALS) to measure brightness at three equally distributed positions on the robot, and three *directed light sensors* (DLS) in the front of the robot to sense the direction towards light sources (with a maximal viewing angle of ± 90 degrees). For simplicity, light can penetrate obstacles freely. All experiments have been simulated with the NERD Toolkit [19] and can be replicated with material from our supplementary page.

4 Experiments

4.1 Experiments with MRS

To demonstrate the method, five experiments with different complexities have been performed under modulated random search. The experiments are typical for early evolutionary robotics experiments and are still used in many learning scenarios. In all experiments, a robot has to learn a simple task from scratch, starting with a plain, specifically designed LMNN. The predefined MSN of the network produces global neuromodulators for undesired behaviors, while the given CSN defines the topology in which solutions can develop. Neuromodulation is global since all synapses of the CSN are sensitive to NM concentration, and they start out disabled, so that the network connectivity develops together with the synaptic weights, while all synapses of the MSN are insensitive to neuromodulation and are therefore static. As such, each experiment can be defined by a robot, a task, an environment and a control and modulatory subnetworks (MSN and CSN, respectively).

Tasks and Environments. The first experiment (E1) is a positive light-tropism task (Fig. 2a). Four light sources are distributed in some distance from the corners of a quadratic arena. At any time, only one light source is switched on. Each light source is bright enough to cover the entire arena. When the robot arrives at that light source, it is switched off and a randomly chosen source is switched on.

The second experiment (E2) focuses on an obstacle-avoidance task (Fig. 2b), where the robot has to navigate in a quadratic environment riddled with round objects and narrow corners. The robot also needs to explore its whole environment. Thus, the arena also comprises a number of light sources each emitting a different, homogeneous light that allow the robot to recognize different locations and hence to monitor its own exploration behavior.

As a combination of the previous experiments, E3 extends the first experiment with four small obstacles placed with a small asymmetric shift near the four light sources (Fig. 2d). Here, the robot has to approach the lights and simultaneously avoid the obstacles next to the light sources.

Table 2 Experimental setups for global neuromodulation

Exp.	τ_{exp}	τ_{temp}	Sensors	NMC modules
E1	120	0.5	2 DLS	Light
E2	240	5	3 DS	Obst, Drive, Explore
E3	720	0.75	2 DS, 2 DLS	Light, Obst
E4	720	0.75	2 DS, 2 DLS	Light, Obst
E5	240	5	1 AS	$2 \times$ TurningAngle

τ_{exp} is the experiment time in simulated minutes, τ_{temp} is the duration in minutes without neuro-modulation production to consider a behavior a successful temporary solution

A more difficult variant is experiment E4. While the task remains the same, there are now larger obstacles in the middle of the arena and one of the corners is more narrow (Fig. 2e). Furthermore, a fifth light source was added in the center of the arena. All lights are now also randomly moved away from their initial positions every time they get switched on. In contrast to E3 the robot now gets confronted with many more different light-obstacle combinations, which makes the task quite difficult.

The pendulum experiment (E5, Fig. 2c) requires the controller to learn to swing with a specific amplitude between the two target angles ±65° with a tolerance of ±5°. The difficulty is that the motors are too weak to get to the target angles without swinging the pendulum up first.

Control Sub-Networks (CSN). Each CSN includes the necessary sensory and motor neurons, a number of intermediate processing neurons and a bias neuron. The latter allows the bias of neurons to be changed using the same technique as used for other synapses. The network substrates vary over the different experiments, ranging from trivial feedforward networks over a layered network with 4 hidden neurons, to fully connected, recurrent networks with 2, 4 and 6 intermediate neurons. The network configurations for the experiments are summarized in Table 2.

Modulatory Sub-Networks (MSN). Each MSN uses *experiment-specific* network structures to detect undesired behavior based on (sensor) activations to produce neuromodulators when needed. As a reaction to the neuromodulators, synapses of the CSN randomly change and explore different topologies and weight distributions. This has an effect on the behavior and, accordingly, on the NM production in the MSN. Similar to the work by Ashby [5], the system is destabilized when an undesired behavior is detected, leading to continuous changes until the system stabilizes again in a new, valid configuration. In this spirit, six different NMCs are used in the experiments (see Table 2).

The Obst cell reacts on the activation of any of the eight force sensors to detect undesired contact with objects. The stimulation is quite rapid so that obstacle contact immediately leads to neuromodulation production to alter the behavior.

The Drive cell gets stimulated when the two motor signals are too low, the robot is moving backwards, or the difference of the motors becomes too large, i.e. the robot is moving in narrow circles. Because the desired behavior also may include moving backwards and especially moving in circles, the stimulation is less rapid and tolerates such movements as long as they do not dominate the behavior.

The Explore cell is stimulated when the robot is not entering the detectable locations frequently (the task E2). Its associated modulating network classifies the signal of one of the ambient light sensors into the nine detectable locations and integrates these signals to determine the duration of each location not being visited. Explore is stimulated if some locations have not been visited for a long time. If a location is entered that has not been visited for a long time, then all integrator neurons for all locations are inhibited, so this potential behavior improvement already leads to a fast decrease in neuromodulator concentration to allow the new configuration to be tested.

The Light cell also uses an auxiliary network that interprets the ambient light sensors to detect whether the robot is getting closer to the light. If not, the NMC is stimulated. This achieved by utilizing neural differentiators of the ambient light sensors activity.

The TurningAngle cell gets persistently, but slowly stimulated over time. However, if the pendulum changes its swinging direction within the desired angle range, then the NMC stimulation decreases rapidly. The desired angular range can be adjusted independently for each of the two NMCs in the pendulum networks.

Table 2 shows which NMCs, with their corresponding auxiliary networks, are used in each experiment. Figure 3 shows the structure of both the CSN and the MSN for experiments E2 and E3, giving also the neural structures for the six auxiliary sub-networks. The experiments here are restricted to a global modulator release with a uniform concentration levels. Table 3 summarizes the parameter choices for the NMCs used across the experiments.

Experiments Setup. Each experiment has been run with five different network substrates for the CSN: a layered network with 4 intermediate neurons (L4) and four fully, recurrently connected networks with 0, 2, 4 and 6 intermediate processing neurons (N0-N6). Due to the differing number of motors and sensors, the total number of synapses varies. An overview can be found in Table 4. All additional settings of the network, specifically the settings for the plastic synapses and the NMC settings, have been fixed at the values given in Table 3.

Each such learning scenario (experiment + network substrate) has been repeated 50 times with identical settings, each starting with a new CSN composed of disabled synapses with zero weights. Thus, the entire network topology and the synaptic weights had to be learned from scratch within the given network substrate.

4.2 Comparative Experiment with MRS and MGW

We compare the two plasticity mechanism on a task that combines light tropism and obstacle-avoidance with no exploration, i.e. the MSN contains NMCs Obst, Drive, and Light. The chosen CSN of this task is similar to the layered architecture L4, but

Table 3 Parameter values for NMCs and the modulated random search in the global modulation experiments

Param.	Obst	Drive	Explore	Light	TurningAngle	Param.	Synapses
S^{min}, S^{max}	0.9,1.0	0.9,1.0	0.4,1.0	0.9,1.0	0.5,1.0	W	0.0001
S^{gain}, S^{drop}	0.01,0.01	0.001,0.001	0.001,0.01	0.0002,0.0001	0.005,1	D	0.00002
T^{prod}, T^{red}	0.95,0.95	0.95,0.95	0.95,0.95	0.99,0.99	0.95,0.95	W^{min}	−1.5
C^{max}	2	1	1	1	1	W^{max}	1.5
C^{gain}, C^{drop}	0.1,0.1	0.001,0.01	0.001,0.01	0.01,0.1	0.001,1	M	1.0

Fig. 3 Two exemplary control subnetworks that result from learning, with their associated modulator subnetworks

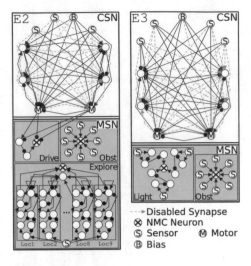

Table 4 Number of plastic synapses in each of the experiments. All configurations include a bias neuron. L4 provides a layered network with 4 neurons, all others are fully connected

	Number of processing neurons				
	N0	N2	N4	N6	L4
E1	14	32	60	96	46
E2	10	28	54	88	42
E3	14	32	60	96	42
E4	14	32	60	96	42
E5	4	15	35	63	32

with few simplifications that decrease the number of plastic synapses considerably. First, the hidden layer is split into two pairs of neurons. One pair is connected to the two distance sensors only, while the other pair is connected to the two directed light sensors. This results in a modular structure that enforces a kind of specialization to each pair. The two modules are also fully-connected to each other, adding eight plastic synapses that are responsible for the fusion of behavior. Furthermore, a symmetry constraint is added to each module. This means that a change of some synapse at the left side of the module would be copied to corresponding synapse at the right side. This constraint is meant to reflect the symmetry in the body morphology of the robot, which would result in a symmetric behavior. No constraints are imposed on the connections between the two modules. As such, the number of plastic synapses in this CSN, including those coming from a bias neuron, are only 22.

The parameters of MRS are chosen as before but with the probability of enabling or disabling a synapse set to zero. The range of weights is restricted to ± 1.5 for both MRS and MGW. For the latter, the variance σ^2 is set to 0.2. Each learning rule was tested on 64 runs, with 8 hours simulation time.

5 Results and Discussion

5.1 Results on Modulated Random Search

For all experiments and with all but one of the different network substrates, solutions have been found within the given time windows. All behaviors discovered in this way have been sufficiently effective and comply with the desired and expected behaviors. However, as can be seen in Fig. 4, by far not all runs did finally end up with a proper behavior network during the limited learning time. Consistent with intuition, the easier the task is, the larger the percentage of successful learning trials.

Therefore, The simple light-tropism task E1 led to successful behaviors in almost all cases, despite its comparably short learning time of up to only two hours. Also, the final solutions have been found very fast (Fig. 5a-E1) without many intermediate temporal solutions (Fig. 5c-E1).

In contrast, the almost similarly short duration of the obstacle-avoidance task E2 with four hours seems to be much too low to consistently find solutions, contrary to our expectation. Therefore, only about half of the experiments were successful. A reason for this may be the relatively slow detection of insufficient exploration behavior with the Explore NMC. This modulator has to react with a larger delay to give the networks a chance to actually do exploration.

So, behaviors violating the exploration condition – while still doing a fine obstacle-avoidance – are detected only after a significant delay. Also, such intermediate solutions get destroyed quite easily when a bad exploration behavior is detected, leading to the destruction – not to a refinement – of the temporary solution. This, obviously, is one of the major limitations of the stochastic search: due to the missing directedness of the learning, temporary solutions are usually not improved, but rather destroyed and replaced by very different networks.

The results for combining light tropism and obstacle avoidance (E3) reflect the increasing difficulty of the task. Even though the experiment was simulated 12 hours

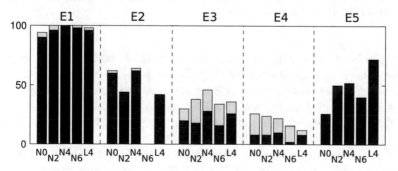

Fig. 4 Percentage of successful experiments with stable solutions. The *gray tips* indicate the number of temporary solutions with a continuous modulator-free behavior during at least 30 min, which would be interpreted as solutions in intermediate-term evaluations

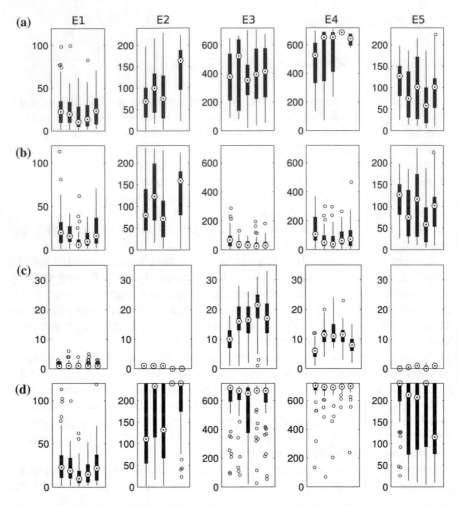

Fig. 5 **a** Time to final solution. **b** Time to first (temporary) solution. **c** Number of (temporary) solutions. **d** Minutes spent in learning mode

per try, only $\approx 20\,\%$ of the runs lead to a fully stable behavior. First temporary solutions have been found quite fast (Fig. 5b-E3), but most light tropism behaviors with only a partial obstacle-avoidance behavior are easily destroyed due to hitting one of the small obstacles close to the light sources. Because the light sources are approached with slightly different angles, at some point a situation is encountered where the obstacle-avoidance behavior briefly fails and the obstacle is hit. This leads to a strong production of NM and the behavior is usually destroyed. This alternation between many temporary solutions (Fig. 5c-E3) and the subsequent network destruction, and thus long phases with enabled plasticity (Fig. 5d-E3), describes the typical way how network configurations are explored with the stochastic search: only if *all requirements* of the behavior are *fully* met with a single mutation burst, the behavior remains stable in the long run. This *all or nothing* approach is another limiting characteristics of the simple stochastic search.

This becomes even more severe in the aggravated variant of this experiment (E4), in which large and more various obstacles enforce the robot to do significant detours against the desired direction towards the light. Here, a proper behavior requires a fine tuning of weights, which makes it much more difficult to accidentally stumble upon a working network. The percentage of final solutions, therefore, is even lower with only about 10 %. However, the number of long-term temporary solutions with a continuous runtime of more than 30 min exceeds the number of stable solutions by a factor of ≈ 2 (Fig. 4-E4). These behaviors would in many evaluations with a short test (e.g. evolutionary algorithms) already be considered solutions, but it shows that even slight weaknesses due to an unfortunate sequence of target light sources can lead to a destruction of such *almost stable* networks in the long run. As in E3, temporary solutions are found quite fast (Fig. 5b-E4), but are destroyed later, so that most of the time is spent trying new network configurations (Fig. 5d-E4).

The pendulum behavior again is an example of a simpler single-goal task. The number of successful runs is, with almost 50 %, quite high and the networks are also found fast within the first 2 hours (of a total of 4 hours). Due to the characteristics of the experiment, there are almost no temporary solutions: if a solution is found, then this solution tends to be stable in the long run, because there are no disturbances in the simple pendulum motion (compare Fig. 5a-E5, 5b-E5, and 5c-E5).

An interesting observation can be made concerning the network complexity. It was expected, that the performance of the experiments primarily depends on the size of the neural substrate, because with an increasing search space the probability of finding a stable solution should drop down significantly. However, at least for the network sizes used in these experiments, there is only a small influence of the network substrate on the performance (Fig. 4). Only in E2 the largest network showed a significant drop in the number of solutions compared to the other substrates in the same experiment. And in E5 it seems that the layered network has an advantage over the fully recurrent neural networks. This may indicate, that – as long as the topology can vary within the substrate – there are similar or equivalent network configurations contained in all substrates and that with an increasing number of synapses, the fraction between feasible and improper network configurations may remain in the same order of magnitude. In forthcoming experiments, larger networks have to be tested to find the actual limiting size for this simple class of robot experiments. In these experiments, anyway, the impact of the chosen experiment complexity has a much higher impact on the performance than the chosen network substrate, so the major effort in designing such experiments should probably be focused on defining a well suited experiment, not on choosing a particularly suited network substrate.

To examine the learning process in more detail, Fig. 6 shows the weight changes and the related neuromodulator concentrations for one of the learning runs in experiment E2. As expected, the weight changes in learning phases are random and undirected. However, from time to time, the system stabilizes in a network configuration, because no neuromodulator is produced as a response to the (partially) working

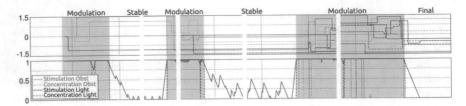

Fig. 6 Example run for the light tropism-behavior, showing the alternation between stable and plastic states during the behavior learning. The *upper graph* shows the individual weights over time, the *lower graph* the stimulation and concentration level of the two NMCs

behavior. It can also be seen in the lower part of Fig. 6 that even during these stable states, the stimulation of the NMCs is not just zero, but that their stimulation level remains active, though not high enough to enter their *production mode*. So, slight violations of the behavior restrictions still take place, but these violations are not strong enough to be interpreted as a failing behavior. But if the stimulation level exceeds the limit to *production mode*, then often one of the first random changes destabilizes the system so much, that other neuromodulators are triggered as side-effect. This leads to a strong relearning, usually destroying the previous temporary solution, until the modulation stops when a new potentially working configuration has been found.

5.2 Comparing MRS and MGW

As the previous section demonstrated, due to its uncontrolled random changes to network structure, MRS leads to the destruction of solutions. In comparison, limiting random changes to small values, as is the case in MGW, results in the preservation of found solutions. In the modular light-tropism/obstacle-avoidance experiment, outlined in Sect. 4.2, MRS has shown 34 temporary solutions that lasted longer than 5 min in simulation time, with an average of 5.7 min per solution. On the other hand, MGW found almost double the number of temporary solutions, with an average of 12.5 min per solution. Also, under MGW, the agent spends more of its time exploiting the found solutions. While temporary solutions that lasted longer than 5 min occupied more than 11.4 % of the experiment time of robots trained by MGW, only 8.2 % of the experiment time is covered by the temporary solutions found by MRS. This means that learning with Gaussian walk is more stable since the learning rule does not result in the sudden destruction of behavior when neuromodulation is released due to minor lapses in behavioral fitness. Further results suggest that MGW refine network structures that are on the verge of becoming a solution by inducing small changes to the networks' synaptic weights. This is demonstrated by the fact that only 40 % of controllers trained by MRS found a temporary solution at all, while MGW lead to 70 % of the runs leading to a temporary solution at some stage of learning.

6 Conclusions

We demonstrated with five typical experiments from the field of robot learning and early evolutionary robotics, that a simple random search on a given network topology is sufficient to find many suitable solutions, as long as the network changes are started and stopped by a reasonable feedback signal. In our case, this feedback is realized with neuromodulators that are triggered as a reaction to the sensed behavior. Because of this, and the simplicity of the implementation, the learning should also work directly on physical robots without external supervision. The tasks show that the feasibility of the method strongly depends on the experiment complexity, not so much on the chosen network substrate. Also, temporary solutions appear and get relearned when the behavior proves ineffective in some situations. These aspects – already available in such a simple approach – are highly desired in the field of robot learning to allow adaptive, self-contained robots with life-long learning capabilities.

Simple random search, however, is not meant to be used as a competitive learning paradigm for real robots. Our results show that by simply replacing the fully-random search with a more confined random walk of synaptic weights lead to a huge increase in the number of solutions and of their stability. This points to the possible benefits of incorporating more directed learning rules and synaptic dynamics to the neuro-modulation framework. Our intention of this study is to provide the mechanism that signals to an autonomous system the need to start learning, i.e. when to learn. The suggested learning mechanism itself, i.e. how to learn, needs to prove superior to the simple random search, as was demonstrated by the Gaussian walk, in order to justify its increased complexity.

Acknowledgments This research was partially funded by the German Research Foundation (DFG) priority program 1527. The contribution of Christian Rempis to this project is gratefully acknowledged. The authors thank Josef Behr, Andrea Suckro, and Florian Ziegler for testing and refining the simulation models in the NERD Toolkit, and particularly the latter for his role in the current study. Thanks to Kevin Koschmieder for implementing the modulated Gaussian walk.

References

1. Dean, J.: Animats and what they can tell us. Trends Cogn. Sci. **2**(2), 60–67 (1998)
2. Meyer, J.A.: The animat approach to cognitive science. In: Roitblat, H. Meyer, J.A. (eds.) Comparative Approaches to Cognitive Science, pp. 27–44. The MIT Press/Bradford Books (1995)
3. Meyer, J.A., Guillot, A.: Biologically inspired robots. In: Siciliano, B., Khatib, O. (eds.) Springer Handbook of Robotics, pp. 1395–1422. Springer (2008)
4. Pfeifer, R., Bongard, J.: How the body shapes the way we think: a new view of intelligence. MIT press (2007)
5. Ross Ashby, W.: Design for a brain: the origin of adaptive behavior (2nd edn). Chapman and Hall, London UK (1960)
6. Di Paolo, E.A.: Organismically-inspired robotics: homeostatic adaptation and teleology beyond the closed sensorimotor loop. In: Murase, K., Asakura, T. (eds.) Dynamical Systems Approach

to Embodiment and Sociality, pp. 19–42. Advanced Knowledge International, Adelaide, Australia (2003)

7. Ziemke, Tom: The embodied self: theories, hunches and robot models. J. Conscious. Stud. **14**(7), 167–179 (2007)

8. Ikegami, T., Suzuki, K.: From a homeostatic to a homeodynamic self. BioSystems **91**(2), 388–400 (2008)

9. Der, R.: Artificial life from the principle of homeokinesis. In: Proceedings of the German Workshop on Artificial Life (2008)

10. Hebb. D.O.: The Organization of Behavior. Wiley, New York (1949)

11. Cooper, L.N., Intrator, N., Blais, B.S., Shouval, H.Z.: Theory of Cortical Plasticity. World Scientific (2004)

12. Turrigiano, G.G., Nelson, S.B.: Homeostatic plasticity in the developing nervous system. Nat. Rev. Neurosci. **5**(2), 97–107 (2004)

13. Rumelhart, D.E., Hinton, G.E., Williams, R.J.: Learning representations by back-propagating errors. Nature **323**(9), 533–536 (1986)

14. Hopfield, J.J.: Neural networks and physical systems with emergent collective computational abilities. Proc. Natl. Acad. Sci. USA **79**(8), 2554–2558 (1982)

15. Jennings, H.S.: Contributions to the study of the behavior of lower organisms. Number 16, Carnegie institution of Washington (1904)

16. Smith, T., Husbands, P., Philippides, A., O'Shea, M.: Neuronal plasticity and temporal adaptivity: Gasnet robot control networks. Adapt. Behav. **10**(3–4), 161–183 (2002)

17. Timmis, J., Neal, M., Thorniley, J.: An adaptive neuro-endocrine system for robotic systems. In: Proceedings of the IEEE Workshop on Robotic Intelligence in Informationally Structured Space, RIISS'09, pp. 129–136 (2009)

18. Moioli, R.C., Vargas, P.A., Husbands, P.: A multiple hormone approach to the homeostatic control of conflicting behaviours in an autonomous mobile robot. In: Proceedings of IEEE Congress on Evolutionary Computation, CEC'09, pp. 47–54 (2009)

19. Rempis, C., Thomas, V., Bachmann, F., Pasemann, F.: NERD—Neurodynamics and Evolutionary Robotics Development Kit. In: Simulation, Modeling, and Programming for Autonomous Robots, pp. 121–132. Springer (2010)

Prediction Capabilities of Evolino RNN Ensembles

Nijolė Maknickienė and Algirdas Maknickas

Abstract Modern portfolio theory of investment-based financial market forecasting use probability distributions. This investigation used an ensemble of genetic algorithm based recurrent neural networks (RNN), which allows to obtain multi-modal distribution for predictions. Comparison of the two different models—scatted points based prediction and distributions based prediction—opens new opportunities to create profitable investment tool, which was tested in real time demo market. Dependence of forecasting accuracy on the number of Evolino recurrent neural networks ensemble was obtained for five forecasting points ahead. This study allows to optimize the cluster based computational time and resources required for sufficiently accurate prediction.

Keywords Distribution of expected returns · Ensembles · Evolino · Financial markets · Prediction · Recurrent neural networks

1 Introduction

Neural networks and their systems are successfully used in forecasting. There are several factors that determine the predictive accuracy of the prediction—input selection, neural network architecture and the quantity of training data.

The paper [1] is to provide a practical introductory guide in the design of a neural network for forecasting economic time series data. An eight-step procedure to design a neural network forecasting model is explained including a discussion of trade offs in parameter selection, prediction dependence on number of iterations. In paper [2], the effects of different sizes of training sample sets on forecasting currency exchange rates are examined. It is shown that those neural networks-given an appropriate

N. Maknickienė (✉) · A. Maknickas
Vilnius Gediminas Technical University, Sauletekio Al. 11, 10223 Vilnius, Lithuania
e-mail: nijole.maknickiene@vgtu.lt
URL: http://www.vgtu.lt

A. Maknickas
e-mail: algirdas.maknickas@vgtu.lt

© Springer International Publishing Switzerland 2016
K. Madani et al. (eds.), *Computational Intelligence*,
Studies in Computational Intelligence 613,
DOI 10.1007/978-3-319-23392-5_26

amount of historical knowledge—can forecast future currency exchange rates with 60 percent accuracy, while those neural networks trained on a larger training set have a worse forecasting performance. More over, the higher-quality forecasts, the reduced training set sizes reduced development cost and time. In the paper [3], the relationship between the ensemble and its component neural networks is analysed, which reveals that it may be a better choice to ensemble many instead of all the available neural networks. This theory may be useful in designing powerful ensemble approaches. In order to show the feasibility of the theory, an ensemble of twelve neural networks (NN) approach named GASEN is presented.

The methodology in paper [4] proposes an architecture-altering technique, which enables the production of highly antagonistic solutions while preserving any weight-related information. The implementation involves genetic programming using a grammar-guided training approach, in order to provide arbitrarily large and connected neural logic network. The ensemble of 1–5 neural networks was researched by [5], resumed that "incorporating more neural networks into the model does not guarantee that the error would be lowered". As it can be seen in the application case study, the model with two neural networks did not perform more satisfactorily than the single neural network.

In paper [6], was proposed a general framework for designing neural network ensembles by means of cooperative coevolution. The proposed model has two main objectives: first, the improvement of the combination of the trained individual networks; second, the cooperative evolution of such networks, encouraging collaboration among them, instead of a separate training of each network. Authors [7] made ensemble of neural predictors is composed of three individual neural networks.The experimental results have shown that the performance of individual predictors was improved significantly by the integration of their results. The improvement is observed even during the application of different quality. In paper [8] a prediction technique was proposed which was called "an ensemble of simple regression models" to improve the prediction accuracy of cross-project prediction. To evaluate the performance of the proposed method, was conducted 132 combinations of cross-project prediction were conducted using datasets of 12 projects from NASA IVV Facility Metrics Data Program. Brezak et al. [9] made a comparison of feed-forward and recurrent neural networks in time series forecasting.The obtained results indicate satisfactory forecasting characteristics of both networks. However, recurrent NN was more accurate in practically all tests using less number of hidden layer neurons than the feed-forward NN. This study once again confirmed a great effectiveness and potential of dynamic neural networks in modelling and predicting highly nonliner processes.

Recurrent Neural networks (RNN) ensembles acts as expert systems, and for its results are using expert methods. One of them, the fuzzy Delphi, was used for sales forecasting by [10] and integrated with artificial NN for stock market forecasting by [11]. New forecasting mechanism was proposed by [12]. It was modelled by integrating Fuzzy Delhi Method (FDM) with Artificial Neural Network (ANN) techniques to manage the demand with incomplete information.

In order to form the investment strategies in financial markets, there is a need for a proper forecasting technique, which can forecast the future profitabilities of assets (stock prices or currency exchange rates) not as particular values but as probability distributions of values. Such approach is analytically meaningful because future is always uncertain and we cannot make any unambiguous conclusion about it. For this reason the adequate portfolio model is used, developed by [13], which is an amplification of Markowitz portfolio model. The adequate portfolio conception is based on the adequate perception of reality that portfolio return possibilities should be expressed as a probability distribution with its parameters. The analysis of the whole probability distribution is especially important taking into account that portfolio return possibilities usually do not conform to Normal probability distribution form and therefore it is not enough to know their mean value and standard deviation. The initial concept of adequate portfolio over time was also applied to the analysis of other complex processes in the scientific works of A.V.Rutkauskas and his co-authors [14–17].

Decision maker is soliciting opinions as data for statistic inference, with the additional complication of strategic manipulation from interested experts [18]. Authors investigated proportion of correct decisions made by different number of agents— $a \in [1, 1000]$.

The aim of the paper is to investigate the influence of the number of neural nets on accuracy of financial markets prediction, to find new conditions of constructing investment portfolios. Knowing how much RNN is enough that the ensemble makes sufficiently accurate forecasting, to allow the saving of time and power resources.

2 Prediction Using Artificial Intelligence

2.1 Prediction

The forecasting we understand the ability to correctly guess a certain amount of unknown data in time with some precision. After all, the predicted data set is compared with a set of known data to evaluate the correlation between these.

Suppose it is known that p is an element of some set of distributions P. Choose a fixed weight w_q for each q in P such that the w_q add up to 1 (for simplicity, suppose P is countable). Then construct the Bayesmix $M(x) = \sum_q w_q q(x)$, and predict using M instead of the optimal but unknown p. How wrong could this be? The recent work of Hutter provides general and sharp loss bounds [19]: Let $LM(n)$ and $Lp(n)$ be the total expected unit losses of the M-predictor and the p-predictor, respectively, for the first n events. Then $LM(n) - Lp(n)$ is at most of the order of $\sqrt{Lp(n)}$. That is, M is not much worse than p. And in general, no other predictor can do better than that. In particular, if p is deterministic, then the M-predictor won't make any more errors. If P contains all recursively computable distributions, then M becomes the celebrated enumerable universal prior. The aim of this paper is to construct a model that can make predictions with a small enough difference $M(t) - p(t)$ for some fixed time t.

2.2 Evolino RNN

Autors [20, 21] propose a new class of learning algorithms for supervised recurrent neural networks—EVOLINO RNN. EVolution of recurrent systems with Optimal LINear Output. EVOLINO-based Long Short-Term Memory (LSTM) recurrent networks learn to solve several previously unlearnable tasks. "EVOLINO-based LSTM was able to learn up to 5 sins, certain context-sensitive grammars, and the Mackey-Glass time series, which is not a very good RNN benchmark though, since even feedforward nets can learn it well" [22].

In Evolino RNN Enforced SubPopulations (ESP) evolves neurons instead of full networks. Neurons are segregated into subpopulations, and networks are formed by randomly selecting one neuron from each subpopulation. A neuron accumulates a fitness score by adding the fitness of each network in which it has participated. The best neurons within each subpopulation are mated to form new neurons. The network shown here is an LSTM network with four memory cells. In Evolino, only the connection weights in the recurrent part of the network are evolved. The weights to the output units are computed analytically during each evaluation.

2.3 Ensembles of NN

Modularity is a feature often found in nature. It can be of two types-(1) when the modules are connected to each other in parallel or sequentially, (2) when the modules are connected by another module inside. We constructed a modular Evolino RNN system connecting them in parallel.

Ensembles acts as expert systems, like group with different opinions. One of experts methods is delphy method that gives a certain priority evaluation for results of the group.

The Delphi method is based on the assumption that group judgements are more valid than individual judgements. Our observations on the Evolino recurrent neural network prediction [23] made clear that some of the predictions are very accurate, but some others are contradictory, unstable, and must be rejected. The Delphi method makes it possible to achieve a certain consensus or clustering of forecasts. The steps of classical Delphi method are:

(1) The group of experts receives a questionnaire and assesses their prognoses using numeric values, argues their assessments, and completes the questionnaire.
(2) The answers are arranged in ascending order and the media μ and quartiles Q_1, Q_3 are calculated. After determining the upper and lower quartiles, the range between the two averages $Q_1\mu$ and $Q_3\mu$ is considered the most desirable interval. The compatibility of the predictions is calculated, such as whether there is a consensus of the experts. The experts are then acquainted with the results and the arguments and prognoses are made again.

(3) The second step is then repeated. Theoretically, the Delphi process can be continuously iterated until a consensus is determined to have been achieved. In practice, the number of iterations is limited by the time available for decision making.

Consensus in expert system based on RNN ensembles is calculating in each step of delphy method. Therefore it is necessary to assess the compatibility of the expert assessments and calculate the interquartile coefficient. The variation of the responses is taken to be the difference between the first and third quartiles, $Q_3 - Q_1$. The interquartile coefficient is the quotient of the variation response by the median:

$$q = \frac{Q_3 - Q_1}{\mu},\qquad(1)$$

The interquartile coefficient ranges from 0 to +1 and is close to zero when the distribution has very little variation.

Forecast getting by RNN ensemble for the one point in the future may be: (1) a single point; (2) the most likely forecast interval; (3) the distribution of expected values.

2.4 Distributions of Expected Values

The form of the distribution can be described using Sharpe characterization plane [24] and main indicators: skewness and kurtosis.

Skewness is indicator used in distribution analysis as a sign of asymmetry and deviation from a normal distribution:

$$\gamma_1 = \frac{\sum_{i=1}^{N}(y_i - \bar{y})^3}{(N-1)\sigma^3},\qquad(2)$$

where $\bar{y}$ is the mean, σ is the standard deviation, and N is the number of data points.

Skewness > 0—Right skewed distribution—most values are concentrated on left of the mean, with extreme values to the right. Skewness < 0—Left skewed distribution—most values are concentrated on the right of the mean, with extreme values to the left. Skewness = 0—mean = median, the distribution is symmetrical around the mean.

Kurtosis is indicator used in distribution analysis as a sign of flattening or "peakedness" of a distribution:

$$\beta_2 = \frac{\sum_{i=1}^{N}(y_i - \bar{y})^4}{(N-1)\sigma^4},\qquad(3)$$

and excess kurtosis:

$$\gamma_2 = \frac{\sum_{i=1}^{N}(y_i - \bar{y})^4}{(N-1)\sigma^4} - 3.\qquad(4)$$

Kurtosis > 3—Leptokurtic distribution, sharper than a normal distribution, with values concentrated around the mean and thicker tails. This means high probability for extreme values. *Kurtosis* < 3—Platykurtic distribution, flatter than a normal distribution with a wider peak. The probability for extreme values is less than for a normal distribution, and the values are wider spread around the mean. *Kurtosis* = 3— Mesokurtic distribution—normal distribution for example. The investment decisions in financial markets are always taken under uncertainty. Therefore, distributions are more informative and reliable than scattered projections.

Our artificial intelligence system is ensembled from EVOLINO RNN. We are solving a problem: How many elements must be in ensemble for accurate prediction and rational decision making.

3 Description of Models Based on Ensembles of Recurrent Neural Networks

Two different models have been developed and tested. Technical feasibility has been a major factor in determining both the creation of models.

3.1 Scattered Points Based Prediction Model

Evolino RNN-based prediction model, which is applied to the average for PC. This model, which uses eight predictors, was investigated with the phython program by the following steps:

Data step. Getting historical financial markets data from Meta Trader—Alpari. We choose for prediction EURUSD (Euro and American Dollar), GBPUSD (Great Britain Pound and American Dollar), exchange rates and their historical data for the first input, and for the second input, two years historical data for XAUUSD (gold prise in USA dollars), XAGUSD (Silver price of USA dollars), QM (Oil price in USA dollars), and QG (Gass price in USD dollars). At the end of this step we have a basis of historical data.

Input Step. The python script calculates the ranges of orthogonality of the last 80–140 points of the exchange rate historical data chosen for prediction, and an adequate interval from the two years historical data of XAUUSD, XAGUSD, QM, and QG. A value closer to zero indicates higher orthogonality of the input base pairs. Eight pairs of data intervals with the best orthogonality were used for the inputs to the Evolino recurrent neural network. Influence of data orthogonality to accuracy and stability of financial market predictions was described in paper [25].

Prediction Sstep. Eight Evolino recurrent neural networks made predictions for a selected point in the future. At the end of this step, we have eight different predictions for one point of time in the future.

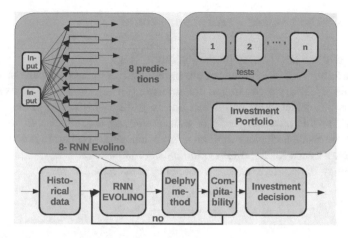

Fig. 1 Scheme of the points based model

Consensus Step. The resulting eight predictions are arranged in ascending order, and then the median, quartiles, and compatibility are calculated. If the compatibility is within the range [0; 0.024], the prediction is right. If not, then step 3 is repeated, sometimes with another "teacher" if the orthogonality is similar. At the end of this step, we have one most probable prediction for the chosen exchange rate.

Decision of trading are making by constructing portfolio of exchange rates with taking into account of predictions—medians, got by described model I (Fig. 1).

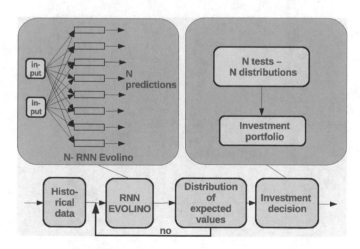

Fig. 2 Scheme of the distribution based model

3.2 Distribution Based Predictions Model

Second—Evolino RNN-based prediction model (Fig. 2).

For calculation of big amount of ensembles software and hardware acceleration were employed. Every predicting neural network from ensemble could be calculated separately. So, calculations could be done in parallel. MPI wrapper mpi4py [26] were used for this purpose. Cycle of each predicting neural network was divided into equal intervals and every interval were calculated on separate processor node. There are not needs for communication between mpi threads, so obtained equal to one efficiency of parallelism, where efficiency is described as folow [27, 28]:

$$S = \frac{1}{P} \times \frac{T_{seq}}{T(P)} \tag{5}$$

where P is number of processors, $T(P)$ is the runtime of the parallel algorithm, and T_{seq} is the runtime of the sequential algorithm. Hardware acceleration were achieved using six nodes of Intel(R) Xeon(R) CPU E5645 @ 2.40 GHz on the cloud www.time4vps.eu. So calculations of ensemble of 300 predicting neural networks are 6.25 h time long.

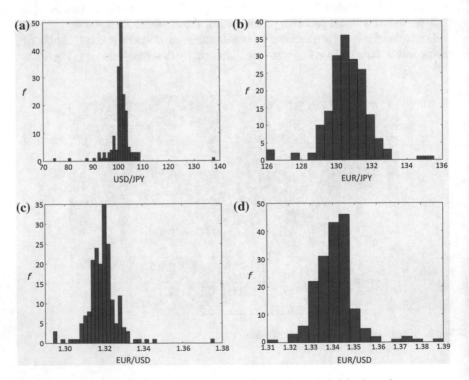

Fig. 3 Examples of distributions: **a** tight; **b** scattered; **c** multimodal; **d** right skewed

Fig. 4 Exess kurtosis versus $skew^2$

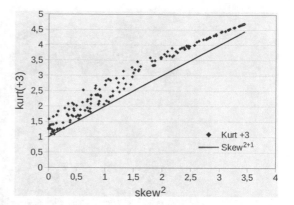

The first two steps—*Data step* and *Input step*—remain the same as in the first model. This is followed by other steps:

Prediction Step. We can choose n neural network forecasting. Neural networks can lead to the number of hours required for a decision. Therefore, it is necessary to select the optimum number of ensemble. When $n > 60$, the forecast assumes the shape of the distribution. At the end of this step, we have a distribution with all parameters of it—mean, median, mode, skewness, kurtosis and et. Decision of trading are making by composed portfolio of exchange rates by analysing the distribution parameters.

The result of prediction is probability distribution, which has form and parameters. In Fig. 3 are shown examples of different distributions: (a) tight—shows clear direction of predicting exchange rate; (b) scattered—shows big riskiness; (c) multimodal—shows some different forces in the financial market; (d) *right* skewed— shows that most probable values are concentrated on the left of the mean.

Distributions are not normal. Shape Characterization Plane [24] was used for testing. Dependency of excess kurtosis from $skewness^2$ was investigated and we got that, when number of NN in our ensemble is from 60–300 its are multimodal distributions (Fig. 4).

It is very important to investigate an accuracy of prediction when new models are testing. The comparison of the performance of the forecasting models was made in terms of the accuracy of the forecasts on the test case domain.

3.3 Comparison of Predictions Accuracy

The test of the accuracy of models on 1–5 steps ahead forecasts was investigated by MAPE. An interval forecast is considered to be correct if the actual value falls in side the predicted 95 % confidence interval. Point estimation accuracy was measured using the Mean Absolute Percentage Error (MAPE) of forecasts:

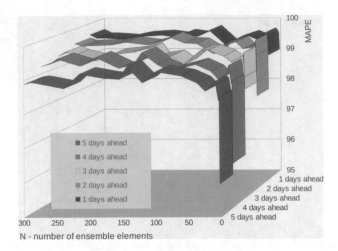

Fig. 5 Dependency forecasting accuracy of the number of RNN EVOLINO: **a** in 1 and 2 days ahead; **b** 3, 4 and 5 days ahead

$$P_{ea} = 100 - \frac{100}{n} \sum \frac{|Y_i - \hat{Y}_i|}{Y_i} \tag{6}$$

where n—number of observations in the test set, Y_i—actual output and $\hat{Y}_i$—forecasted output. Test from 5 observations was made in 20/01/2012—15/03/2012 (Fig. 5). Accuracy of predictions obtained in the interval 94–99,6%. Increase of accuracy depends on number of networks and forecasting becomes more stable. This investigation shows that in some cases more is not always better—with a lot of predictions EVOLINO RNN require more calculating processes time and resources. An interval of number of EVOLINO RNN N [1; 120] has hight accuracy, but is not stable. Distribution of predictions has not form of clear shape and parameters are not informative. An interval [120; 200] is accurate and stable, so it not require too many time and resources. Distribution of predictions is sufficiently informative. An interval of N [200; 300] is good for investigation, but require to many calculation time—the investment decision in finance market so could be too late.

First and second points ahead forecasts are accurate and stable, and 3, 4 and 5 points ahead forecasts stability is reached only when the ensemble consists of over 64 RNN. In time series forecasting, the magnitude of the forecasting error increases over time, since the uncertainty increases with the horizon of the forecast.

And what happens if the N will be much bigger? The ensemble of 1008 Evolino RNN elements require more resources or more time. After 3, 5 days we got such distributions (Fig. 6). The shape of distributions are clear, accuracy is hight but not more, then with 200–300 elements. The application of this distributions of probabilities in the investment portfolio needs further investigation.

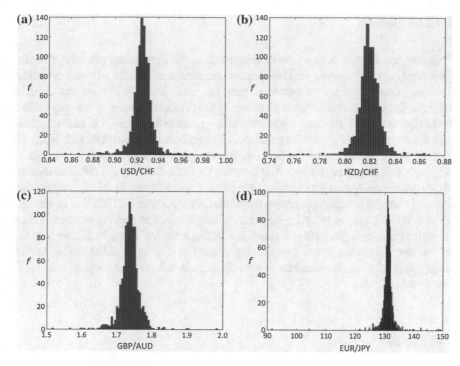

Fig. 6 Distributions, when N = 1008, 02/09/2013

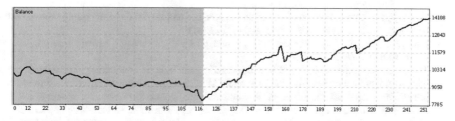

Fig. 7 Trading in demo Forex results: *gray*-single point prediction; *white*-distribution prediction

The best experimental validation of predictive models is testing in a Forex market in real time. Daily trading in exchange market, using EURUSD, USDJPY, GBPUSD and EURJPY exchange rates from 02/01/2013 to 01/03/2013 by single predictions and from 01/03/2012 to 29/10/2013 by distribution forecasting model is shown in Fig. 7. For first input was used historical data of exchange rates and for second input—historical data of gold, silver or oil prices. Number of Evolino RNN in ensemble is equal 176, invested funds are distributed equally to each currency pair.

Investment by using 1008 Evolino RNN or using optimised investment portfolio needs further investigation.

4 Conclusions

Neural network architecture is very important in the forecasting process. The single neural network system provides a point forecast that accuracy is very unstable. Ensemble from eight neural networks provides more accurate forecasting point in the expected range. The decision to invest in the financial markets are always taken under uncertainty. Therefore, distributions are more informative and more reliable than the scatter projections. When number of neural networks exceeds 120, obtained multi-modal distribution of predictions, which opens up opportunities to use it as profitable investment portfolio tool. However, more is not always better. The ensemble for prediction requires more calculating time and resources. Stable and not feather growing prediction accuracy, gotten by increasing the number of RNN in ensemble, when $n > 120$, allows to optimize the investment decision-making process. Those ensembles makes it possible to expect prediction accuracy of up to 5 days into the future. Several attempts was made using ensemble with $n > 1000$ RNN. At this stage application of distributions of probabilities in the investment portfolio needs further investigation.

References

1. Kaastra, I., Boyd, M.: Designing a neural network for forecasting financial and economic time series. Neurocomputing **10**(3), 215–236 (1996)
2. Walczak, S.: An empirical analysis of data requirements for financial forecasting with neural networks. J. Manag. Inf. Syst. **17**(4), 203–222 (2001)
3. Zhou, Z.H., Wu, J., Tang, W.: Ensembling neural networks: many could be better than all. Artif. Intell. **137**(1), 239–263 (2002)
4. Tsakonas, A., Dounias, G.: An architecture-altering and training methodology for neural logic networks: Application in the banking sector. In: Madani, K. (ed.) Proceedings of the 1st International Workshop on Artificial Neural Networks and Intelligent Information Processing, ANNIIP 2005, Barcelona, Spain, NSTICC Press (Sep 2005) 82–93 In conjunction with ICINCO 2005
5. Nguyen, H., Chan, C.: Multiple neural networks for a long term time series forecast. Neural Comput. Appl. **13**(1), 90–98 (2004)
6. Garcia-Pedrajas, N., Hervas-Martinez, C., Ortiz-Boyer, D.: Cooperative coevolution of artificial neural network ensembles for pattern classification. IEEE Trans. Evolut. Comput. **9**(3), 271–302 (June, 2005)
7. Siwek, K., Osowski, S., Szupiluk, R.: Ensemble neural network approach for accurate load forecasting in a power system. Int. J. Appl. Math. Comput. Sci. **19**(2), 303–315 (2009)
8. Uchigaki, S., Uchida, S., Toda, K., Monden, A.: An ensemble approach of simple regression models to cross-project fault prediction. In: 13th ACIS International Conference on Software Engineering, Artificial Intelligence, Networking and Parallel Distributed Computing (SNPD), pp. 476–481, Aug 2012
9. Brezak, D., Bacek, T., Majetic, D., Kasac, J., Novakovic, B.: A comparison of feed-forward and recurrent neural networks in time series forecasting. In: IEEE Conference on Computational Intelligence for Financial Engineering Economics (CIFEr), pp. 1–6, March, 2012
10. Kuo, L.L.R., Lee, C.: Integration of artificial neural networks and fuzzy delphi for stock market forecasting. In: IEEE International Conference on Systems, Man, and Cybernetics, vol. 2, IEEE, pp. 1073–1078 (1996)

11. Chang, P., Wang, Y.: Fuzzy delphi and back-propagation model for sales forecasting in pcb industry. Expert Syst. Appl. **30**(4), 715–726 (2006)
12. Kabir, G., Sumi, R.: Integrating fuzzy delphi method with artificial neural network for demand forecasting of power engineering company. Manag. Sci. Lett. **2**(5) (2012)
13. Rutkauskas, A.V.: Formation of adequate investment portfolio for stochasticity of profit possibilities. Prop. Manag. **4**(2), 100–115 (2000)
14. Rutkauskas, A.V., Miečinskienė, A., Stasytytė, V.: Investment decisions modelling along sustainable development concept on financial markets. Technol. Econ. Dev. Econ. **14**(3), 417–427 (2008)
15. Rutkauskas, A.V., Lapinskaitė-Vvohlfahrt, I.: Marketing finance strategy based on effective risk management. In: The 6th International Scientific Confcrence Business and Managemen, vol. 1, pp. 162–169, Technika (13–14 May 2010)
16. Rutkauskas, A.V., Stasytytė, V.: Optimal portfolio search using efficient surface and three-dimensional utility function. Technol. Econ. Dev. Econ. **17**(2), 305–326 (2011)
17. Rutkauskas, A.V.: Using sustainability engineering to gain universal sustainability efficiency. Sustainability **4**(6), 1135–1153 (2012)
18. Riley, B.: Practical statistical inference for the opinions a.of biased experts blake riley. In: The 50th Annual Meeting of the MVEA, vol. 2., Missouri Valley Economic Association (2012)
19. Hutter, M.: On universal prediction and bayesian confirmation. Theor. Comput. Sci. **384**(1), 33–48 (2007)
20. Schmidhuber, J., Wierstra, D., Gomez, F.: Modeling systems with internal state using evolino. In: Proceedings of the conference on genetic and evolutionary computation (GECCO). Washington, ACM Press, New York, NY, USA, pp. 1795–1802 (2005)
21. Wierstra, D., Gomez, F.J., Schmidhuber, J.: Evolino: Hybrid neuroevolution/optimal linear search for sequence learning. In: Proceedings of the 19th International Joint Conference on Artificial Intelligence (IJCAI), pp. 853–858, Edinburgh (2005)
22. Schmidhuber, J., Wierstra, D., Gagliolo, M., Gomez, F.: Training recurrent networks by evolino. Neural Comput. **19**(3), 757–779 (2007)
23. Maknickienė, N., Maknickas, A.: Financial market prediction system with evolino neural network and delphi method. J. Bus. Econ. Manag. **14**(2), 403–413 (2013)
24. Wheeler, D.: Problems with skewness and kurtosis (2013)
25. Maknickas, A., Maknickienė, N.: Influence of data orthogonality to accuracy and stability of financial market predictions. In: 4th International Conference on Neural Computation Theory and Applications (NCTA 2012), pp. 616–619, Barselaona, Spain, October (2012)
26. Dalcin, L.: https://code.google.com/p/mpi4py/ (2012/12/01)
27. Fox, G., Johnson, M., Lyzenga, G., Otto, S., Salmon, J., Walker, D.: Solving problems on concurrent processors **1**, 1373849 (1988)
28. Kumar, V., Gmma, A., Anshul, G.: Introduction to parallel computing: Design and analysis of algorithms (1994)

Gene Ontology Analysis on Behalf of Improved Classification of Different Colorectal Cancer Stages

Monika Simjanoska, Ana Madevska Bogdanova and Sasho Panov

Abstract The colorectal cancer is a serious cause of death worldwide. Diagnosing the current colorectal cancer stage is crucial for early prognosis and adequate treatment of the patients. Even though the scientists have developed various techniques, determining the real colorectal cancer stage is still critical. In this paper we utilize Gene Ontology analysis information to address this issue. We compose a set of special genes that are used to obtain two main results—we show the distinction between the carcinogenic and healthy tissue by difference in the range of their DNA gene expressions, and we propose a novel methodology that improves the colorectal cancer stages classification.

Keywords Gene ontology · Colorectal cancer stages · Gene expression · Bayes' theorem

1 Introduction

In 2008, the World Health Organization (WHO) conducted a research on the cancer's incidence, mortality and prevalence. The results showed that the colorectal cancer (CRC) deserves serious attention since it causes approximately 608,000, which is 8 % of total cancer deaths [1]. The incidence and prevalence results showed that 60 % of the 1,234,000 new cases occur in the developed regions, from which 663,000 at man and 571,000 at women.

M. Simjanoska (✉)
Computer Science and Engineering, Ss. Cyril and Methodius University, Skopje, Macedonia
e-mail: m.simjanoska@gmail.com

A.M. Bogdanova
Natural Sciences and Mathematics, Ss. Cyril and Methodius University, Skopje, Macedonia
e-mail: ana.madevska.bogdanova@finki.ukim.mk

S. Panov
Institute of Biology, Skopje, Macedonia
e-mail: sasho@mt.net.mk

© Springer International Publishing Switzerland 2016
K. Madani et al. (eds.), *Computational Intelligence*,
Studies in Computational Intelligence 613,
DOI 10.1007/978-3-319-23392-5_27

Recently, the scientists provide intensive gene expression profiling experiments in order to compare the malignant to the healthy cells in a particular tissue. The advantage of the microarray technologies enables simultaneous observation of thousands of genes and allows the researchers to derive conclusions whether the disorder is a result of the abnormal expression of a subset of genes.

In our previous work we used gene expression experiments from Affymetrix Human Genome U133 Plus 2.0 Array to perform analysis of colorectal carcinogenic and healthy tissues [2]. During the research we developed methodology for biomarkers detection based on the two types of tissues, carcinogenic and healthy. The obtained set of biomarkers was then used to build a machine learning based classifier capable of distinguishing between carcinogenic and healthy patients. Since the classification analysis resulted in very high accuracy when classifying both CRC and healthy patients, we proceeded to inspect whether the biomarkers we discovered play important biological role in the colorectal cancer development [3]. For that purpose, we provided gene ontology (GO) analysis and inspected the molecular functions and the biological processes of a particular set of genes that showed to be overrepresented among all biomarkers. Considering the colorectal cancer significance of the biomarker genes, we confirmed few biomarkers to be tightly related to the disease: $CHGA$, $GUCA2B$, $MMP7$, $CDH3$ and PYY.

Consequently, since gene expression profiling by microarrays is expected to advance the progress of personalized cancer treatment based on the molecular classification of subtypes [4], we used the same set of biomarkers to model the different CRC stages (I–IV) [5]. The modelling resulted in an accurate Bayesian classifier that showed satisfying results when diagnosing tissues in the critical stages, I and IV, and, II and III, which, as presented in Sect. 2, are often found to be problematic for prognosis.

Even though, we exceeded the problems of distinguishing between CRC stage I and IV, and, II and III, that remained common problem in the literature, we decided to go deeper in the problem in order to improve our classification results. In this paper we conduct a research that follows two threads of our previous work, the GO analysis of the biomarkers [3] and the classification of the different CRC stages [5]. In this research we preform GO analysis for each of the different CRC stages probed with the same Affymetrix platform. Our aim is to compare the stages that are critical for diagnosing and also the neighbouring stages, in order to derive conclusions on their common biological and molecular functions (enriched genes). Obtaining the enriched genes involved in the common GO functions and inspecting their range of DNA expression is very important for determining the distinguishing functions between the CRC stages. Once we discovered the enriched genes, we were able to remodel the prior probabilities of the different CRC stages and we got significantly improved classification results.

The rest of the paper is organized as follows. In Sect. 2 we briefly present the latest work related to our point of interest of this paper. In Sect. 3 we describe the methods for biomarkers selection and GO analysis. The results from the analysis are presented in Sect. 4 and eventually, we derive our conclusions and present our plans for future work in the final Sect. 5.

2 Related Work

In this section we present a work related to CRC stages analysis and GO appliance in the research of various diseases.

Recently, the classification of different CRC stages has been in the focus of many researches. Even though, the authors developed many procedures for diagnosis and survival prediction [6, 7], the analysis showed that an accurate classification of intermediate-stage cases, II and III, as well as stage I and stage IV, is problematic [8, 9].

The microarray data used in this paper, has also been used for distinguishing patterns in different CRC stages.

Laibe et al. [10] profiled both stage II and stage III carcinomas. They realized that expression profile of stage II colon carcinomas distinguishes two patterns, one pattern very similar to that of stage III tumors, based on a 7-gene signature. The function of the discriminating genes suggests that tumors have been classified according to their putative response to adjuvant targeted or classic therapies. Tsukamoto et al. [11] performed gene expression profiling and found that the over expression of OPG gene may be a predictive biomarker of CRC recurrence and a target for treatment of this disease. Hong et al. [12] aimed to find a metastasis-prone signature for early stage mismatch-repair proficient sporadic CRC patients for better prognosis. Their best classification model yielded a 54 gene-set with an estimated prediction accuracy of 71 %. Another problem of limited discrimination for Dukes stage B and C disease is presented by Jorissen et al. [13]. They conclude that metastasis-associated gene expression changes can be used to refine traditional outcome prediction, providing a rational approach for tailoring treatments to subsets of patients. Finally, three of the five microarray data sets used in this paper, have also been used by Schlicker et al. [14]. They model the heterogeneity of CRC by defining subtypes of patients with homogeneous biological and clinical characteristics and match these subtypes to cell lines for which extensive pharmacological data is available, thus linking targeted therapies to patients most likely to respond to treatment.

Regarding ontology and classification analysis related to colorectal cancer, authors in [15] sum up the biomarkers results from 23 different researches. Even though most of them show diversity in the significant genes revealed, the authors in their research take into account the unique biomarkers, which are nearly 1000, and perform ontology analysis using various tools. Similarly, in [16] the researchers use Affymetrix microarray data from 20 patients to reveal significant gene expression, which resulted in 1469 biomarkers. From the ontology analysis they ranked top 10 most important pathways. Since the non overlapping between the biomarkers sets discovered in different scientific papers is very common, a new meta-analysis model of colorectal cancer gene expression profiling studies is proposed in [17]. As the authors ranked the biomarker genes according to various parameters, the gene CDH3 which we found to play role in the colorectal cancer [3], is also found by their meta-analysis model. Another interesting approach maintained with classification analysis is presented in [18], where the authors constructed disease-specific gene networks and used them to identify significantly expressed genes. A particular attention is given to five biomarkers, from which one of them, IL8, was also detected by our methodology [3], but it was not considered important in our research since no specific connection to the colorectal cancer was found in the literature.

3 Methods and Methodology

In this section we define the methodology that we developed to detect the genes whose expression is statistically and biologically markable among the different CRC stages and the healthy tissues. We also present the GO procedure that we used to obtain the genes involved in the common biological and molecular functions of the all four CRC stages. Eventually, we present the modified classification procedure that was used for obtaining the new improved results.

3.1 The CRC Stages

Colorectal stages systems are designed to enable physicians to stratify patients in terms of expected predicted survival, to help select the most effective treatments, to determine prognoses, and to evaluate cancer control measures [19]. All data is organized into four CRC stages [20]:

1. *Stage I*—In this stage cancer has grown through the superficial lining, i.e., mucosa of the colon or rectum, but has not spread beyond the colon wall or rectum.
2. *Stage II*—In this stage cancer has grown into or through the wall of the colon or rectum, but has not spread to nearby lymph nodes.
3. *Stage III*—In this stage cancer has invaded nearby lymph nodes, but is not affecting other parts of the body yet.
4. *Stage IV*—In this stage cancer has spread to distant organs.

3.2 Choosing the Biomarkers

In the process of CRC stages biomarkers selection, instead of using the whole genome data, we use the same set of biomarkers which ability to distinguish carcinogenic and healthy patients is previously confirmed by classification and GO analysis [2, 3].

Once we obtained the initial set of biomarkers, B, we repeated the procedure for biomarkers selection for each stage S_i, where i is the current CRC stage, versus *Healthy* tissues, in order to produce subsets of biomarkers, B_i. The process for revealing the biomarkers consists of the following steps:

1. *Quantile normalization.* Since our aim is to unveil the difference in gene expression levels between the carcinogenic and healthy tissues, we proposed the Quantile normalization (QN) as a suitable normalization method [21].
2. *Low entropy filter.* We used low entropy filter to remove the genes with almost ordered expression levels [22], since they lead to wrong conclusions about the genes behaviour.
3. *Paired-sample t-test.* Knowing the facts that both carcinogenic and healthy tissues are taken from the same patients, and that the whole-genome gene expression follows normal distribution [23], we used a paired-sample t-test.
4. *FDR method.* False Discovery Rate (FDR) is a reduction method that usually follows the t-test. FDR solves the problem of false positives, i.e., the genes which are considered statistically significant when in reality there is not any difference in their expression levels.
5. *Volcano plot.* Both the t-test and the FDR method identify different expressions in accordance with statistical significance values, and do not consider biological significance. In order to display both statistically and biologically significant genes we used volcano plot visual tool.

3.3 Gene Ontology Analysis

The analyses of single markers have been in the focus of the genome-wide association studies. However, it often lacks the power to uncover the relatively small effect sizes conferred by most genetic variants. Therefore, using prior biological knowledge on gene function, pathway-based approaches have been developed with the aim to examine whether a group of related genes in the same functional pathway are jointly associated with a trait of interest [24].

The goal of the Gene Ontology Consortium is to produce a dynamic, controlled vocabulary that can be applied to all eukaryotes even as knowledge of gene and protein roles in cells is accumulating and changing [25]. The GO project provides ontologies to describe attributes of gene products in three non-overlapping domains of molecular biology [26]:

1. Molecular Function describes activities, such as catalytic or binding activities, at the molecular level. GO molecular function terms represent activities rather than the entities that perform the actions, and do not specify where, when or in what context the action takes place.
2. Biological Process describes biological goals accomplished by one or more ordered assemblies of molecular functions.
3. Cellular Component describes locations, at the levels of subcellular structures and macromolecular complexes.

There are many tools based on Gene Ontology resource; however, in this research we use the freely accessible Gene Ontology Enrichment Analysis Software Toolkit, GOEAST. It is a web based tool which applies appropriate statistical methods to identify significantly enriched GO terms among a given list of genes. Beside the other functions, GOEAST supports analysis of probe set IDs from Affymetrix microarrays. It provides graphical outputs of enriched GO terms to demonstrate their relationships in the three ontology categories. In order to compare GO enrichment status of multiple experiments, GOEAST supports cross comparisons to identify the correlations and differences among them [27].

In this paper we define few test cases to compare the ontologies of the critical and the neighbouring stages:

1. *Test case 1*—Compare Stage I and Stage II
2. *Test case 2*—Compare Stage II and Stage III
3. *Test case 3*—Compare Stage III and Stage IV
4. *Test case 4*—Compare Stage I and Stage IV

3.4 Remodelling the Prior Distributions

Previously revealed biomarkers showed high precision while diagnosing both carcinogenic and healthy patients [2]. In order to produce improved CRC stage classification, we used the developed procedure [5], and introduced a powerful key subprocedure that enables reshaping the probability distributions of the training and test set:

1. *Round-up threshold method*
2. *Normalization*
3. *Smoothing method*
4. **Boosting the enriched biomarkers**: as we have analysed the common biological and molecular functions of all four CRC stages from the GO analysis, we introduce an additional method which as presented in Sect. 4, produced an improved prior distributions modelling of the CRC stages. We chose special biomarker genes that play role in the common biological functions among the CRC stages. In order to increase the importance of the special genes, we multiplied the set in ratio 3 : 1, so that the new set is now a leading factor in the distributions shape.
5. *Hypothesis testing*.

3.5 Multiclass Bayesian Classification

As we remodelled the prior distributions of all four CRC stages, we are now able to use them in the Bayes' theorem and to calculate the posterior probability for each patient to belong to each of the four classes. Given the prior distributions we can calculate the class conditional densities, $p(x|C_i)$, as the product of the continuous probability distributions of each gene from x distinctively:

$$p(x|C_i) = \prod f_1 f_2 \dots f_n \tag{1}$$

Since we have unequal number of patients in all four classes, considering the total number of 657 tissues, we defined the prior probabilities $P(C_i)$, to be $P(C_1) = 0.2085$, $P(C_2) = 0.3912$, $P(C_3) = 0.2770$ and $P(C_4) = 0.1233$. Therefore, we calculate the posterior probability $P(C_i|x)$, as:

$$p(C_i|x) = \frac{p(x|C_i) * P(C_i)}{\sum_1^4 p(x|C_i) * P(C_i)} \qquad (2)$$

The tissue x is classified according to the rule of maximizing the a posteriori probability (MAP):

$$C_i = \max p(C_i|x) \qquad (3)$$

4 Experiments and Results

4.1 Gene Expression Data

In order to unveil the biomarker genes in the initial biomarkers set B, discussed in Sect. 3.2, we used the microarray experiment retrieved from Gene Expression Omnibus database [28] with GEO accession ID $GSE8671$, where 32 carcinogenic and 32 adjacent normal tissues were probed with the $Affymetrix\ Human\ Genome$ $U133\ Plus\ 2.0\ Array$.

The microarray experiments used for CRC stages biomarkers detection are retrieved by using the following GEO accession IDs: $GSE37892$, $GSE21510$, $GSE9348$, $GSE14333$ and $GSE35896$. The experiments have been performed using the same $Affymetrix$ platform. All data is organized into four CRC stages:

- *Stage I* contains gene expression from 137 patients.
- *Stage II* contains gene expression from 257 patients.
- *Stage III* contains gene expression from 182 patients.
- *Stage IV* contains gene expression from 81 patients.

4.2 Gene Ontology Results

According to the methodology we defined in Sect. 3.2, for each CRC stage we created new subset of biomarkers, B_i, for each stage $i = 1, .., 4$:

- $B_1 = 70$
- $B_2 = 72$
- $B_3 = 73$
- $B_4 = 66$

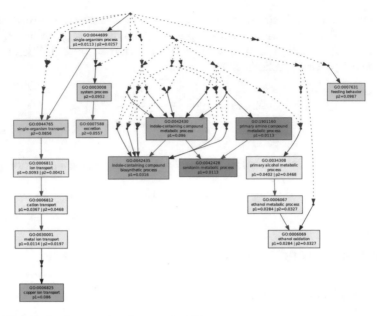

Fig. 1 Biological processes: stage I versus stage II

Thus, instead of analysis all the 138 initial biomarkers as we did in our previous work [3], we analyse the subsets of biomarkers that represent the current CRC stage.

For each subset B_i we performed GO-stage analysis using the GOEAST online tool previously discussed in Sect. 3.3. In order to compare multiple GO results, we used the Multi-GOEAST tool and produced three types of ontologies to describe: Biological processes, Molecular function and Cellular component. The different colour saturation degrees in the graphs present the enrichment significance of each GO term, defined by the p-value. In the graphical output of Multi-GOEAST results, each set is represented with different colour. Therefore, red and green boxes represent enriched GO terms only found in one of the biomarkers set, whereas yellow boxes represent commonly enriched GO terms in both experiments.

Since all ontologies refer to the same problem, in this paper we present only the *Biological processes* view.

Figure 1 presents the comparison of the ontology analysis between the neighbouring stage I and stage II. As we can see, those stages have 7 biological processes in common. Considering the critical stages II and III, Fig. 2 depicts their common biological processes, most of them overlapping with the common processes between stage I and stage II. Figure 3 presents the common biological processes in the neighbouring stages III and IV. They have nearly the same common biological processes as in the previous test cases.

Finally, we compared the critical stages I and IV (Fig. 4). As a result of these comparisons, we choose the following processes for further analysis, since they are common in the all four cases:

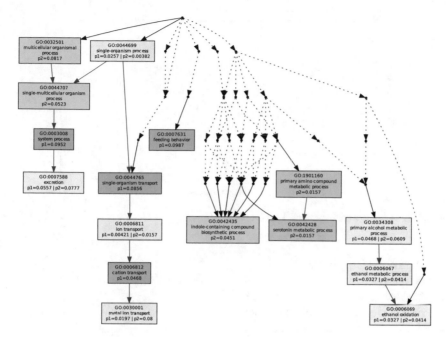

Fig. 2 Biological processes: stage II versus stage III

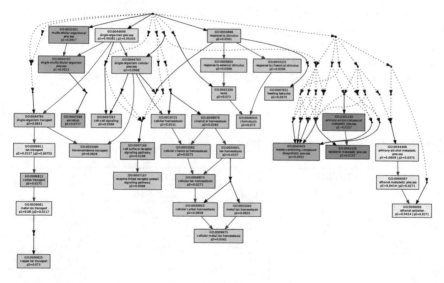

Fig. 3 Biological processes: stage III versus stage IV

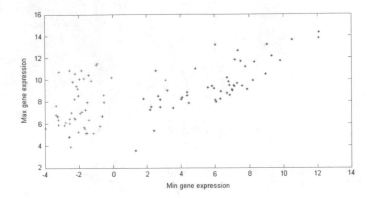

Fig. 4 Biological processes: stage I versus stage IV

- Single organism process;
- Ion transport;
- Metal ion transport;
- Primary alcohol metabolic process;
- Ethanol metabolic process and
- Ethanol oxidation.

Considering the common biological processes, we extracted 50 gene probes, which we refer to as *special genes*, that are directly involved in these processes. In Table 1 we present the gene symbols that are equal to the gene probes. An important fact is that four ($CHGA$, $GUCA2B$, $CDH3$ and PYY) of the five biomarkers we found to be highly correlated with the CRC phenomena, are found in the common biological processes.

Table 1 Biomarkers from common biological processes

Biomarker genes				
UNC5C	TRPM6	TPH1	SST	SLC6A19
SLC30A10	SLC26A3	SLC25A34	SCNN1B	SCN9A
SCN7A	RSPO2	PYY	PRKAA2	PLP1
NRXN1	NEUROD1	LGI1	INSM1	INSL5
GUCA2B	GREM2	GCNT2	GCG	FCRLA
FAM5C	CXCL13	CP	CLDN8	CHGB
CHGA	CHAD	CDH3	CDH19	CCL23
BEST4	BCHE	ASCL2	ANGPTL1	AFF3
ADH1B				

4.3 Classification Results

As we finished the GO-stage analysis and obtained the enriched set of genes (the special genes), we inspected the ranges of the gene expression of the special genes at both carcinogenic and healthy tissues from the data in our disposition. Figure 5 presents the ranges of expression at test patients which were not involved in the biomarkers selection process and we can conclude that the special genes clearly distinguish the carcinogenic and the healthy patients. Following this first result, we proceeded with further experiments in order to improve the classification results in CRC stages classification.

Hereupon, we applied the methodology in Sect. 3.4 to remodel the gene expression distributions of each CRC stage. A key point in remodelling the probability distributions was to boost the enriched set of genes **3** times, thus the special genes are now a leading factor in the distributions shape. Using the boosting method, we additionally avoid the overlap between the probability distributions of the critical CRC stages. Therefore, as a results we got a set of 238 biomarkers which produced the distributions of the training sets depicted in Fig. 6.

As we remodelled the probability distributions, we used them in the Multiclass Bayesian classifier developed in Sect. 3.5 and achieved the improvements presented in Table 2. The improvement of recognition in the first three stages is significant, so we can decide to use this procedure, even the fourth-stage classification has decreased for few points. The *Old results* refer to the results published in [5].

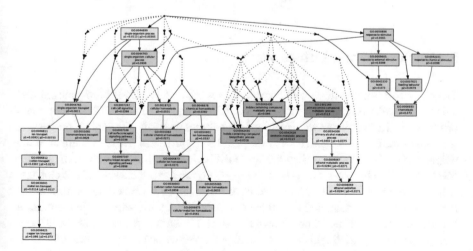

Fig. 5 Gene expression ranges of the special genes in carcinogenic and healthy testing tissues

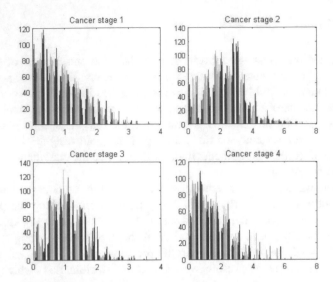

Fig. 6 Probability distributions of the training sets

Table 2 Classification results

Class	Old results (%)	New results (%)
Stage I	73.72	80.29
Stage II	53.69	63.42
Stage III	72.52	92.30
Stage IV	64.19	58.02

5 Conclusions and Future Work

The paper follows two threads of our previous work on CRC DNA chip gene expressions: the developed classificator of the four CRC stages [5] and GO CRC biomarkers analysis [3]. In this paper we developed GO-stages analysis (Biological processes) for each of the cancer stages (I–IV) using the 138 biomarkers. We have compared the neighbouring GO-stages as they are more difficult to distinguish one from another $(GO_1-GO_2, GO_2-GO_3, GO_3-GO_4, GO_1-GO_4)$ and extracted several conclusions from this analysis. In this process we have differentiated 6 functions, common for the all 4 stages. We deducted the 50 gene probes directly included in the 6 common functions. The experiments has shown that the DNA chip expressions of the 50 special genes, clearly distinguish the carcinogenic versus healthy patients. Following this important result, we upgraded the existing methodology for preprocessing the multiclass Bayesian classification by adding the boosting method which multiplied the existing enriched biomarkers. As a result we achieved new probability distributions applicable for Bayesian modelling. The novel preprocessing procedure was confirmed by the experiments.

This result of distinguishing the genes responsible for the common functions for all 4 CRC stages, is very important for the further analysis of the colorectal cancer stages. In the future work, GO stages will help us to determine the distinguishing functions between the CRC stages, that will enable us to further improve the stages classification process and pinpoint the biomarkers that are responsible for the different CRC stages.

References

1. GLOBOCAN: (2008)
2. Simjanoska, M., Bogdanova, A.M., Popeska, Z.: Bayesian posterior probability classification of colorectal cancer probed with Affymetrix microarray technology. In: Information & Communication Technology Electronics & Microelectronics (MIPRO), 2013 36th International Convention on, IEEE, pp. 959–964 (2013)
3. Simjanoska, M., Bogdanova, A.M., Panov, S.: Gene ontology analysis of colorectal cancer biomarkers probed with Affymetrix and Illumina microarrays. In: Proceedings of the 5th International Joint Conference on Computational Intelligence, IJCCI, 2013, IJCCI, pp. 396–406 (2013)
4. Jain, K.: Applications of biochips: from diagnostics to personalized medicine. Curr. Opin. Drug Discov. Devel. 7(3), 285–289 (2004)
5. Simjanoska, M., Bogdanova, A.M., Popeska, Z.: Bayesian multiclass classification of gene expression colorectal cancer stages. In: ICT Innovations 2013, pp. 177–186. Springer (2014)
6. Ahmed, F.E.: Artificial neural networks for diagnosis and survival prediction in colon cancer. Molecular cancer 4(1), 29 (2005)
7. Frederiksen, C.M., Knudsen, S., Laurberg, S., Ørntoft, T.F.: Classification of dukes' b and c colorectal cancers using expression arrays. J. Cancer Res. Clin. Oncol. 129(5), 263–271 (2003)
8. Eschrich, S., Yang, I., Bloom, G., Kwong, K.Y., Boulware, D., Cantor, A., Coppola, D., Kruhøffer, M., Aaltonen, L., Orntoft, T. F., et al.: Molecular staging for survival prediction of colorectal cancer patients. J. Clin. Oncol. 23(15), 3526–3535 (2005)
9. Salazar, R., Roepman, P., Capella, G., Moreno, V., Simon, I., Dreezen, C., Lopez- Doriga, A., Santos, C., Marijnen, C., Westerga, J., et al.: Gene expression signature to improve prognosis prediction of stage ii and iii colorectal cancer. J. Clin. Oncol. 29(1), 17–24 (2011)
10. Laibe, S., Lagarde, A., Ferrari, A., Monges, G., Birnbaum, D., Olschwang, S.: The COL2 Project: A seven-gene signature aggregates a subgroup of stage ii colon cancers with stage iii. OMICS: J. Integr. Biol. 16(10), 560–565 (2012)
11. Tsukamoto, S., Ishikawa, T., Iida, S., Ishiguro, M., Mogushi, K., Mizushima, H., Uetake, H., Tanaka, H., Sugihara, K.: Clinical significance of osteoprotegerin expression in human colorectal cancer. Clin. Cancer Res. 17(8), 2444–2450 (2011)
12. Hong, Y., Downey, T., Eu, K.W., Koh, P.K., Cheah, P.Y.: A metastasis-prone signature for early-stage mismatch-repair proficient sporadic colorectal cancer patients and its implications for possible therapeutics. Clin. Exp. Metastasis 27(2), 83–90 (2010)
13. Jorissen, R.N., Gibbs, P., Christie, M., Prakash, S., Lipton, L., Desai, J., Kerr, D., Aaltonen, L.A., Arango, D., Kruhøffer, M., et al.: Metastasis-associated gene expression changes predict poor outcomes in patients with dukes stage b and c colorectal cancer. Clin. Cancer Res. 15(24), 7642–7651 (2009)
14. Schlicker, A., Beran, G., Chresta, C.M., McWalter, G., Pritchard, A., Weston, S., Runswick, S., Davenport, S., Heathcote, K., Castro, D.A., et al.: Subtypes of primary colorectal tumors correlate with response to targeted treatment in colorectal cell lines. BMC Med. Genomics 5(1), 66 (2012)

15. Lascorz, J., Chen, B., Hemminki, K., Försti, A.: Consensus pathways implicated in prognosis of colorectal cancer identified through systematic enrichment analysis of gene expression profiling studies. PLoS ONE **6**(4), e18867 (2011)
16. Xu, Y., Xu, Q., Yang, L., Liu, F., Ye, X., Wu, F., Ni, S., Tan, C., Cai, G., Meng, X., et al.: Gene expression analysis of peripheral blood cells reveals toll-like receptor pathway deregulation in colorectal cancer. PLoS ONE **8**(5), e62870 (2013)
17. Chan, S.K., Griffth, O.L., Tai, I.T., Jones, S.J.: Meta-analysis of colorectal cancer gene expression profiling studies identifies consistently reported candidate biomarkers. Cancer Epidemiol. Biomarkers Prev. **17**(3), 543–552 (2008)
18. Jiang, W., Li, X., Rao, S., Wang, L., Du, L., Li, C., Wu, C., Wang, H., Wang, Y., Yang, B.: Constructing disease-specific gene networks using pair-wise relevance metric: application to colon cancer identifies interleukin 8, desmin and enolase 1 as the central elements. BMC Syst. Biol. **2**(1), 72 (2008)
19. O'Connell, J.B., Maggard, M.A., Ko, C.Y.: Colon cancer survival rates with the new American joint committee on cancer sixth edition staging. J. Natl. Cancer Inst. **96**(19), 1420–1425 (2004)
20. MayoClinic: Colon cancer (2013)
21. Wu, Z., Aryee, M.: Subset quantile normalization using negative control features. J. Comput. Biol. **17**(10), 1385–1395 (2010)
22. Needham, C., Manfield, I., Bulpitt, A., Gilmartin, P., Westhead, D.: From gene expression to gene regulatory networks in arabidopsis thaliana. BMC Syst. Biol. **3**(1), 85 (2009)
23. Hui, Y., Kang, T., Xie, L., Yuan-Yuan, L.: Digout: viewing differential expression genes as outliers. J. Bioinf. Comput. Biol. **8**(supp01), 161–175 (2010)
24. Wang, K., Li, M., Hakonarson, H.: Analysing biological pathways in genome-wide association studies. Nat. Rev. Genetics **11**(12), 843–854 (2010)
25. Ashburner, M., Ball, C.A., Blake, J.A., Botstein, D., Butler, H., Cherry, J.M., Davis, A.P., Dolinski, K., Dwight, S.S., Eppig, J.T., et al.: Gene ontology: tool for the unification of biology. Nat. genetics **25**(1), 25 (2000)
26. Harris, M., Clark, J., Ireland, A., Lomax, J., Ashburner, M., Foulger, R., Eilbeck, K., Lewis, S., Marshall, B., Mungall, C., et al.: The gene ontology (go) database and informatics resource. Nucleic acids research 32(Database issue) (2004) D258
27. Zheng, Q., Wang, X.J.: Goeast: a web-based software toolkit for gene ontology enrichment analysis. Nucleic acids research **36**(suppl 2), W358–W363 (2008)
28. Gene Expression Omnibus: (2013)

Artificial Curiosity Emerging Human-Like Behavior: Toward Fully Autonomous Cognitive Robots

Kurosh Madani, Christophe Sabourin and Dominik M. Ramík

Abstract This chapter is devoted to autonomous cognitive machines by mean of the design of an artificial curiosity based cognitive system for autonomous high-level knowledge acquisition from visual information. Playing a chief role as well in visual attention as in interactive high-level knowledge construction, the artificial curiosity is realized through combining visual saliency detection and Machine-Learning based approaches. Experimental results validating the deployment of the investigated system have been obtained using as well simulation facilities as a real humanoid robot acquiring visually knowledge about its surrounding environment interacting with a human tutor. As show the reported results and experiments, the proposed cognitive system allows the machine to discover autonomously the surrounding world in which it may evolve, to learn new knowledge about it and to describe it using human-like natural utterances.

1 Introduction and Problem Stating

If nowadays machines and robotic bodies are fully automated outperforming human capacities, nonetheless, none of them can be called truly intelligent or pretend defeating human's cognitive skills. The fact that human-like machine-cognition is still beyond the reach of contemporary science only proves how difficult the problem is. Somewhat, it is due to the fact that the science is still far from fully understanding the human cognitive system. On the other hand, it is so because if contemporary machines are often fully automatic, they linger rarely fully autonomous in their

K. Madani (✉) · C. Sabourin · D.M. Ramík
Images, Signals and Intelligence Systems Laboratory (LISSI/EA 3956),
University PARIS-EST Creteil (UPEC), Senart-FB Institute of Technology,
Bât.A, Av. Pierre Point, 77127 Lieusaint, France
e-mail: madani@univ-paris12.fr

C. Sabourin
e-mail: sabourin@univ-paris12.fr

D.M. Ramík
e-mail: dominik.ramik@univ-paris12.fr

© Springer International Publishing Switzerland 2016
K. Madani et al. (eds.), *Computational Intelligence*,
Studies in Computational Intelligence 613,
DOI 10.1007/978-3-319-23392-5_28

501

knowledge acquisition. Nevertheless, the concepts of bio-inspired or human-like machine-cognition remain foremost sources of inspiration for achieving intelligent systems (intelligent machines, intelligent robots, etc…).

Emergence of cognitive phenomena in machines has been and remains active part of research efforts since the rise of Artificial Intelligence (AI) in the middle of the last century. Among others, [1] provides a survey on cognitive systems. It accounts on different paradigms of cognition in artificial agents markedly on the contrast of emergent versus cognitivist paradigms and on their hybrid combinations. It is also worth of mentioning the work of [2] which brings an in-depth review on a number of existing cognitive architectures such those which adheres to the symbolic theory and reposes on the assumption that human knowledge can be divided to two kinds: declarative and procedural. Another discussed architecture belongs to class of those using "If-Then" deductive rules dividing knowledge again on two kinds: concepts and skills. In contrast to above-mentioned works, the work of [3] focuses the area of research on cognition and cognitive robots discussing purposes linking knowledge representation, sensing and reasoning in cognitive robots. However, there is no cognition without perception (a cognitive system without the capacity to perceive would miss the link to the real world and so it would be impaired) and thus autonomous acquisition of knowledge from perception is a problem that should not be skipped when dealing with cognitive systems.

Prominently to the machine-cognition's issue is the question: "what is the compel or the motivation for a cognitive system to acquire new knowledge?" For human cognitive system Berlyne states, that it is the curiosity that is the motor of seeking for new knowledge [4]. Consequently a few works have been since there dedicated to incorporation of curiosity into a number of artificial systems including embodied agents or robots. However the number of works using some kind of curiosity motivated knowledge acquisition with implementation to real agents (robots) is still relatively small. Often authors view curiosity only as an auxiliary mechanism in robot's exploration behavior. One of early implementations of artificial curiosity may be found in [5]. According to the author, the introduction of curiosity further helps the system to actively seek similar situations in order to learn more. On the field of developmental and cognitive robotics a similar approach may be found in [6] where authors present an approach including a mechanism called "Intelligent Adaptive Curiosity". Two experiments with AIBO robot are presented showing that the curiosity mechanism successfully stimulates the learning progress. In a recent publication, authors of [7] implement the psychological notion of surprise-curiosity into the decision making process of an agent exploring an unknown environment. Authors conclude that the surprise-curiosity driven strategy outperformed classical exploration strategy regarding the time-energy consumed in exploring the delved environment. On the other hand, the concept of surprise, relating closely the notion of curiosity, has been exploited in [8] by a robot using the surprise in order to discover new objects and acquire their visual representations. Finally, the concept of curiosity has been successfully used in [9] for learning affordances of a mobile robot in navigation task. The mentioned works are attempting to respond the question: "how an autonomous cognitive system should be designed in order to exhibit the behavior and functionality close to its human users".

That is why even though *curiosity killed a cat*,[1] taking into consideration the enticing benefits of curiosity, we have made it our principle foundation in investigated concept. The present paper is devoted to the description of a cognitive system based on artificial curiosity for high-level human-like knowledge acquisition from visual information. The goal of the investigated system is to allow the machine (such as a humanoid robot) to observe, to learn and to interpret the world in which it evolves, using appropriate terms from human language, while not making use of a priori knowledge. This is done by word-meaning anchoring based on learning by observation stimulated (steered) by artificial curiosity and by interaction with the human. Our model is closely inspired by juvenile learning behavior of human infants [10, 11].

In Sect. 2, we detail our approach by outlining its architecture and principles. We explain how the machine generates its beliefs about the world from observing the surrounding environment and the role of human-robot interaction in the learning process. Section 3 focuses the validation of the proposed approach using as well simulation facilities as a real robot evolving in real environment. Finally Sect. 4 discusses the achieved results and outlines the future work.

2 Brief Overview of Multi-level Cognitive Concept

Accordingly to Berlyne's theory of human curiosity [4], two cognitive levels contribute to human's desire of acquiring new knowledge. The first is so-called "perceptual curiosity", which leads to increased perception of stimuli. It is a lower level cognitive function, more related to perception of new, surprising or unusual sensory input. It contrasts to repetitive or monotonous perceptual experience. The other one is called "epistemic curiosity", which is more related to the "desire for knowledge that motivates individuals to learn new ideas, eliminate information-gaps, and solve intellectual problems" [12]. It also seems that it acts to stimulate long-term memory in remembering new or surprising (e.g. what may be contrasting with already learned) information [13]. By observing the state of the art (including the referenced ones), it may be concluded that the curiosity is usually used as an auxiliary mechanism instead of being the fundamental basis of the knowledge acquisition. To our best knowledge there is no work to date which considers curiosity in context of machine cognition as a drive for knowledge acquisition on both low (perceptual) level and high ("semantic") level of the system. Without striving for biological plausibility whilst by analogy with natural curiosity, we founded our system on two cognitive levels ([14, 15]). Depicted in Fig. 1, the first ahead of reflexive visual attention plays the role

---

[1]In 'Different', Eugene O'Neill, 1920: BENNY—(with a wink): *"Curiosity killed a cat! Ask me no questions and I'll tell you no lies."*

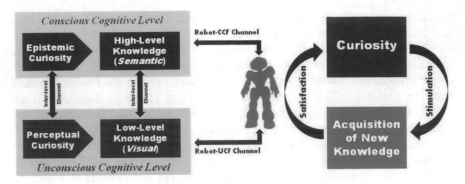

Fig. 1 General bloc-diagram of the proposed curiosity driven architecture (*left*) and principle of curiosity-based Stimulation-Satisfaction mechanism for knowledge acquisition (*right*)

of perceptual curiosity and the second coping with intentional learning-by-interaction undertakes the role of epistemic curiosity.

2.1 From Observation to Interpretation

The problem of autonomous learning conveys the inbuilt problem of distinguishing the pertinent sensory information from the impertinent one. The solution to this task is natural for human, it remain very far from being obvious for a robot. In fact, when a human points to one object among many others giving a description of that pointed object using his human natural language, the robot still has to distinguish, which of the detected features and perceived characteristics of the object the human is referring to. To achieve correct anchoring, the proposed architecture adopts the following strategy. By using its perceptual curiosity, realized thanks to artificial salient vision and adaptive visual attention (described in [16–18]), the robot extracts features from important objects found in the scene along with the words the human used to describe the objects. Then, the robot generates its beliefs about which words could describe which features. Using the generated beliefs as organisms in a genetic algorithm, the robot determines its "most coherent belief". To calculate the fitness, a classifier is trained and used to interpret the objects the robot has already seen. The utterances pronounced by the human for each object are compared with those the robot would use to describe it based on its current belief. The closer the robot's description is to that given by the human, the higher the fitness is. Once the evolution has been finished, the belief with the highest fitness is adopted by the robot and is used to interpret occurrences of new (unseen) objects. Figure 2 depicts through an example important parts and operations of the proposed system.

Let us suppose a robot equipped by a sensor observing the surrounding world and interacting with the human. The world is represented as a set of features $I = \{i_1, i_2, \ldots, i_k\}$, which can be acquired by robot's sensor. Each time the robot makes

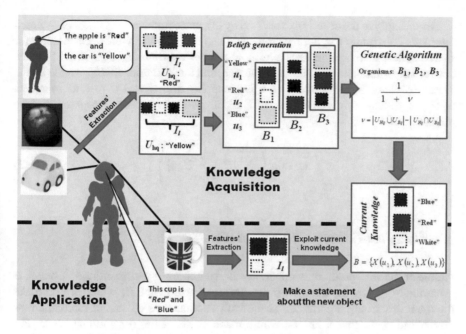

Fig. 2 Example showing main parts of the system's operation in the case of autonomous learning of *colors*

an observation o, its epistemic curiosity stimulates it to interact with the human asking him to gives a set of utterances U_H describing the found salient objects. Let us denote the set of all utterances ever given about the world as U. The observation o is defined as an ordered pair $o = \{I_l, U_H\}$, where $I_l \subseteq I$, expressed by (1), stands for the set of features obtained from observation and $U_H \subseteq U$ is the set of utterances (describing O) given by human in the context of that observation. i_p denotes the pertinent information for a given u (i.e. features that can be described semantically as u in the language used for communication between the human and the robot), i_i the impertinent information (i.e. features that are not described by the given u, but might be described by another $u_i \in U$) and sensor noise ε. The goal is to distinguish the pertinent information from the impertinent one and to correctly map the utterances to appropriate perceived stimuli (features). Let us define an interpretation $X(u) = \{u, I_j\}$ of an utterance u as an ordered pair where $I_j \subseteq I$ is a set of features from I. So, the belief B is defined accordingly to (2) as an ordered set of interpreting utterances u from U.

$$I_l = \bigcup_{U_H} i_p(u) + \bigcup_{U_H} i_i(u) + \varepsilon \qquad (1)$$

$$B = \{X(u_1), \ldots, X(u_n)\} \text{ with } n = |U| \qquad (2)$$

Accordingly to the criterion expressed by (3), one can calculate the belief B which interprets coherently the observations made so far: in other words, by looking for such a belief, which minimizes across all the observations $o_q \in O$ the difference between the utterances U_{Hq} made by human, and those utterances U_{Bq}, made by the system by using the belief B. Thus, B is a mapping from the set U to I: all members of U map to one or more members of I and no two members of U map to the same member of I.

$$\arg\min_{B} \left(\sum_{q=1}^{|O|} |U_{Hq} - U_{Bq}| \right) \tag{3}$$

2.2 The Most Coherent Interpretation Search

Although the interpretation's coherence is worked out by computing the belief B accordingly to Eq. (3), the system has to look for a belief B, which would make the robot describing a particular scene with utterances as close and as coherent as possible to those that a human would made on the same (or similar) scene. For this purpose, instead performing the exhaustive search over all possible beliefs, we propose to search for a suboptimal belief by means of a Genetic Algorithm (GA). For doing that, we assume that each organism within it has its genome constituted by a belief, which, results into genomes of equal size $|U|$ containing interpretations $X(u)$ of all utterances from U.

In our genetic algorithm, the genomes' generation is a belief generation process generating genomes (e.g. beliefs) as follows. For each interpretation $X(u)$ the process explores whole the set O. For each observation $o_q \in O$, if $u \in U_{Hq}$ then features $i_q \in I_q$ (with $I_q \subseteq I$) are extracted. As described in (1), the extracted set of features contains as well pertinent as impertinent features. The coherent belief generation is done by deciding, which features $i_q \in I_q$ may possibly be the pertinent ones. The decision is driven by two principles. The first one is the principle of "proximity", stating that any feature i is more likely to be selected as pertinent in the context of u, if its distance to other already selected features is comparatively small. The second principle is the "coherence" with all the observations in O. This means, that any observation $o_q \in O$, corresponding to $u \in U_{Hq}$, has to have at least one feature assigned into I_q of the current $X(u) = \{u, I_q\}$.

To evaluate a given organism, a classifier is trained, whose classes are the utterances from U and the training data for each class $u \in U$ are those corresponding to $X(u) = \{u, I_q\}$, i.e. the features associated with the given u in the genome. This classifier is used through whole set O of observations, classifying utterances $u \in U$ describing each $o_q \in O$ accordingly to its extracted features. Such a classification results in the set of utterances U_{Bq} (meaning that a belief B is tested regarding the qth observation). The fitness function evaluating the fitness of each above-mentioned organism is defined as "disparity" between U_{Bq} and U_{Hq} (defined in previous subsection) which is computed accordingly to the Eq. (4), where v is the number of utterances that are not present in both sets U_{Bq} and U_{Hq} (e.g. either missed or arc superfluous utterances interpreting the given features). The globally best fitting organism is chosen as the belief that best explains observations O made (by robot).

$$D(v) = \frac{1}{1+v} \quad \text{with} v = |U_{Hq} \cup U_{Bq}| - |U_{Hq} \cap U_{Bq}| \tag{4}$$

Figure 3 gives the bloc diagram of the designed evolutionary process. It is important to note that here the above-described GA based evolutionary process doesn't operates as only an optimizer but it generate the machines (e.g. robot's) most coherent belief about the observation accomplished by this robot and about the way that the same robot will autonomously construct a human-like description of the observed reality. In other words, it is the GA based evolutionary process that drives the robot's most coherent semantic understanding of the observed reality. It plays also a key role in implementation of the epistemic curiosity because the drop of the search for the most coherent belief, due to leakage of knowledge about the observed reality, makes the robot interacting with its human counterpart and thus drives its epistemic curiosity.

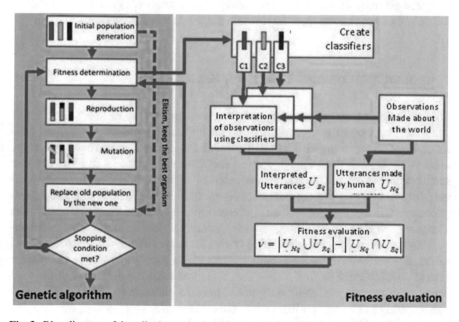

Fig. 3 Bloc diagram of described genetic algorithm's workflow. The *left part* describes the genetic algorithm itself, while the *right part* focuses on the fitness evaluation workflow

2.3 Role of Human-Robot Interaction

Human beings learn both by observation and by interaction with the world and with other human beings. The former is captured in our system in the "best interpretation search" outlined previous subsections. The latter type of learning requires that the robot be able to communicate with its environment and is facilitated by learning by observation, which may serve as its bootstrap. In our approach, the learning by interaction is carried out in two kinds of interactions: human-to-robot and robot-to-human. The human-to-robot interaction is activated anytime the robot interprets wrongly the world. When the human receives a wrong response (from robot), he provides the robot a new observation by uttering the desired interpretation. The robot takes this new corrective knowledge about the world into account and searches for a new interpretation of the world conformably to this new observation. The robot-to-human interaction may be activated when the robot attempts to interpret a particular feature classified with a very low confidence: a sign that this feature is a borderline example. In this case, it may be beneficial to clarify its true nature. Thus, led by the epistemic curiosity, the robot asks its human counterpart to make an utterance about the uncertain observation. If the robot's interpretation is not conforming to the utterance given by the human (robot's interpretation was wrong), this observation is recorded as a new knowledge and a search for the new interpretation is started.

3 Implementation and Validation Results

The validation of the proposed system has been performed on the basis of both simulation of the designed system as by an implementation on a real humanoid robot. A video capturing different parts of the experiment may be found online on: http://youtu.be/W5FD6 zXihOo. As real robot we have considered NAO robot (a small humanoid robot from Aldebaran Robotics) which provides a number of facilities such as onboard camera (vision), communication devices and onboard speech generator. The fact that the above-mentioned facilities been already available offers a huge save of time, even if those faculties remain quite basic in that kind of robots.

Although the usage of the presented system is not specifically bound to humanoid robots, it is pertinent to state two main reasons why a humanoid robot is used for the system's validation. The first reason for this is that from the definition of the term "humanoid", a humanoid robot is aspired to make its perception close to the human's one, entailing a more human-like experience of the world. This is an important aspect to be considered in context of sharing knowledge between a human and a robot. The second reason is that humanoid robots are specifically designed to interact with humans in a "natural" way by using e.g. a loudspeaker and microphone set in order to allow for a bi-directional communication with human by speech synthesis and speech analysis and recognition. This is of importance when speaking about a natural human-robot interaction during learning.

3.1 Simulation Based Validation and Results

The simulation based validation finds its pertinence in assessment of the investigated cognitive-system's performances. In fact, due to difficulties inherent to organization of strictly same experimental protocols on different real robots and within various realistic contexts, the simulated validation becomes an appealing way to ensure that the protocol remains the same. For simulation based evaluation of the behavior of the above-described system, we have considered color names learning problem. In everyday dialogs, people tend to describe objects, which they see, with only a few color terms (usually only one or two), although the objects in itself contains many more colors. Also different people can have slightly different preferences on what names to use for which color. Due to this, learning color names is a difficult task and it is a relevant sample problem to test our system.

In the simulated environment, images of real-world objects were presented to the system alongside with textual tags describing colors present on each object. The images were taken from the Columbia Object Image Library (COIL) contains 1000 color images of different views of 100 objects database. Five fluent English speakers were asked to describe each object in terms of colors. We restricted the choice of colors to "Black", "Gray", "White", "Red", "Green", "Blue" and "Yellow", based on the color opponent process theory [19] (Schindler 1964). The tagging of the entire set of images was highly coherent across the subjects. In each run of the experiment, we have randomly chosen a tagged set.

The utterances were given in the form of text extracted from the descriptions. The object was accepted as correctly interpreted if the system's and the human's interpretations were equal. The rate of correctly described objects from the test set was approximately 91 %. Figure 4 gives the result of interpretation by the system of the colors of the WCS table. Figure 5 shows the learning rate versus the increasing number of exposures of each color.

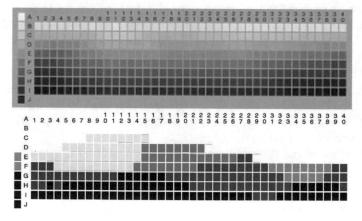

Fig. 4 Original WCS table (*upper image*), its system's made interpretation (*lower image*)

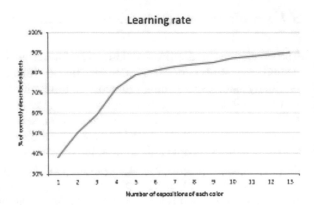

Fig. 5 The rate of correct learning versus the number of presented examples (of the same object) to the system

3.2 Implementation on Real Robot

The designed system has been implemented on NAO robot (from Aldebaran Robotics). It is a small humanoid robot which provides a number of facilities such as onboard camera (vision), communication devices and onboard speech generator. The fact that the above-mentioned facilities are already available offers a huge save of time, even if those faculties remain quite basic in that kind of robots. If NAO robot integrates an onboard speech-recognition algorithm (e.g. some kind of speech-to-text converter) which is sufficient for "hearing" the tutor, however its onboard speech generator is a basic text-to-speech converter. It is not sufficient to allow the tutor and the robot conversing in natural speech. To overcome NAO's limitations relating this purpose, the TreeTagger tool[2] was used in combination with robot's speech-recognition system to obtain the part-of-speech information from situated dialogs. Standard English grammar rules were used to determine whether the sentence is demonstrative (e.g. for example: "This is an apple."), descriptive (e.g. for example: "The apple is red.") or an order (e.g. for example: "Describe this thing!"). To communicate with the tutor, the robot used its text-to-speech engine.

The core of the implementation's architecture is split into five main units: Communication Unit (CU), Navigation Unit (NU), Low-level Knowledge Acquisition Unit (LKAU), High-level Knowledge Acquisition Unit (HLAU) and Behavior Control Unit (BCU). Figure 6 illustrates the bloc-diagram of the implementation's architecture. The aforementioned units control NAO robot (symbolized by its sensors, its actuators and its interfaces in Fig. 6) through its already available hardware and software facilities. In other words, the above-mentioned architecture controls the whole robot's behavior.

The purpose of NU is to allow the robot to position itself in space with respect to objects around it and to use this knowledge to navigate within the surrounding environment. Capacities needed in this context are obstacle avoidance and

---

[2] Developed by the ICL at University of Stuttgart, available online at: http://www.ims.uni-stuttgart.de/projekte/corplex/TreeTagger.

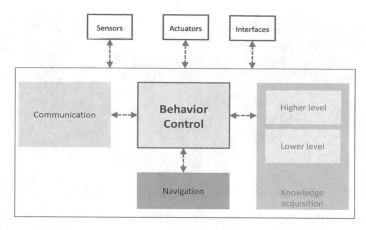

Fig. 6 Block diagram of the implementation's architecture

determination of distance to objects. Its sub-unit handling spatial orientation receives its inputs from the camera and from the LKAU. To get to the bottom of the obstacle avoidance problem, we have adopted a technique based on ground color modeling. Inspired by the work presented in [20], color model of the ground helps the robot to distinguish free-space from obstacles.

The LKAU ensures gathering of visual knowledge, such as detection of salient objects and their learning (by the sub-unit in charge of salient object detection) and sub-recognition (see [18, 21]). Those activities are carried out mostly in an "unconscious" manner, i.e. they are run as an automatism in "background" while collecting salient objects and learning them. The learned knowledge is stored in Long-term Memory for further use.

The HKAU is the center where the intellectual behavior of the robot is constructed. Receiving its features from the LKAU (visual features) and from the CU (linguistic features), this unit processes the beliefs' generation, the most coherent belief's emergence and constructs the high-level semantic representation of acquired visual knowledge. Unlike the LKAU, this unit represents conscious and intentional cognitive activity. In some way, it operates as a baby who learns from observation and from verbal interaction with adults about what he observes developing in this way his own representation and his own opinion about the observed world [22].

The CU is in charge of robots communication. It includes an output communication channel and an input communication channel. The output channel is composed of a Text-To-Speech engine which generates human voice through loud-speakers. It receives the text from the BCU. The input channel takes its input from a microphone and through an Automated Speech Recognition engine (available in NAO) the syntax and semantic analysis (designed and incorporated in BCU) it provides the BCU labeled chain of strings representing the heard speech.

The BCU plays the role of a coordinator of robot's behavior. It handles data flows and issues command signals for other units, controlling the behavior of the robot and

Fig. 7 Robot and the subset of collected objects used for learning (*left-side picture*) and the recognition by robot of two objects among those learned by the robot in different posture and in different location (*right-side picture*)

its suitable reactions to external events (including its interaction with humans). BCU received its inputs from all other units and returns its outputs to each concerned unit including robot's devices (e.g. sensors, actuators and interfaces) [22]. The human-robot interaction is performed by this unit in cooperation with HLAU. In other words, driven by HLAU, a part of the robot's epistemic curiosity based behavior is handled by BCU.

3.3 Experimental Validation

The total of 25 every-day objects was collected for purposes of the experiment. The collected set has been randomly divided into two sets for training and for testing (Fig. 7). The learning set objects were placed around the robot and then a human tutor pointed to each of them calling it by its name. Using its 640×480 monocular color camera, the robot discovered and learned the objects around it by the salient object detection approach we have described in [18]. Here, this approach has been extended by detecting the movement of the human's hand to achieve joint attention. In this way, the robot was able to determine what object the tutor is referring to and to learn its name. The right-side picture in Fig. 7 shows the recognition by robot of two objects among those learned by the robot in different posture and in different location.

During the experiment, the robot has been asked to learn a subset among the 25 considered objects: in term of associating the name of each detected object to that object. At the same time, a second learning has been performed involving the interaction with the tutor who has successively pointed the above-learned objects describing (e.g. telling) to the robot the color of each object. Extracted from the video of the experimental validation, Fig. 8 shows the robot observing and learning

Fig. 8 Experimental setup showing tutor pointing different objects from learning set and robot learning those objects

different objects chosen by the human tutor. Here-bellow an example of the Human-Robot interactive learning is reported:

- **Human** [*pointing a red aid-kit*]: "This is a first-aid-kit!"
- **Robot**: "I will remember that this is a first-aid-kit."
- **Human**: "It is red and white".
- **Robot**: "OK, the first-aid-kit is red and the white."

After learning the names and colors of the discovered objects, the robot is asked to describe a number of objects including as well some of already learned objects but in different posture (for example the yellow box presented in reverse posture) as a number of still unseen objects (as for example a red apple or a white teddy-bear). The robot has successfully described, in a coherent linguistics, the presented seen and unseen objects. Extracted from the video of the experimental validation, Fig. 9 shows the human tutor asking the robot to describe the pointed object (which is a red apple) in term of colors (left-side picture of Fig. 9) and the ground truth detected objects as the robot perceives them. Finally, Fig. 10 shows two examples of observed objects' interpretation by the robot. Here-bellow is the Human-Robot interaction during the experiment:

- **Human** [*pointing the unseen white teddy-bear*]: "Describe this!"
- **Robot**: "It is white!"
- **Human**: [*pointing the already seen, but reversed, yellow box*]: "Describe this!"
- **Robot**: "It is yellow!"
- **Human**: [*pointing the unseen apple*]: "Describe this!"
- **Robot**: "It is red!"

Fig. 9 Experimental setup showing tutor pointing a red apple which has not been seen before, (by the robot) asking the robot to describe that object in term of colors (*left-side picture*) and the ground truth detected objects as the robot perceives them (*right-side picture*)

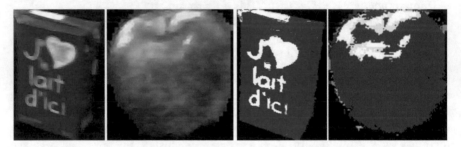

Fig. 10 Two objects observed and interpreted by the robot: the original image provided by robot's camera (*left-side pictures*) and the interpretation of those objects by the robot (*right-side pictures*). For the "apple", the robot's given description was "the object is red". For the box, the description was "the object is blue and white"

4 Conclusion and Perspectives

This chapter has presented, discussed and validated a cognitive system for high-level knowledge acquisition based on the notion of artificial curiosity. Driving as well the lower as the higher levels of the presented cognitive system, the emergent artificial curiosity allow such a system to learn in an autonomous manner new knowledge about unknown surrounding world and to complete (enrich or correct) its knowledge by interacting with a human. Experimental results, performed as well on a simulation platform as using the NAO robot show the pertinence of the investigated concepts as well as the efsfectiveness of the designed system. Although it is difficult to make a precise comparison due to different experimental protocols, the results we obtained show that our system is able to learn faster and from significantly fewer examples, than the most of more-or-less similar implementations.

Based on obtained results, it is thus justified to say, that a robot endowed with such artificial curiosity based intelligence will necessarily include autonomous cognitive capabilities. With respect to this, several appealing perspectives are pursuing to push further the presented work. The current implemented version allows the robot to work with a single category or property at a time (e.g. for example the color in utterances like "it is red"). We are working on extending its ability to allow the learning of multiple categories at the same time and to distinguish which of the used words are related to which category. While, concerning the middle-term perspectives of this work, they will focus aspects reinforcing the autonomy of such cognitive robots. The ambition here is integration of the designed system to a system of larger capabilities realizing multi-sensor artificial machine-intelligence. There, it will play the role of an underlying part for machine cognition and knowledge acquisition.

References

1. Vernon, D., Metta, G., Sandini, G.: A survey of artificial cognitive systems: implications for the autonomous development of mental capabilities in computational agents. IEEE Trans. Evol. Comput. **11**(2), 151–180 (2007)
2. Langley, P., Laird, J.E., Rogers, S.: Cognitive architectures: research issues and challenges. Cogn. Syst. Res. **10**(2), 141–160 (2009)
3. Levesque, H.J., Lakemeyer, G.: "Cognitive robotics", Handbook of Knowledge Representation, Dagstuhl: Schloss Dagstuhl—Leibniz-Zentrum fuer Informatik (2010)
4. Berlyne, D.E.: A theory of human curiosity. British J. Psychol. **45**(3), 180–191 (1954)
5. Schmidhuber, J.: Curious model-building control systems. In: Proceedings of International Joint Conference on Neural Networks (IEEE-IJCNN 1991), Vol. 2, pp. 1458–1463 (1991)
6. Oudeyer, P.-Y., Kaplan, F., Hafner, V.V.: Intrinsic motivation systems for autonomous mental development. IEEE Trans. Evol. Comput. **11**(2), 265–286 (2007)
7. Macedo, L., Cardoso, A.: The exploration of unknown environments populated with entities by a surprise-curiosity-based agent. Cogn. Syst. Res. **19–20**, 62–87 (2012)
8. Maier, W., Steinbach, E.G.: Surprise-driven acquisition of visual object representations for cognitive mobile robots, pp. 1621–1626. In: Proceedings of IEEE International Conference on Robotics and Automation, Shanghai (2011)
9. Ugur, E., Dogar, M.R., Cakmak, M., Sahin, E.: Curiosity-driven learning of traversability affordance on a mobile robot. In: Proceedings of IEEE 6th International Conference on Development and Learning, pp. 13–18 (2007)
10. Yu, C.: The emergence of links between lexical acquisition and object categorization: a computational study. Connection Sci. **17**(3–4), 381–397 (2005)
11. Waxman, S.R., Gelman, S.A.: Early word-learning entails reference, not merely associations. Trends Cogn. Sci. (2009)
12. Litman, J.A.: Interest and deprivation factors of epistemic curiosity. Personality Individ. Differ. **44**(7), 1585–1595 (2008)
13. Kang, M.J.J., Hsu, M., Krajbich, I.M., Loewenstein, G., McClure, S.M., Wang, J.T.T., Camerer, C.F.: The wick in the candle of learning: epistemic curiosity activates reward circuitry and enhances memory. Psychol. Sci. **20**(8), 963–973 (2009)
14. Madani, K., Sabourin, C.: Multi-level cognitive machine-learning based concept for human-like artificial walking: application to autonomous stroll of humanoid robots. Neurocomputing, 1213–1228 (2011)

15. Ramik, D.-M., Sabourin, C., Madani, K.: From visual patterns to semantic description: a cognitive approach using artificial curiosity as the foundation. Pattern Recogn. Lett. **34**(14), 1577–1588 (2013). Elsevier
16. Ramík, D.M., Sabourin, C., Madani, K.: A real-time robot vision approach combining visual saliency and unsupervised learning. In: Proceedings of 14th International Conference on CLAWAR, Paris, France, pp. 241–248 (2011)
17. Ramík, D.M., Sabourin, C., Madani, K.: Hybrid salient object extraction approach with automatic estimation of visual attention scale. In: Proceedings of Seventh International Conference on Signal Image Technology & Internet-Based Systems, Dijon, France, pp. 438–445 (2011)
18. Ramik, D.M., Sabourin, C., Moreno, R., Madani, K.: A machine learning based intelligent vision system for autonomous object detection and recognition. J. Appl. Intell. (2013). Springer, doi:10.1007/s10489-013-0461-5
19. Schindler, M., Von Goethe, J.W.: Goethe's theory of colour applied by Maria Schindler. New Knowledge Books, East Grinstead, Eng (1964)
20. Moreno, R., Ramik, D.M., Graña, M., Madani, K.: Image segmentation on the spherical coordinate representation of the RGB color space. IET Image Process. **6**(9), 1275–1283 (2012)
21. Madani, K., Sabourin, C.: Multi-level cognitive machine-learning based concept for human-like "artificial" walking: application to autonomous stroll of humanoid robots. Neurocomuting **74**, 1213–1228 (2011)
22. Ramik, D.M., Sabourin, C., Madani, K.: Autonomous knowledge acquisition based on artificial curiosity: application to mobile robots in indoor environment. J. Robot. Auton. Syst. **61**(12), 1680–1695 (2013)

Author Index

© Springer International Publishing Switzerland 2016
K. Madani et al. (eds.), *Computational Intelligence*,
Studies in Computational Intelligence 613,
DOI 10.1007/978-3-319-23392-5

517

Printed in the United States
By Bookmasters